AF520900

Chemistry of the Soil

Second Edition

Related Books

- Agriculture Chemical Users Manual — *Geoff Cowles*
- Agrotechnology for Dryland Farmaing *2nd Revised Ed* — *A.M. Dhopte*
- Allelopathy in Soil Sickness — *S.S. Narwal*
- Biofertilizer Technology — *S. Kannaiyan*
- Comprehensive assessment of soil health *3rd Ed* — *SIPS*
- Crop Production in Salt Affected Soils — *I.C. Gupta*
- Crop Production in Waterlogged Saline Soils — *S.K. Gupta*
- Crop Protection Strategies under Climate Change Scenarios — *P.P. Reddy*
- Drainage Engineering: Principles and Practices — *S.K. Gupta*
- Dryland Resources and Technology (Vol. 8) — *L.L. Somani*
- Dryland Technology *2nd Ed* — *M.L. Jat*
- Fertilizer Manual — *IFDC*
- Fundamentals of Soil Science — *A. Rathinasamy*
- Genesis and Management of Sodic (alkali) Soils — *S.K. Gupta*
- Handbook of Saline and Alkali Soils Diagnosis Reclamation and Management — *S.K. Gupta*
- Indiras Objective Agriculture — *R.L. Arya*
- Manual on Fertilizer Distribution** — *FAO*
- Methods of Analysis for Soils of Arid and Semi Arid Regions** — *I.I. Bashour*
- Methods of Soil Analysis: Microbiological and Biochemical Properties Part 2** — *R.W. Weaver*
- Nanofertilizers: Challenges and Prospects** — *J.C. Tarafdar*
- Nanotechnology — *J.C. Tarafdar*
- New Vistas of Organic Farming *2nd Ed* — *Mukund Joshi*
- Organic Farming Theory & Practice — *S.P. Palaniappan*
- Rainwater Management Theory and Practice — *M.L. Jat*
- Salinity Tolerance in Plants: Methods, Mechanisms and Management *2nd Ed* — *B.K. Garg*
- Sasya Vigyan Ke Sidhant (Principles of Agronomy) (Hindi) — *R.L. Arya*
- Soil and Plant Analysis — *C.S. Piper*
- Soil and Water Conservation in Semi-Arid Areas — *N.W. Hudson*
- Soil Chemical Analysis — *M.L. Jackson*
- Sustainable Agriculture — *R.S. Meena*
- Sustainable Dryland Farming** — *A.M. Dhopte*
- The Use of Saline Waters for Crop Production — *FAO*
- Tillage and Crop Production *2nd Ed* — *N.R. Das*
- Water Measurement Manual *3rd Ed* — *USDI*

Chemistry of the Soil

Second Edition

Edited by

Firman E. Bear

Editor-in-Chief, Soil Science
Former Chairman, Department of Soils
Rutgers, Tha State University of New Jersey

United Book Prints

Reprinted in India by:

UNITED BOOK PRINTS
(Imprint of Scientific Publishers, INDIA)
4806/24, Ansari Road, Daryaganj,
New Delhi 110 002, INDIA
E-mail: info@scientificpub.com
Website : http://www.scientificpub.com

Print : 2020

ISBN: 978-81-92686-00-4
eISBN: 978-93-88148-12-2

Printed in India

General Introduction

American Chemical Society's Series of Chemical Monographs

By arrangement with the Interallied Conference of Pure and Applied Chemistry, which met in London and Brussels in July, 1919, the American Chemical Society was to undertake the production and publication of Scientific and Technologic Monographs on chemical subjects. At the same time it was agreed that the National Research Council, in cooperation with the American Chemical Society and the American Physical Society, should undertake the production and publication of Critical Tables of Chemical and Physical Constants. The American Chemical Society and the National Research Council mutually agreed to care for these two fields of chemical progress. The American Chemical Society named as Trustees, to make the necessary arrangements of the publication of the Monographs, Charles L. Parsons, secretary of the Society, Washington, D. C.; the late John E. Teeple, then treasurer of the Society, New York; and the late Professor Gellert Alleman of Swarthmore College. The Trustees arranged for the publication of the ACS Series of (a) Scientific and (b) Technological Monographs by the Chemical Catalog Company, Inc. (Reinhold Publishing Corporation, successor) of New York.

The Council of the American Chemical Society, acting through its Committee on National Policy, appointed editors (the present list of whom appears at the close of this sketch) to select authors of competent authority in their respective fields and to consider critically the manuscripts submitted.

The first Monograph of the Series appeared in 1921. After twenty-three years of experience certain modifications of general policy were indicated. In the beginning there still remained from the preceding five decades a distinct though arbitrary differentiation between so-called "pure science" publications and technologic or applied science literature. By 1944 this differentiation was fast becoming nebulous. Research in private enterprise had grown apace and not a little of it was pursued on the frontiers of knowledge. Furthermore, most workers in the sciences were coming to see the artificiality of the separation. The methods of both groups of workers are the same. They employ the same instrumentalities, and frankly recognize that their objectives are common, namely, the search for new knowledge for the service of man. The officers of the Society therefore combined the two editorial Boards in a single Board of twelve representative members.

Also in the beginning of the Series, it seemed expedient to construe rather broadly the definition of a Monograph. Needs of workers had to be recognized. Consequently among the first hundred Monographs appeared works in the form of treatises covering in some instances rather broad areas. Because such necessary works do not now want for publishers, it is considered advisable to hew more strictly to the line of the Monograph character, which means more complete and critical treatment of relatively restricted areas, and, where a broader field needs coverage, to subdivide it into logical subareas. The prodigious expansion of new knowledge makes such a change desirable.

These Monographs are intended to serve two principal purposes: first, to make available to chemists a thorough treatment of a selected area in form usable by persons working in more or less unrelated fields to the end that they may correlate their own work with a larger area of physical science discipline; second, to stimulate further research in the specific field treated. To implement this purpose the authors of Monographs are expected to give extended references to the literature. Where the literature is of such volume that a complete bibliography is impracticable, the authors are expected to append a list of references critically selected on the basis of their relative importance and significance.

Preface

So many advances have been made in our knowledge of soils and in the development of the techniques employed in their study since "Chemistry of the Soil" was first published in 1955 that it seemed desirable to develop a second edition to bring the book up to date.

Unfortunately, of the original authors, George K. Fraser and F. G. Merkle are deceased, Arthur L. Prince is retired, Adolf Mehlich and Kirk Lawton are on extended leave on foreign assignments, and D. Wynne Thorne is now fully engaged in administrative work.

Since new authors had to be sought to replace these men and since some additional phases of the subject needed to be developed, it was decided to reorganize the presentation, to change the titles and coverage of some of the previous chapters, and to add chapters on the biochemistry of soils, and radioisotopes and soils.

We are very fortunate that Marion L. Jackson, James L. Mortensen, Ian L. Stevenson, Harold B. Peterson, Ellis R. Graham, and William J. Hanna agreed to replace the original contributors who are presently unavailable and to add new material.

Dr. Mortensen died before he had finished his chapter. We are very much indebted to Frank L. Himes, his associate, who volunteered to complete this assignment and did so in a highly satisfactory manner.

A list of selected references follows each chapter. To assist the reader, the references in this edition have been keyed into the text.

Much credit is due Ruth M. Field, managing editor of *Soil Science*, for her careful attention to details in editing the manuscripts and preparing them for publication.

FIRMAN E. BEAR

April 1, 1964

Contents

Introduction

What is soil? Veatch of Michigan defined it as "a mass of inorganic material that holds inorganic and organic colloids, dead and living plant and animal material, water, and gases in variable but balanced proportions."

Robinson of Wales added this important point: "An adequate conception of the soil can be obtained only from a study of all the horizons or soil strata down to the parent material. And whilst, for some purposes, it is sufficient to study the properties of samples taken from individual horizons of a profile, a complete picture cannot be obtained until the laboratory studies are supplemented by observations on natural profiles in the field."

Earlier, Cameron of the United States Department of Agriculture had pointed out that "in studying the soil as a medium for crop production we must consider the plant itself, or at least that part of it that enters the soil, namely: the root, the solid particles of the soil, the soil water or the aqueous solution from which the plant draws all the materials for its sustenance excepting the carbon dioxide by its aerial portions, the soil atmosphere, and the biological processes taking place. The one common characteristic of all these things is that they are continually in a state of change. Therefore the soil is essentially dynamic."

The soil chemist is interested in knowing, in detail, the chemistry of the original rocks and minerals that constituted the materials from which soils were derived. But he is much more concerned with tracing the changes that have occurred and that are still taking place as a consequence of the action of meteorological, geological, and biological agencies on these rocks. The resulting soils, by reason of the great variation in the composition of the rocks and minerals of which they are the residue and the wide diversity in the nature and mode of operation of the several agencies affecting their formation, are highly variable products. They vary both horizontally and vertically over the surface of the earth.

Concepts of soil chemistry, as of all other divisions of chemistry, have changed greatly over the years. Davy of England is sometimes credited with having been the first soil chemist. He was particularly interested in soil as a possible source of alkalies and alkali-earths, but in a series of lectures on agricultural chemistry he dealt with soils in relation to plant nutrition.

The basic importance of soil in food production required that attention be directed toward the study of the chemical composition of soil in relation to crop needs. Of the pioneer chemists in this field, the best known are Liebig of Germany, Boussingault of France, and Gilbert of the Rothamsted Experimental Station in England. These men analyzed soils for their total contents of the several mineral elements as the starting point in evaluating the fertility status. They also analyzed crop plants to determine what the plants removed from soils.

In the United States, this type of soil chemistry reached its peak in the analytical program carried on under the direction of Hopkins of Illinois during the early part of the twentieth century. At the start, complete mineral analyses were made of large numbers of soil samples. Later, the analytical work was confined to total amounts of the major mineral nutrients and nitrogen in the plow depth of soils, representing what came to be termed the "bank account" against which "checks" were drawn in the process of growing crops. Such analyses showed that soils varied greatly from area to area in their total contents of the several nutrient elements. Still greater variations were found on a regional basis. However, serious doubts began to develop as to the value of the thousands of such analyses that were being accumulated and recorded in the literature.

Meanwhile, an effort was being made to measure the availability to plants of the nutrient elements contained in the soil complex. Hilgard of California was of the opinion that zeolites constituted the primary source of the available mineral nutrients. Consequently, he thought it necessary to substitute strong-acid digestion for fusion in analyzing soils. At his suggestion studies were made of the solubility of soils in three concentrations of hydrochloric acid. It was found that maximum solubility was induced by use of acid of a specific gravity of 1.115. Hilgard then wrote: "This being the most readily obtainable by simple steam distillation of acid of any other strength, the writer adopted it as best suited for the purpose of soil analysis." The Association of Official Agricultural Chemists made this solvent and concentration, with certain time and temperature stipulations, an integral part of the "official" method for the United States.

In due time, however, it was shown that zeolites were not present in soils to any appreciable extent. Instead, it was demonstrated, first by Way of England and later by a number of other chemists, notably Gedroiz of Russia, Mattson of Sweden, Hendricks of the United States Department of Agriculture, and Kelley of California, that the available supplies of the nutrient cations were adsorbed on the clay minerals and colloidal humus in exchange forms, from which they could readily be released by use of neutral salt solutions. This led to studies of the cation-exchange capacities of soils and of the quantities and ratios of the several adsorbed cations in soils of widely diverse origins. Some of the nutrient anions were found to be similarly adsorbed by soils, and a considerable amount of research was undertaken to explore these phenomena.

Meanwhile, a great deal of work was being done on soil solutions. This involved a variety of procedures for separating extracts from soils. Various kinds of weak solvents were used for extraction. Some of the earliest work was done by Dyer in England, who used a 1 per cent solution of citric acid for simulating the supposed dissolving action of plant roots on the soil. Truog of Wisconsin, Bray of Illinois, and Morgan of Connecticut were among the earliest workers in this field of study in the United States. Morgan's method of extraction, in which use was made of a highly buffered solution of sodium acetate, was widely employed. Later Purvis and Hanna of New Jersey perfected a quick-test technique for electrodialyzing soils in a weak solution of boric acid. Considerable work was also done on measuring the nutrient-supplying capacities of soils by analyzing plants grown on them.

During all this period much effort was being put into field studies of soils in an effort to find a more dependable basis for classification. This ultimately resulted in the discovery by Dokuchaev and his co-workers in Russia that climate and related vegetation were dominant factors in determining the nature of mature soils. On this basis, the soils of the earth were finally divided into a number of Great Soil Groups. Four outstanding examples of these are the gray-red Desert soils; the black Chernozems of warm regions of moderate rainfall; the gray-brown Podzols of cool, moist regions; and the red Latosols of high-rainfall tropical areas.

Once this concept had been developed, it soon became apparent that differences in these Great Soil Groups, as evidenced superficially by color and other readily observable characteristics of the several horizons of their profiles, merited careful study. Such studies revealed that there had been a downward movement of soluble and colloidal substances in the soil profile, the extent of which was determined by precipitation in relation to loss of water by evaporation, temperature and time. The essential points of the chemist's findings in this connection are illustrated by the salinity and alkalinity of Desert soils, the calcium carbonate deposits in the B horizons of Chernozems, the acidity of Podzols and the accumulation of iron and aluminum oxides in their B horizons, and the loss of silicate silica and the accumulation of iron and aluminum oxides in Latosols.

These studies, which were highly illuminating in themselves, provided a starting point for more careful investigation of the clay minerals on which the physical and chemical properties of soils so largely depend. As would be expected, the character of these clay minerals was shown also to be related to the nature of the climate and of the soils resulting therefrom. Montmorillonite was found to be the dominant type of clay mineral in the Chernozems, illite in the Podzols, and kaolinite in the Latosols.

The producing capacities of soils are profoundly affected by the organic matter naturally contained in them or added to them during farming operations. Thaer of Germany was a pioneer in the study of soil organic matter. Among the many other workers in this field of research were Mülder of Germany, Deherain of France, Schreiner and Shorey of the United States Department of Agriculture, Jenny of California, and Waksman of New Jersey.

Continued study of soils in relation to the needs of crop plants finally revealed that the nutrient elements which are required only in very small amounts form a list of considerable size. Included in this group of "trace" elements are manganese, zinc, copper, boron, molybdenum and chlorine. The quantities of these trace elements contained in soils and required by plants were found to be so small as to present some very difficult analytical problems.

The tools of the soil chemist now include all the modern instruments employed in qualitative and quantitative analytical procedures in the chemical laboratory. They also include x-ray diffraction and differential thermal units, petrographic and electron microscopes, the spectrograph, and a considerable number of other instruments. And the soil chemist finds it essential to examine soils in place in the field as well as in the laboratory.

This introductory statement is designed merely to pave the way for the presentation of the more detailed discussions by the men who, because of their special qualifications, were asked to write the several chapters in this monograph. It is hoped that those who read and study this volume will find in it the material required for a reasonably complete understanding of the chemistry of the soil as it is now known.

FIRMAN E. BEAR

CHAPTER 1

Chemistry of Soil Development

ISAAC BARSHAD

Department of Soils and Plant Nutrition
University of California, Berkeley

Soil is an independent, dynamic body of nature that acquires properties in accordance with the forces which act upon it. Recognition of this fact establishes Hilgard and Dokuchaev as the founders of pedology. These men were the first to recognize that many soil properties result from certain physical and chemical reactions which are conditioned not only by the rocks or rock materials from which soils are derived but also by the factors of climate, vegetation, topography, and time of reaction. Through their efforts, and the efforts of the many scientists who followed them, the science of pedology has become an important branch of the earth sciences.*

From the broadest point of view, pedology is a study of the development of soil material. *Soil material* may be defined as the material occupying the outermost part of the earth's crust and possessing distinct morphological, mineralogical, chemical, and physical properties resulting from certain inheritance factors, from its unique position on the surface of the earth, and from the environmental factors existing at that position. One can thus study the development of a given soil property of a soil material from a fixed position with respect to the surface as it varies from locality to locality, such as nitrogen content of the surface 6 inches as it is affected by a variation in temperature, or one can study the variation of the soil material as it is affected by its position with respect to the surface at any one site. Recognition of the variation of the soil material with depth and with longitude and latitude as a result of developmental processes is one of the most unique contributions of pedology.

*** See references 1, 14, 18, 19, 22, 26, 27, 34, 38, 40, 45, 46, 56, 57, 60, 62, 67–69, 72, 73, 79, 80, and 82.**

Many pedologists believe that a *soil* should be defined by the unique variation in its properties with depth and lateral extension, and that all soil materials extending from the surface downward to a point where the properties remain uniform should be considered as one body and be termed the *soil profile* or the *pedon* (74); the individual layers of soil material that exhibit distinct variations with depth are termed *soil horizons*.

To evaluate the development of a soil profile it is necessary to define and evaluate the initial state of the soil material at each horizon. This initial state is referred to as the *parent material*. Profiles which developed from uniform parent material are said to contain the parent material at the bottom of the profile, but in reality this latter material is only the preserved parent material. In such profiles the horizons above the parent material are collectively termed the *solum*. Many soil profiles exist, however, which have developed from parent material that was not uniform with depth and, as a result, evaluation of soil development in such profiles is more difficult.

It must be recognized that the parent material of a soil profile may consist of a variety of materials: it may be a solid rock, it may be a decomposed rock, it may be a soil material—such as an alluvial deposit, or it may consist of a prior soil profile. Soil development can be evaluated on each of these materials as long as the initial state can be determined. It would appear, therefore, that one of the most important aspects of soil development studies is to establish the initial states of the soil material at each position of the soil profile. Once the initial state is established, the degree of development can readily be determined in terms of the mineralogical, chemical, physical, and morphological changes that occur in the transformation of the parent material to soil material. It is the purpose of this chapter to outline methods by which these changes can be measured quantitatively, and also to consider the nature of the chemical and physical reactions responsible for the transformation of a parent material to a soil material and a soil profile.

ANALYSES FOR EVALUATING SOIL DEVELOPMENT

In measuring changes resulting from a chemical reaction one must clearly distinguish between the reactants and the resulting products. The extent of the reaction is determined by the amounts of reactants and products at the beginning and at the end of the reaction. As simple as this appears to the chemist in studying chemical reactions, it did not appear so to early soil pedologists or geologists in evaluating rock decomposition. Thus, for example, most early chemical analyses of soils consisted of total chemical analysis of the soil as a whole, which included both products and reactants.

Such chemical analysis, although valuable in indicating gross changes, cannot reveal precisely the changes that occurred in the reactants and the kind of products obtained. One of the main reasons for failure to analyze separately the reactants and the products of a soil lies in the difficulty of separating and clearly distinguishing between them. Even when such a separation can be achieved, one finds that many reactants undergo chemical changes in their solid state, and, in a sense, these reactants consist of both reactants and products which are inseparable. But for the purpose of evaluating soil development, with the type of soil analyses available at present, it could be assumed that the nonclay fractions represent the reactants and the clay fractions represent the products.

To obtain chemical and mineralogical data that could be used for evaluating soil profile development most accurately, the techniques to be employed in soil analyses should yield results which would differentiate the three main types of inorganic substances in soils: (*a*) the primary minerals, inherited from the original igneous and metamorphic rocks; (*b*) the secondary minerals, formed from the primary minerals—such as the clay minerals; and (*c*) the uncombined oxides and carbonates—such as Fe_2O_3, Al_2O_3, SiO_2, $CaCO_3$.

As a rule, the largest proportion of the primary minerals is confined to the coarser fractions of the soil—those with a diameter larger than about 2μ. The secondary minerals, on the other hand, are confined mainly to the finer fractions of the soil—those with a diameter smaller than 2μ—the clay fraction. But the coarse fractions may contain a considerable number of partly altered minerals, and the finer fractions may contain a considerable number of unaltered primary minerals, both of which should be recognized in a complete analysis of a soil. The uncombined oxides and carbonates are present in both the coarse and the fine fractions in the form of an adsorbed coating on the surfaces of the primary and secondary minerals, and also as precipitates acting as cementing materials to bind the primary and the secondary minerals together to form aggregates, concretions, or even hardpans—which are consolidated soil material. Determination, therefore, of the uncombined oxides requires their removal from the soil by bringing them into solution but without dissolving either the primary or the secondary minerals. The analytical techniques available for this purpose have not been entirely satisfactory, but considerable progress is being made in that direction. Even with the best method that could be devised for analysis of the free oxides and carbonates, there would always be, the possibility that some primary and secondary minerals might have suffered some dissolution during removal of the uncombined oxides. For evaluating profile development in such soils, this difficulty may be overcome by simply assuming that such solution is insignificant and that the amounts of uncombined oxides and carbonates determined by a given method are truly representative.

In light of the preceding discussion, soil analyses to obtain data for evaluating profile development should include the following steps:

1. Removal and the determination of organic matter.

2. Removal and determination of uncombined oxides, carbonates, and soluble salts.

3. Extraction of the clay fraction and its determination in soil freed from organic matter and uncombined oxides.

4. Total chemical analysis of the clay fraction freed from organic matter and uncombined oxides and saturated with a known exchangeable cation.

5. Total chemical analysis of the whole soil also freed of organic matter and uncombined oxides and saturated with the same exchangeable cation as the clay fraction.

6. Cation and anion exchange properties: exchangeable bases, exchange acidity, and total cation and anion exchange and fixation capacities for various cations and anions.

7. A mineralogical analysis of the soil, which should differentiate clearly between the clay and the nonclay fractions. As in chemical analyses, the soil should be freed from organic matter and uncombined oxides and carbonates. The techniques used by mineralogists are applicable to the nonclay fractions, but special techniques must be used for the analysis of the clay fraction. These include specialized x-ray techniques, differential thermal analysis, thermogravimetric analysis, surface determination analysis, cation and anion exchange analysis, and a separate evaluation of the total chemical analysis.

8. A morphological analysis of the soil should include a macroscopic and microscopic description of the various minerals, aggregates, and voids as to shape, size, and arrangements with respect to each other and with respect to soil depth and lateral extension. The techniques employed in soil survey are applicable to the macroscopic description (74), and the petrologist technique of a direct microscopic examination of soil peds and of *thin sections* of the naturally occurring and undisturbed soil material is applicable to the microscopic description of *structure* and *fabric*. The thin section is also useful for qualitative and quantitative mineralogical analysis (13, 43, 44).

CRITERIA FOR ESTABLISHING UNIFORMITY OF PATENT MATERIAL

The first task in evaluating soil profile development is to establish the degree of uniformity of the parent material of the whole soil profile at the start of soil formation.

Because of the great variety of parent materials of soils there are also a variety of criteria for establishing uniformity.* These criteria are as follows:

1. Total mineralogical analysis, with particular attention to the heavy mineral suite.
2. Nature of particle-size distribution of the resistant minerals—either the heavy or the light—of the nonclay fractions.
3. Nature of the ratio of two resistant minerals in any one fraction of the nonclay fraction, preferably the fine sand or coarse silt.
4. Particle-size distribution of the whole nonclay fraction.
5. Nature of clay distribution with depth.
6. Nature of the change in chemical composition of the nonclay fraction.

Total Mineralogical Analysis

A qualitative mineralogical analysis is very useful in differentiating materials from different sources. It is not sensitive enough, however, for differentiating small differences that may exist in any one of the materials from the common source. It was applied successfully in soils to detect stratification of parent material and to determine the source of parent materials (25).

Particle-size Distribution of Resistant Minerals

Use of the pattern of particle-size distribution of resistant minerals to identify uniformity of a material with depth is based on the principle that, since these minerals do not undergo any significant change during the course of soil formation, their particle-size distribution pattern would remain undisturbed. Consequently, constancy of the particle-size distribution pattern of such a mineral throughout the soil profile indicates that the soil profile was formed from a uniform parent material; but appearance of two or more patterns of distribution indicates that the parent material was stratified and the point of contact of the strata is at the point where the change in particle-size distribution occurs. The minerals which were found useful for this purpose are zircon, tourmaline, garnet, anatase, rutile, quartz, albite, and microcline.†

Tables 1.1, 2.1, and 3.1 demonstrate the effect of the change in particle-size distribution of resistant versus nonresistant minerals upon weathering. It is seen clearly that the particle-size distribution of the biotite and hornblende, the nonresistant minerals, changes with depth, whereas that of the quartz and albite, the resistant minerals, remains nearly constant. In fact, the change in particle-size distribution of a mineral in soils with an increasing degree of weathering can be used as a criterion for establishing its degree

* See references 3, 11, 25, 28, 33, 47, 48, and 52.

† See references 3, 25, 28, 47, 48, and 52.

TABLE 1.1 PARTICLE-SIZE DISTRIBUTION OF HORNBLENDE (>5μ) IN THE SHERIDAN SOIL PROFILE

Depth (in.)	% of Total hornblende in 100 g soil, for 6 fractions						
	2.0–1.0 mm	1.0–0.5 mm	0.5–0.25 mm	0.25–0.10 mm	0.10–0.05 mm	0.05–0.005 mm	Total
0–6	3.62	8.15	4.91	9.91	22.28	51.13	100.00
12–18	2.96	9.30	6.61	13.80	41.93	25.40	100.00
18–24	7.24	19.91	8.87	18.04	34.13	11.81	100.00
24–30	9.39	23.22	13.78	18.33	23.70	11.58	100.00

TABLE 1.2. PARTICLE-SIZE DISTRIBUTION OF MICA (>5μ) IN THE SHERIDAN SOIL PROFILE

Depth (in.)	% of Total mica in 100 g soil, for 6 fractions						
	2.0–1.0 mm	1.0–0.5 mm	0.5–0.25 mm	0.25–0.10 mm	0.10–0.05 mm	0.05–0.005 mm	Total
0–6	0.00	0.00	0.00	0.00	0.50	99.50	100.00
12–18	0.00	1.10	4.40	16.10	15.80	62.60	100.00
18–24	0.18	6.66	8.86	29.45	27.60	27.35	100.00
24–30	0.17	6.23	13.09	26.05	28.61	25.85	100.00

TABLE 1.3. PARTICLE-SIZE DISTRIBUTION OF QUARTZ + ALBITE IN THE SHERIDAN SOIL PROFILE

Depth (in.)	% of Total quartz + albite in 4 fractions				
	2–1 mm	1–0.5 mm	0.5–0.25 mm	0.25–0.10 mm	Total
0–6	23.05	33.10	21.28	22.57	100.00
12–18	20.30	32.55	22.40	24.75	100.00
18–24	23.73	34.01	21.80	20.46	100.00
24–30	23.55	33.82	19.60	23.03	100.00

of resistance. The quartz and albite distribution, as shown in Table 1.3, is such an example—in this case the particle-size distribution of the two minerals remains constant, indicating that in this soil profile both are equally resistant.

In Tables 1.4 and 1.5, two examples of the particle-sized distribution pattern of zircon (48), a heavy resistant mineral, are given which indicate stratification. Thus, in the Grundy profile (Table 1.4), there are four distribution patterns: one extends through horizons 1 to 3, another through 4 to 6, and still another through 7, 8, and 9; a fourth is present in horizon

TABLE 1.4. PARTICLE-SIZE DISTRIBUTION OF ZIRCON IN A GRUNDY SILT LOAM PROFILE

Horizon no.	% of Total zircon in 3 fractions			Total zircon as % of nonclay fraction	2 μ as % of the whole soil
	0.125–0.046 mm	0.046–0.025 mm	0.025–0.011 mm		
1	33	35	32	0.091	15.58
2	35	36	29	0.077	21.78
3	33	37	30	0.073	38.78
4	26	34	40	0.062	41.03
5	24	38	38	0.062	36.07
6	22	36	40	0.062	34.21
7	22	46	32	0.074	28.33
8	21	44	35	0.073	27.23
9	24	41	35	0.077	26.25
10	50	32	18	0.104	34.39

TABLE 1.5. PARTICLE-SIZE DISTRIBUTION OF ZIRCON AND OF THE WHOLE SOIL IN A TILSIT SILT LOAM PROFILE

Depth	Zircon fractions (%)*			Whole soil fractions (%)*		
	0.1–0.05 mm	0.05–0.02 mm	0.02–0.01 mm	0.1–0.05 mm	0.05–0.02 mm	0.02–0.01 mm
0–8 in.	5	83	12	11	52	37
8–17 in.	3	81	16	8	52	40
17–28½ in.	5	85	10	11	51	38
- - - - -	- - -	- - -	- - -	- - -	- - -	- - -
28½–35 in.	17	76	7	36	46	18
Weathered sandstone	25	74	1	51	42	7
Fresh sandstone	31	67	2	79	18	3

The dashed line denotes the depth at which stratification occurs.
* % of the total zircon or the total soil in fraction.

10. In the Tilsit (25) there are two distinct distribution patterns (Table 1.5): one extends through horizons 1, 2, and 3, and another through horizons 4, 5, and 6.

The distribution patterns of quartz and feldspar—particularly albite or microcline in moderately weathered soils—are just as useful for indicating stratification as the heavy resistant minerals as shown in Tables 1.6, 1.7, and 1.8. Thus, the quartz distribution in the Hagerstown soil profile (33) shows (Table 1.6) one distinct pattern in horizon A_0 through B_1 and another

TABLE 1.6. PARTICLE-SIZE DISTRIBUTION OF QUARTZ IN AN HAGERSTOWN SOIL PROFILE

Horizon	Depth (in.)	% of Total quartz in fractions*		Total quartz as % of whole soil*
		0.25–0.10 mm	0.10–0.05 mm	
A_0	0–1.25	18.85	81.15	16.20
A_1	1.25–3.5	16.30	83.70	19.07
A_2	3.5–8.5	15.00	85.00	22.14
A_3	8.5–12.75	15.20	84.80	21.56
B_1	12.75–16.25	17.70	82.30	17.86
B_2	16.25–25.00	28.30	71.70	9.62
B_3	25–36	25.70	74.30	10.35
B_4	36–51	27.60	72.40	7.39

* Calculated from the original data of Jeffries and White (33).

TABLE 1.7. PARTICLE-SIZE DISTRIBUTION OF QUARTZ IN THE RESIDUAL GRANITE PROFILE

Sample*	Depth (in.)	Total quartz as % of whole soil†	% of Total quartz in fractions†				
			>0.5 mm	0.5–0.25 mm	0.25–0.05 mm	0.05–0.02 mm	0.02–0.005 mm
1	0–4	47.36	3.72	1.60	2.93	35.35	56.40
2	4–10	39.61	2.17	1.24	1.59	33.40	61.60
3	10–18	25.66	1.40	1.40	1.70	30.50	65.00
4	18–24	22.05	4.50	1.77	3.13	39.00	51.60
5	24–30	25.71	16.72	4.16	2.45	57.77	18.90
6	26	24.08	13.95	5.00	4.03	52.72	24.30
7	24–30	24.92	26.49	6.19	5.21	52.60	9.51
8	26	24.50	38.10	9.74	7.10	38.52	6.54

* 5—Decomposed hull of weathered parent material; 6—partially decomposed hull enclosing unaltered granite; 7—crushed fragments of weathered parent material; and 8—crushed fragments of unaltered granite.

† Calculated from the original data of Humbert and Marshall (28).

in B_2 through B_4 ; the feldspar and the muscovite of this profile show a similar distribution pattern. The distribution of the quartz as well as that of the feldspar of a granitic profile (28) also shows (Tables 1.7 and 1.8) two distinct patterns of distribution, one extending through samples 1, 2, 3, and 4 and another through samples 5, 6, 7, and 8. (The increase in the relative amounts of coarse fraction in samples 7 and 8 is undoubtedly due to incomplete fractionation of the minerals of these samples due to the absence of decomposition.)

TABLE 1.8. PARTICLE-SIZE DISTRIBUTION OF FELDSPAR IN THE RESIDUAL GRANITE PROFILE

Sample	Depth (in.)	Total feldspar as % of whole soil*	% of Total feldspar in fractions				
			>0.5 mm	0.5–0.25 mm	0.25–0.05 mm	0.05–0.02 mm	0.02–0.005 mm
1	0–4	28.31	1.77	1.18	1.45	33.40	62.20
2	4–10	27.71	.58	.83	.87	28.60	69.12
3	10–18	25.51	.63	.86	.94	18.70	78.87
4	18–24	33.19	3.42	2.31	2.37	18.20	73.80
5	24–30	43.51	14.10	5.78	4.47	24.55	51.10
6	26	42.39	11.05	5.79	5.55	22.01	55.60
7	24–30	64.52	24.80	8.15	4.95	28.20	33.90
8	26	69.04	35.40	13.70	9.00	20.10	21.80

* Calculated from the original data of Humbert and Marshall (28).

Ratio of Two Resistant Minerals in a One-particle-size Fraction

Determination of uniformity of parent material by the constancy of a ratio of two resistant minerals in any one-size fraction is based on the same logic as the constancy of particle-size distribution of any one resistant mineral, namely, that in the absence of any decomposition, the resistant minerals will remain unaffected by soil formation, and their ratio in any one fraction throughout the profile would, therefore, also be unaffected and would reflect the degree of uniformity of the parent material (13, 47, 48, 52).

Table 1.9 illustrates three such examples. In one soil, the Bethel silt loam, the zircon/tourmaline ratio of the coarse silt fraction is an example

TABLE 1.9. RATIOS OF RESISTANT MINERALS IN ANY ONE FRACTION IN SOIL MATERIALS OF TWO PROFILES

Horizon no.	Bethel silt loam	Sheridan sandy clay loam		
	Zircon/tourmaline	Albite/quartz		SiO_2/Al_2O_3
	0.05–0.02 mm	2–1 mm	1–0.5 mm	0.25–0.10 mm
1	7.4	2.0	1.1	6.7
2	7.4	2.1	1.1	6.8
3	—	2.0	1.1	6.6
4	7.4	1.9	1.1	6.8
5	7.8	1.9	1.1	6.6
6	7.9	—	—	—

of a heavy resistant mineral ratio, demonstrating uniformity of parent material, whereas in the Sheridan sandy clay loam profile the albite/quartz ratio of two fractions and the SiO_2/Al_2O_3 ratio of one fraction (which includes only the quartz and the albite) demonstrate the value of the light resistant minerals as a means for determining uniformity of parent material as well as the resistance of the minerals themselves. The SiO_2/Al_2O_3 ratio, in the case of the fine sand fraction of the quartz plus albite minerals, is an example of the use of chemical instead of mineralogical analysis to demonstrate the constancy of the ratio of the two resistant minerals, since Al_2O_3 is present only in the albite.

Particle-size Distribution of the Whole Nonclay Fraction of the Soil

Since both quartz and albite, and in many soils possibly other feldspars, are unaffected or only slightly affected by soil formation, and since these minerals constitute most of the nonclay fraction, their particle-size distribution could also serve as an indicator of uniformity of parent material. This is particularly true when the particle-size distribution is calculated on the basis of the nonclay fraction alone. Indication of stratification may be marked much more strongly in some fractions than in others; and even in uniform parent materials the particle-size distribution may not be absolutely constant because of experimental difficulties in fractionation. This method of testing the uniformity of parent material was found to be as reliable as any of the methods so far (3, 52) and is particularly useful for sedimentary parent materials. Tables 1.5, 1.10, 1.11, and 1.12 are examples which clearly demonstrate the usefulness of such analyses.

Nature of Clay Distribution with Depth

The nature of the clay distribution with depth may at times indicate clearly the occurrence of stratification. This is particularly true when a nonpedogenic type of distribution occurs. A pedogenic type of clay distribution may have several patterns. In very young soils developed from a uniform alluvial parent material, the clay content is constant with depth. In moderately weathered soils, on granitic or basaltic rocks, the clay content would be highest at the surface soil and gradually decrease with depth. In fairly well-developed soils the clay content increases with depth to a maximum; below this depth it decreases until it either remains constant or completely disappears, as it would in solid rock. After it passes the maximum horizon of accumulation, at no time does the clay content increase again through pedogenic processes. If such an increase occurs, it is certain that the material of this horizon is of a different nature from the overlying material. In several soils in which this condition was found to occur, the

TABLE 1.10. PARTICLE-SIZE DISTRIBUTION OF SOILS WITH UNIFORM DISTRIBUTION WITH DEPTH OF BOTH CLAY AND NONCLAY

Depth (in.)	% of Whole soil						
	Final gravel	Sand				Silt	Clay <5 μ
		Coarse	Medium	Fine	Very fine		
Greenfield Sand							
0–15	8	19	11	24	21	10	6
15–40	9	13	23	23	24	12	6
40–72	8	20	11	23	20	12	6
Hanford Sandy Loam							
0–16	8	16	8	18	20	18	11
10–72	8	16	7	19	19	19	11
Montazuma Clay Loam							
0–14		0.0	1.3	1.2	8.7	32.7	28.1
14–32		0.0	0.1	0.1	8.5	35.2	29.1
32–40		0.7	0.1	1.0	8.1	34.8	29.0

TABLE 1.11. PARTICLE-SIZE DISTRIBUTION OF SOILS WITH UNIFORM CLAY BUT STRATIFIED NONCLAY DISTRIBUTION WITH DEPTH

Depth* (in.)	% of Whole soil						
	Fine gravel	Sand				Silt	Clay <5 μ
		Coarse	Medium	Fine	Very fine		
Hesperia Sand							
0–12	31	31	22	6	1	9	7
12–36	6	26	18	25	13	3	7
36–72	13	24	26	20	10	2	6
Yolo Gravelly Loam							
0–10	30	4	10	27	29	16	12
10–72	7	11	21	34	10	17	11

* Dashed lines denote stratification.

TABLE 1.12. PARTICLE-SIZE DISTRIBUTION OF SOILS WITH PEDOGENIC TYPE OF CLAY PROFILE BUT WITH A NONCLAY DISTRIBUTION WHICH INDICATES STRATIFICATION OF PARENT MATERIAL

Depth* (in.)	% of Nonclay fraction alone						% of Whole soil
	Fine gravel	Sand				Silt	Clay <5 μ
		Coarse	Medium	Fine	Very fine		
McClusky Sandy Loam							
0–16	0.7	7	29	32	21	11	11
13–30	0.0	11	19	33	22	15	11
30–60	0.0	0	4	30	36	30	40
60–72	0.0	0	1	4	24	71	29
Delano Loamy Sand							
0–19	10	16	10	28	20	15	11
10–31	6	17	11	28	17	20	10
31–38	8	16	11	26	19	20	11
38–55	3	7	4	13	17	58	24
55–66	5	9	6	14	16	48	21

* Dashed lines denote stratification.

lack of uniformity was always confirmed by the variation in the particle-size distribution of the nonclay fractions, as illustrated in Table 1.4 by horizon 10.

Chemical Composition of the Nonclay Fraction

The change in chemical composition of the nonclay fraction with depth, particularly in moderately weathered soil profiles or in profiles in which the most pronounced pedogenic process involved clay migration rather than clay formation, may reveal the state of uniformity of the parent material. This is particularly true for the elements that are present in the resistant minerals or even in the moderately resistant minerals. An example can be cited for the Grundy profile in which the total zircon was determined for each horizon. When the total zircon is expressed not on the basis of the whole soil but on the basis of the nonclay fraction alone (Table 1.4), it is seen that horizons 7, 8, and 9 contain a distinct quantity, which changes abruptly in horizon 6 to another quantity, which again persists in horizons

5 and 4 but which, in horizon 10, changes to another quantity. The constancy of the zircon content in horizons 4, 5, and 6, and its change to another constant value in horizons 7, 8, and 9, would clearly indicate stratification of the parent material at positions 4, 7, and 10. It was already shown, by means of the particle-size distribution, that stratification occurs at these positions in the profile. Another example is given in Table 1.9 in which the ratio of SiO_2/Al_2O_3 in the 0.25 to 0.10 mm fraction remains constant with depth indicating uniformity of parent material as demonstrated by the mineralogical index.

CALCULATING CHANGES DUE TO PROFILE DEVELOPMENT

For many years, geologists have calculated gains and losses involved in the chemical weathering of rocks by assuming that the oxide of one element remains constant during the course of weathering (51). Various oxides, such as Al_2O_3, Fe_2O_3, and SiO_2, have been used for this purpose; the choice depends on the condition of the weathering process and upon the nature of the rock being weathered. The weakness of this method in studying the development of a soil profile lies in the fact that, during the process of soil profile formation, colloidal particles, which may contain all the elements present in the whole rock, move from one point of the profile to another and thereby destroy the constancy of any element that may be chosen for calculating gains and losses of the other elements. It will be shown presently, however, that this problem can be solved by two methods. One is based on the mineralogical, and the other on the chemical, composition of the nonclay fractions of the soil.

Evaluating Profile Development by Mineralogical Analysis

Marshall proposed a method for measuring gains and losses or other changes that may occur during soil formation. This method is based upon a quantitative determination, in every part of the soil profile, of a mineral that is resistant to weathering and at the same time is immobile, that is, it must be contained in the nonclay fraction of the soil. Such a mineral, Marshall termed the *index mineral*. If the resistant mineral to be determined contains an element which is present only in that mineral, as zirconium in zircon, then instead of making a mineralogical analysis one can determine the element chemically, as shown in Table 1.9 for SiO_2 and Al_2O_3 and by Beavers (11) for zirconium and calcium. Marshall recommended zircon, tourmaline, garnet, anatase, or rutile for use as index minerals (47). To this list, the writer (3) would add quartz, albite, and microcline, or the sum of any of these, and in some soils even a whole-size fraction. Choice of the index mineral may vary from one soil to another, according to the mineralogical composition of the parent material of the individual soil. Thus Marshall and Haseman used zircon for a Grundy profile (48); Mikelson used tourmaline, garnet, and zircon for three Ohio profiles (32), and the writer has used the sum of quartz and albite for one profile and the whole fine sand fraction for another profile.

Theoretically, it may appear that determination of an index mineral in any one fraction might be sufficient, but because of experimental errors in fractionation, as well as the possibility that the index mineral may be present in very small amounts, it is advisable to determine the index mineral in two or more particle-size fractions.

The calculations involved in determining total gain or loss and the gain or loss of any one constituent of a soil at any position in a profile are identical to those used for an oxide such as Al_2O_3. Thus, if the percentages by weight of the index mineral of the soils of the various horizons are expressed as R_1, R_2, R_3, . . . R_p, where R_p is that of the parent material, the original weight (X) of, let us say, 100 g of soil from horizon 1 can be calculated from the following expression:

$$XR_p = 100 \times R_1$$

and, therefore,

$$X = 100 \frac{R_1}{R_p} \tag{1}$$

From the value obtained, the total gain or loss can readily be calculated.

The change in volume that may have occurred during soil formation can be determined by dividing the calculated original weight of a soil confined to the present volume by the bulk density of the parent material.

The most serious error that may enter into the evaluation of the changes occurring during the transformation of a parent material to a soil profile is in the determination of the precise nature of the parent material, particularly with respect to the clay content. So far, in most of the soils for which index minerals have been determined, it appears that the soil material, sampled to represent the parent material, either gained clay through migration or, because of stratification, failed to represent the parent material of the whole profile.

If these calculations are carried out not on the basis of 100 g of soil but on the basis of weight of analyzed soil present in a given volume which is bounded by unit surface area and the thickness of each horizon, then one obtains an expression similar to that of Marshall and Haseman as applied to the Grundy profile (48). To make such a calculation, the total amount of the analyzed material present in a given volume must be known. If the bulk density, as commonly measured, truly represents the soil analyzed, and the thickness of each of the horizons of a profile is known, the amount of soil analyzed equals the product of the bulk density times the volume. But in soils which contain voids and materials which are not being analyzed, such as gravel or larger rocks, indurated nodules, and large roots, the bulk density as commonly measured is inadequate. In such soils the total amount of soil material actually analyzed in a given volume must be determined directly. The bulk density of this material can be determined after subtracting from the total volume sampled the volume of the voids and of the materials not analyzed and dividing the weight of material analyzed by this volume.

The method proposed here for calculating profile development rests on the assumption that clay is formed from the nonclay fraction and that in any one horizon the amount of clay formed is proportional to the loss in the nonclay fraction. This proportionality factor may be determined in various ways, depending on the nature of the parent material, as will be shown presently. In parent materials in which losses in the nonclay fraction may be caused by losses other than those due to clay formation, such as losses of $CaCO_3$ and $MgCO_3$, the index mineral should be calculated on the basis of the soil freed of $CaCO_3$ and $MgCO_3$, or the nonclay fraction that yields the clay should be expressed as an entity apart from the $CaCO_3$ and $MgCO_3$ or other substances which are not involved in clay formation. This nonclay fraction in the C horizon is considered to be the *parent material* for clay formation.

The problems that may be encountered in calculating clay formation, clay migration, and other changes by use of index minerals can best be illustrated by the actual application of this method to various profiles.

Changes due to soil development can be evaluated with the greatest certainty in profiles in which the clay distribution with depth indicates that the assumed parent material, or what appears to be the preserved parent material, has gained no clay through migration, and that any clay found in the parent material was there when soil formation began. Such profiles are usually those in which the clay content of the soil material below the B horizon remains constant, as may occur in sedimentary parent materials, or in which the soil material below the B horizon grades into solid rock. In this latter instance, the decomposed rock material could be assumed to represent the parent material, with the further assumption that all the clay found in this material was formed in place.

Another prerequisite for calculating clay formation and migration with least uncertainty is the assumption that the profile was not subjected to erosion or deposition during its development. Furthermore, the calculations

TABLE 1.13. A QUANTITATIVE EVALUATION, BY THE INDEX MINERAL METHOD, OF CLAY FORMATION, CLAY MIGRATION, AND VOLUME CHANGE DURING DEVELOPMENT OF THE SOIL MATERIALS OF SHERIDAN CLAY LOAM PROFILE

Horizon and depth (in.)	Present volume per horizon (cc)*	Bulk density (g/cc)	Present weight of soil	Weight of index mineral†	Present weight of $<5\,\mu$ clay	Present weight of $>5\,\mu$ nonclay	Original weight of $>5\,\mu$	Loss in $>5\,\mu$ nonclay‡ (g)	Original weight of $<5\,\mu$ clay per horizon (g)
			g/horizon						
A_1, 0–6	14.6	1.50	21.9	12.6	6.2	15.7	23.6	−7.9	1.7
A_2, 6–12	14.6	1.56	22.8	13.0	6.7	16.1	24.3	−8.2	1.8
B_1, 12–18	14.6	1.76	25.7	12.9	7.5	18.2	24.1	−5.9	1.8
B_2, 18–24	14.6	1.84	26.9	11.8	5.6	21.2	22.1	−0.9	1.6
B_3, 24–30	14.6	1.80	26.3	12.2	3.5	22.8	22.8	0.0	1.7
C, 30–36	14.6	1.80	26.3	13.1	1.8	24.5	24.5	0.0	1.8
Totals					31.3			−22.9	10.4

Total clay formed = 31.3 − 10.4 = 20.9 g.
Proportionality factor for calculating clay formation per horizon = 20.9 ÷ 22.9 = 0.912.

Horizon	Clay formed per horizon (g)	Total clay in absence of migration (g)	Loss or gain of clay	Weight of parent material	Original volume	Change in volume	Relative clay formation (g)§	Relative clay loss or gain (g)**	Relative change in volume (cc)††
			g/horizon		cc/horizon				
A_1	7.2	8.9	−2.7	25.3	14.1	+0.5	30.5	−19.2	+3.5
A_2	7.5	9.3	−2.6	26.1	14.5	+0.1	30.8	−17.9	+0.7
B_1	5.4	7.2	+0.3	25.9	14.3	+0.3	22.4	+2.1	+2.1
B_2	0.8	2.4	+3.2	23.7	13.2	+1.4	3.6	+24.2	+10.6
B_3	0.0	1.7	+1.8	24.5	13.6	+1.0	0.0	+13.2	+7.6
C	0.0	1.8	0.0	26.3	14.6	0.0	0.0	0.0	0.0

* Per 1 cm^2 of surface × thickness of horizon in cm.
† Quartz + albite.
‡ Due to soil formation.
§ Per 100 g nonclay of parent material.
** Per 100 cc of parent material.
†† Per 100 cc of parent material. If it were assumed that the change in volume occurs only in the vertical direction, then the change in volume would be a measure of the change in thickness of the horizons.

must be made on the basis of amounts of the analyzed soil restricted to volumes bounded by unit surface area and thicknesses of the horizons. Calculations for such a profile are shown in Table 1.13 and involve the following steps:

1. Calculating the amount of analyzed soil present in the volumes bounded by unit surface area and the thickness of each horizon by means

of the bulk densities of the soil material at each horizon or by direct determination during sampling.

2. Calculating the amounts of index mineral, nonclay fraction, and clay fraction at each horizon.

3. Calculating the amounts of nonclay and clay originally present at each horizon with the aid of the index mineral.

4. Calculating the loss in nonclay at each horizon.

5. Calculating the total clay formed in the profile by equating it to the difference between the total clay now present and the total clay originally present.

6. Calculating the clay formed at each horizon by dividing the total clay formed by the total loss of the nonclay of the whole profile and multiplying by the loss of the nonclay at each horizon. (It is assumed that the clay formed is proportional to the loss of nonclay.)

7. Calculating the total clay that would have been present at each horizon had no clay migrated, that is, it is equal to the sum of the clay formed and the clay originally present.

8. By subtracting the total clay from the actual clay present, calculating the loss or gain of clay at each horizon due to migration.

9. Calculating changes in volume, as expressed by the change in thickness of the original soil material at each horizon, by dividing the sum of the original nonclay and clay by the bulk density of the parent material and comparing the result with the present soil thickness.

10. Calculating relative clay formation at each horizon by dividing the clay formed by the original amount of nonclay and multiplying by 100.

11. Calculating relative clay loss or gain at each horizon by dividing the loss or gain in clay by the original volume of each horizon and multiplying by 100.

12. Calculating relative changes in volume or thickness (assuming the change in volume is in only one direction) by dividing the change in thickness of each horizon by the original thickness and multiplying by 100.

The foregoing method for calculating clay formation and clay originally present at each horizon cannot be applied to profiles in which part of the clay of the assumed parent material is believed to have been gained through migration but in which the nonclay fraction represents the original nonclay. In this example it is again assumed that no erosion or deposition has occurred. Calculations for such a profile, shown in Table 1.14 involve the following steps:

1. By means of the bulk densities of the soil material, calculating the amount of soil present in the volumes bounded by unit surface area and the thickness of each horizon.

TABLE 1.14. QUANTITATIVE EVALUATION, BY THE INDEX MINERAL METHOD, OF CLAY FORMATION AND CLAY MIGRATION DURING DEVELOPMENT OF THE SOIL MATERIALS OF BETHEL SILT LOAM PROFILE

Horizon and depth (in.)	Present volume per horizon (cc)*	Bulk density (g/cc)	Present weight of soil	Weight of index mineral†	Present weight of < 2 μ clay	Present weight of > 2 μ nonclay	Original weight of > 2 μ	Loss in > 5 μ‡ (g)	Weight of clay formed (loss × 0.866) (g)
			g/horizon						
A_1, 0–3	7.6	1.08	8.2	0.00123	1.2	6.2	7.8	−1.6	1.4
A_2, 3–7	10.0	1.22	12.4	0.00186	1.9	10.3	11.8	−1.5	1.3
A_3, 7–14	17.8	1.34	23.8	0.00319	4.3	18.2	20.2	−2.0	1.7
B_2, 14–24	25.4	1.43	36.3	0.00376	14.2	21.4	23.8	−2.4	2.1
B_3, 24–31	17.8	1.51	26.8	0.00268	9.4	16.9	16.9	0.0	0.0
Totals					31.0		80.5		6.5

Total clay originally present = 31.0 − 6.5 = 24.5.

Horizon	Original weight of 2 μ clay (g)	Total clay in absence of migration (g)	Loss or gain of clay per horizon (g)	Relative clay formation (g)§	Relative clay loss or gain (g)**	Present weight of clay < 0.05 μ per horizon (g)
A_1	2.4	3.8	−2.6	17.9	−34.2	0.26
A_2	3.6	4.9	−3.0	11.0	−30.0	0.35
A_3	6.1	7.8	−3.5	8.4	−19.7	1.17
B_2	7.2	9.3	+4.9	8.8	+19.3	5.03
B_3	5.2	5.2	+4.2	0.0	+23.6	3.26

* Per cm^2 of surface thickness of horizon in cm.
† Tourmaline.
‡ Due to soil formation.
§ Per 100 g nonclay of parent material.
** Per 100 cc of present soil.

2. Calculating the amount of the index mineral and nonclay fraction present at each horizon.

3. Calculating the amount of nonclay originally present at each horizon.

4. Calculating the loss in the nonclay fraction at each horizon.

5. Calculating the amount of clay formed at each horizon by multiplying the amount of loss of the nonclay by a proportionality factor which relates the real density of the clay formed to the real density of the clay-forming minerals of the nonclay fractions. It may be necessary to establish this factor for each soil studied, but for many soils it was found that an average real density of the clay is approximately equal to 2.40 g/cc and for the nonclay minerals about 2.77 g/cc; the proportionality factor is equal, therefore, to 2.40/2.77 or 0.866 = 0.87. Use of this factor is believed to be justified on the basis that during the transformation of the crystal structure

of the nonclay minerals to that of the clay minerals, the change in the real volumes is negligible, since the volume of both structures is determined essentially by closely packed oxygens characteristic of the aluminosilicate minerals. On the basis of this assumption, therefore, the following relation holds:

$$\frac{\text{weight loss of nonclay}}{\text{real density of nonclay}} = \frac{\text{weight of clay formed}}{\text{real density of clay}}$$

Consequently

$$\text{weight of clay formed} = \text{weight loss of nonclay} \times \frac{\text{real density of clay}}{\text{real density of nonclay}}$$

6. Calculating the total amount of clay originally present in the parent material for the whole profile by subtracting the total clay formed from the total clay present.

7. Calculating the amount of clay originally present at each horizon by dividing the total clay originally present by the total nonclay originally present and multiplying by the amount of nonclay originally present at each horizon.

8. The calculations of the amounts of clay lost or gained at each horizon are exactly the same as outlined previously.

9. Changes in volume cannot be calculated for such a profile since the bulk density of the assumed preserved parent material is not that of the true parent material.

10. Relative clay formation is calculated as indicated previously, but relative loss or gain in clay is calculated on the basis of the present volume of the soil, since the original volumes are not known.

A third method for calculating the loss or gain of clay at each horizon consists of a combination of the two methods previously described: (*a*) the clay originally present at each horizon is calculated on the basis of the assumption that the clay content of the preserved parent material represents the original clay content of the parent material, and (*b*) the amount of clay formed at each horizon is equal to the loss of the nonclay multiplied by a proportionality factor that relates the real density of the clay and the clay-forming minerals of the nonclay fractions. This method is particularly useful for profiles that may have suffered erosion during their development or a gain of clay by deposition from some outside source other than through formation from the nonclay fractions. With the results obtained by this method, the extent of erosion during soil profile development can be calculated in terms of the thickness of the soil eroded away. An example of this method of calculation is shown in Table 1.15.

TABLE 1.15. QUANTITATIVE EVALUATION, BY THE INDEX MINERAL METHOD, OF CLAY FORMATION, CLAY MIGRATION, AND VOLUME CHANGE DURING DEVELOPMENT OF THE SOIL MATERIALS OF ADELANTO LOAMY SAND PROFILE

Horizon and depth (in.)	Present volume per horizon (cc)	Bulk density (g/cc)	Present weight of soil	Weight of index mineral*	Present weight of <5 μ clay	Present weight of >5 μ nonclay	Original weight of >5 μ nonclay	Loss in >5 μ† (g)	Weight of clay formed (loss × 0.866) (g)
			g/horizon						
A_1, 0–9	23.0	1.49	34.3	8.3	4.4	29.9	37.6	−7.7	6.7
A_2, 9–16	15.4	1.50	23.1	5.6	2.1	21.0	25.3	−4.3	3.7
B_1, 16–30	35.8	1.73	62.0	9.9	17.4	44.5	44.9	−0.4	0.3
B_2, 30–43	33.3	1.58	52.7	9.7	8.8	43.9	43.9	0.0	0.0
C, 43–63	51.2	1.55	79.6	15.5	9.4	70.2	70.2	0.0	0.0

Horizon	Original weight of <5 μ per horizon (g)	Total clay in absence of clay migration (g)	Loss or gain of clay	Weight of parent material	Original volume	Change in volume	Relative clay formation (g)‡	Relative clay loss or gain (g)**	Relative change in volume (cc)**
			g/horizon		cc/horizon				
A_1	5.0	11.7	−7.3	42.6	27.4	−3.4	17.8	−26.6	−12.4
A_2	3.4	7.1	−5.0	28.7	18.5	−3.1	14.6	−27.0	−16.8
B_1	6.0	6.3	+11.1	50.9	32.7	+3.1	0.7	+34.0	+9.5
B_2	5.9	5.9	+2.9	49.8	32.0	+1.3	0.0	+9.1	+4.1
C	9.4	9.4	0.0	79.6	51.2	0.0	0.0	0.0	0.0

* The fine sand fraction.
† Due to soil formation.
‡ Per 100 g of nonclay of parent material.
** Per 100 cc of parent material (see footnote †† of Table 1.13).

Note: Surface soil eroded away during profile formation can be calculated as follows: The amount of clay gained by the B_1 and B_2 horizons in excess of that lost by the A_1 and A_2 horizon, 1.7 g (14.0 − 12.3 = 1.7 g), is assumed to have been lost from a soil similar to that of A_1 horizon. Since a volume of 23.0 cc of the A_1 horizon lost 7.3 g of clay, the excess of 1.7 g of clay gained by the B_1 and B_2 horizon must have been lost by a volume of (23.0/7.3) × 1.7 = 5.4 cc. In terms of horizon thickness per cm² of surface, this loss would amount to 5.4 cm.

Changes in parent material during formation of a soil profile, other than changes in clay content, such as in the size distribution of particles other than clay, mineralogical composition, and chemical composition, can also be determined by the proposed methods. Examples of such calculated changes in the mineralogical and chemical composition of 100 g of parent material during its transformation to soil material of the A horizon of the Sheridan clay loam profile are shown in Tables 1.16 and 1.17.

For comparison of the relative intensity of soil development of different soil materials, with either depth or location, the calculated changes on a profile basis, that is, per square centimeter and thickness of each horizon, are not adequate, since the thicknesses of the horizons and the bulk densities may vary both with depth and from profile to profile. This difficulty, however, may be resolved by expressing the changes not only on a profile-volume basis but also on an equal-weight basis, such as 100 g of parent material or 100 g of nonclay, or on an equal-volume basis, such as 100 cc of parent material or present soil material (see Tables 1.13, 1.14, and 1.15).

Since the relative intensity of soil development varies with depth, developments of different profiles should be compared strictly on an equivalent-depth basis. For example, the relative intensity of clay formation of two profiles should be based on the amount of clay formed per 100 g of nonclay of the parent material, let us say, at the 0- to 6-inch level, or the comparison may be made on a strictly equivalent-volume basis, for ex-

TABLE 1.16. CHANGE IN MINERALOGICAL COMPOSITION OF 100 G OF PARENT MATERIAL DURING ITS TRANSFORMATION TO SOIL MATERIAL OF THE A_1 HORIZON OF THE SHERIDAN CLAY LOAM PROFILE

Material	Total (g)	Quartz + albite (g)	Hornblende (g)	Biotite (g)	Clay <5 μ (g)	Uncombined oxides (g)	Organic matter (g)	Loss in mineral matter (g)	Loss due to clay migration (g)	Loss due to clay formation (g)
Parent	100	49.0	24.1	18.7	6.9	1.0	0.3			
Soil	95.0	49.0	7.2	4.9	24.2	3.4	6.1			
Loss or gain*	−5.0	0.0	−16.9	−13.8	+17.3	+2.4	+5.8	−11.0	−10.7	−0.3

* Due to soil formation.

TABLE 1.17. CHANGE IN CHEMICAL COMPOSITION OF 100 G OF PARENT MATERIAL DURING TRANSFORMATION TO THE INORGANIC SOIL MATERIAL OF A_1 HORIZON OF THE SHERIDAN CLAY LOAM PROFILE

Material	% of Weight of material after ignition								
	Total	SiO_2	TiO_2	Fe_2O_3	Al_2O_3	CaO	MgO	K_2O	Na_2O
Parent material	100.0	57.40	1.50	14.02	16.33	4.80	2.80	0.50	2.70
Soil material	86.6	53.15	1.06	9.64	14.10	3.84	1.88	0.26	2.61
Total loss due to soil formation	13.4	4.25	0.44	4.38	2.23	0.96	0.92	0.24	0.09
Loss in oxides due to clay migration	10.7	4.98	0.27	2.51	2.23	0.21	0.28	0.09	0.04
Gain or loss in oxides due to clay formation	−2.7	+0.73	−0.17	−1.87	0.00	−0.75	−0.64	−0.15	−0.05

ample, a volume bound by a surface of 10 cm^2 and a depth of 10 cm. Either basis of comparison leads to the conclusion that, for evaluating soil development, soil profiles should be sampled on a completely uniform basis, that is to say all soil profiles should be sampled in an identical manner or the results extrapolated to a uniform-depth basis. Since the thickness of equivalent horizons of different profiles may vary, it would be necessary either to sample each profile twice, once on a strictly horizon basis and once on a predetermined depth basis, or if sampled once, on a horizon basis, the results obtained should also be expressed on a predetermined depth basis.

Evaluating Profile Development by Chemical Analysis

The methods proposed for evaluating clay formation and clay migration based on the determination of index minerals are, at present, of limited value because very few data are available to which the methods can be applied. It is hoped, however, that the elucidation of these methods may stimulate mineralogical analysis of soils for index minerals and chemical analysis for index elements which could be applied to a determination of soil development.

As pointed out earlier, the methods of evaluating clay formation rest on the assumption that the clay fractions are the products of weathering, whereas the nonclay fractions are the reactants; the amount of clay formed must, therefore, be proportional to the loss in those minerals of the nonclay fraction which alter to clay. Since chemical composition of a material reflects its mineralogical composition, the degree of change in the chemical composition of the nonclay fraction of the parent material as a result of soil development should also reflect the amount of clay formed. An accurate measurement, therefore, of the degree of change in the chemical composition of the nonclay fraction of the parent material, as a result of soil development, could be utilized to calculate the amount of clay formed. Such a method would be very valuable, since chemical analyses of many soils are available and could be utilized for evaluating soil profile development under various conditions.

The proposed method rests on the availability of data on the chemical composition of the whole soil and of the clay, and on the amount of clay at each horizon of a soil profile. With such data at hand the steps involved in calculating clay formation and clay migration are as follows:

1. *Calculating the chemical composition of the nonclay fraction.* This can be accomplished with the aid of the following relation:

$$\text{Composition of soil} \times 100 = \text{composition of clay} \times \text{per cent of clay} + \text{composition of nonclay} \times (100 - \text{per cent clay})$$

Since the composition of the nonclay is the only unknown, this equation can readily be solved. (The word "composition" in this equation represents the amount of any elemental oxide in 100 g of material.)

2. *Calculating the amount of clay formed in any horizon from 100 g of nonclay of the parent material.* Let this amount of clay be designated by f. Changes in chemical composition of the nonclay fraction of the parent material during soil formation may be due to a loss in elemental oxides strictly associated with clay formation, and to a loss of elemental oxides or carbonates not involved in clay formation; a change in chemical composition may result from a gain in certain elemental oxides such as water and CO_2. Therefore, in calculating the changes in chemical composition of the nonclay fraction of the parent material brought about by clay formation, it is necessary to choose those elements in which the change is strictly the result of clay formation. These elements must be chosen separately for each soil and according to the nature of the chemical composition of the clays. The most common elements (expressed as oxides) found useful for this purpose are SiO_2, Al_2O_3, Fe_2O_3, and occasionally MgO and K_2O. Although the loss of a given elemental oxide of the nonclay fraction may be strictly proportional to the amount of clay formed, the amount left in the nonclay fraction, when expressed on a percentage basis of the altered nonclay fraction, could also be affected by the loss, or possibly a gain, of other elemental oxides that are not involved in clay formation. This change in composition can be expressed as a proportionality factor (k), which is related to the amount of the nonclay fraction left unaffected by losses or gains due to causes other than to those associated with clay formation.

In the light of the foregoing statements, the change in the SiO_2 content, for example, of the nonclay fraction of the parent material as a result of clay formation, as well as other changes may be expressed by the following equation:

$$a_{\cdot} \times 100 - a' \times f = a_2(100 - f)k \qquad (2)$$

where $a_{\cdot}$ = per cent of SiO_2 of the nonclay fraction of the parent material; a' = per cent of SiO_2 of the clay formed, let us say, in the A horizon; a_2 = per cent of SiO_2 of the nonclay fraction of the A horizon; f = amount of clay formed in the A horizon from 100 g of nonclay of the parent material; and k = the proportionality factor affecting the SiO_2 content of the nonclay of the A horizon due to losses or gains other than to clay formation.

Since there are two unknowns, f and k, to be determined, another equation, which would also contain these two unknowns must be set up. Such an equation can readily be set up for another element, let us say Fe_2O_3, the change in the content of which was also due to clay formation and to losses or gains not involved in clay formation. This second equation is as follows:

$$b_{\prime} \times 100 - b' \times f = b_2(100 - f)k \qquad (3)$$

where $b_{\prime}$ = per cent of Fe_2O_3 in the nonclay fraction of the parent material; b' = per cent of Fe_2O_3 of the clay formed; b_2 = per cent of Fe_2O_3 of the nonclay fraction of the A horizon; and f and k the same unknowns as in equation (*2*).

The solution for f and k, therefore, is as follows. Divide equation (*2*) by (*3*) and solve for f, thus:

$$\frac{a_{\prime} \times 100 - a' \times f}{b_{\prime} \times 100 - b' \times f} = \frac{a_2(100 - f)k}{b_2(100 - f)k} = \frac{a_2}{b_2} \qquad (4)$$

and let $a_2/b_2 = B$
then

$$f = \frac{a_{\prime} - Bb_{\prime}}{a' - Bb'} \times 100$$

Once f is known, numerically, the solution for k can readily be found by either equation (*2*) or equation (*3*).

To test the validity of choice of the elements for the foregoing problem k should be calculated from both equation (*2*) and equation (*3*). If k is the same for the two calculations, it is proof that the choice of the elements is correct. But if k is not the same it indicates that the choice was incorrect, and the calculation for f should be repeated with another element until k is the same for both equations.

A source of difficulty in these calculations is encountered in profiles in which the chemical composition of the clay varies greatly with depth. In such profiles the chemical composition of the clay present in the C or the lower B horizon was found to be a more satisfactory choice than the clay present in the horizon for which clay formation is calculated. The reason for this choice is based on the assumption that the composition of the clay in these horizons reflects more truly the composition of the clay initially formed. Whereas the clay present in the surface horizons might have undergone considerable change after formation, the clay present in the C or lower B horizon would have suffered the least amount of change after its formation.

3. *Calculating the amount of clay formed in the A horizon which resulted in the present amount of nonclay in 100 g of soil.* Let this quantity be designated by F. If the amount of clay in 100 g of soil in the A horizon is equal to d g then the nonclay is equal to $(100 - d)$ g.

When f g of clay is formed from 100 g of nonclay of the parent material, the amount of nonclay left would be equal to $(100 - f)$ g, if no other change has occurred in the nonclay. But in the preceding calculations it was assumed that a change in the nonclay, other than a change to clay formation, does occur and that this change is measured by the proportionality factor k. The amount of the nonclay left in the A horizon upon the formation of f g of clay, which is of the same composition as the nonclay of the present A horizon, would therefore equal $(100 - f) \times k$. Since 100 g of soil of the A horizon contains $(100 - d)$ g of nonclay, the amount of clay formed which would yield this $(100 - d)$ g of nonclay must be

$$\frac{f}{(100 - f)k} \times (100 - d) \quad \text{or} \quad F \text{ g}$$

And the original amount of nonclay after F g of clay is formed, but before other changes have occurred, is equal to $(100/f - 1) \times F$; and the original amount of nonclay of the parent material, before clay formation occurred in the A horizon, is equal to $(100/f) \times F$, which we shall designate O.

4. *Calculating the amount of clay originally present in the A horizon.* Let this quantity be designated by N. There are two methods for calculating the amount of clay originally present in the A horizon, or any other soil horizon, depending on whether or not the clay content of the preserved parent material represents the original clay content.

If the clay content of the preserved parent material represents the original clay content, then the calculations are as follows: Let c represent the original clay content of 100 g of parent material. Then the nonclay associated with it is equal to $(100 - c)$ g, and since $(100/f) \times F$ g of nonclay of parent material yields the nonclay of the soil material (for example, $(100 - d)$ g in the A horizon), then the original clay associated with it must have been

$$\frac{c}{(100 - c)} \times \frac{100 \times F}{f} = N$$

If part of the clay associated with the nonclay in the preserved parent material is believed to have been gained through accumulation, to calculate the original clay content the bulk densities or the total amount of analyzed soil and also the thickness of each horizon must be known. Also, it must be assumed that no erosion or deposition has occurred. The steps involved in the calculation are as follows:

(*a*) Calculating the amount of analyzed soil material present in a volume of each horizon bounded by 1 cm^2 and the thickness of the horizon.

(*b*) Calculating the amount of clay formed for the amounts of soil found in step (*a*): the amount of clay formed per 100 g of soil (*F*) is divided by 100 and then multiplied by the amount of soil in the desired volume.

(*c*) Calculating the original nonclay of the parent material for the amounts of soil found in step (*a*): the calculated original nonclay per 100 g of soil (*O*) is divided by 100 and multiplied by the amount of soil in the desired volume.

(*d*) The amount of clay originally present at each horizon is then calculated by subtracting the total amount of clay formed in the profile from the total amount of clay actually present in the profile, and the result (which is the total clay originally present in the profile) is divided by the total nonclay originally present in the profile and multiplied by the amount of nonclay originally present at each horizon.

5. *Calculating the loss or gain in clay at each horizon during soil profile development.* The loss or gain in clay is calculated by adding the clay formed to the clay originally present at each horizon, either for 100 g or for the amount of soil in a desired volume, and subtracting the sum from the actual amount of clay present at each horizon.

6. *Calculating the change in volume of each horizon during soil development.* This calculation can be carried out if the bulk densities at each horizon are known and the bulk density of the preserved parent material represents the bulk density of the parent material.

The calculation is as follows: the *original nonclay* and the *original clay* at each horizon for the desired volume are added together, and the sum, which is the original weight of whole parent material, is divided by the bulk density of the parent material. The result represents the original volume of the horizon; a comparison with the present volume of the horizon indicates the change during soil development. If it is assumed that the change in volume is only in the vertical direction, then the change in volume would represent the change in thickness of the horizons.

7. *Calculating the change in the total chemical composition of the soil at each horizon during soil profile development.* This can be determined by calculating the total amount of the elemental oxides present in the original amount of the parent material at each horizon—for the amount of soil in a desired volume, for 100 g of present soil, for 100 g of parent material, or for 100 cc of parent material at each horizon—and comparing these values with those of the present soil.

Examples of the application of the proposed methods for calculating clay formation, clay migration, changes in volume, and changes in chemical composition during development of the A and B horizons of a Dayton silt loam profile (59) are shown in Tables 1.18 and 1.19.

The proposed method for evaluating clay formation based on the change

TABLE 1.18. QUANTITATIVE EVALUATION, BY THE CHEMICAL ANALYSIS (SiO_2 AND Fe_2O_2 OF NONCLAY AND CLAY) METHOD OF CLAY FORMATION, CLAY MIGRATION, AND VOLUME CHANGES DURING THE DEVELOPMENT OF THE SOIL MATERIAL OF THE A_2 AND B HORIZONS OF DAYTON SILT LOAM PROFILE

Horizon	Depth (in.)	Present $<2\mu$ clay* (g)	f (g)	k	F (g)	N (g)	$F+N$ (g)	Loss or gain of clay* (g)	O (g)	$O+N$ (g)
A_2	6–10	16.7	18.5	0.92	20.7	33.4	54.1	−37.4	112.0	145.4
B	18–22	48.7	9.9	0.96	5.9	17.8	23.7	+25.0	59.6	77.4
C	42–50	23.0	0.0	1.00	0.0	23.0	23.0	0.0	77.0	100.0

Horizon	Present volume per horizon (cc)	Bulk density (g/cc)	Weight of present soil (g/horizon)	Weight of parent material (g/horizon)	Original volume (cc/horizon)	Change in volume (cc/horizon)	Relative clay loss or gain (g) †	Relative clay loss or gain (g) ‡	Relative change in volume§ (cc)
A_2	10.2	1.22	13.4	19.4	14.0	−3.8	−35.7	−25.7	−27.1
B	10.2	1.56	15.9	12.3	8.8	+1.4	+45.2	+32.3	+15.9
C	20.3	1.39	28.2	28.2	20.3	0.0	0.0	0.0	0.0

For explanation of the symbols f, k, F, N, and O, see text.
* Per 100 g soil.
† Per 100 cc of parent material.
‡ Per 100 g parent material.
§ Per 100 cc of parent material (see footnote †† of Table 1.13).

in the chemical composition of the nonclay fraction was applied to a large number of soils to determine the effect on clay formation and clay migration of the factors of soil formation (7). This lead to the following conclusions:

1. In all soil profiles, regardless of their classification, clay formation decreases with increasing distance of the parent material from the surface of the soil with one exception, that is, maximum clay formation occurs not at the immediate surface but in a layer below the surface at a depth ranging from 2 to 20 inches.

2. A comparison of the quantities of clay formed with depth in any one profile with the quantities of clay present leaves no doubt that clay migration is the most important process in the occurrence of a horizon of clay accumulation, claypan formation, and the large changes in chemical composition of the whole soil.

3. In the transformation of a parent material to a soil material, clay formation is mainly responsible for the changes in the mineralogical composition, but clay migrations, reflected by losses or gain of clay, are mainly responsible for the changes in chemical composition.

TABLE 1.19. CHANGE IN CHEMICAL COMPOSITION OF THE MAJOR ELEMENTAL OXIDES OF PARENT MATERIAL DURING TRANSFORMATION TO SOIL MATERIAL OF THE A_2 AND B HORIZONS OF THE DAYTON SILT LOAM PROFILE

Material	Total	SiO_2	Al_2O_3	Fe_2O_3	TiO_2	MgO	CaO	K_2O	Na_2O
	g/100 g parent material								
A_2 Horizon									
Parent material	100.0	65.44	18.06	7.31	1.26	2.13	2.18	1.92	1.30
Soil material	68.8	51.35	9.58	3.08	0.90	0.63	0.67	1.56	0.92
Total loss due to soil formation	31.2	14.09	8.48	4.23	0.36	1.50	1.51	0.36	0.38
Loss in oxides due to clay migration	25.7	14.38	6.14	3.52	0.34	0.67	0.36	0.30	0.00
Gain or loss of oxides due to clay formation	−5.5	+0.29	−2.34	−0.71	−0.02	−0.83	−1.15	−0.06	−0.38
B Horizon									
Parent material	100.0	65.44	18.06	7.31	1.26	2.13	2.18	1.92	1.30
Soil material	129.2	83.20	24.97	11.53	1.75	2.47	1.91	1.85	1.16
Total gain due to soil formation	29.2	17.76	6.89	4.22	0.49	0.34	−0.27	−0.07	−0.14
Gain in oxides due to clay migration	32.3	18.06	7.65	4.43	0.43	0.84	0.45	0.36	0.0
Gain or loss of oxides due to clay formation	−3.1	−0.30	−0.76	−0.21	+0.06	−0.50	−0.72	−0.43	−0.14

4. Clay formation affects mostly the content of the alkali and alkaline earth metals, whereas clay migration affects mostly the silica and sesquioxide content of the parent material in its transformation to a soil.

5. An increase in both temperature and rainfall increases total clay formation of a given parent material. Often this increase in clay formation is most sharply expressed in the lower horizons of the profile rather than in the surface. This increase is not only in the amount of clay formed at each horizon but also in the increased depth to which the parent material weathers. Among the Great Soil Groups belonging to the Pedalfer category, clay

formation was found to be in the following increasing order: Podzols < Gray Brown Podzols < Brown Earth < Prairies < Red and Yellow Latosols < Laterites.

6. Topographical conditions which decrease the effective drainage of a soil appear to enhance rate of clay formation.

7. Grass-type vegetation is more effective in promoting clay formation than tree-type vegetation.

8. The following properties of the parent material were found to affect clay formation: (*a*) mineralogical composition, (*b*) texture of the minerals, (*c*) chemical composition, (*d*) porosity and density, (*e*) structure and fabric, and (*f*) degree of consolidation. Whether the parent materials are igneous, metamorphic, or sedimentary rocks is important only so far as these properties are affected.

Clay formation is enhanced by a larger content in the parent material of nonresistant minerals, the finer texture of these materials, and higher base and Al_2O_3 content in them. High porosity and a structure and fabric of parent materials which enhance the water-holding capacity and permeability to water and gas were found also to enhance clay formation.

9. Even under the most intense weathering conditions, as in Latosols, the annual rate of clay formation is very small, ranging from 0.0001 to 0.002 g per 100 g parent material.

FACTORS DETERMINING RELATIVE STABILITY OF MATERIALS

In the discussion of methods for evaluating soil development, it was pointed out that the determination of clay formation rests on the assumption that the clay is formed from the nonclay. The chemical and mineralogical composition of the aluminosilicate minerals of the nonclay fraction and the factors that determine their stability or chemical reactivity should, therefore, be of primary interest in a study of soil development.

The aluminosilicate minerals may be viewed as aggregates of oxygen ions, the interstices of which are occupied by the cations that hold the structure together (12). The size of the interstice determines the size of the cation that can occupy it, and the stability of the structure depends largely on how well the cations fit these interstices. The most important interstice is in the center of the oxygen tetrahedron, since this unit of structure is the primary building block of all the aluminosilicates. If oxygens are considered as rigid spheres with a radius of 1.32 Å, a sphere that just fits the inside of the interstice of a tetrahedron would have a radius of 0.30 Å. Only cations with radii approximating this radius may be present in such an interstice without distorting the tetrahedron. Only silicon ion, with a radius of 0.39 Å, fits this interstice without undue distortion, and such a structure is, therefore, the most stable one. An Al^{3+}, with a radius of 0.57 Å in such an interstice distorts the tetrahedron considerably. Consequently, the presence of Al^{3+} in tetrahedral positions should impart to an aluminosilicate mineral a degree of instability proportional to the number of such ions. There is considerable evidence in support of this reasoning.

Another important interstice is in the octahedral positions of the micas, amphiboles, and the clay minerals. The radius of a sphere that would just fit in an octahedral position is calculated to be 0.70 Å. Among the ions commonly found in this position, the Mg^{2+} with a radius of 0.78 Å and Fe^{3+} with a radius of 0.67 Å appear to fit without distortion, but Al^{3+} with a radius of 0.57 Å and Fe^{2+} with a radius of 0.83 Å are either too small or too large. As expected, structures with Mg^{2+} and Fe^{3+} are, therefore, more stable than structures with Al^{3+} and Fe^{2+}. Furthermore, structures with octahedrons in which the interstices are smaller than in the mica-type structure, as in gibbsite with an interstice just large enough to accommodate a sphere of 0.55 Å radius, form a much more stable structure. This is indicated by the fact that gibbsite may be a product of weathering of biotite, muscovite, or many other aluminosilicates.

The most important structural units of the aluminosilicate minerals are the silica and alumina tetrahedrons. Their linkage to form the framework of the minerals has been used as a basis for classification of the aluminosilicates (12, 63).

The following classification of the structure based on the method of linkage of the tetrahedrons is commonly adopted.

1. Independent tetrahedrons sharing no oxygen atoms with neighboring tetrahedrons; the tetrahedrons are linked together with basic cations—the orthosilicate-type structure. Examples are olivine, zircon, and garnet.

2. Separate tetrahedrons sharing one or more corners with neighboring tetrahedrons: (*a*) Si_2O_7—double tetrahedrons; (*b*) five tetrahedrons in open group; (*c*) rings of three tetrahedrons; and (*d*) rings of six tetrahedrons.

3. Infinite chains of tetrahedrons—the metasilicate-type: (*a*) single-chain linkage (pyroxenes); and (*b*) double-chain linkage (amphiboles).

4. Infinite sheets of tetrahedrons, two-dimensional linkage (talc, micas, and clay minerals).

5. Continuous framework of linked tetrahedrons—sharing all four oxygen atoms with the neighboring tetrahedrons (feldspars, feldspathoids, zeolites, and quartz).

The foregoing classification of aluminosilicate minerals was found extremely useful in detecting the sequence of crystallization under igneous conditions and also the sequence of resistance to decomposition by weathering conditions (23, 63, 77).

The order of crystallization from a structural point of view was found to be as follows: independent tetrahedrons → single chains → double chains → sheets. From the point of view of chemical composition, the larger the number of independent tetrahedrons, the more basic is the mineral, that is the larger is the ratio of basic cations to Si^{4+}. Similarly, the larger the number of alumina tetrahedrons, the more basic is the mineral. The most basic minerals crystallize first, followed by minerals with a decreasing degree of basicity. The order of crystallization accordingly has been pictured as follows:

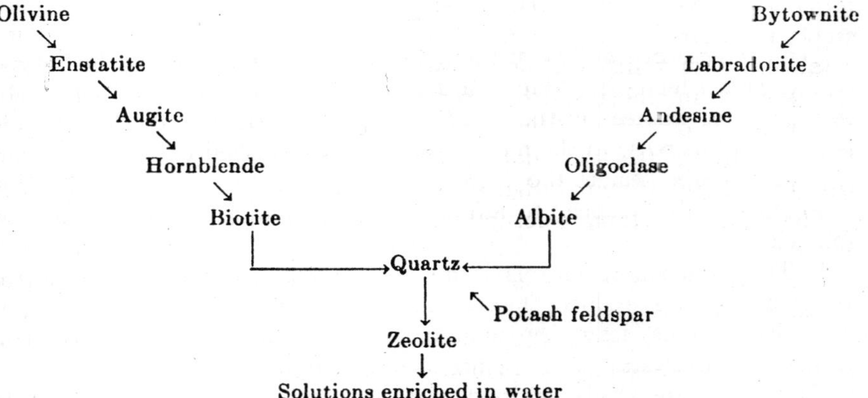

The increasing basicity of the minerals in the branch rising to the left of quartz is due to a decreasing degree of tetrahedral linkage, whereas in the branch to the right, it is due to an increased number of alumina tetrahedrons. During the course of crystallization indicated above, the temperature falls. Apparently, crystallization of minerals with a sheet structure is favored by a lowering base content and a decreasing temperature.

The relative stability of the minerals to weathering and to decomposition with acids also appears to be related to the degree of basicity, degree of linkage of the tetrahedrons, relative number of alumina and silica tetrahedrons, and other factors that induce a lowering of the basicity of the mineral and a destruction of bonds linking the tetrahedrons. Various listings of the relative stability of minerals in weathering follow closely the order of crystallization, that is the less basic the mineral the more stable it is. One such list is:

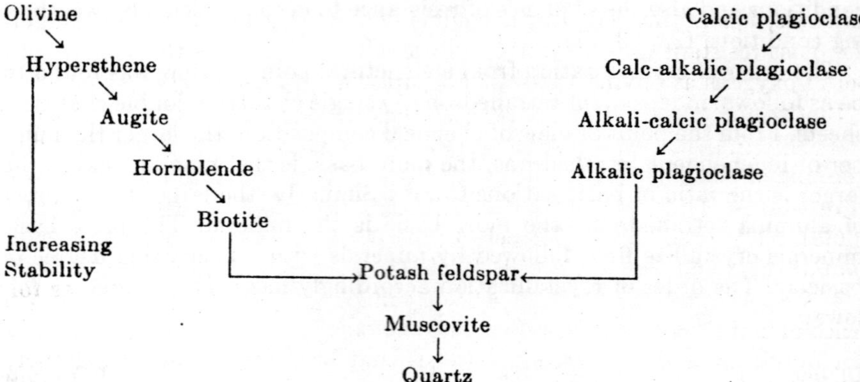

The decrease in stability of the minerals in the branch rising to the left of the potash feldspar is due to a decrease in the degree of linkage of the tetrahedrons, whereas in the branch rising to the right of it, the decrease is due to an increase in the number of alumina tetrahedrons.

Other factors, besides the degree of linkage and the number of alumina tetrahedrons that appear to affect the stability of the minerals are the following:

1. The presence of ferrous iron or other cations that oxidize during weathering greatly reduces the structural stability, for upon oxidation some other cation must leave the structure to maintain the electrostatic neutrality of the crystal lattice. Since such a cation may be involved in the linkage of the tetrahedrons, its departure would weaken the structure. In fact, the presence of Fe^{2+} in minerals is one of the most important factors contributing to their instability in weathering.

2. The more closely the oxygens are packed about the cation other than in tetrahedral positions, that is, the smaller the volume occupied by a given number of oxygens, the more stable is the mineral. To demonstrate this relation, it is necessary to compare minerals with an identical type of tetrahedral linkage and with the same number of alumina tetrahedrons. Two examples can be cited, one with the feldspar-type linkage and one with the olivine-type linkage.

The greater resistance of microcline as compared with orthoclase, and of albite as compared with soda-orthoclase, may be attributed to the smaller volumes occupied by the oxygens of the microcline and albite, due to their triclinic symmetry as contrasted with the monoclinic symmetry of the orthoclase.

The contrast in resistance of olivine and zircon is a more striking example of the effect of tightness of packing. Thus olivine, one of the least resistant minerals, has a unit cell with a volume of 291 Å^3, whereas zircon, one of the most resistant minerals, has a unit cell which contains the same number of oxygens as olivine but which occupies a volume of only 231 Å^3.

Biotite and muscovite, minerals with the same type of tetrahedral linkage and with the same number of alumina tetrahedrons, can be cited as an example of the contribution of both volume and composition to difference in resistance. Thus, biotite, which is the less resistant of the two minerals, contains Fe^{2+} and has a unit cell of 489 Å^3, whereas muscovite does not contain Fe^{2+} and has a unit cell of 459 Å^3.

3. Closely related to tightness of packing of the oxygens is the occurrence of empty positions in certain parts of the structure. Empty positions not only reduce the electrostatic forces that bind the structure together, they also serve as ports of entry and departure to the interior of a crystal particle and thereby serve to accelerate reactions, involving cations in the interior of the particle, which alter the particle.

An example of structures in which empty positions are common is the micas; these positions are the cavities in the center of the hexagonal rings of oxygens on the bases of the tetrahedrons. Such empty positions facilitate greatly the exchange of K^+, normally found in these positions, with other cations such as Ca^{2+}, Mg^{2+}, Na^+, or H^+ which alter the structure to that of vermiculite (6). Such empty positions were found to convert a muscovite from one of the most stable minerals to only a moderately stable one.

The presence of large channels in their structure is recognized as the primary cause of the low stability and of the cation-exchange properties of zeolites.

4. Since the reactions of crystal particles during weathering occur in the unit cells exposed at their surfaces, it would be expected that the smaller the percentage of unit cells of a particle exposed at the surface, the more stable is the mineral. This percentage is determined by the size and shape of the individual particle and can be calculated as follows:

Let the dimensions of a particle be measured in terms of unit cells instead of Å, then the width would equal a, the breadth b, and the thickness c. The total number of unit cells in such a particle is therefore equal to $a \times b \times c$. The total number of unit cells that are not exposed to the surface must be equal to the number of unit cells in a particle with dimensions just two unit cells smaller on each edge than in the one just described. The number of unit cells in such a particle must be equal to $(a - 2)(b - 2)(c - 2)$, and therefore the total number of unit cells which are exposed at the surface of the particle with the a, b, and c dimensions is equal to:

$$a \times b \times c - (a - 2)(b - 2)(c - 2)$$

and the percentage of unit cells that are exposed at the surface is equal to:

$$\frac{a \times b \times c - (a - 2)(b - 2)(c - 2)}{a \times b \times c} \times 100$$

Using this method of calculation, one can readily determine the relation between particle size and percentage exposure and between particle shape and exposure.

For simplicity's sake, let particle size be expressed in terms of number of unit cells in a given particle, and let the shape be defined by the ratio of the width a, in unit cells, to thickness c, in unit cells. Thus, if we let the width a equal the breadth b, then the shape of the particle may be defined as follows: if the ratio $a{:}c$ is smaller than 1, the particle is a rod; if the ratio is equal to 1, it is a cube; and if the ratio is greater than 1, it is a plate.

Table 1.20 shows the percentage exposure of particles of different size and shape. For a given-size particle, the plate-shaped particles have the largest percentage exposure, the rod-shaped next, the cube-shaped the least, and the greater the "*platiness*" or the "*roddiness*", the larger the percentage exposure. Increase in exposure with a decrease in particle size is greatest in the plate-shaped particles, less in the rod-shaped, and least in the cube-shaped.

It should be noted, however, that a consideration of the percentage of unit cells of a given particle is significant only in minerals that do not have channels in their structure or defects that enable solutions or ions to penetrate to the interior of the particle.

5. The relative stability of a given mineral may also be affected by the nature of the other minerals associated with it through their effect on the composition of the solution in contact with the mineral or through their direct contact with the mineral.

6. Recent research of the writer (unpublished) with montmorillonite and vermiculite clay minerals indicates that the total cation exchange

Table 1.20. Effect of Particle Size and Shape on the Percentage of Total Crystal-Lattice Unit Cells Exposed at the Surface

Particle size*	Particle shape†				
	1:20	1:10	1:1	10:1	20:1
1000	80	69	49	100	100
10,000	45	37	25	48	70
100,000	22	18	12	23	34

* Measured by number of unit cells.

† Measured by width to thickness ratio ($a{:}c$).

capacity strongly affects their stability in an acid environment. The higher the cation exchange capacity, the more readily and rapidly does adsorbed H^+ interchange with Al^{3+} and Mg^{2+} of octahedral, or Al^{3+} of tetrahedral positions.

ALTERATION OF MINERALS OF THE PARENT MATERIAL

One of the most important changes in the minerals of the parent material during soil development is their diminution in particle size. This diminution is brought about partly by physical reactions, involving thermal expansion and contraction or freezing and thawing of adsorbed water in fractures, and partly by chemical reactions, involving the breakdown of unit cells, bordering fractures or changes in certain planes of the crystal structure that induce adsorption of water and organic substances, which in turn may cause enough swelling to disrupt the particle.

This diminution in particle size induces greater chemical activity in the minerals, since all the chemical reactions involve primarily the unit cells exposed at the surface of a crystal particle. In fact, one of the methods of determining the nature of the reactions that occurred in a specific mineral during soil formation is to determine the changes in the mineral as a function of particle size.

The reactions to be considered in the present discussion are as follows: ion exchange, hydrolysis, solution, diffusion, oxidation-reduction, and adsorption and swelling. Before each of these reactions is discussed in detail, it should be noted that: (*a*) Water is essential in all of these reactions: it functions as the solvent, it is an important source for H^+ and OH^- ions for several of the reactions, and it is the most important polar liquid involved in adsorption, swelling, and gelation. Without water, very few chemical reactions would occur. (*b*) The reactions do not necessarily occur in the sequence listed, and several of them may be involved in what might appear to be one reaction. (*c*) The relative importance of the various reactions in the alteration of different minerals might be considerably different. (*d*) A given change in a mineral can be viewed as resulting from any one of several reactions; for example, the loss of K^+ from biotite may be viewed as resulting from an exchange reaction; thus

$$\text{K biotite} + H^+ \rightarrow \text{H biotite} + K^+$$

or, as resulting from a hydrolysis reaction; thus

$$\text{K biotite} + H_2O \rightarrow \text{H biotite} + K^+ + OH^-$$

but since part of the biotite—the K^+—goes into solution, the foregoing reaction may also be viewed as a solution reaction. These three reactions,

TABLE 1.21. THE DEGREE OF HYDROLYSIS OF THE READILY EXCHANGEABLE K^+ (K^+ WHICH IS EXCHANGEABLE WITH NH_4^+) OF GROUND BIOTITE DURING FRACTIONATION INTO VARIOUS-SIZED PARTICLES WITH DISTILLED WATER

Particle size (μ)	meq of K^+ present*		K^+ Hydrolized during fractionation	
	Before fractionation	After fractionation	(meq)	(% of Total)
<1	73.9	27.3	46.6	63.1
1–5	57.3	23.6	33.7	58.8
5–50	60.6	25.6	35.0	57.8
>50	72.4	39.0	33.4	46.3

* Milliequivalents per 100 g.

therefore, should be considered as one, and it is one of the most important reactions involved in the loss of bases from the minerals. Table 1.21 illustrates how readily K^+ is lost from a ground biotite upon continuous contact with distilled water, as during fractionation into various particle sizes. Loss of the exposed K^+ ranged from 46 to 63 per cent.

In considering the breakdown or alteration of the various minerals, one must consider the weakest bond in each structure under study, for it is this bond that determines the degree of stability of the whole structure. Generally, the better a cation fits into its coordination, that is the size of the cation relative to size of the interstice in which it is present, the more strongly it is linked to the surrounding structure and the higher will be the energy barrier that must be overcome before the ions can move into another coordination or leave the structure altogether.

The weakest bonds in some of the important structures are as follows: (*a*) In structures with independent silica tetrahedrons, or with only a small number of linked tetrahedrons, the weakest bond is the one which binds the tetrahedrons together; the larger the coordination number and the smaller the charge of the cation, the weaker the bond. (*b*) In metasilicates containing single or double chains of linked tetrahedrons (pyroxenes and amphiboles), the weakest bond binds the tetrahedron chains together. (*c*) In structures with two-dimensional tetrahedral linkage—the sheet structure as in the micas—the weakest bond binds the bases of the tetrahedrons together, as the K^+ in mica. (*d*) In the structure with three-dimensional linkage of the tetrahedrons, the weakest bond binds the cations that balance the charge of the alumina tetrahedrons, as the K^+, Na^+, or Ca^{2+} in the feldspars.

The second weakest bond in the foregoing structures is in the alumina tetrahedrons, because Al^{3+} fits poorly in a fourfold coordination. Therefore, the larger the number of alumina tetrahedrons in a structure, the weaker it will be; and upon disintegration, the breakage should occur at the oxygen tetrahedrons containing the Al^{3+}.

The disintegration, therefore, of the aluminosilicate minerals could be expected to follow the sequence of the strength of the bonds of the structure. In the feldspars, for example, K^+, Na^+, and Ca^{2+} would be expected to leave the structure first, then Al^{3+}, and finally Si^{4+}; in the mica structure, the K^+ would be expected to leave first, Al^{3+} in tetrahedral positions next, and finally the structure would break up into units containing essentially only silica tetrahedrons and Al^{3+} or Mg^{2+} octahedrons.

The structure of the silicates is essentially ionic in nature. To maintain, therefore, the electric neutrality during disintegration or alteration, the departure of a cation must be accompanied either by departure of an anion or by adsorption of another cation. The nature of the adsorbed cation determines whether the structure would disintegrate and go into an ionic and a molecular solution or whether it would alter only enough to accommodate the coordination of the new cation. Adsorption of H^+ ions brings about the former change, whereas adsorption of other cations such as Na^+, K^+, Ca^{2+}, and Mg^{2+}, brings about the latter change. The collapse of a structure upon adsorption of H^+ is induced by two factors, the relative weakness of the H^+ bond and the appearance of large interstices previously occupied by the departed cations but which remain nearly empty because of the smallness of the H^+.

Since H^+ is one of the most abundant ions in a large number of soil solutions, it must also be the cause of the major breakdown of the minerals during weathering. In other words, solution reactions in an acid medium play a very important role in mineral disintegration. For this reason, the manner of decomposition of aluminosilicate minerals upon acid treatment is of particular interest in a study of soil development. Minerals that are attacked by acid can be divided into two groups, those which go completely into solution that gelatinizes upon standing or concentration, and those which go only partly into solution and leave behind a residue of insoluble silica.

The relation between the structure of the minerals and their behavior toward the acid treatment is summarized by Murato (54, 55) in the following manner. The minerals that completely dissolve and whose solutions gelatinize have the following structures:

1. Those minerals containing silicate radicals of small molecular weight, namely, orthosilicates, pyrosilicates, and possibly silicates containing ring structures of three silicon atoms.

2. Those minerals with large continuous silicon-oxygen networks that will disintegrate into units of low molecular weight.

(*a*) Disilicates containing appreciable ferric iron in the silicon-oxygen sheets.

(*b*) Minerals of the silica type with three-dimensional networks that contain aluminum in the ratio of at least two aluminum atoms to three silicon atoms.

Minerals that separate insoluble silica, instead of gelatinizing, upon being treated with acid, are characterized by silicon-oxygen structures of large dimensions that do not disintegrate into small units under acid attack. These are SiO_3 chains, Si_4O_{11} double chains, Si_2O_5 sheets not containing large amounts of ferric ion replacing silicon, and three-dimensional networks having an aluminum content less than the ratio of two aluminum atoms to three silicon atoms.

Under normal weathering conditions the rate of breakdown of the minerals is low, and most of the products do not remain long in solution. Both the rate of solution and the fate of the resulting products are strongly affected by such conditions as temperature, degree of acidity, particle size, and crystal structure.

Adsorption through cation exchange of ions other than H^+ plays a very important role in the alteration of the mica minerals. Thus it was found that the adsorption of ions such as Ca^{2+}, Mg^{2+}, or Na^+ causes the surfaces of the mica to hydrate and also to expand the crystal lattice in the direction of the C axis. This expansion results in a crystal structure similar to that found in vermiculite or in the montmorillonite clay minerals (5, 6).

There is ample evidence that a part of the montmorillonite and vermiculite clay minerals in soils results from this reaction.

Collapse of the structural units along fracture lines throughout the body of the primary mineral particles, through solution or exchange, is probably responsible for their reduction to clay-size dimensions.

The diffusion of cations into and out of the interior of crystal particles plays an important role in cation exchange and also in the cation interchange reaction (8, 9) whereby adsorbed H^+ interchanges with tetrahedral Al^{3+} or with octahedral Al^{3+}, Mg^{2+}, or other ions present there. The diffusion reaction is responsible for converting a mica-type structure to a vermiculite-type, or vice versa. Empty positions in a structure, exchange of monovalent for divalent cations and of large for small cations, and adsorption of water during diffusion accelerate the diffusion process.

The interchange, through diffusion, of adsorbed H^+ with Al^{3+}, Mg^{2+} or other cations determines the nature of soil acidity, the exchangeable Mg^{2+} to Ca^{2+} ratio, and the nature of clay formation, as will be discussed later.

The diffusion of adsorbed H^+ to the interior of the crystal lattice suggests a mechanism by which the chemical alteration and breakdown of the silicate minerals proceeds beyond the first stage of hydrolysis. If the H^+ is replaced by Mg^{2+} or other alkaline or alkali cations, then they can again be replaced, through hydrolysis or exchange, by other H^+ ions, and this process can be repeated until all such cations are depleted in the crystal.

If the H^+ ions are replaced by Al^{3+}, and the pH of the soil system is buffered, as by CO_2, organic substances or a high degree of base saturation (particularly by Na^+), the Al^{3+} may precipitate as $Al(OH)_3$ or become hydroxylized to various degrees and thereby polymerize and remain as strongly adsorbed polyvalent cations on the exchange sites of the soil minerals and impart to them their exchange acidity (8, 9, 31, 32). These polymerized aluminum ion species are believed also to be the first step in clay formation. They were found to react readily with soluble polymerized silica to form a gelatinous amorphous precipitate which, upon aging, may crystallize to the various clay minerals under appropriate reaction conditions (Barshad, unpublished).

If the hydroxylized polyvalent aluminum species grow in number and become equal to the number of exchange sites of a soil, the interchange reaction comes to a stop, thereby depleting the soil of exchangeable basic cations, reducing their base cation-exchange capacity, and stopping further weathering. It is for this reason that even under intense weathering conditions many of the soil clay minerals remain stable. The process of weathering, however, resumes upon the removal of the polyvalent aluminum ions either by the formation of gibbsite or the aluminosilicate clay minerals, by interaction with soluble polymerized silica, or by chelation with organic acids and removal by water movement to another site.

If Fe^{2+} iron is present in the crystal lattice of a mineral, it too would be replaced by the interchange reaction, but as it reaches the surface of the crystal and comes in contact with dissolved oxygen, it becomes oxidized to Fe^{3+}. The Fe^{3+} iron, due to its chemical behavior, will become hydroxylized even more readily than Al^{3+}. In the pH range of most soils, the Fe^{3+} will precipitate as $Fe(OH)_3$ and will, therefore, rarely participate in clay mineral formation but will remain as free Fe_2O_3. The replacement of Fe^{2+}, therefore, would be expected to proceed even faster than that of Al^{3+}, and, indeed, minerals which contain large amounts of Fe^{2+} weather more rapidly than those which do not.

Oxidation reactions are of primary importance in minerals containing ferrous iron. Because the electrostatic neutrality of a crystal structure must be maintained, the oxidation of ferrous to ferric iron brings about the departure of some other cation; the more ferrous iron, the more cations must depart. Such a departure leaves empty positions, which cause the structure to collapse, to become more susceptible to acid attack and dissolution, or to alter to other species. The ferrous-bearing minerals, therefore, are among the first to decompose upon weathering.

The conversion of biotite to vermiculite and/or montmorillonite is a good illustration of the effect of oxidation on structure alteration. The loss of K^+ due to oxidation leaves empty positions in the interlayer planes and also weakens the electrostatic bonds which hold these oxygen planes together. These two changes in the crystal lattice facilitate the exchange between the remaining adsorbed K^+ and the soluble Ca^{2+}, Mg^{2+}, and Na^+ in a soil solution to bring about the conversion to vermiculite or montmorillonite as determined by the total charge left on the planal surfaces. Thus, a high charge, greater than 150 me per 100 g, results in vermiculite, and a low charge, less than 130 me per 100 g, results in montmorillonite (3). From a consideration of the oxidation-reduction potentials of Fe^{2+} to Fe^{3+} it may be predicted that the oxidation reaction would take place much more readily in an alkaline environment than in acid environment (3). This leads to the conclusion that biotite mica would be converted, when weathering in an alkaline medium, to montmorillonite, whereas in an acid environment it would be converted to vermiculite. Numerous clay mineral analyses of soils in recent years confirm this conclusion.

Adsorption reactions are important in the breakdown or alteration of minerals because they accelerate other reactions, such as the exchange, the interchange, and the oxidation. The exchange reaction itself may be considered partly an adsorption, and partly a desorption reaction. The adsorption of water by the interplanar surfaces of the mica structure with the resulting swelling enables the exchange reaction to go to completion. The adsorption of OH^- by exchangeable Al^{3+}, Fe^{2+}, Fe^{3+}, and other acidic metallic cations which result from the interchange reaction enables this reaction to be sustained until all of these ions are completely hydroxylized. During this stage of the reaction both H^+ and OH^- are simultaneously adsorbed by the minerals; H^+ is adsorbed by the negative exchange sites and OH^- by the acidic exchangeable cations. Since these OH^- ions in turn were found to be capable of undergoing exchange reactions with other anions, such as phosphates, molybdates, borates, sulfates, and chlorides, they may be considered the seat of the anion exchange and fixation property of most soil clay minerals (32). The oxidation reaction of ferrous to ferric could not proceed in either an alkaline or an acid medium without adsorbed water, for water itself is one of the reactants.

Water, H^+ and OH^- ions; phosphate, molybdate, borate, sulfate and chloride ions; various organic substances; and the oxides or hydroxides of iron, aluminum, silicon, manganese, and titanium are some of the important adsorbates.

Adsorption of water is termed *hydration*, whereas the adsorption of H^+ and OH^- is termed *hydroxylation*. Adsorption of H^+ and OH^- ions at broken edges of a particle is the first step in releasing the elements as oxides from the crystal particles. Adsorption of the released oxides by the surfaces of other crystal particles as an aid to clay formation is discussed in a later section.

The importance of adsorption of organic substances, such as amino acids and proteins, lies in the fact that they cause large particles to break into smaller ones through swelling and exfoliation. This is particularly true of crystals in which the organic substance and water penetrate interplanar positions and may cause an expansion that exceeds the interlayer attractive forces holding the structure together.

REACTIONS AMONG PRODUCTS OF WEATHERING AND BETWEEN PRODUCTS AND REACTANTS

The initial nature of the products of weathering is determined by the reactants, the pH of the medium in which the reactions occur, and the presence of chelating substances.

As indicated previously, some of the reactants are completely soluble, whereas others are only partly soluble. Some silica separates, for instance, as insoluble silica. Some of the soluble products may be removed from the site of the reaction by percolating water, but some remain at the site and react with one another or with the reactants. Thus the soluble cations Na^+, K^+, Ca^{2+}, and Mg^{2+} may be adsorbed through exchange reactions by the clay minerals or can enter into the structure of the primary minerals or combine with other products to form the clay minerals.

During soil development, it is doubtful whether products like SiO_2 and Al_2O_3 remain soluble for long after their separation from the crystal lattice of the nonresistant minerals. Of these products, SiO_2 is probably the most soluble, particularly when separated from structures with independent silica tetrahedrons or from structures that break up into smaller units. Under excessive leaching and in soils ranging in pH from 6 to 8, some of this SiO_2 may be completely removed from the site of decomposition, as in Latosols, but a large proportion will remain in place and will, together with Al_2O_3 and Fe_2O_3, form the framework of the clay minerals. Because of the pH of the soil solution, generally ranging from 4.5 to 7.4, the Al_2O_3, Fe_2O_3, and TiO_2 will separate from the decomposing minerals as polymerized complex hydroxy ions and colloidal hydroxides. Since colloidal SiO_2 is negatively charged, and colloidal Al_2O_3 and Fe_2O_3 or $Al(OH)_3$ and $Fe(OH)_3$ are positively charged, and particularly the polymerized hydroxy ions, it is postulated that they interact to form the clay minerals. The kind of clay mineral to form would depend on the ratio of SiO_2 to Al_2O_3 and Fe_2O_3, the pH of the medium, and the presence in solution of basic cations such as Ca^{2+} and Mg^{2+} and K^+. Thus, in a medium in which the proportion of colloidal SiO_2 to Al_2O_3 and Fe_2O_3 ranges from 2 to more than 4, with a pH of 7 or higher, and which contains relatively high amounts of exchangeable and soluble Ca^{2+} and Mg^{2+}, montmorillonite-type clay minerals tend to form. In such a medium when K^+ instead of Ca^{2+} or Mg^{2+} are high, the illite type form. On the other hand, in a medium in which the proportion of colloidal SiO_2 to Al_2O_3 and Fe_2O_3 is 2 or less, with a pH of 7 or less, and which is low in exchangeable and soluble bases, kaolin-type clay minerals tend to form (41, 49, 50, 78). But in such a medium when Mg^{2+} is high, vermiculate-type minerals tend to form (Barshad, unpublished).

Any condition which prevents colloidal silica and the various forms of aluminum and iron from reacting with each other also prevents clay mineral formation and induces the separation of each element as a distinct species, as occurs frequently in Podzols and Latosols. Thus, in Podzols the aluminum and iron species, through the action of organic acid compounds, become soluble or peptized and are moved by water from the A_2 to the B horizons, causing the silica to remain in an amorphous form in the A_2 horizon and the aluminum and iron to accumulate as gibbsite and hematite in the B horizon. On the other hand, in Latosols, due to the rapid decomposition of organic matter, the silica is fairly soluble and is, therefore, moved out of the soil entirely, leaving behind the aluminum and iron to accumulate as gibbsite and hematite.

Numerous analyses of the clay mineral species in various soils* confirm the foregoing conclusions that the kind of clay mineral which is being formed is determined by the prevailing condition in the soil environment, particularly as relating to the acid-base status, composition of the soil solution, and its concentration. It is for this reason that in many soil profiles a different kind of clay mineral forms in different horizons. This is to be expected, since in profiles the reaction conditions frequently differ in the different horizons due to variation in the degree of leaching and in the action of plant growth and microorganisms. Similarly, since the degree of leaching of a soil depends greatly on the amount of rainfall, temperature, and runoff, it is not surprising to have found a good correlation between the kind of clay minerals being formed and the amount of rainfall or drainage conditions, irrespective of the nature of the parent materials of the soil (Barshad, unpublished). Thus, in soils in which $CaCO_3$ either forms or accumulates, due to low rainfall or due to their position in a landscape, montmorillonite predominates. But in soils forming under conditions in which $CaCO_3$ cannot accumulate, due either to high rainfall, to a highly porous parent material, or to low temperature, vermiculite and the kaolin minerals predominate. For a given mean annual temperature (50 to 60°C), the higher the mean annual rainfall (in the range between 25 and 80 inches), the higher is the ratio of the kaolin-type minerals to the vermiculite-type, and for a given mean annual rainfall (in the range between 35 and 40 inches), the lower the mean annual temperature (in the range between 60 and 35°C),

* See references 10, 20, 21, 24, 30, 66, 74, and 75.

the lower is the ratio of the kaolin-type to the vermiculite-type minerals. The formation of gibbsite and hematite or other iron species is favored by either a low temperature, as in Podzols, or by a high temperature, as in Latosols, as described previously. The formation, however, of illite appears to be determined not so much by the leaching conditions as by the parent material: only those high in K_2O content favor the formation of illite. Thus, in granites or other rocks with primary micas, illite is formed as a result of a reduction in particle size of these micas, but in rocks without primary micas, as in obsidian or in acid volcanic ash (Mono Crater area, California) or volcanic ash high in leucite (Mt. Vesuvius area, Italy), illite is synthesized in the same manner as kaolin-type or vermiculite-type minerals (Barshad, unpublished).

The mechanism by which the colloidal silica reacts with colloidal alumina and iron to form the crystalline clay minerals is not precisely known. Various theories have been suggested (78), but so far artificial preparation of clay minerals under conditions similar to those encountered in soils has not been reported.

FORMATION OF CLAY MINERALS

Formation of the clay minerals in soils from soluble and colloidal products is believed to be catalyzed by the other aluminosilicate minerals of the soil through the adsorption of these products on their surfaces. The surfaces of the existing clay minerals, or of primary minerals, may be considered as nuclei which initiate the crystallization process. Crystal lattice growth may be said to begin when only a few silica tetrahedrons are adsorbed on an OH surface of a kaolinite particle or on the surface of other minerals covered by adsorbed hydroxy aluminum ions, and when a few molecules of alumina or hydroxy aluminum ion are adsorbed on an oxygen surface of a mica, a montmorillonite, or a kaolinite particle. These adsorbed molecules would tend to become oriented in a pattern similar or complementary to that of the substrate and become condensed, through olation and dehydration, into two-dimensional sheets of linked tetrahedrons or octahedrons. At any one time, crystal-lattice growth may proceed in the direction of any one of the crystallographic axes, depending on the available reactants and the conditions favoring adsorption and condensation, such as pH, moisture, presence of complexing, chelating, or peptizing substances, and possibly other conditions that need to be studied.

One condition which is believed to be of significance in clay formation is that growth of the crystal lattice of the clay minerals is very slow; even in well-developed soil profiles the amount of clay formed, as calculated by the methods previously proposed, is relatively small. For example, in some of the well-developed soils of the Midwest, the clay formed in the surface 0- to 6-inch layer is about 20 g per 100 g of parent material. If we assume further that the soil is ten thousand years old, then the amount of clay formed annually is 0.002 g per 100 g of parent material. The surface area of such an amount of clay, if it consisted of a monolayer of unit cells, can be calculated to be 1.8 square meters. If we calculate clay formation in terms of surface area of monolayers of unit cells, then the rate would be 1.8 square meters a year, or about 2.5 cm^2 per hour, and even at this rate clay formation is extremely high for most soils (7).

These results clearly indicate that any parent material would have sufficient surface area to induce clay formation by means of adsorption and condensation even if the surfaces capable of such adsorption were restricted to a relatively few minerals.

Because of the low rate of clay formation, it may be postulated that the various clay minerals can form simultaneously in the same soil or soil profile, according to the nature of the adsorbing surface, its position in the profile, and the supply of the reactants at the site of adsorption. Furthermore, growth of any one type of crystal lattice on any one surface may vary periodically as a result of seasonal or other variations in the factors that may control the supply of the reactants and affect adsorption and condensation. It is not surprising, therefore, to find that in a large number of soils, several clay minerals are always present together, either as simple mixtures or as interleaved crystals.

In light of the foregoing discussion, it would be expected that any condition which promotes adsorption and condensation would accelerate clay mineral formation, and any condition which hinders adsorption and condensation would retard clay formation. Thus, it might be expected that with a given supply of reactants, clay minerals would form faster in a parent material with a relatively large adsorptive surface, such as one that already contains clay minerals, than in one with a small adsorptive surface. In two such soils, therefore, one should find a larger amount of uncombined oxides in the one with the lower adsorptive surface.

Depicting the formation of clay minerals by a process of adsorption and condensation leads to two interesting conclusions: (*a*) Their crystal lattices should contain no Al^{3+} in tetrahedral positions, since colloidal alumina, which is positively charged, would not be adsorbed on the same surface as colloidal silica, which is negatively charged; moreover, existence of independent Al^{3+} tetrahedrons after they have separated from a crystal lattice of a primary mineral is most unlikely, because of the unstable state of the Al^{3+} in a tetrahedral coordination. (*b*) Crystalline particles of the clay minerals of soils should generally have poorly defined shapes, for the conditions of crystallization as outlined are not conducive to the formation of well-defined crystal faces. Electron micrographs of soil clay minerals verify this conclusion.

That the crystal structure of the kaolin minerals contains no Al^{3+} tetrahedrons is well recognized. A number of the montmorillonite and vermiculite minerals, however, contain Al^{3+} in tetrahedral positions. Their presence is believed to result from the two distinct processes by which montmorillonite and vermiculite minerals in a soil may have been formed. One involves adsorption and condensation, which does not favor the presence of Al^{3+} tetrahedrons. The other involves the alteration of the mica minerals through exchange of K^+ with Ca^{2+}, Na^+, and Mg^{2+}, and, since mica contains Al^{3+} in tetrahedral positions, they remain in the clay that is formed. If it were possible to distinguish the portion of montmorillonite and vermiculite clays with tetrahedral Al^{3+} from the portion without tetrahedral Al^{3+}, then the relative amount of the montmorillonite and vermiculite clays formed by each process would also be known. Schwertmann (71) suggests such a possibility.

FORMATION OF PROFILES OF SOIL PROPERTIES

The specificity of many properties of a given soil material is the result of its unique position with respect to the surface and the substratum, and its development cannot be well understood except in relation to its position as it occurs in nature. Consequently, in an evaluation of the development of a soil material, its position must be defined with respect to the surrounding soil materials. The whole assemblage of soil materials as a function of depth is defined, as indicated previously, as the *soil profile* or the pedon, and the term *soil development* commonly refers to the development of the *soil profile* or the pedon (74).

The technique of investigating the variation of soil material with depth is to dissect the profile into layers of various thicknesses and then study each layer separately. The method of dissection depends on the properties chosen for differentiating the soil materials. Generally, those properties are chosen which show the greatest variation with depth, the most common of which are texture, color, and structure. The object in dissection is to separate the profile into layers, each of which consists of soil material of maximum uniformity. Many profiles consist of such layers differentiated naturally by the soil-forming processes. These layers, termed *horizons*, are designated as follows: the surface soil horizon (A), the subsoil horizon (B), and the preserved parent material (C). In sampling, each horizon might be further dissected into thinner layers, depending on the variation of the soil within it; consequently each of these thinner layers is designated by a subscript number, for example, A_1, A_2, A_3, B_1, B_2, B_3, C_1, C_2, etc. A standardized scheme of horizon designation for all the possible types of horizons which occur in soils has been proposed (74). In soil profiles in which differentiation into natural horizons is not readily recognized, the profile may be dissected arbitrarily according to the objective of an investigation. Such an arbitrary sampling may also be necessary, as pointed out earlier, in contrasting the development of one soil profile with that of another.

The assumed, or measured, thickness of a profile or a horizon depends to some extent upon the soil properties chosen as a guide for sampling and upon the emphasis placed on the role of soil developmental processes in determining the profile thickness.

If the concept that the thickness is determined by the soil-forming processes is accepted, then the soil profile should include only soil materials that show a difference in one or more properties from the preserved parent material. Accordingly, the preserved parent material is not a part of the soil profile but must be included in sampling, because it is the reference material by which the degree of soil development can be measured. In many soils, the point at which the soil profile stops and the preserved parent material begins may depend somewhat on a choice to facilitate the evaluation of soil development. Thus, in soil profiles which developed from consolidated rocks, one may argue that the consolidated rock is the parent material, but, on the other hand, it is more advantageous analytically to choose the disintegrated rock in contact with the consolidated rock as the parent material, since it is more amenable to quantitative mineralogical analysis. This proposed concept has been accepted (74). The disintegrated rock is designated by the letter C and the consolidated rock by the letter R.

Similarly, in a soil profile which developed from an unconsolidated or partly consolidated calcareous loess, the choice of the point of contact between the soil profile and the preserved parent material is subject to some degree of arbitrariness in localities where lime was completely removed from a very thick layer of the parent material. The problem here is whether to choose as the original parent material the calcareous loess, which may be found at a depth of, let us say, 10 or more feet below the surface, or the altered loess, which lost lime but otherwise remained unchanged and which is found, let us say, at about 4 feet below the surface. Obviously, the first choice would lead to a profile thickness of 10 feet, whereas the second one would lead to a profile thickness of 4 feet. The thickness of the soil profile is particularly important in contrasting the degree and nature of the soil profile development of different soils.

Since soils are compared by their properties, the profile thickness may be defined not in terms of the sum of all the properties but rather in terms of each soil property. Consequently, any given profile may have different thicknesses, depending on the property under study. Thus the thickness of the profile with respect to organic matter may be 4 feet; with respect to clay, 3 feet; with respect to the nature of the nonclay, 2 feet; with respect to uncombined Fe_2O_3, 5 feet; and with respect to lime distribution, 10 feet. The sampling, therefore, of profiles for the purpose of contrasting their development should be to a depth below which the material remains completely uniform with respect to all properties to be studied.

The following section deals with processes that lead to the formation of profiles of soil properties. The most important of these processes are plant growth and microbiological and animal activity, water movement, diffusion, dispersion and flocculation, wetting and drying, expansion and contraction, and cementation and decementation.

Plant Growth

Plant growth, as such, effects profile differentiation by furnishing organic matter, by absorbing plant nutrients and water, by producing roots, and by supplying a cover to the surface. Indirectly plants affect profile differentiation by being subjected to occasional burning and windfalls.

Organic Matter. The most pronounced means of profile differentiation during the initial stages of soil development is by organic matter. The nature of the organic matter profile formed depends on the method of supply and the nature of the organic matter itself, which in turn depends on the species of plants grown. Thus, trees supply a much larger percentage of organic matter through their vegetative parts above ground than through those below ground, whereas grasses supply a considerable proportion of the organic matter by the vegetative parts below ground. In consequence, a much larger percentage of the organic matter is at or near the surface in a profile developed under tree-type vegetation than in a profile developed under grass-type vegetation (45, 79). For this reason, organic matter is one of the most distinguishing and the most readily formed features of a soil profile. In fact, in the early stages of soil development the only measurable difference between the parent material and the soil material may lie in the organic matter or result directly from its presence.

The nature of the growing organic matter, with respect to its organic and inorganic composition, as affected by plant species and environment, is an important factor in the kind of profile differentiation. The important factors of composition (42) are the nature of the ash, particularly with respect to CaO, Fe_2O_3, and SiO_2; the acid and base buffer content; the carbon and nitrogen ratio; and the exact organic composition with respect to sugars, starches, carbohydrates, pentosans, pectins and other hemicelluloses, true cellulose, lignins and tannins, fats, waxes, oils, sterols, fatty acids, and proteins and their derivatives. The course of decomposition and the nature of the intermediate and final products are affected greatly by the composition of the fresh organic matter.

Absorption of Nutrients. The active absorption of nutrients during plant growth is recognized as one of the most important factors contributing to the alteration of the primary minerals, and possibly also to the formation of the clay minerals, for it affects the composition of the soil solution with respect to all the cations and anions and thus its pH and buffering capacity, and it keeps the solution saturated with CO_2. The continued absorption of nonexchangeable K^+ by plants is a classical example of the power of a plant to replace a cation from its crystalline environment (65). The circulation of many of the elements in the profile and the accumulation of some in the surface horizons, as P_2O_5 and many of the minor elements, are direct results of active plant absorption. Without plant absorption of the inorganic nutrients and their deposition on the surface, the upper part of a soil profile under humid conditions would quickly be depleted of exchangeable bases. Plant growth, therefore, is greatly responsible for the observed differences in the cation-exchange profiles of most soils. The seasonal variation in composition of the soil solution resulting from plant absorption (15) may, by the processes of adsorption and condensation, affect clay mineral formation, as follows: during active plant growth, when the soil solution is being depleted of its bases and enriched with H^+, the kaolin mineral would tend to form, whereas during periods of inactive plant growth, when the soil solution becomes enriched with bases and OH^-, the montmorillonite minerals would tend to form.

A comparison of the annual uptake of mineral nutrients by plants with possible annual rates of clay formation (7) suggests that even if only a very small portion of the *ash* of the plants is converted to colloidal constituents upon their decay in the soil, these constituents would be an important contribution to clay formation, at least to several important constituents of the clay fraction, such as free SiO_2, Fe_2O_3, and Al_2O_3. It is also visualized that the SiO_2 which is circulated through the plants and later released to the soil solution is more polymerized and soluble than SiO_2 which has not circulated through the plant and, therefore, it can react more readily with polymerized hydroxyaluminum ions, present either on the absorption complex of the soil minerals or in solution, to initiate clay-mineral formation.

Absorption of Water. The absorption of water by plants is the most important means by which soil material 6 or more inches below the surface can be dehydrated to a moisture content below its field capacity. This dehydration by plants is particularly important in soils developed under limited moisture supply, and during a period of drought in regions with more moisture. Dehydration affects profile differentiation in several ways. It concentrates the soil solution to the point of saturation with respect to some of the dissolved ions and thereby causes their precipitation, as with $CaCO_3$, $CaSO_4$, and SiO_2 in semiarid and arid regions. It may also concentrate suspensions of colloidal particles of iron, aluminum, and silicon oxides and thereby cause them to flocculate and aggregate and thus prevent their downward movement. It brings about nodulation and cementation. In semiarid and arid regions it limits movement of water to a relatively shallow depth, by preventing accumulation of enough water for downward percolation, and, as a result, limits the depth of profile formation. It brings about contraction and degelatinization of the soil colloidal material and thereby contributes to structure and fabric formation. It also helps aeration by emptying pores fully or partly occupied by water.

Growth and Movement of Roots. Apart from their functions in absorbing nutrients and water and supplying organic matter through their growth and movement, roots contribute to profile differentiation in two distinct ways. First, they help maintain a high permeability to water and air, which are essential for plant growth and the movement of various substances to the subsoil or completely out of the profile. Second, they help in structure formation through the pressure they exert on a wet soil mass as they increase in size. This pressure brings about orientation not only of clay particles around points resisting flow, such as sand and silt particles, but also of the whole soil material around the roots themselves. Such orientation is believed to be a prerequisite for aggregate formation upon drying (14).

The decay of large roots leaving channels which are often filled with surface soil material is an important contribution to a disturbance of horizontal stratification of the soil horizons and to a unique morphology to many pedons.

Surface Cover. The cover that plants supply to the surface of a soil affects considerably the temperature and moisture regime and thereby affects the rate of reactions in the soil (21). It also limits erosion.

Burning. Burning of vegetation by naturally occurring fires may often disturb the natural plant ecology of an area and thus initiate a *new cycle* of soil formation. Burning, particularly of forests, suddenly leaves behind large quantities of ash, some of the elements of which are less soluble and less available as plant nutrients than those of ash of unburned vegetation. This sudden deposit of ash may affect the pH of the soil and thereby the pH-dependent reactions. Under humid conditions the base reserve of the soil, as represented by the ash content, may be lost through leaching. Burning greatly depletes the soil available nitrogen and thereby affects plant growth and the distribution of plant species which reoccupy the burned area.

Severe burning may alter the color of the surface horizon and, if large roots have burned, leave channels and charcoal which may eventually affect the morphology of the pedon.

Windfalls. Uprooting of trees which is often associated with windfalls causes a mixing of soil horizons by depositing subsoil horizons on the surface of the soil and causing the surface soil to fall into the holes left behind by the roots. Such mixing disturbs the natural development of horizons of eluviation and illuviation.

Microbiological and Enzymatic Activity

The activity of the microorganisms in profile formation is as important as plant growth, for together they form the organic matter cycle of the profile: plant growth furnishes the organic matter and microorganism activity destroys it. Without the destruction of organic matter, all the CO_2 of the atmosphere would soon be bound in the undecomposed plants, and without CO_2, plant growth would cease. Besides destroying fresh plant material to produce organic matter in the soil, the microorganisms fix most of the nitrogen necessary for plant growth and thus indirectly affect the nature and amount of plant material produced. The organic matter of the soil is not merely plant residue; it reflects the activity of the microorganisms as well as their residue. Through the activity of microorganisms, not only carbonic acid but such inorganic acids as nitric, sulfuric, and phosphoric acids, and such organic acids as humic, fulvic, crenic, apocrenic, oxalic, and lichenic acids (42, 70) are produced, which enhance the destruction or alteration of minerals of the soil or parent material. Since a major portion of the acids are produced in the surface soil, they are most effective there in altering the nonclay fraction and consequently contribute greatly to the differentiation of the nonclay fraction profile.

The production or liberation from organic matter of substances with highly chelating or peptizing properties for iron and aluminum, such as fulvic or humic acids, is, in part, responsible for the movement of uncombined iron and aluminum from the surface to the subsoil horizon.* On the other hand, an increase in pH, in soluble Ca^{2+} and Mg^{2+}, and possibly the destruction by the microorganisms of the chelating properties of these organic substances in the subsoil horizon, could explain the deposition and accumulation of iron and aluminum oxides in these horizons (42).

* See references 2, 42, 53, 70, 81.

These activities of the microorganisms, therefore, enable the formation of profiles of uncombined iron and aluminum oxides, two of the most conspicuous profile properties to distinguish one soil from another. Microorganisms contribute also to the formation of organic matter with a high cation-exchange capacity. The presence of such organic matter is one of the most important means of preserving the base status of soils and affecting the nature of the cation-exchange properties of the soil profile.

Soil microorganisms play an important role in the formation of soil structure by the enmeshing of soil particles by mycelia and mold and by producing slimy, and other, substances which bind soil particles together to form aggregates.

Lichens, mosses, and algae play a very important role during the initial stages in rock weathering, and in rock weathering at very high altitudes. These organisms, by secreting powerful chelating and oxidizing substances, dissolve the substrate rock to make available to themselves the essential elements for their growth. By enmeshing the individual minerals with mycelia and mold and by producing substances which adhere strongly to the surface of the minerals and their own structures, these lithophylic organisms initiate the physical breakdown of rocks (42, 61, 70).

Animal Activity

The activity of animals which inhabit the soil influences greatly its morphology and often disturbs the horizontal pattern of horizon stratification. It has been suggested that worms, through their ingesting of soil material, contribute greatly to the formation of stable soil aggregates (40).

Water Movement

Downward water movement is probably one of the most important processes in profile differentiation, for water carries both the dissolved and the colloidal substances that may be moved out of the soil profile completely or moved from the surface to the subsoil. The amount of water supplied to a soil is not so important as the amount that penetrates the soil and moves downward. The factors, therefore, which affect water penetration, movement, and runoff also affect profile differentiation.

In pedology, the term *eluviation* has been applied to the loss of material from the surface horizon, and the term *illuviation*, to the gain of material by the subsoil horizon. Consequently, the surface horizon is designated as the *eluvial horizon* and the subsoil as the *illuvial horizon*.

The results presented earlier in this chapter clearly show that large amounts of material are moved from the surface horizon and deposited in the subsoil horizon. This movement of material—particularly that of clay—has been termed *migration*.

The profiles of the individual exchangeable cations are the result of water movement accompanied by the cation exchange and the cation interchange reactions, by the circulation of elements through plant growth and decomposition of organic matter, and by clay formation through adsorption and condensation (9). The general pattern of distribution is such that the most strongly adsorbed cations, such as Ca^{2+}, H^+, or hydronium, and the various complex hydroxyaluminum ions tend to accumulate in the surface horizon in greater amounts than do the other cations; the ions that are less strongly adsorbed, such as Na^+ and Mg^{2+}, either are reduced to extremely low values throughout the profile, as is Na^+, or increase with depth, as does Mg^{2+} in many soils.

Little water moves upward unless a perched water table is present within about 5 feet of the surface and, at the same time, the surface is subjected to desiccation. Such movement may bring about accumulation of salts in the surface horizon and on the surface, and may result in the formation of alkali and alkaline soils.

Diffusion

The upward or downward movement of soluble salts in a profile, without the movement of water, is termed *diffusion*. The contribution of such movement of salts to soil development has been studied but little. There is some evidence that the alkali conditions in certain soils of India have developed through salt movement by diffusion.

Plant absorption of exchangeable and soluble cations during periods of relatively low moisture content—in the range below the field capacity and above the permanent wilting percentage—may initiate salt diffusion to restore the equilibrium of the soil system.

The exchange of cations in the interlayer planes of montmorillonite and vermiculite clay particles and interlayer mica K^+ with cations in solution also results from a diffusion reaction.

Dispersion and Flocculation

Data presented earlier showed that the changes of greatest magnitude in a parent material during its development into a soil material are brought about by a loss or gain in clay through the process of migration. Furthermore, there is ample evidence that the clay involved in migration consists of particles of 0.1μ or less in diameter, (for example, the data of Table 1.14 show that the $<0.05\mu$ clay content of the B horizons is several times higher than in the A horizons, and also that the gain in clay in the B horizons is nearly equal to its <0.05 clay content, relations which suggest that this clay must have been involved in the migration), and that in many soils a large proportion of this clay consists of minerals of the montmorillonite type admixed with uncombined Fe_2O_3 and Al_2O_3. One of the previous sections noted that clay is displaced by the downward movement of water. But such displacement can occur only if the clay being moved is in a suspended state and the clay which accumulates is being flocculated. In other words, it is necessary to postulate the existence of conditions that lead to a state of dispersion in the surface horizon and of flocculation in the subsoil horizon, or of conditions that lead first to a state of dispersion and later to a state of flocculation.

Conditions that favor dispersion of colloids are a high state of hydration, a low electrolyte content, a pH far from the isoelectric point, and absence of oppositely charged colloids in the same system. On the other hand, conditions that favor flocculation are dehydration, a high electrolyte content, a pH at the isoelectric point, and the presence of an oppositely charged colloid.

In a soil the silicate clay minerals and colloidal silica are negatively charged, whereas colloidal iron and aluminum hydroxides or oxides are positively charged. On the other hand, colloidal organic matter may be either positively or negatively charged, according to the isoelectric point of the system. Conditions of hydration are induced by prolonged wetness of the soil, and conditions of dehydration are induced by evaporation and transpiration. The electrolyte content of a soil solution continuously fluctuates as a result of plant growth, plant decomposition, and water movement. Thus, active plant growth and excessive water movement induce a low electrolyte content, whereas active plant decomposition and very little or no water movement induce a high electrolyte content. The presence of substances that chelate or peptize iron and aluminum, such as oxalates, citrates, tartrates, humates, and fulvates, cause the iron and aluminum colloids to go into solution as complex ions or become peptized, with the result that the clay minerals in a wet state become dispersed. On the other hand, complete destruction by microorganisms of such chelating and peptizing substances, or merely destruction of their chelating or peptizing properties, releases the iron and aluminum and induces conditions which would be favorable to flocculation of the clay minerals.

In natural soil profiles, one or more of the conditions that induce dispersion or flocculation may exist simultaneously and may occur regularly during the course of a year.

Conditions conducive to dispersion in the surface soil and to flocculation in the subsoil horizon occur as a result of the manner of downward movement of water. Studies dealing with the composition of the *soil solution*

(15) have demonstrated that, regardless of the water content of a soil, water or a solution applied to the surface of a soil column does not distribute itself evenly throughout the column but instead displaces the soil solution from the surface downward. Consequently, the downward-moving soil solution under such conditions consists of two distinct phases: one, in the upper part of the column, comprising the added water or solution, and the other, in the lower part of the column, comprising the soil solution present prior to the addition of water. This condition would persist until all the original soil solution was displaced from the entire profile. A condition favoring dispersion in the surface soil would occur, therefore, after a period of sustained rainfall during which the surface soil solution with a high electrolyte content was being displaced to the lower part of the profile by a solution of low electrolyte content. Further rainfall on such a soil might bring about clay movement, but this movement would cease upon the cessation of the downward movement of water. Upon dehydration, whether through evaporation or transpiration, flocculation would occur and further movement of the clay would not resume until the soil was again hydrated and dispersed. The flocculation of the downward-moving clay may also occur without dehydration (*a*) by encountering, in the subsoil, colloidal iron and aluminum hydroxides or oxides brought down in the form of chelated or peptized complexes but released there by the action of microorganisms, or (*b*) by encountering a fairly high electrolyte content brought there by displacement. Any condition, therefore, leading to a low electrolyte content or a removal of the colloidal iron and aluminum would lead to dispersion, and any condition leading to a high electrolyte content and the presence of colloidal iron and aluminum hydroxides or oxides would lead to flocculation. High rainfall and active plant growth, both of which lower the electrolyte concentration, induce dispersion and favor clay migration, whereas low rainfall, poor plant growth, and active plant decomposition, conditions that increase the electrolyte content, induce flocculation and hinder clay migration. In any soil, therefore, clay migration takes place during some periods, and during other periods flocculation sets in and clay migration ceases. Furthermore, the iron and the clay minerals may move independently. Thus iron and aluminum hydroxides or oxides may move downward after a period of active plant decomposition, as a result of the enrichment of the soil solution with organic chelating, peptizing, or reducing substances that tend to bring the iron and aluminum oxides into solution, whereas during the same period, because of the high electrolyte content of the surface soil, the clay minerals would remain immobile.

It is well established that even under excessive water supply the downward movement of clay proceeds to a relatively shallow depth, ranging from 3 to 5 feet below the surface. This limited movement appears to result from the flocculating conditions that nearly always exist in the lower part of the profile (because of the nature of the movement of the soil solution) and from the ever-decreasing pore-size distribution of the subsoil through which the clay must pass. Clay moves downward farthest probably during the early stages of profile differentiation. With increasing clay content, through migration and formation, the depth to which clay can move decreases gradually to a point where movement may completely cease, as it probably does in some of the most highly differentiated profiles containing dense claypans.

The foregoing discussion of clay movement appears plausible upon consideration of the rate of clay movement, the concentration of clay in suspension which is necessary to bring about the observed migration, and the relation between the concentration of clay and the solutes in such suspensions. The first part of this chapter showed that the clay lost from the A and gained by the B horizons of a few profiles ranged from about 5 to 15 g/cm^2 of surface, and the annual loss, assuming the soil is ten thousand years old, would range from 0.0005 to 0.0015 g/cm^2 per horizon. The concentration of the suspensions involved in the movement of these amounts of clay would be estimated by considering the annual rainfall on these soils and the proportion of rainwater that moves through the A horizon. Thus, if we assume an average annual rainfall of about 20 $in./cm^2$ of soil and that only about a fourth of the water supplied is involved in clay movement, the volume of water in ml would be equal to $(20/4) \times 2.54$ or 12.5 ml of water. The concentrations of the suspensions, in 100 ml of water would therefore be $(0.0005/12.5) \times 100$ or 0.004 g, and equal to $(0.0015/12.5) \times 100$ or 0.012 g, or 40 and 120 ppm. To determine whether conditions of dispersion or flocculation would exist in such suspensions, it is necessary to consider the relative concentration of the solutes in solution and the total exchangeable cations on the clay. Several workers (15, 64) have found that the total concentration of solutes in displaced soil solutions ranges from 8 to 1000 g equivalents per million. Since the exchange capacities of clays range from 10 to 100 meq per 100 g, the concentration of exchangeable cations in the suspensions under consideration would range from 4 to 120 meq, or 0.004 to 0.120 g equivalents per million. The soluble salts, therefore, are several thousand times as concentrated as the exchangeable cations. It is well known that such conditions cause flocculation. Since, in soils, such conditions are the rule rather than the exception, it is not surprising to find that clay movement is limited with respect to both depth and time. The foregoing discussion substantiates the earlier suggestion that clay movement can occur only after a sustained period of rainfall and plant growth, conditions that induce a lowering of the concentration of the soil solution and therefore also a possible state of dispersion.

Wetting and Drying

The mere wetting and drying of the soil material, besides affecting all reactions involving water and water movement, plays an important role, as discussed previously, in dispersion and flocculation and in expansion and contraction reactions which effect profile differentiation. The differences in degree and in frequency of wetting and drying of the surface and subsoil horizons also play an important role in profile differentiation. Because of the manner of soil water movement and water supply, the surface soil is subjected more frequently to wetting and drying than is the subsoil. Moreover, the degree of drying of the surface soil is much greater than that of the subsoil, the difference being due to the mechanism of drying. At the surface, the soil is dried by both evaporation and transpiration, and as a result, the soil may be reduced to air dryness. The subsoil, however, is dried mainly by transpiration, and consequently the soil moisture rarely reaches a degree of dryness below the permanent wilting percentage, a water content much higher than that of air dryness.

The degree of dryness attained by the surface soil plays an important role in enhancing both clay formation through adsorption and condensation and primary mineral alteration through diffusion and exchange. The adsorption and condensation of colloidal silica, alumina, and iron by the surfaces of both the primary and the clay minerals are enhanced greatly by drying, since formation and elimination of water are involved in both of these reactions. Diffusion and exchange are accelerated as a result of an increase in concentration of the soluble salts in the soil solution.

Wetting, which leads to expansion and gelation, and drying, which leads to contraction and degelation, are extremely important in structure formation and, hence, in profile differentiation with respect to structure. Wetting and drying and expansion and contraction affect structure formation because they affect orientation of the clay particles with respect to one another, with respect to the sand and silt particles, and with respect to the roots. Drying, besides affecting orientation, also induces a high degree of irreversibility with respect to the manner and the degree of swelling of the clay minerals, particularly of the montmorillonite group in their Ca^{2+}, Mg^{2+}, H^{+} or Al^{3+} saturated state.

Wetting and drying associated with expansion and contraction have an important effect on permeability of the soil profile to both water and air.

Expansion and Contraction

Expansion and contraction associated with wetting and drying, freezing and thawing, and particularly with gain and loss of material through migration, leaching, and primary mineral alteration, contribute greatly to the formation of the bulk density profile. In consequence, the bulk density profile is one of the few soil profile properties that sum up the important soil-forming reactions.

Cementation and Decementation

The term *cementation* as used in the present discussion refers to the consolidation of soil material by means of cementing materials such as $CaCO_3$, Fe_2O_3, Al_2O_3, and/or SiO_2, whereas *decementation* refers to disaggregation of a material resulting from removal of these cementing materials.

Cementation and decementation are two of the most important processes in the differentiation of the textural profile, as well as in the formation of the unique profile feature of a hardpan in its subsoil horizon.

Removal of $CaCO_3$ from the surface soil and its accumulation in the subsoil plays a very important role in profile formation in semiarid soils; it also illustrates one of the important means by which clay is gained or lost by a soil or parent material without migration or alteration of primary minerals.

A large number of parent materials of sedimentary origin—sandstones, shales, loesses—are essentially cemented soil material and therefore may contain various amounts of clay in a cemented state. Clay formation from such parent materials is essentially a decementation reaction. Consequently, the amount of clay formed in such parent materials appears to be high when the method for calculating clay formation presented earlier in this paper is applied. Soil formation from consolidated or partly consolidated sedimentary materials is most strongly affected by the process of decementation. On the other hand, partial cementation of the soil material of subsoil horizons by $CaCO_3$, Fe_2O_3, Al_2O_3, or SiO_2 is very widespread. It is caused by an irreversible precipitation or flocculation and may occur under almost any soil-forming condition. It is most prevalent, however, in semiarid and arid conditions, and in humid conditions subject to periods of intense dehydration, as is common in the tropics.

One of the most interesting results of cementation is found in a number of soils in California (58): these contain a horizon, or a layer, which is completely cemented and which is termed a *hardpan layer*. It has been suggested that the presence of such a hardpan does not represent the normal course of profile development but rather a reflection of the initiation of new soil-forming processes which lead to formation of the hardpan after the normal soil profile features have already formed. In other words, the soil is believed to be the product of *polygenesis* and not of *monogenesis*—terms applied to changing soil-forming conditions as against uniform soil-forming conditions.

FACTORS OF SOIL FORMATION

In the discussion presented so far, emphasis has been placed on the kinetics of soil formation, that is, on the type of reactions that may occur during the transformation of a parent material to a soil material and to a soil profile. But in studying the development of a particular soil material or soil profile and in contrasting its development with that of others, we must consider not only the kinetics of soil development but also the conditions that affect the direction and rate of the soil-forming reactions, as well as their duration.

These conditions are inherent in, and determined by, what pedologists have termed the *factors of soil formation*, *parent material*, *topography*, *climate*, and *organisms* (16, 18, 19, 34). These may be defined as follows: *parent material* is the material from which the soil profile or soil material developed, that is, the initial state; *topography* is the configuration of the soil surface, including slope, exposure, position in the landscape, and position from surface; *climate* includes all the conditions from which it is made up, such as temperature, rainfall, and wind; *organisms* comprise all plant, animal, and microbiological life.

The most important conditions that affect the nature, direction, and rate of a chemical reaction are the nature of the reactants and products, their concentrations and amounts, the manner in which the reactants are supplied, and the products removed from the site of the reaction, as well as temperature, pressure, and volume. In a soil system each of these conditions is affected by each of the factors of soil formation. To clarify this statement, let us consider a few reaction conditions and show how each of the factors contributes to its determination.

For example, let us consider the reactants: the most important contributor is, of course, the *parent material* factor, for it contributes minerals and other materials; it also affects infiltration, permeability, and drainage, and thereby the water and gaseous supply. The contribution of *topography* to the reactants is indirect, in that it affects the runoff and drainage conditions and, thereby, affects the amount of water that reaches the site of the reaction and various substances that might be carried by the water moving through or over the soil. The *climate* factor determines the water supply, through precipitation and evaporation, and the various solutes that may be present in rainwater. The *organisms* supply many products, such as the CO_2, various organic substances, and inorganic salts, bases, and acids; and organisms also affect the water supply through transpiration and through their effect on runoff, drainage, and permeability.

Another good example is the *temperature* and *heat supply*. The *climate* factor is the most important contributor in determining the temperature and total quantity of heat supplied through solar radiation. The *parent material* affects the temperature at the site of the reaction by its color, heat capacity, and conductivity. *Topography*, through its exposure and slope, affects the amount of heat reaching a given area through radiation and air movement. *Organisms* affect the nature of the cover over the surface as well as transpiration, evaporation, and air movement.

The water supply also represents the contribution of all the factors to a given reaction condition. *Parent material* thus contributes, through its water-holding capacity, infiltration, permeability, and drainage; *topography*, through runoff and drainage; *climate*, through precipitation and evaporation; and *organisms*, through transpiration and effect on evaporation, infiltration, permeability, and drainage. In a similar manner it can be shown that nearly every reaction condition that occurs in the soil is affected by all the factors of soil formation; any one reaction condition could, consequently, result from a number of combinations of the various factors of soil formation. Since the conditions of a reaction, including time, determine the nature and amount of the products, it follows that a given product and its amount may result from a number of combinations of the various factors of soil formation.

If, in contrasting the development of one soil with another, only those profile properties are chosen which are truly the products of weathering, such as amount and nature of organic matter, nature of exchangeable cations, pH, nature and amount of clay minerals and free oxides, bulk density profile, and soil color profile (which is the result of many other properties), then a number of soils will be found to be similar and this similarity will be found to result from a combination of the factors of soil formation which induced similar reaction conditions. Such a grouping is essentially the basis for the classification of soils into what pedologists term the great soil groups. These groups include latosol, podzol, gray-brown podzol, gray forest, brown forest, brown, prairie, chernozem, chestnut, and several other groups.* Because of common properties, these groups constitute the most widely used scheme of soil classification. This principle is also the basis for the new comprehensive scheme (the 7th Approximation) for the classification of the soils of the United States.

Another scheme, however, which has been used successfully in comparing soils or individual properties of different soils, singles out the effect of any one factor of soil formation on a given soil property. This scheme was most successfully applied by Jenny (34) and is based on the principle of comparing a given soil property among soils with identical factors of soil formation except one.

* See references 1, 22, 29, 40, 44, 62, 67–69, 72, 73, 80, and 82.

This relation is expressed by Jenny as follows:

$$s = f_{cl}\ (\text{climate})_{o\ ,\ r\ ,\ p\ ,\ t\ldots}$$

$$s = f_{o}\ (\text{organisms})_{cl\ ,\ r\ ,\ p\ ,\ t\ldots}$$

$$s = f_{r}\ (\text{topography})_{cl\ ,\ o\ ,\ p\ ,\ t\ldots}$$

$$s = f_{p}\ (\text{parent material})_{cl\ ,\ o\ ,\ r\ ,\ t\ldots}$$

$$s = f_{t}\ (\text{time})_{cl\ ,\ o\ ,\ r\ ,\ p\ldots}$$

where s = any soil property, such as organic matter, clay content; or free iron oxide content; f = a function of; cl = climate; o = organisms; r = topography; p = parent material; and t = time. The three dots after the last terms indicates additional, unspecified factors, for example fires and windstorms. Thus the first equation states that the s property of a group of soils is affected only by variation in the climate factor; the second equation by variation in organisms only, and so forth.

The foregoing equations were recently modified by Jenny (39) to bring out the principle that, in the development of a given soil property, the factor within the parenthesis is the dominant one, whereas the factors outside the parenthesis are not really constant, as the equations imply, but are ineffective, that is, the partial derivative of the property with respect to these factors is nearly zero. Jenny proposed, therefore, that in writing these equations all the factors be placed within the parenthesis, but that the dominant factor be listed first and printed in a distinguishing type, such as bold face, to indicate its dominance, thus:

$$s = f\ (\mathbf{cl}, o, r, p, t, \ldots)\ \text{climofunction}$$

$$s = f\ (\mathbf{o}, cl, r, p, t, \ldots)\ \text{biofunction}$$

$$s = f\ (\mathbf{r}, cl, o, p, t, \ldots)\ \text{topofunction}$$

$$s = f\ (\mathbf{t}, cl, o, r, p, \ldots)\ \text{chronofunction}$$

A few functions are illustrated in Figures 1.1 to 1.7.

One serious objection to the factorial method of studying soil development is an implied assumption, not so stated, that soils are monogenetic, that is, they were developed under constantly uniform conditions since their initial state. Such a condition is true only for very young soils, those not older than the last ice age. Numerous investigators at present are arriving at the conclusion that most of the well-developed Great Soil Groups are very old—hundreds of thousands or a few million years—and therefore, the climatic and possibly the topographic factors must have changed several times during the course of soil development. It is, consequently, surprising to find a correlation as good as that between certain soil properties and present climatic conditions seen in the examples of Figures 1.1 to 1.7. An explanation for this "apparent anomaly" can be found by a proper definition of the "initial state" or the "parent material" for each "individual" soil property under study rather than for the "whole" soil. Thus for those soil properties which reflect primarily the influence of the biotic factor, as C/N, and total organic matter, the initial state should be considered the state of the soil at the time the present climatic condition began rather than the "preserved" parent material of the C horizon. This state may even be a well-developed soil profile with respect to clay formation and migration.

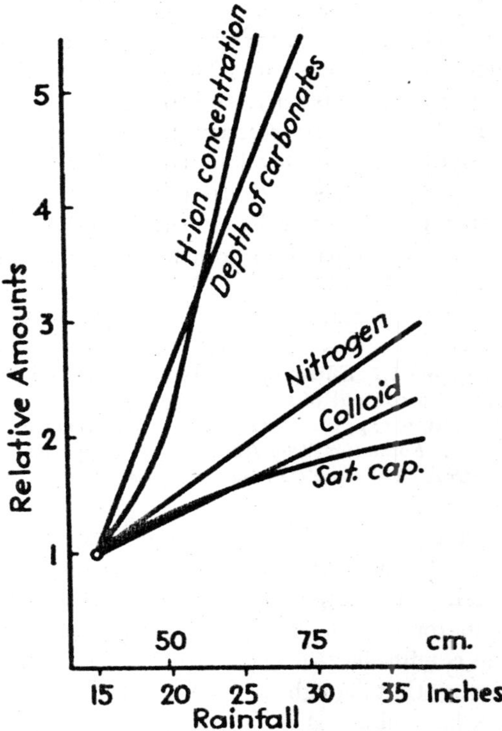

Figure 1.1. Effect of the rainfall factor on various soil properties: $s = f$ (*rainfall*, temperature, $o, r, p, t \ldots$). (*By permission, from "Factors of Soil Formation," by Hans Jenny. Copyright, 1941. McGraw-Hill Book Co., Inc., New York.*)

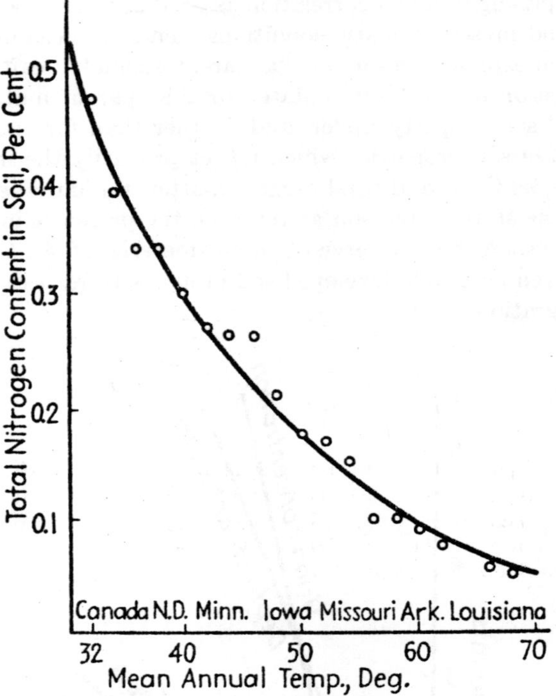

Figure 1.2. Effect of the temperature factor on nitrogen content of soils: nitrogen = *f* (*temperature*, rainfall, *o*, *r*, *p*, *t* . . .). (*By permission, from "Factors of Soil Formation," by Hans Jenny. Copyright, 1941. McGraw-Hill Book Co., Inc., New York.*)

Such an assumption, however, still does not explain the fact that even such properties as "clay formation," with respect to both quantity and quality, which reflect the total "lifetime" of a soil in relation to the "preserved" parent material of the C horizon, show a good correlation with the present climatic conditions. An explanation for this "apparent anomaly" may lie in assuming that though the absolute values of climate, as rainfall, at each soil site, have changed with time, the relative value between sites remained essentially constant. For some soil properties, even if small variations in the relative values have occurred, they would have been affected by these variations very little. This is particularly true for soil properties which show a correlation with a variation in the factor only within certain limits above or below which the state factor becomes ineffective, as was shown by Jenny for nitrogen content in relation to pre-

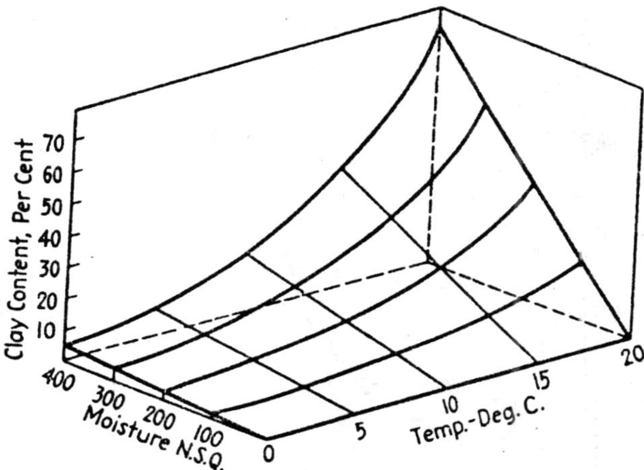

Figure 1.3. Effect of the temperature and rainfall (climate) factors on clay content of soils: clay content = f (*climate, o, r, p, t . . .*). (*By permission, from "Factors of Soil Formation," by Hans Jenny. Copyright, 1941. McGraw-Hill Book Co., Inc., New York.*)

cipitation in northern India (37). The pH of a soil is also such an example: it tends to be neutral or slightly higher over a wide range in rainfall, ranging from 0 to possibly 15 inches per year, and about 5.0 ± 0.3 for rainfall ranging between 35 and 100 inches. Fluctuations within these ranges of rainfall would hardly affect the pH and, consequently, the pH-dependent reactions such as clay mineral species formation.

Only for certain soil profile characteristics does the polygenetic nature of a soil profile become significant, as for the formation of an indurated hardpan, as it occurs in some soils in the western United States, and for certain morphological features related to color, structure, and fabric (44).

This method of correlating certain soil properties with the state factors, in spite of the uncertainties indicated, enables the detection of the most important reaction condition that plays a role in their development. It also enables evaluation of these properties in some soils without a detailed examination. Most important, it would enable prediction of changes in soil properties under known reaction conditions.

SOIL-FORMING PROCESSES

In conclusion, it is necessary to consider whether development of a soil undergoing a specific set of soil-forming reactions would lead to a state of equilibrium or a steady state. Theoretically, as for any other chemically

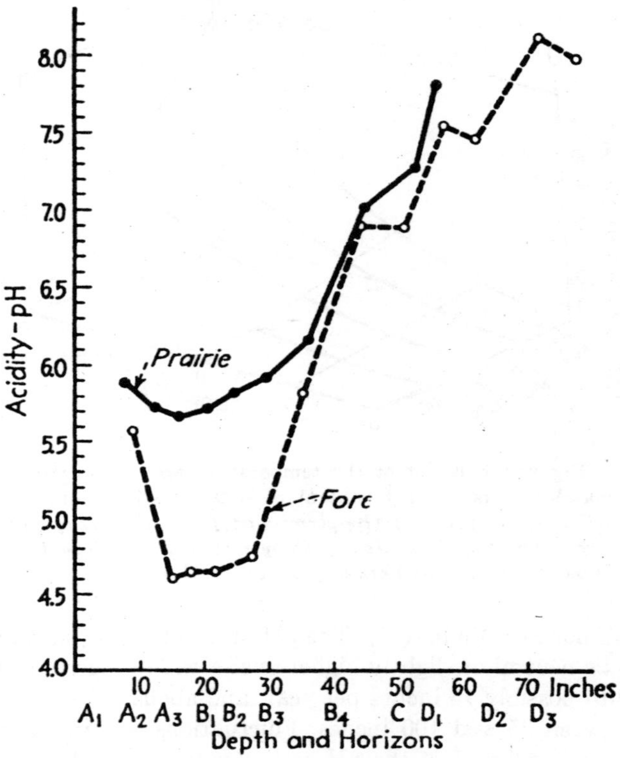

Figure 1.4. Effect of the organism (vegetation) factor on pH of soils: pH = *f* (*vegetation, cl, r, p, t . . .*). (*By permission, from "Factors of Soil Formation," by Hans Jenny. Copyright, 1941. McGraw-Hill Book Co., Inc., New York.*)

reactive system, this would have been so had soils been "closed systems" and had the reacting conditions, the "state factors," remained constant. Neither of these conditions holds true for soils. Soils are "open systems"; both energy and matter flow continuously in and out of them, and for most soils, due to their age, the reaction conditions have changed numerous times by changing climatic conditions and topographical features. The latter is clearly demonstrated by soils developed from sedimentary parent materials. In such soils a number of properties reflect the previous weathering and soil-forming conditions—such properties as amount and nature of the clay minerals, degree of base saturation, color as affected by free Fe_2O_3, and possibly other properties (29). This is also true for soils developing on primary igneous rocks, even though it is not so obvious.

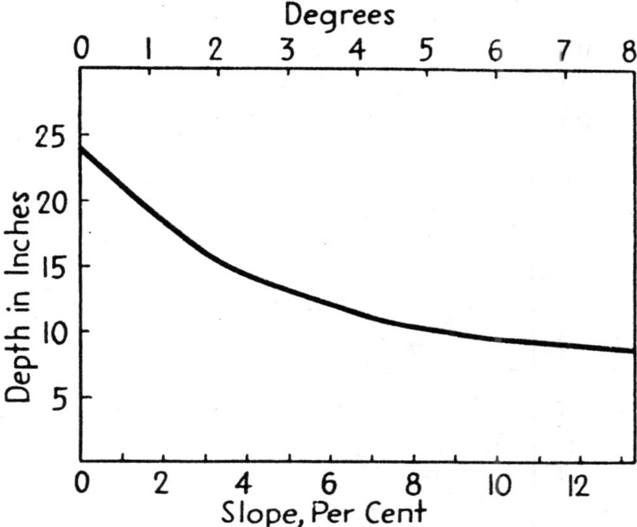

Figure 1.5. Effect of the topography factor on thickness of the A horizon of timbered soils: thickness of A = *f* (*slope, cl, o, p, t* . . .). (*By permission, from "Factors of Soil Formation," by Hans Jenny. Copyright, 1941. McGraw-Hill Book Co., Inc., New York.*)

The appearance, nevertheless, of a number of distinct soil groups possessing certain common features and occupying broad geographical areas—as those occupied by the great soil groups—indicates clearly that some of their features do reach a certain degree of equilibrium with the current environment, particularly for those features which are dependent on the biotic factor. Because such features are relatively few, soil classifiers who have attempted to develop a comprehensive scheme of soil classification found it more feasible to base the classification on soil properties rather than on the so-called "genetic" principles (74).

The presence of distinctly different soil types does not imply that the natures of the soil-forming reactions themselves are different in each soil, but rather that the relative importance of different soil-forming reactions is different in each soil. For example, the soil-forming reactions that occur to produce profile differences in uncombined Fe_2O_3 could consist of the following:

1. Release of Fe^{3+} from the primary minerals;
2. Movement of Fe^{3+} from the surface to the subsoil by water percolation;

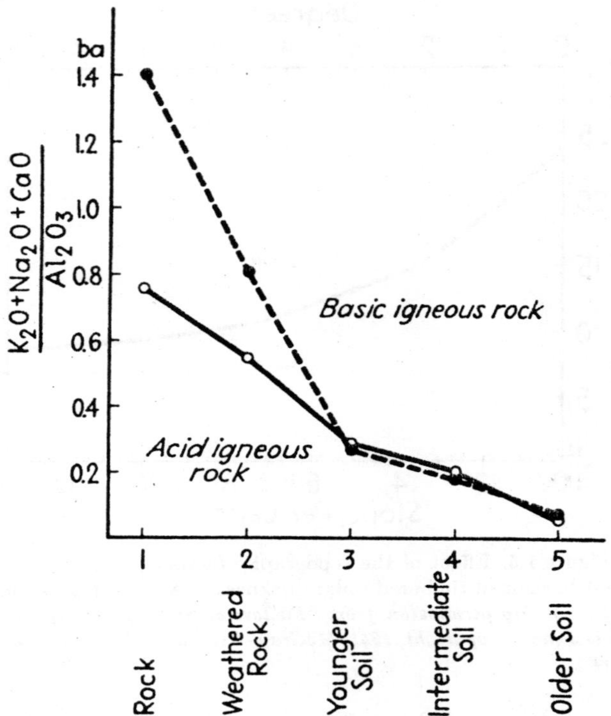

Figure 1.6. Effect of the parent material factor on relative base status of soils: base status = *f* *(parent material, cl, o, r, t . . .)*. *(By permission, from "Factors of Soil Formation," by Hans Jenny. Copyright, 1941. McGraw-Hill Book Co., Inc., New York.)*

3. Recombination of Fe^{3+} with silica and alumina to form clay minerals;
4. Precipitation of Fe^{3+} as $Fe(OH)_3$ and its conversion to Fe_2O_3 ;
5. Movement of iron from the subsoil back to the surface soil by plant absorption, and its release there through plant decomposition.

All five reactions occur in all soils but with such variation in intensity and relative importance as to produce different profile features. Thus, reactions 1 and 2 might be of equal intensity, but reactions 3, 4, and 5 could so vary in intensity as to produce entirely different profile features.

On the other hand, the presence of similar features in soils which are distinctly different does not imply that these features are the result of similar reaction conditions. A good example is the occurrence of gibbsite and hematite in both podzols and latosols. In both of these soils the forma-

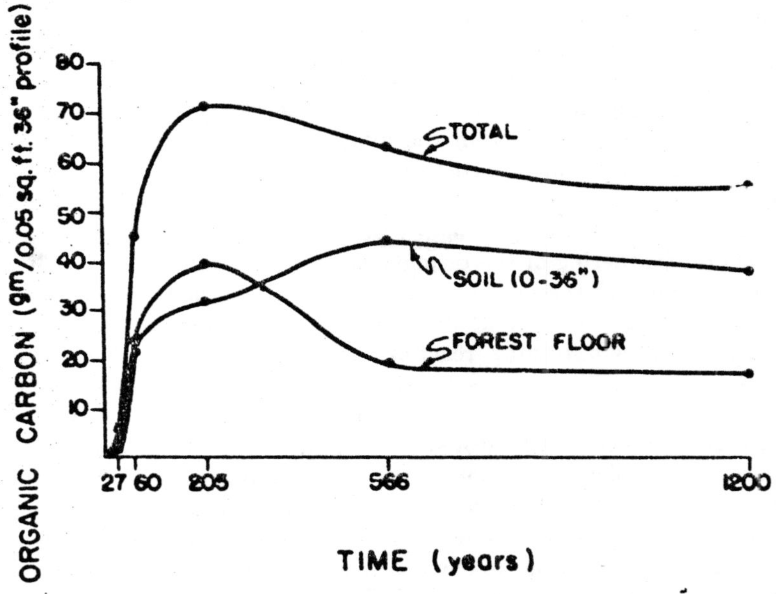

Figure 1.7. Effect of time of reaction, the time factor, on organic carbon of soils: organic carbon = f (*time, cl, o, r, p* . . .). (*Courtesy B. A. Dickson and R. L. Crocker and the Clarendon Press, Oxford, publishers of Journal Soil Science.*)

tion of these constituents is the presence of conditions which prevent silica from reacting with Al^{3+} and Fe^{3+}. In podzols the occurrence of acid conditions renders the silica insoluble and the Al^{3+} and Fe^{3+} "soluble," and, through water movement, the Al^{3+}, or complexes of Al^{3+}, and Fe^{3+} are separated from the silica in the A_2 horizon and deposited in the B horizon where they form gibbsite and hematite; whereas in latosols the occurrence of neutral or only slightly acid conditions renders the silica soluble and the Al^{3+} and Fe^{3+} insoluble, and, again through water movement, the silica is separated from the Al^{3+} and Fe^{3+} which then forms gibbsite and hematite in the A horizon, or throughout the profile over prolonged periods of time. If podzols have suffered erosion and the original surface horizons have eroded away, the present-day surface soil, which represents the original B horizon, contains gibbsite and hematite, a feature in common with latosols. This condition has often been mistaken to represent latosolic- or lateritic-forming conditions, but, in reality, as was indicated, they do not.

The foregoing example is also true for several other clay-mineral products which occur in common in different kinds of soils but which were formed through diverse chemical and physical reactions. The clay mineral vermiculite is a good example. In soils derived from granites or other rocks containing primary mica minerals, the occurrence of vermiculite is generally believed to have resulted from the alteration of the micas, as discussed previously; but, in soils derived from basalt or other basic rocks which do not contain primary micas, the occurrence of vermiculite is believed to have formed by synthesis through the interaction of silica, alumina, and Mg^{2+}—products of weathering of the primary nonresistant minerals.

Due to the inheritance of many soil features throughout its "life history" which remain stable, it is doubtful whether the designation of a given combination of soil-forming reactions as *podzolization* or *laterization* as distinct soil-forming processes has any validity other than in relatively young soils forming under current climatic and topographic conditions.

Since the factors of soil formation are responsible for conditioning the soil-forming reactions, it is not surprising to find that a good correlation exists between these factors and several characteristic profile features of the soils at their equilibrium or steady state, and that some of these features alter readily upon a change in climate or vegetation, or both, and that some profile features must have a polygenetic origin.

REFERENCES

1. Afanasiev, Ya. N., "Zonal Systems of Soils," Reprint from "Memoirs of the Gorki Agricultural Institute," pp. 1–83, 1922.
2. Atkinson, H. J., and Wright, J. R., *Soil Sci.* **84**, 1 (1957).
3. Barshad, doctoral dissertation, University of California, 1944.
4. Barshad, I., *Soil Sci.* **61**, 723 (1946).
5. Barshad, I., *Am. Mineralogist* **33**, 655 (1948).
6. Barshad, I., *Soil Sci.* **78**, 57 (1954).
7. Barshad, I., in "Clays and Clay Minerals," 6th Conf., p. 110, New York, Pergamon Press, 1959.
8. Barshad, I., *Science* **131**, 988 (1960).
9. Barshad, I., *Trans. Intern. Congr. Soil Sci. 7th Congr. Madison, Wisconsin* **2**, 435 (1961).
10. Barshad, I., Unpublished, 1963.
11. Beavers, A. H., *Trans. Intern. Congr. Soil Sci. 7th Congr. Madison, Wisconsin* **2**, 1 (1961).
12. Bragg, W. L., "Atomic Structure of Minerals," Ithaca, New York, Cornell University Press, 1937.
13. Brower, R., *Trans. Intern. Congr. Soil Sci. 7th Congr. Madison, Wisconsin* **1**, 1 (1961).
14. Brush, A. J., *Pochvovedenye* **710** (1935).
15. Burd, J. S., and Martin, J. C., *Soil Sci.* **18**, 151 (1924).
16. Crocker, R. L., *Quart. Rev. Biol.* **27**, 139 (1952).
17. Dickson, B. A., and Crocker, R. L., *J. Soil Sci.* **4**, 142 (1953).
18. Dokuchaev, V. V., trans. in *St. Petersburg Soc. Nat.* **10**, 64 (1879).

19. Dokuchaev, V. V., *Akad. Nauk S.S.S.R.* 6, 381 (1951).
20. Fields, M., and Swindale, L. D., *New Zealand J. Sci. Technol.* 36, 140 (1954).
21. Fields, M., and Taylor, N. H., *New Zealand J. Sci.* 4, 679 (1961).
22. Glinka, K. D., (1914) trans., C. F. Marbut, Ann Arbor, Mich., Edwards Bros., 1927.
23. Goldich, S. S., *J. Geol.* 46, 17 (1938).
24. Grim, R. E., "Clay Mineralogy," London, McGraw-Hill Publishing Company Ltd., 1953.
25. Haseman, J. F., and Marshall, C. E., "Missouri Agr. Expt. Sta. Research Bull." 387, 1945.
26. Hilgard, E. W., "Report of the Geology and Agriculture of the State of Mississippi," Jackson, Mississippi, State Printer, 1860.
27. Hilgard, E. W., "Soils," New York, The Macmillan Company, 1906.
28. Humbert, R. P., and Marshall, C. E., "Missouri Agr. Expt. Sta. Research Bull." 359, 1943.
29. Jackson, M. L., and Sherman, G. D., *Advances in Agron.* 5, 219 (1953).
30. Jackson, M. L., "Clays and Clay Minerals," 6th Conf., p. 133, New York, Pergamon Press, 1959.
31. Jackson, M. L., *Trans. Intern. Congr. Soil Sci. 7th Congr. Madison, Wisconsin* 2, p. 445, 1961.
32. Jackson, M. L., *Soil Sci. Soc. Am. Proc.* 27, 1 (1963).
33. Jeffries, C. D., and White, J. W., *Soil Sci. Soc. Am. Proc.* 2, 133 (1937).
34. Jenny, H., "Factors of soil formation, a system of quantitative pedology," New York, McGraw-Hill Book Company, Inc., 1941.
35. Jenny, H., *Soil Sci.* 61, 375 (1946).
36. Jenny, H., *Ecology* 39, 5 (1958).
37. Jenny, H., and Raychaudhuri, S. P., "Effect of Climate and Cultivation on Nitrogen and Organic Matter Reserves in Indian Soils" New Delhi, India, Indian Council Agr. Research, 1960.
38. Jenny, H., "E. W. Hilgard and the Birth of Modern Soil Science," Pisa, Italy, Agrochimica Publishing Company, and P.O. Box 564, Berkeley, California, Farallon Publisher, 1961.
39. Jenny, H., *Soil Sci. Soc. Am. Proc.* 25, 385 (1961).
40. Joffe, J. S., "Pedology," New Brunswick, New Jersey, Pedology Publications, 1949.
41. Kelley, W. P., *Soil Sci.* 56, 443 (1943).
42. Kononova, M. M., "Soil Organic Matter, its Nature, its Role in Soil Formation and Soil Fertility," trans. by T. Z. Nowakawski and G. A. Greenwood, New York, Pergamon Press, 1961.
43. Kubiena, W. L., "Micropedology," Ames, Iowa, Collegiate Press, 1938.
44. Kubiena, W. L., "The soils of Europe," London, Thomas Murby and Company, 1953.
45. Lutz, H. J., and Chandler, R. F., "Forest Soils," New York, John Wiley & Sons, Inc., 1946.
46. Marbut, C. F., "Soils of the United States, Atlas of American Agriculture, Part III," Washington, D. C., U.S. Government Printing Office, 1935.
47. Marshall, G. E., *Soil Sci. Soc. Am. Proc.* 5, 100 (1941).
48. Marshall, G. E., and Haseman, J. F., *Soil Sci. Soc. Am. Proc.* 5, 448 (1943).
49. Mattson, S., *Soil Sci.* 28, 179 and 373 (1929).
50. Mattson, S., *Soil Sci.* 30, 75 (1939).

51. Marill, G. P., "A treatise on rocks, rock-weathering and soils," New York, The Macmillan Co., 1897.
52. Mickelson, G. A., Reprint *Abstr. Doctoral Dissertation*, **40**, Columbus, Ohio, The Ohio State University Press, 1943.
53. Mortenson, J. L., *Soil Sci. Soc. Am. Proc.* **27**, 179 (1963).
54. Murata, K. J., *Am. Mineralogist* **28**, 545 (1943).
55. Murata, K. J., *U.S. Geol. Survey Bull.* **1950**, 25 (1946).
56. Nikiforoff, C. C., *Sigma Xi Quart.* **30**, 36 (1942).
57. Nikiforoff, C. C., *Am. J. Sci.* **240**, 847 (1942).
58. Nikiforoff, C. C., and Alexander, L. T., *Soil Sci.* **53**, 157 (1942).
59. Nikiforoff, C. C., and Drosdoff, M., *Soil Sci.* **55**, 459; **56**, 43 (1943).
60. Nikiforoff, C. C., *Soil Sci.* **67**, 219 (1949).
61. Polynov, B. B., *Pochvovedeniye* 3–13 (1948) [trans. in *Soils and Fertilizers*, **14**, 95 (1951)].
62. Ramann, E., "The Evolution and Classification of Soils," London, W. Heffer and Sons, Ltd., 1928.
63. Rankama, K., and Sahama, Th., "Geochemistry," Chicago, University of Chicago Press, 1950.
64. Reitemeier, R. F., *Soil Sci.* **6**, 195 (1946).
65. Reitemeier, R. F., Brown, I. C., and Holmes, R. S., *U.S.D.A. Bull.* **1049**, 1951.
66. Ross, C. S., and Hendricks, S. B., *U.S. Geol. Survey*, Professional paper **205-B**, Washington, D. C., U.S. Government Printing Office, 1945.
67. Remezov, N. P., *Pochvovedeniye* **178** (1932).
68. Robinson, G. W., "Soils: Their Origin, Constitution and Classification," 2nd. ed., London, Thomas Murby and Co., 1949.
69. Rode, A. A., *Pochvovedeniye* **400** (1946).
70. Schatz, A., Chernis, N. D., Schatz, V., and Trelawny, G. S., *Penn. Acad. Sci. Proc.* **28**, 44 (1954).
71. Schwertmann, U., *Z. Pflanzenernaehr. Dueng. Bodenk.* **97**, 9 (1962).
72. Sibirtzev, N. M., *Zapiski Nov.-Aleksandr. Agr. Inst.* **9**, 1 (1895).
73. Sigmond, A. A. J. de, "The principles of Soil Science," (trans., A. W. Jacks), London, Thomas Murby and Co., 1938.
74. Soil Survey Staff, U.S. Department of Agriculture, "Soil Classification, A Comprehensive System, 7th approximation," Washington, D. C., U.S. Government Printing Office, 1960.
75. Soil Conservation Service, U.S.D.A., *Virginia Agr. Expt. Sta. Southern Regional Bull.* **61**, Blacksburg, Virginia, Virginia Polytechnic Institute, 1959.
76. Van der Merwe, C. R., *Union of South Africa Dept. of Agr., Forestry, and Chem. Ser.*, **165**, Pretoria, Union of South Africa Government Printer, 1940.
77. Van Hise, Ch. R., "U.S. Geol. Survey Monogr." **47**, 1904.
78. Van Schuylenborgh, J., and Sanger, A. M. H., *Landbouwk. Tijdschr.* **62**, 347 (1950).
79. Vilenskii, D. G., "Soil science," Moscow, U.S.S.R., State Teacher's College Publishing House, 1957 (trans. A. Birron and Z. S. Cole, Jerusalem, Israel, S. Morison, 1960).
80. Wilde, S. A., "Forest Soils Their Properties and Relation to Silviculture," New York, The Ronald Press Company, 1958.
81. Wright, J. R., and Schnitzer, M., *Soil Sci. Soc. Am. Proc.* **27**, 171 (1963).
82. Zakharov, S. A., Russian Pedological Investigations II, pp. 1–47, Leningrad, Academy of Science, 1927.

CHAPTER 2

Chemical Composition of Soils

MARION L. JACKSON

Department of Soil Science
University of Wisconsin, Madison

Soil composition concerns the geochemistry of that portion of the mantle of the earth currently exposed to the atmosphere and subjected to biotic influences. The earth's mantle contributes mineral matter, while organisms introduce an organic matter component into soils, giving the two broad classes of materials which make up the solid phases of soils—mineral and organic matter. Varying proportions of mineral matter are present in the various soils, ranging in amount up to over 99 per cent in sandy soils and 95 to 97 per cent in many productive soils of intermediate texture. The mineral portion drops to about 92 per cent in well-drained dark-colored soils, and drops down into the range of 80 to 60 per cent in many poorly drained soils in which the organic matter is protected by the lowering of the oxygen supply in sites where water fills most of the pore space. Peat and muck soils (organic soils) may contain from 60 per cent to as little as 5 per cent or less of mineral matter, depending on the quantities of aeolean and water-laid sediments received as the organic matter accumulated in marshes. Because the organic phase of soils is the subject of other chapters, the concern of this chapter is confined to the chemical composition of the mineral portion of soils.

The solid phases (mineral and organic) frequently make up only about 50 per cent of the soil volume, the other half being filled by the soil solution and air (Figure 2.1). Since the density of organic matter is only about 0.5 g/cm^3, while that of mineral matter is approximately 2.7 g/cm^3, 5 per cent by weight of organic matter in a soil occupies nearly 25 per cent of the volume of the solids. The liquid and gas phases together often make

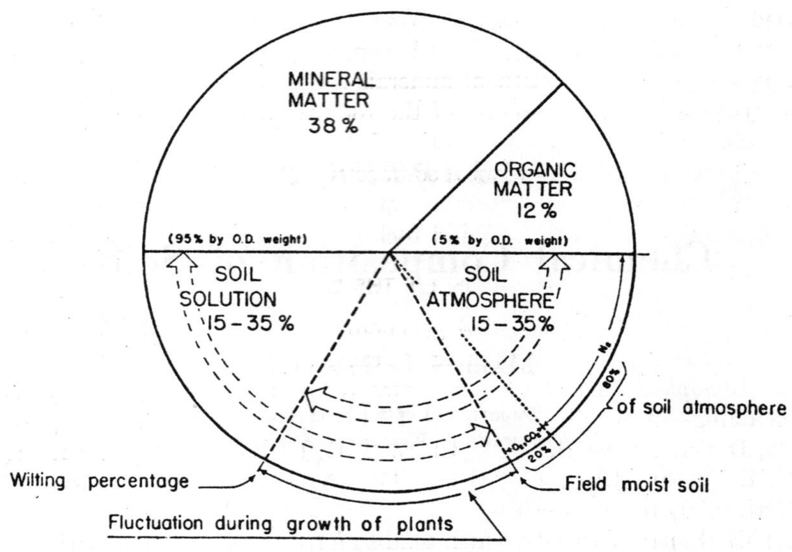

Figure 2.1. Percentage composition of soil (volume basis).

up about half the soil volume, varying rapidly in relative proportions of each as the liquid phase (water) drains away or is used by plants.

The gas phase of soils, which often occupies about a fourth of the soil volume, has a composition of approximately 80 per cent nitrogen, as in ordinary air, the other 20 per cent (by volume) being mainly oxygen. The soil air in well-drained soils contains less than 0.1 per cent CO_2. When aeration is poor, the percentage of CO_2 may rise to 5 or 10 per cent and, in extreme cases, for example, under frozen soil horizons, it may rise to 20 per cent and thus replace all the oxygen. Traces of the rare gas elements, Ne, Ar, Kr, and Xe, which occur in atmospheric air, also occur in the soil atmosphere.

Soils are a dynamic part of the earth's geomorphic cycle of surficial weathering, erosion, deposition, sinking, diagenesis, metamorphosis, uplift, and mountain building. The mineralogical composition of soils influences that of the sedimentary rocks of which eroded soil material becomes a part, just as the rocks which serve as parent materials of soils in turn influence the chemical and mineralogical composition of soils. Other factors of soil formation, namely climatic, biotic, topographic, and time factors, may greatly influence the chemical composition of soils by controlling the weathering reactions which alter the minerals received from the parent rock. Under control of these factors, chemical elements are translocated and deposited in deeper soil horizons; sometimes, these elements are returned to the upper horizons by biocycling, and sometimes they are removed from soil by leaching. Nonetheless, most soils reveal by the chemical composition of their constituent minerals the nature of their original parent rock because of the interaction of the rock composition, texture, and susceptibility to leaching with the other factors of soil formation. In gaining an understanding of the chemical composition of soils, it is useful, therefore, to review the chemical composition of the earth's crust, considering both the elements and the mineral and rock association of the elements.

COMPOSITION OF THE EARTH'S CRUST

The elements are distributed geochemically into five main groups (20, 77, 78) according to their bonding character.

(*a*) *Lithophile elements*, which ionize readily or form stable oxyanions, including—O, Si, Ti, Fe, Mn, Al, H, Li, Na, K, Rb, Cs, Be, Mg, Ca, Sr, Ba, B, Ga, Ge, Sn, Sc, Y, F, Cl, Br, I, C, Hf, Th, P, V, Nb, Ta, Cr, W, U, Zr, (Mo), (Cu), (Zn), (Pb), (Tl), (As), (Sb), (Bi), (S), (Se), (Te), (Ni), (Co), and rare earths.

(*b*) *Chalcophile elements*, which tend to form covalent bonds with sulfide, including—S, Se, Te, (Fe), Ni, Co, Cu, Zn, Pb, Mo, Ag, Sb, (Sn), Cd, In, Tl, Pb, As, Bi, Re, (Mn), (Ga), and (Ge).

(*c*) *Siderophile elements*, which readily form metallic bonds, including—Fe, Ni, Co, Ru, Rh, Pd, Pt, Ir, Os, and Au.

(*d*) *Atmosphile elements*, which tend to remain in atmospheric gases, including—N, (O), He, Ne, Ar, Kr, Xe.

(*e*) *Biophile elements*, which tend to be associated with organisms (and thus collect in the horizons most affected by organisms in soils), including—C, H, O, N, P, S, Cl, I, B, Ca, Mg, K, Na, Mn, Fe, Zn, Cu, Ag, Mo, Co, Ni, Au, Be, Cd, Se, Tl, Sn, Pb, As, and V.

The elements shown in parentheses fall in the additional group or groups because bonding of the element changes with temperature, pressure, oxidation or reduction state, or specific biochemical affinity. For example, oxygen appears in the *atmosphile* group because of its release by photosynthesis, and iron falls in the *chalcophile* group when it occurs in a reduced environment. The *biophile* list includes mainly *lithophile* elements.

The earth's crust consists mainly of the *lithophile* elements, which are common in silicates and in soils. The 80 elements that make up the majority of the rocks in the earth's crust are distributed in about two thousand characteristic compounds or *minerals*, and, of these minerals, a few dozen make up the bulk of surficial rocks. The elements are distributed according to ionic radii and charge as the crystals act like a sorting or "chemical sieving" mechanism.

Despite the large-scale and small-scale heterogeneity seen in rock out-

TABLE 2.1. AVERAGE CHEMICAL COMPOSITION OF THE EARTH'S CRUST

Chemical composition of crust*			
Element	%	Oxide	%
O	46.5	SiO_2	59.07
Si	27.6	Al_2O_3	15.22
Al	8.1	Fe_2O_3	3.10
Fe	5.1	FeO	3.71
Ca	3.6	CaO	5.10
Mg	2.1	MgO	3.45
Na	2.8	Na_2O	3.71
K	2.6	K_2O	3.11
Ti	0.6	TiO_2	1.03
P	0.12	P_2O_5	0.30
Mn	0.09	MnO	0.11
S	0.06	H_2O	1.30
Cl	0.05		
C	0.04		

* Based on 95 per cent igneous rocks, 4 per cent shales, 0.75 per cent sandstone, and 0.25 per cent limestone [after Clarke and Washington (48)].

crops in the outermost portion of the earth's crust, thousands of rock analyses show that a few chemical elements are predominant. According to Clarke (47) the abundance of elements in the earth's crust 10 miles thick decreases in the order: O, Si, Al, Fe, Ca, Na, K, Mg, Ti, P, Mn, S, Cl, and C (Table 2.1). Only the eight elements O, Si, Al, Fe, Ca, Mg, Na, and K surpass 1 per cent in the average composition of earth's crust. The eleven elements Ti, H, P, Mn, F, S, Sr, Ba, C, Cl, and Cr constitute on the order of 0.2 to 1 per cent each; each of the other elements makes up less than 0.2 per cent (20). One element, oxygen, makes up over 90 per cent by volume of the earth's crust, and thus the geochemistry of the crust is mainly concerned with cations interspersed in this oxygen volume. Of the elemental composition of the earth's crust, SiO_2 makes up 59 per cent and Al_2O_3, 15 per cent, and similar percentages of these constituents are present in many sediments and soils, as will be discussed below.

In approaching the study of soil chemical composition, it is useful to note the mineralogical composition of the various rock types in which the chemical elements are held in the mineral portion of soils and soil parent materials. The chemical compositions of the main individual mineral components of rocks, following a brief review of the genesis and mineralogy of the main rock types, are given below.

Igneous Rocks

Igneous rocks, which vary in composition from feldspathic to mafic (Figure 2.2), have resulted from the cooling of *magma* or silicate melts of the *lithophile* elements. If cooling has taken place slowly (as occurs at depth in the earth's crust), coarser-grained, *plutonic* rocks resulted; if an eruption occurred, cooling was rapid, and finer-grained, *volcanic* rocks resulted. Among the plutonic rocks, *granites* and *granodiorites* are far more plentiful than all the others combined. Among volcanic rocks, *basalts* and *basic andesites* greatly predominate. When the melt was deficient in silica, *nepheline* formed from melts high in Na and K, while *olivine* formed at the other extreme when the proportions of Fe and Mg were very high. These extremes of composition can be produced by separation of the high temperature, more mafic phases by sedimentation and structural movements, followed by subsequent crystallization of the lower-temperature phases.

The association of minerals in these rock types results from the fractional crystallization sequences deduced by Bowen and subsequent workers (17, 23):

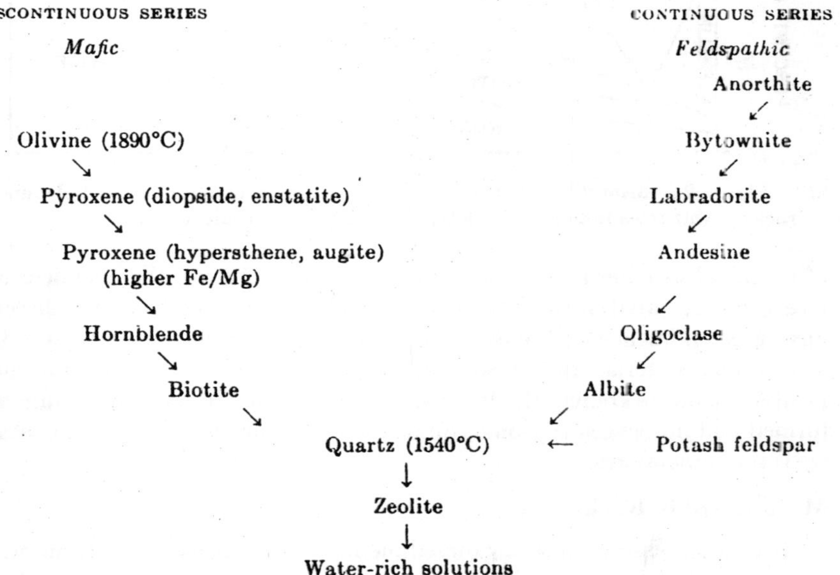

The minerals of high melting point crystallize first, yielding large and euhedral crystals termed *phenocrysts*. At the other extreme, when cooling of volcanic extrusives has been very rapid, crystallization of minerals is prevented and *glasses* result. Higher-temperature crystals formed before eruption of volcanic rocks are frequently cemented together by glass formed after eruption. Volcanic rocks vary in texture from *massive* to *ash*, depending on the degree of fracturing that occurred, as by gas expansion or by chilling in water. Such texture differences make enormous differences in the rate

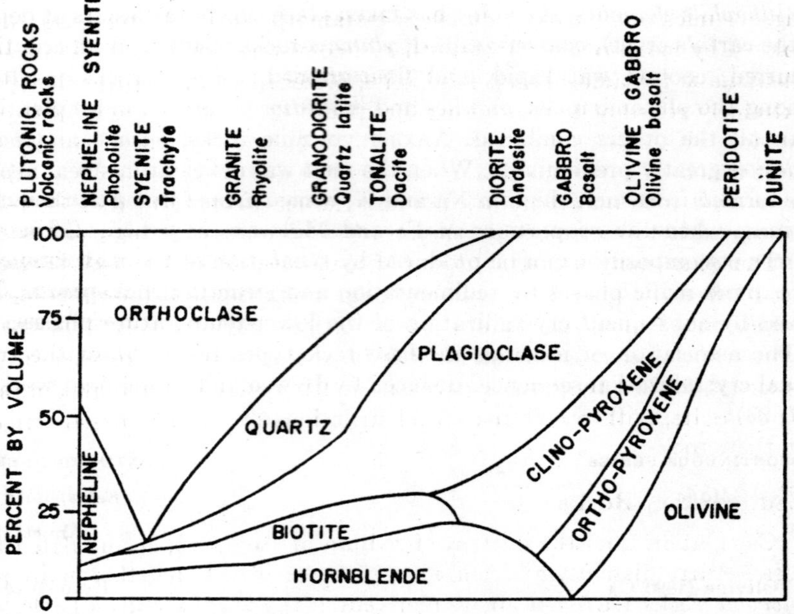

Figure 2.2. Mineralogical composition of common igneous rocks. Volcanic rocks with names in lower-case letters consist of fine-grained minerals.

of chemical weathering during soil formation. Moreover, the chemical differences associated with these various rock types directly and indirectly have great influence on the elemental and mineralogical composition of the soil parent material and of soils developed thereon. The elemental composition of rocks frequently controls the nature of the secondary minerals formed and influences the ionic composition of aqueous solutions in contact with the mineral grains.

Metamorphic Rocks

Heat and pressure operating on sedimentary and igneous rocks form *metamorphic* rocks. The mineralogical composition and texture of the resulting rocks greatly affect the composition of soils formed on them when they are exposed to soil formation.

Gneiss is a laminated or banded metamorphic rock of mineralogical composition similar to granite. *Mica schists* are foliated, usually fine-grained, rocks. Mica schists weather rapidly through the laminous cleavages, giving rise to soil high in vermiculite and quartz. The *slates* are produced by compression and alteration of shales and mud stones, giving fine-grained rocks high in mica and quartz. *Marble* is metamorphosed limestone, frequently colored by inclusions of various other minerals and resistant to weathering. *Quartzite* is metamorphosed sandstone, enriched in interstitial silica; it is extremely resistant to weathering. Frequently this rock may be 99 per cent or more SiO_2. Veins may be filled with dickite or other layer silicate minerals which are formed from the minor quantities of various ionic impurities associated with quartzite rock because the silica phase of the quartzite is inhospitable to cations other than Si.

There are many additional types of metamorphic rocks, of less-extensive occurrence. Related to metamorphosed mafic and dolomitic rocks are *amphibolite* (consisting mainly of hornblende and plagioclase), *amphibole schist* or *hornblende schist*, and *chlorite schist* or greenstone. Another example is *tactite*, formed by granitic invasion and injection of ions into limestone or dolomite, with the formation of inclusions of ferromagnesian minerals in these rocks.

Sedimentary Rocks

Calculation from the content of sodium in the ocean, accumulated from weathering (dissolution) from *igneous* rocks [which together with *metamorphic* rocks make up about nine-tenths the earth's crust (Table 2.2)], gave Clarke (47) an estimate of 5 per cent *sedimentary* rocks that should have formed from the quantity of igneous rocks that weathered. Later estimates of the proportion of sedimentary rocks tend to raise this 5 per cent figure several-fold (71). Sedimentary rocks tend to accumulate near the interface of the crust with the hydrosphere and atmosphere, and thus sedimentary rocks cover about three-fourths the land area of the earth, a much higher proportion of the parent material of soils than of the 10-mile crust. This geomorphic fact is of great importance to soils, because the higher specific surface of the mineral grains and the general nature of sedimentary rocks—less consolidated than igneous rocks—permit a more rapid biotic invasion and development of deeper, more-productive soils on sedimentary rocks. It is probable that the accumulation of hydrous minerals in sedimentary muds favored and conditioned the evolutionary development of life forms on earth. Particularly, the capacity of fine sediments (primordial soils) to sorb water and readily exchangeable ions in coastal margins encouraged the evolutionary movement of life from sea to land, and thus introduced the biotic factor in soil formation.

Sedimentary rocks of marine origin are classified as *shale*, *sandstone*, and *limestone* (including dolomite). Of the sediments of the earth's crust, 80 per cent are shales, 15 per cent are sandstones, and 5 per cent are limestones (Table 2.2). Terrestrial sediments include glacial drift, loess, dunes,

TABLE 2.2 ELEMENTAL COMPOSITION OF ROCKS OF THE EARTH'S CRUST

Elements	% Composition of rocks*			
	Igneous† (95%)	Shale (4%)	Sandstone (0.75%)	Limestone (0.25%)
SiO_2	59.12	58.11	78.31	5.19
Al_2O_3	15.34	15.40	4.76	0.81
Fe_2O_3	3.08	4.02	1.08	0.54
FeO	3.80	2.45	0.30	
TiO_2	1.05	0.65	0.25	0.06
CaO	5.08	3.10	5.50	42.57
MgO	3.49	2.44	1.16	7.89
MnO	0.12	Tr	Tr	0.05
K_2O	3.13	3.24	1.32	0.33
Na_2O	3.84	1.30	0.45	0.05
CO_2	0.10	2.63	5.04	41.54
P_2O_5	0.30	0.17	0.08	0.04
SO_3	—	0.65	0.07	0.05
S	0.05	—	—	0.09
H_2O	1.15	4.99	1.63	0.77
Total	*99.65*	*99.15*	*99.95*	*99.98*

* Per cents in parentheses indicate per cents of rocks in the earth's crust [After Clarke and Washington (48)].

† And metamorphic.

volcanic ash, and stream and lake sediments. The soil itself is sometimes thought of as a sediment. While igneous and metamorphic rocks contain 60 per cent feldspars (Table 2.3), physical and chemical weathering causes the feldspar content to drop to 30 and 12 per cent in shale and sandstone, respectively. Quartz (SiO_2) grains, which are more resistant to weathering, increase from 12 to 22 and 67 per cent, respectively. Calcium carbonates ($CaCO_3$) and to some extent dolomite ($CaCO_3 \cdot MgCO_3$) accumulate in limestones of sedimentary rocks mainly through chemical and biochemical precipitation in the ocean waters, but also as erosion-sedimentation products from chemical precipitates (evaporites) in lakes and in soils.

Soil Minerals

The inorganic constituents of soils occur mainly in a rather limited number of minerals or inorganic compounds of definite crystal structures, as mentioned for rocks (above). Quantitative determination of minerals in soils is based on assignment of chemical elements to the mineral structures revealed by X-ray diffraction analysis. Compositions of a given soil mineral can vary greatly because of isomorphous substitution of different ions in a given crystal structural position. Each mineral structure tends, there-

TABLE 2.3. AVERAGE MINERALOGICAL COMPOSITION OF IGNEOUS AND SEDIMENTARY ROCKS*

Mineral constituent	Igneous rock (%)	Shale (%)	Sandstone (%)
Feldspars	59.5	30.0	11.5
Amphiboles and pyroxenes	16.8	—	†
Quartz	12.0	22.3	66.8
Micas	3.8	—	†
Titanium minerals	1.5	—	†
Apatite	0.6	—	†
Clay	—	25.0	6.6
Limonite	—	5.6	1.8
Carbonates	—	5.7	11.1
Other minerals	5.8	11.4	2.2

* Data from Clarke (47).
† Present in small amounts.

fore, to form an *isomorphous series of minerals*. Many of the essential trace elements occur mainly as isomorphous substitutions in host minerals, and this prevalence of a variety of ions in minerals of various rocks enabled plants to evolve with a dependence on a wide range of trace elements in addition to elements present in larger amounts. Thus plants need C, H, O, P, K, N, S, Ca, Fe, Mg, Mn, Cu, B, Zn, Co, Mo, Cl, Na, and probably several others yet to be confirmed. This list of elements essential to plants can be remembered by the phrase: "C. HOPKiNS CaFe, Mighty good, Many CurB Zones ComMonly Close, Naturally."

A small but important fraction of the total of various elements in soil minerals occurs as readily *exchangeable* or adsorbed ions. This form of the elements is of vital importance to the nutrition of plants growing on different soils. The supply of exchangeable ions is slowly renewed by weathering of ions from within the crystal structure of various minerals. The different mineral species in soils have a controlling influence on the release of structural ions and on the chemical activity of exchangeable ions; the mineralogical composition of soils thus has an important bearing on soil productivity. The rate of release of ions slows down as the soil minerals become more highly weathered, often becoming too slow to support intensive crop production. Elements are then added after chemical processing of selected rocks into commercial fertilizer form. The chemical reactions of fertilizer compounds with soil minerals then determines the efficacy of the fertilizer elements in providing available plant nutrients.

The minerals of soils can be classified into (*a*) *primary minerals* formed at elevated temperatures and inherited from the igneous and metamorphic rocks (sometimes through a sedimentary cycle), and (*b*) *secondary minerals* formed by low-temperature reactions and inherited by soils from sedimentary rocks or formed in soils by weathering.

PRIMARY MINERALS IN SOILS

Primary minerals, inherited from the parent rocks, make up the main part of the sand and silt fractions of most soils. For a given soil, the sedimentation history of the parent material determines the content of sand (particles ranging from 1 to 0.05 mm in diameter), and the combined effect of weathering and sedimentation determine the content of silt (particles ranging from 0.05 to 0.002 mm in diameter). The combined percentage of sand and silt in a given soil mainly governs the percentages of primary minerals in that soil. Primary minerals occur in the clay fraction (particles less than 0.002 mm in diameter) of little-weathered soils, for example those derived from fine-grained rock flour derived by glacial grinding of rocks.

The most abundant minerals and mineral groups in sand and silt of soils, the world over, are *quartz*—SiO_2 (silica), and *feldspars*—$MAlSi_3O_8$ (silicate salts of basic cations, M^+), just as these minerals dominate in the rocks making up the earth's crust. Pyroxenes, amphiboles, olivine, and other accessory minerals of primary origin, though usually present in smaller percentages, are important to the chemistry of soils.

Silica

Free silica occurs in soils mainly in the form of *quartz*—SiO_2, consisting of a continuous framework of silica tetrahedra. All four oxygens of the tetrahedra are shared and there are no cleavage planes. Free silica also occurs in some soils as *opal*—$SiO_2 \cdot nH_2O$, including opal pseudomorphic after plant cells (phytoliths), *chalcedonite*, *agate*, *flint*, and *chert*. It also occurs in some soils as *cristobalite*—SiO_2. Substituted or so-called "stuffed" cristobalite occurs with varying substitution—$(NaAl, Si)O_2$, in which the Na is exchangeable. In certain highly podzolized soils of New Zealand, cristobalite may make up nearly 50 per cent of the clay, the other half being quartz (194); exchange capacity of the clay arises almost entirely through the substitution in cristobalite. Free silica also occurs as amorphous sheets and as aggregates with amorphous iron and aluminum oxide gel. While devoid of nutrients that can be released by weathering, free silica particles provide the framework of soil structure and influence soil aeration.

Quartz occurs in the majority of soils, both alluvial and those developed from various types of sedimentary rocks and residual crystalline rocks. Quartz frequently makes up from 50 to 90 per cent of the sand and coarser silts of many soils. Quartz dikes often introduce quartz into basic parent rock formations in which quartz would not normally be expected. Soils developed from quartz-free basic igneous rocks occasionally are found, and yield a quartz-free sand.

The main process of concentration of free silica is *podzolization*, under which organic matter both chelates the sesquioxidic ions and reduces ferric iron and then mobilizes these ions out of the bleached A_2 horizons. Fine-grained silica is itself easily hydrated and moderately soluble, and silica is removed from soils in abundant quantities under heavy leaching. The tendency therefore is for SiO_2 to accumulate as coarser grains of sands, silt, and even coarse clay in highly podzolized horizons. The most common and conspicuous light gray or white A_2 horizons of many podzolic soils consist of sand-sized SiO_2 and other aluminosilicate particles, but a high content of bleached silt and (rarely) of coarse clay does occasionally occur in the A_2 horizons. Bleached horizons occur widely not only in the cool temperate climates but also in the ground-water podzolic soils of tropical areas, and in temperate climates under the mor-forming *Kauri* tree of New Zealand.

Feldspars

Feldspars are anhydrous aluminosilicates of K, Na, and Ca, and occasionally of other large cations such as Ba. The feldspars make up an average of about 60 per cent by weight of igneous rocks and large fractions of sedimentary rocks (Table 2.3).

The feldspar structure consists of tetrahedra which are linked together by sharing each oxygen atom between adjacent tetrahedra. The result is a 3-dimensional *framework* or *tektosilicate* structure. The tetrahedra contain mainly Si ions with considerable Al substitution. For each Al ion substituted, one equivalent of a larger cation is present to maintain electrical neutrality. The smaller cations, such as Mg and Fe, of rocks segregate into minerals (Figure 2.2) which provide 6-fold coordination and do not occur as essential constituents of feldspars.

Potassium feldspars—$KAlSi_3O_8$ include *orthoclase*, *microcline*, *adularia*, and *sanidine*, with the elemental compositions given in Table 2.4. Orthoclase and microcline are more common in the plutonic (Figure 2.2) and metamorphic rocks. They are common in granites, syenites, gneiss, and some pegmatites. Sanidine is associated with volcanic rocks, particularly rhyolites and trachytes. Adularia is a low-temperature, lath-shaped form common in hydrothermal veins. Solid solution of $KAlSi_3O_8$ with $NaAlSi_3O_8$ occurs at high temperatures; the sodium-rich member is *anorthoclase*, and the potassium-rich member is included with the sanidines. The potassium feldspars occur commonly in the silts and sands of soils. They are

TABLE 2.4. CHEMICAL COMPOSITION OF SOME REPRESENTATIVE SOIL MINERALS*

Mineral	Composition (%)											
										H_2O		
	SiO_2	Al_2O_3	Fe_2O_3	FeO	TiO_2	CaO	MgO	K_2O	Na_2O	110–350°C	350–540°C	540–900°C
Oxides and Hydroxides												
Quartz, SiO_2	100	—	—	—	—	—	—	—	—	—	—	—
Gibbsite, $Al(OH)_3$	—	65.4	—	—	—	—	—	—	—	31.2	3.4	—
Boehmite, AlOOH	—	85.0	—	—	—	—	—	—	—	—	15.0	—
Goethite, FeOOH	—	—	89.9	—	—	—	—	—	—	10.1	—	—
Limonite, FeOOH·½H_2O	—	—	85.5	—	—	—	—	—	—	14.5	—	—
Magnetite, Fe_3O_4	—	—	69.0	31.0	—	—	—	—	—	—	—	3.5†
Hematite, Fe_2O_3	—	—	100	—	—	—	—	—	—	—	—	—
Ilmenite, $TiFeO_3$	—	—	—	47.3	52.7	—	—	—	—	—	—	5.7†
Rutile, TiO_2	—	—	—	—	100	—	—	—	—	—	—	—
Anatase, TiO_2	—	—	—	—	100	—	—	—	—	—	—	—
Allophane, $2SiO_2$·Al_2O_3·3.28 H_2O**	42.7	36.3	—	—	—	—	—	—	—	15.3	4.0	1.7
Feldspars												
Microcline, $KAlSi_3O_8$	64.7	18.4	—	—	—	—	—	16.9	—	—	—	—
Albite, $NaAlSi_3O_8$	68.7	19.5	—	—	—	—	—	—	11.8	—	—	—
Anorthite, $CaAl_2Si_2O_8$	43.2	36.6	—	—	—	20.2	—	—	—	—	—	—
Layer Silicates												
Pyrophyllite, Si_8Al_4·$O_{20}(OH)_4$	66.7	28.3	—	—	—	—	—	—	—	—	—	5.0
Muscovite, $K_2Al_2Si_6$·$Al_4O_{20}(OH)_4$	45.3	38.4	—	—	—	—	—	11.8	—	—	—	4.5
Biotite, $K_2Al_2Si_6(Mg_2$·$Fe^{2+})_3O_{20}(OH)_4$	40.2	11.4	—	16.0	—	—	18.0	10.5	—	—	—	4.0
Chlorite, $AlMg_4Fe^{2+}$·$(OH)_{12}$·$Al_2Si_6(Al$·$Mg_4Fe^{2+})O_{20}(OH)_4$	30.7	17.4	—	12.2	—	—	27.5	—	—	—	—	12.3
Kaolinite, $Si_4Al_4O_{10}$·$(OH)_8$	46.5	39.5	—	—	—	—	—	—	—	—	14.0	—
Halloysite, $Si_4Al_4O_{10}$·$(OH)_8$·4H_2O	40.8	34.6	—	—	—	—	—	—	—	12.2‡	12.3	
0.78 H_2O	44.2	37.5	—	—	—	—	—	—	—	2.2	14.8	1.3

* Values for end-members; in practice solid solution and isomorphous substitution cause variations explained in the text [after Jackson (106)].

† Weight gain resulting from oxidation of ferrous iron at 540 to 700°C; Fe^{2+} may be replaced by Mg, thus reducing gain.

‡ Fully hydrated basis rather than 110°C basis.

** See Choyo soil (Table 2.5) for percentage composition with a SiO_2/Al_2O_3 ratio of about unity.

abundant in the clay-size fraction of soils which have undergone little weathering. As finer silt and clay sizes they serve as an important source of available soil potassium, but are generally less important than the micas as mineral sources of potassium in soils.

The *plagioclase* feldspars are a series consisting of a solid solution of *albite*—$NaAlSi_3O_8$ and *anorthite*—$CaAl_2Si_2O_8$. Classically, the *plagioclase* series is named by compositional increments: *Albite*, Ab_{100}-Ab_{90} ; *oligoclase*, Ab_{90}-Ab_{70} ; *andesine*, Ab_{70}-Ab_{50} ; *labradorite*, Ab_{50}-Ab_{30} ; *bytownite*, Ab_{30}-Ab_{10} ; and *anorthite*, Ab_{10}-Ab_0 (An_{100}).

Once thought to be an ideal continuous isomorphous series because of the similarity of ion size between Na and Ca, the series is found to involve solid solution at high temperatures and near the end-members at ordinary temperatures. Structural discontinuities often develop on rock cooling, giving extremely fine-grained "unmixing" of species when the limits of solid solution have been exceeded. Microscopically or macroscopically visible exsolution intergrowths of sodic plagioclase and potassic feldspar crystals are known as *perthites*.

Albite and oligoclase occur commonly as pegmatites and in metamorphic rocks, but do occur in some igneous rocks. Their weathering rates are slow and they occur commonly in soils in the temperate climatic zones.

The bytownite and labradorite end of the series becomes more abundant in the less-siliceous rocks (Figure 2.2), such as basalts, and may be important sources of available calcium in less-weathered soils. Anorthite is formed through contact metamorphism of limestones. Anorthite in soils weathers out rapidly, even in temperate climates. The chemistry of the plagioclase series of feldspars provides an interesting and important relationship of degree weathering to soil fertility. Initially, given a full range of plagioclases, as the weathering intensity and time factors increase, the calcium supply can progressively be obtained from the more albitic plagioclase. In many well-weathered tropical and subtropical soils, the presence or absence of plagioclases in the parent rock makes the difference between productive and unproductive soils.

Accessory Minerals of Soils

The *accessory* minerals of soils include a wide variety of minerals occurring in small but significant amounts. Most of the accessory minerals occur in the sand and silt fractions of soils, but some of them occur in the clay fraction, especially as inherited from glacial rock flour in soils which have not been weathered severely. In soil genesis studies, the accessory minerals are separated out in the heavy specific gravity fractions, most of them falling in the gravity fraction greater than 2.95 g/cm^3. The proportions of the different heavy mineral species are used to deduce the rock origin of the soil parent material.

Pyroxenes and Amphiboles. The pyroxenes and amphiboles are two groups of ferromagnesian minerals the structure of which consists of long chains of linked silica tetrahedra. The pyroxenes consist of a single chain (2 oxygens shared in each tetrahedron) and the amphiboles of double chains (alternately 2 and 3 oxygen atoms shared of successive tetrahedra). These cham silicates are sometimes termed *inosilicates.* The chains are arranged in parallel and bound together by Fe, Mg, Ca, and sometimes Na, Al, Li, Mn, Ti, and other cations. The calcium is surrounded by 8 oxygen atoms and the magnesium and iron by 6. The variety of substitution possible in pyroxenes and amphiboles makes them excellent host minerals for trace elements in soils as well as the main constituent cations Ca, Mg, and Fe. Their weathering rate is adequate to provide significant amounts of these elements in forms available for plant nutrition.

The *pyroxene* compositions are represented by *enstatite*—$MgSiO_3$, *hypersthene*—$(Mg, Fe)SiO_3$, *diopside*—$CaMgSi_2O_6$, and *augite*—$Ca(Mg, Fe, Al)(Si, Al)_2O_6$, of which augite is the most important in amounts as well as compositional significance to soils. The silicate group containing SiO_3 in its formula is known in classical silicate chemistry as *metasilicates.* Augite is a dark-colored ferromagnesian mineral common in basalt and other igneous rocks and occurs as dark mineral grains in soils.

The *amphibole* composition may be represented by the isomorphous series between the end-members, *tremolite*—$Ca_2Mg_5Si_8O_{22}(OH)_2$ and *actinolite*—$Ca_2Fe_5Si_8O_{22}(OH)_2$ or $Ca_2(Mg, Fe)_5Si_8O_{22}(OH)_2$. Tremolite is characteristic of metamorphosed dolomites; actinolite is also characteristic of metamorphic rocks. The series can occur in the fibrous form of *asbestos* used, for example, in laboratory filtration.

The amphibole group is widely represented in the *hornblende* series—$(Na, Ca)_2(Mg, Fe, Al)_5(Si, Al)_8O_{22}(OH)_2$, which is widely distributed in igneous rocks from syenite and granite to gabbro (Figure 2.2), and in metamorphic rocks such as gneiss, hornblende schist, and amphibolites. Hornblende weathers fairly rapidly, and its extent of weathering is a measure of the intensity and time of weathering of soils in cool, humid regions.

Olivines. Olivines—$(Mg, Fe)_2SiO_4$ are olive-green minerals forming an isomorphous series between the end-members *forsterite*—Mg_2SiO_4 and *fayalite*—Fe_2SiO_4. The structure consists of independent silica tetrahedra in which oxygen ions are not shared and Al does not substitute for Si, and alkali ions are excluded. The magnesium and ferrous ions are each surrounded by 6 oxygen ions of these tetrahedra, which are arranged with each alternate tetrahedron inverted. In the older silicic acid chemistry, these minerals were considered *orthosilicate* salts of orthosilicic acid—

H_4SiO_4. They are included with the *nesosilicates*, and they have also been designated *unitetrahedral* silicates. Forsterite and olivine are incompatible in magma with free silica, reacting with it to give pyroxene. Fayalite does not react with quartz and occurs occasionally in granites. The distribution of olivine in igneous rocks is given in Figure 2.2. Large crystals or *phenocrysts* of olivine are frequently found in basic rocks. On weathering in soils, these crystals yield nontronite and free iron oxides. The relative ease of weathering of olivine contributes to the fertility of the soils in which it is present.

Other Accessory Minerals. The accessory mineral, *apatite*—$Ca_{10}(F, OH, Cl)_2(PO_4)_6$, is the most common primary mineral carrier of *phosphorus* in rocks and little-weathered soil parent materials. Apatites occur commonly as fine mineral grains in basalt and other mafic rocks, and less commonly in more siliceous igneous rocks. They occur also in calcareous sediments. *Fluorapatite*—$Ca_{10}F_2(PO_4)_6$ is the more common rock form, but *chlorapatite* occurs as a member of the series. *Hydroxyapatite*—$Ca_{10}(OH)_2(PO_4)_6$ is the common form in bones and teeth and enters sediments through this route.

Tourmaline—$Na(Mg, Fe)_3Al_6(BO_3)_3Si_6O_{18}(OH)_4$ contains the *cyclosilicate* (closed chain) structure represented by the Si_6O_{18} part of the formula. Tourmaline is of interest to soil chemistry because of its content of the plant-essential element boron. Since the mineral is highly insoluble, the available boron of soils is related mainly to the non-tourmaline fractions including boron held in sorbed, organic, and micaceous (67, 178) forms.

Epidote—$Ca_2(Al, Fe^{3+})_3Si_3O_{12}(OH)$ represents a mineral group in which some Ca may be substituted by rare earths and some Mg may substitute for Fe. Minerals of the epidote group characteristically occur in metamorphic rocks. Similar in composition to anorthite, the epidotes apparently are more stable and may account for the rarity of metamorphic anorthite.

Other accessory minerals used for tracing the origin of soil parent materials include *zircon*—$ZrSiO_4$; *topaz*—$Al_2SiO_4(OH, F)_2$; *garnets*—$X_3Y_2(SiO_4)_3$ in which X may be Ca, Mg, Fe^{2+}, or Mn^{2+} and Y may be Al, Fe^{3+}, or Cr^{3+}, and sometimes Ti or Mn^{3+}; *magnetite*—Fe_3O_4, *ilmenite*—$FeTiO_3$, *sphene*—$CaTiSiO_5$, *rutile* and *anatase*—TiO_2; and *kyanite*, *sillimanite*, and *andalusite*—Al_2SiO_5 and similarly structured *staurolite*—$2Al_2SiO_5 \cdot Fe(OH)_2$.

CARBONATE AND SULFUR-BEARING MINERALS IN SOILS

Carbonates

The most abundant carbonate mineral of soils is *calcite*—$CaCO_3$, which forms in the C_{ca} horizon in subsoils of subhumid and more arid regions, often at the lower boundary of the depth of rooting. Calcite in soil is also inherited from limestone, marl, and chalk rocks, as well as from marble parent materials. In some instances the isomer, *aragonite*—$CaCO_3$, occurs in sediments and mantle rocks and may, therefore, be found in some soils.

Depending on the CO_2 pressure and pH of the soil solution, some $CaCO_3$ is dissolved as $Ca(HCO_3)_2$ and this dissolution process gives a source of available calcium in soils containing $CaCO_3$. The reaction results in the complete depletion of $CaCO_3$ from the more highly leached soils of humid regions. Calcite is applied to acid soils in large quantities in the form of ground limestone.

Only in rare environments can *magnesite*—$MgCO_3$ persist in surficial soil and rock layers, because of its relatively high solubility in water. In magnesium-rich environments, some Mg frequently substitutes for Ca in the calcite lattice of pedogenic deposits.

Dolomite—$CaCO_3 \cdot MgCO_3$ is inherited in soil parent materials from dolomitic limestone and mantle rocks derived therefrom. Its dissolution by carbonic acid supplies both available calcium and magnesium. Ground dolomitic rocks are frequently applied to soils as liming materials. Pedogenic dolomite has been reported in the Red River Valley of Minnesota by Sherman and co-workers (185).

Siderite—$FeCO_3$ is formed by decomposition of iron-bearing minerals in environments suitable for carbonate accumulation. It is more resistant than $CaCO_3$ to the action of carbonic acid.

Sulfur-bearing Minerals

Gypsum—$CaSO_4 \cdot 2H_2O$ accumulates in soils of semiarid and arid regions, frequently in a horizon situated somewhat beneath the horizon in which $CaCO_3$ is accumulated. It sometimes occurs as indurated horizons and as crystals disseminated in surface soil horizons. Gypsum occurs in certain subsoils as a weathering product of *pyrite*—FeS_2, even in soils of humid regions. In arid regions, *epsomite*—$MgSO_4 \cdot 7H_2O$ and other soluble sulfates may occur in soils. Minor amounts of the insoluble minerals *sphalerite*—ZnS, *chalcopyrite*—$CuFeS_2$, and *cobaltite*—CoAsS, may occur in soils.

Pyrite and *marcasite*—FeS_2 ("fool's gold") are produced in a reduction phase of the geomorphic cycle of sedimentation through the reduction of the iron of ferric oxides and the sulfur of sulfate. These sulfides frequently occur in shales, underclays of coal seams, and in other sedimentary rocks, including some limestones (Table 2.2). Pyrite also occurs as an accessory mineral in igneous and metamorphic rocks.

The sulfate reduction reaction is extremely fundamental inasmuch as it utilizes iron of ferric oxide as a cation to replace the Na of Na_2SO_4 and to liberate free alkali into the soil solution, lake, or ocean:

$$2\ Fe_2O_3 + 4\ Na_2SO_4 + 4\ H_2O + \frac{9}{n}C_n \rightarrow 4\ FeS + 8\ NaHCO_3 + CO_2$$

a reaction occurring under anaerobic conditions, in which C_n represents organic matter. In time the FeS transforms to FeS_2. The sulfate ion is reduced under the influence of the almost universally present sulfur-reducing bacteria. The sulfate reduction reaction is believed (9, 186, 217) to be responsible for the development of alkalinity in poorly drained saline soils.

These sulfide systems become subjected to oxidation when good drainage is established under many soil-forming situations. Drainage of some tidal marshes or exposure of acid-forming underclays of coal beds in outcrops causes H_2SO_4 to form, which results in extremely acid soils (pH 2.5 is common). Drained tidal marshes frequently become excessively acid and agriculturally unproductive. In certain areas careful regulation of the water table of tidal marshes to prevent excessive oxidation has permitted paddy culture of rice in pyrite-containing marshes.

Calcium is removed from calcium-bearing minerals by the action of the H_2SO_4, and gypsum crystals are formed. The reaction of the sulfuric acid with iron-bearing minerals in soils yields *jarosite*—$KFe_3(OH)_6(SO_4)_2$ (177), and with aluminum-bearing minerals often yields *alunite*—$KAl_3(OH)_6(SO_4)_2$. In many sites, the pyrite oxidation cycle may have been completed in nature, the iron having formed oxides and the sulfide having been oxidized, consumed, or leached out; in agricultural practice, however, bringing about the completion of an uncompleted cycle by alternating drainage with sea-water flooding is feasible only at especially favorable locations and elevations.

LAYER SILICATE MINERALS OF SOILS

Layer silicates or *phyllosilicates* of soils, of which mica is a representative familiar to the most casual observer, play a prominent role in most soils in determining the physical properties, such as plasticity and structure, and the chemical properties, such as cation exchange, ion release, and ion fixation, when fertilizer is added to soil. So important are the layer silicates in determining the properties of most clays that the term "clay mineral" has sometimes been anomalously restricted to mean exclusively fine-grained layer silicates. Actually, many clays contain other hydrous silicate and oxide components of great importance in affecting soil properties. In addition, not only are layer silicates important components of soil clays, they are also important components of silts and sands and even gravels of soils; thus, the layer silicate names should not be restricted in meaning to clays.

Crystal Structure of Layer Silicates

Two ionic coordination groups are basic to the design of the crystal structures of layer silicates. The first consists of the silicon tetrahedron in which a Si^{4+} ion lies equidistant from 4 oxygen ions—$SiO_4{}^{4-}$. To some extent, Al^{3+} can substitute for Si^{4+}. The excess negative charge of the tetrahedral unit is decreased by an oxygen ion being shared by two adjacent tetrahedra. The sharing in this way of three basal oxygen ions forms a tetrahedral sheet. The apical oxygens point mainly in a single direction, but occasionally a reversed tetrahedron is possible.

The second coordination group consists of an Al, Fe, or Mg ion surrounded by 6 equally spaced oxygen or hydroxyl ions which form an octahedron, for example $Al(OH)_6{}^{3-}$. Adjacent octahedra share three hydroxyl or oxygen ions (share octahedral faces), forming an *octahedral sheet* which is two hydroxyl ions in thickness. With trivalent ions, only two-thirds of the octahedral positions are normally filled (a *dioctahedral* structure), while with divalent ions all of them are filled (a *trioctahedral* structure).

In a series of classical structures deduced by Pauling in 1930, the layer silicate structures consist of tetrahedral silica sheets attached to octahedral hydroxyl sheets, with each apical oxygen of the tetrahedral sheet replacing one hydroxyl position of the octahedral sheet. Different combinations of these two general structural units, the octahedral and tetrahedral sheets, yield the structures of the various layer silicates of importance in soils, including mica, vermiculite, montmorillonite, attapulgite, chlorite, kaolinite, and halloysite, as well as various interstratified and intergradient forms of these layer silicates.

When one silica sheet is attached to each side of the octahedral sheet, the structure with this two-to-one ratio is designated as the 2:1 layer silicate structure. The unit cell formula of 2:1 layer silicates can be represented by the formula:

$$XSi_8M_6O_{20}(OH)_4 \cdot n$$

in which the X term represents interlayer charge (satisfied by cations or charged hydroxy structural units); M represents small divalent cation positions (two thirds of the 6 may be occupied by 4 trivalent cations), and n the number of interlayer solvating molecules per unit cell.

The unsubstituted, uncharged, idealized end-members for the 2:1 layer silicates are *pyrophyllite*—$Si_8Al_4O_{20}(OH)_4$, and *talc*—$Si_8Mg_6O_{20}(OH)_4$. These structures have one tetrahedral silica sheet attached to each side of an octahedral hydroxyl sheet. Both of these end-members are rare in soils. Substitution of $(KAl)_2$ for Si_2 of the tetrahedral sheets of pyrophyllite yields the dioctahedral, *muscovite* mica structure discussed in the next section. Substitution of $(KAl)_2$ for Si_2 in talc, and substitution of some Fe^{2+} for Mg, gives the trioctahedral, *biotite* mica structure. A variety of such substitutions yields the several isomorphous series of 2:1 layer silicates common in soils.

Another series of layer silicates has a tetrahedral silica sheet attached to only one side of the octahedral hydroxyl sheet. These are the 1:1 layer silicates. The unsubstituted dioctahedral end-member is *kaolinite*—$Si_4Al_4O_{10}(OH)_8$ to be discussed in a later section. The trioctahedral series of 1:1 layer silicates have much Mg and Fe^{2+} with Al in the octahedral layer, and make up the *serpentines* of the *antigorite* and *chrysotile* series.

In the expansible 2:1 layer silicates, montmorillonite and vermiculite, substitutions of various cations in 2:1 layer silicate structures provide a lower charge than in micas, which results in intercalation of water and exchangeable cations between the 2:1 layers. These minerals, which vary widely in composition, are the plastic clays and silts of high cation exchange capacity in soils. Substitution of a complete, charged hydroxyl sheet between mica-like 2:1 layers gives a 2:1:1 layer structure or the 2:2 layer structure of *chlorites*.

Much of soil chemistry is rooted in the isomorphous substitutions and resultant electrostatic charges of the layers and interlayers of these various layer silicate structures, which occur abundantly in the majority of soils.

Micas

Micas occur extensively in soils. They originate in soils primarily by inheritance from the parent rock from which the soil was formed, although an occasional suggestion has been made that, to a limited extent, micas may be formed in soils. *Well-ordered* micas are inherited by soils mainly from igneous and metamorphic rocks. The idealized end-member micas [(161), Figure 2.3] are dioctahedral *muscovite*—$K_2Al_2Si_6Al_4O_{20}(OH)_4$, and its isomorphous analogue *paragonite*—$Na_2Al_2Si_6Al_4O_{20}(OH)_4$, and trioctahedral *biotite*—$K_2Al_2Si_6(Fe^{2+}, Mg)_6O_{20}(OH)_4$ and *phlogopite*—$K_2Al_2Si_6Mg_6O_{20}(OH)_4$. A variety of additional ion substitutions, beyond those indicated for the end-members, such as Li for Mg, more Al for Si, F or O for OH, and possibly OH for O, can occur in micas. Layer charge, represented equivalent to K_2 in the end-members, decreases in some biotites because of substitution of Al for Mg, or Fe^{3+} for Fe^{2+} in the trioctahedral layer. The layer charge is extremely high in *margarite*—$Ca_2Al_4Si_4Al_4O_{20}(OH)_4$, which is representative of the "brittle" micas.

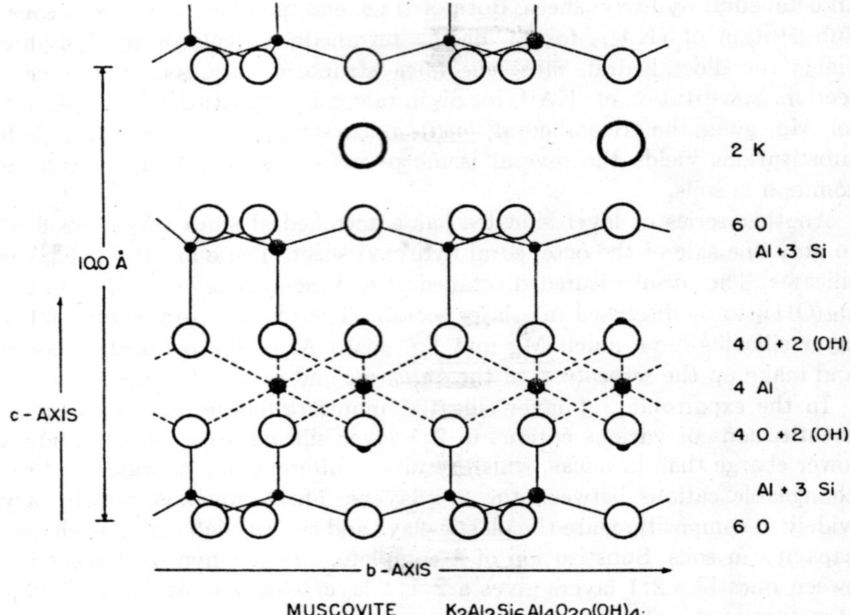

Figure 2.3. Some crystal structure features of muscovite, a dioctahedral mica.

Soils also inherit *imperfectly ordered* micas from rocks, frequently from sedimentary rocks in which authigenesis and diagenesis have not completely perfected the mica structure. Moreover, weathering in soils of well-ordered micas creates disordering of the structure. The imperfectly ordered micas from either of these sources contain less potassium and more water than the well-ordered micas. Study of weathered biotite (55) has shown gain in Al, Si, and H_2O content, oxidation of Fe, and loss in Mg and K. Although disordered micas are most abundant in the clay fraction of soils, they occur as well in silts, sands, and gravels. The broad group of "micas of argillaceous sediments," containing some domains of 10A sequences of structural layers, are often termed hydromica (90) or *illite* (79). According to Grim (80), "all gradations can exist between illite and well-crystallized muscovites on the one hand and montmorillonite on the other hand." According to Yoder (220), "illite" is a rock term which refers to random interstratification of mica and montmorillonite. Jackson and coworkers (88, 101, 103, 146) have presented extensive evidence that illites are micas with some expanding layers intermixed. The extra H and the deficiency of K are explained by the fact that many of the illite-type specimens as well as

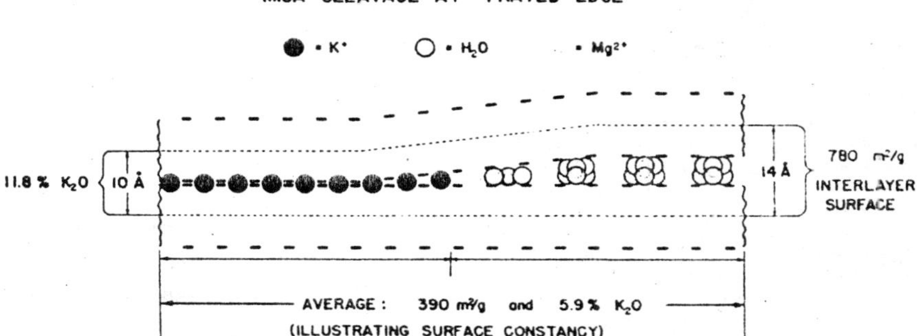

Figure 2.4. "Frayed-edge" type of interlayering caused by mica weathering, an example with 50 per cent expanded interlayers and 50 per cent residual mica interlayers. The analytical data illustrate the tendency for interlayer specific surface constancy (112). (*Courtesy Pergamon Press, publishers of Clays and Clay Minerals.*)

the disordered micas of soils exhibit some mixed-layering with phases of vermiculite, montmorillonite, chlorite, and intergrades of several of these species. The latter phases have hydrated exchangeable cations, water, and hydroxides intercalated between some of the expanded silicate layers, and these are distributed among mica sequences of layers, giving the extra H_2O and OH found on analysis. This type of phenomena is considered further in the section on intergradient 2:1 to 2:2 layer silicates.

Another series of explanations has been offered for the extra H and the corresponding deficiency of K in illites as compared to the ideal micas. The presence of excess H_2O, OH_3^+, or OH in the mica structure proper by the replacement of K by H_2O (209) or by OH_3^+ (35, 138) in some of the interlayer cavities of illite, or proton addition (103, 144, 170) to the apical tetrahedral (octahedrally shared) oxygen of mica, has been proposed. The chemistry of exchangeable hydronium shows that it tends to react with layer silicate lattices to liberate octahedral cations. This property tends to militate against its being stable as a substitute for K in a layer silicate.

There is a tendency for interlayer constancy of the sum of interlayer sorption surface and K-occupied interlayers (Figure 2.4) in mica-vermiculite mixed-layer systems (146). This constancy relationship suggests that under ordinary weathering conditions any given mica interlayer segment tends to remain completely filled with K or else to become completely affected by interlayer swelling. In acid soils (of pH 4.5 to 5), there is a tendency for hydroxyaluminum-iron "islands" or interlayers to occupy a portion of the expanded spacings, as will be discussed below. Among highly acid (pH 2.5 to 3.5) underclay illites, such as the commercial Fithian and Morris

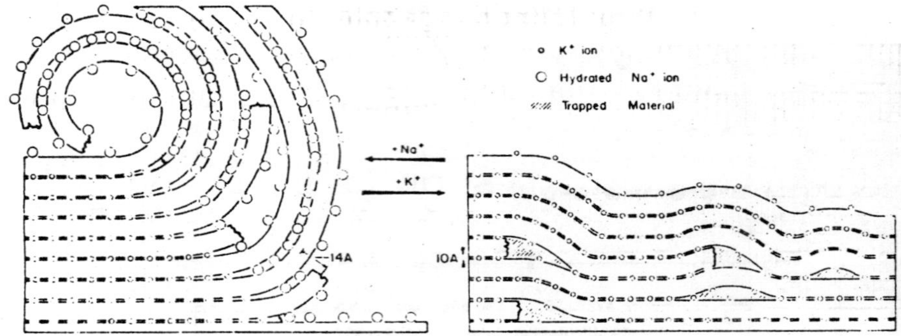

Figure 2.5. Curling of layers by exchange of hydrated Na ions for K ions at the mica cleavage surface as revealed by the electron microscope (204). Drafting by S. B. Jackson.

illites of Illinois and acid gley soils, the hydroxyaluminum-iron interlayers are dissolved out and freely expansible layers are interstratified with mica.

In mica weathering in soils, potassium and other interlayer cations slowly diffuse out of the interlayer spaces formed by cleavage between the mica layers into the soil solution, which results in cleavage at the weathering edges of mica (Figure 2.4). The K release from micas is hastened by removal (92, 139) of the resulting exchangeable or soluble K^+ ions by roots, by leaching, or by chemical precipitation in the laboratory. As the interlayer K is subjected to depletion through chemical equilibrium with the soil solution, there is gain that occurs in readily exchangeable ions and interlayer water, according to the equation of Jackson and others (103):

$$\text{Mica} \rightleftarrows \text{illite} \rightleftarrows \text{vermiculite} \rightleftarrows \text{montmorillonite} \qquad (1)$$

The shifting of equation (*1*) to the right (to give lattice expansion) by the lowering of K^+ activity has been demonstrated in laboratory studies (54, 181, 213). Shift to the right has also been accomplished by replacement with strongly sorbed alkyl ammonium ions (212). Drying (or freezing) the mica of soils hastens K release (84, 123) if the exchangeable K level is low. The levering action of hydrated Na ions in curling back mica layers is revealed (204) by mica surface morphology (Figure 2.5).

Addition of a high level of soluble potassium ions to soils by mass action tends to reverse (151) equation (*1*). Drying or freezing also facilitates a shift of equation (*1*) to the left (66) if the soluble and exchangeable potassium level in the soil is high, because K ions strongly bond (178) to the two crystallographic cavities kept aligned (112) by the mica core (left side of Figure 2.4). The cleavage process is abruptly reversed (204), causing unrolling (Figure 2.5), by addition of NH_4 or K ions that fix in the crystal-

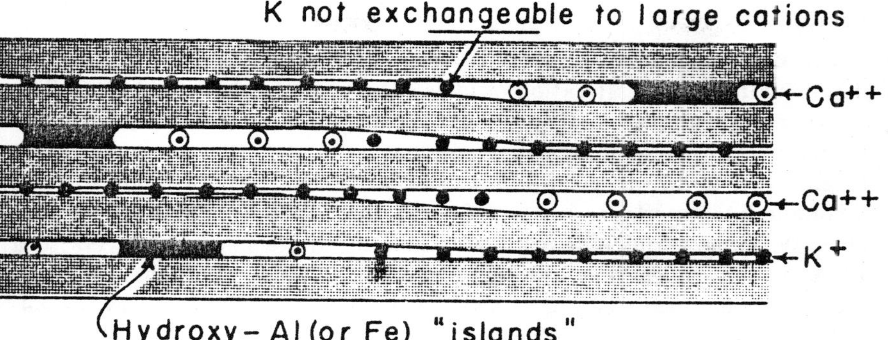

Figure 2.6. Potassium exchange of larger hydrated cations is sterically hindered at the cleavage site in weathering mica crystals (167). *Courtesy Williams and Wilkins Co., publishers of Soil Science.)*

lographic cavities normally occupied by K of mica layers. Equation (*1*) has been amplified to include the chemical process of hydroxy sesquioxide interlayer or "2:2 lattice building" (Figure 2.6). Cations of smaller hydrated radius, such as NH_4^+, Ag^+, OH_3^+, diffuse into the wedge-shaped area to replace K more readily than larger hydrated cations (167). Mica weathering eventually leads to the formation of a layer silicate clay complex (175) consisting of unexpanded mica cores, vermiculite, montmorillonite, and 2:1 to 2:2 intergrades (Figure 2.7).

Figure 2.7. Principal features of a dominantly montmorillonite layer silicate clay complex formed by weathering of mica. Different proportions of the various components occur in mica-derived layer-silicate clays of different soils and other sediments. Drafting by G. A. Borchardt.

Vermiculite

Vermiculite occurs extensively in soils. It forms as a product of weathering or hydrothermal alteration of micas and possibly also of chlorites. The weathering or alteration of micas has replaced the potassium by predominently exchangeable magnesium and has expanded the interlayer space (Figures 2.4 to 2.7) to yield a unit cell *c*-spacing of 14 Å or more, so as to accommodate interlayer water (83) and interlayer exchangeable cations (14). Washing vermiculite with a KCl solution collapses the layers to a 10 Å spacing (mica-like), through replacement by K of the hydrated exchangeable cations. Heating to 300°C is used as a criterion to test the collapsibility of vermiculite, since at this temperature the water of hydration of the interlayer cations is largely expelled.

The layer structure of vermiculite tends to resemble closely that of the mica species from which the vermiculite was derived. Thus there is a trioctahedral series and a dioctahedral series of vermiculites. Trioctahedral vermiculite of soils may be represented by the formula $X_{1.1}(Al_{2.3}Si_{5.7})(Al_{0.5}Fe^{3+}_{0.7}Mg_{4.8})O_{20}(OH)_4 \cdot nH_2O$ (Figure 2.8). Dioctahedral vermiculite of soils may be represented by the formula, $X_{1.1}(Al_{2.3}Si_{5.7})(Al_3Fe^{3+}_{1.2}Mg_{0.3}) \cdot O_{20}(OH)_4 \cdot nH_2O$. The letter X represents the exchangeable cations, or net negative charge, neutralized by 1.1 Na ions or 0.55 Mg ions per unit cell in this example. The determined cation exchange capacity (14) of vermiculite is generally lower than the layer charge of micas. Yet more tetrahedral Al is generally deduced from the elemental analyses of vermiculite than the 2.0 units of fully charged mica, represented in this formula as 2.3 but reported by various workers in the range of 2.2 to 2.6. The tetrahedral charge excess over exchange capacity ($2.3 - 1.1 = 1.2$, in the example) is neutralized by excess ions in octahedral layers. This internal balancing of charge typically occurs in chlorites and to some extent in biotites.

Several possible ionic substitutions have been suggested to account for vermiculite analyses and properties, such as (*a*) oxidation of the octahedral ferrous to ferric iron (83, 103); (*b*) substitution of additional positive ions in the vacant octahedral sites of dioctahedral minerals; (*c*) addition of a proton to the apical oxygen of the silica tetrahedron in the structure to extend the completeness of the hydroxylation of the octahedral layer, giving, for example, $O_{18}(OH)_6$; (*d*) substitution of four protons for Al^{3+} in the tetrahedral layer (144), giving $(OH)_4$; (*e*) substitution of oxygen in excess of the theoretical O_{20} for hydroxyl (27) in amounts equivalent to ferric iron [in the above example, $Fe^{3+}_{0.7}O_{20.7}(OH)_{3.3}$] as occurs in some chlorites, leaving only Al to balance the excess negative charge of the layer; and (*f*) formation of scattered positively charged hydroxyaluminum interlayer "islands" or polymers (36, 57, 59, 121, 164) similar to the more ex-

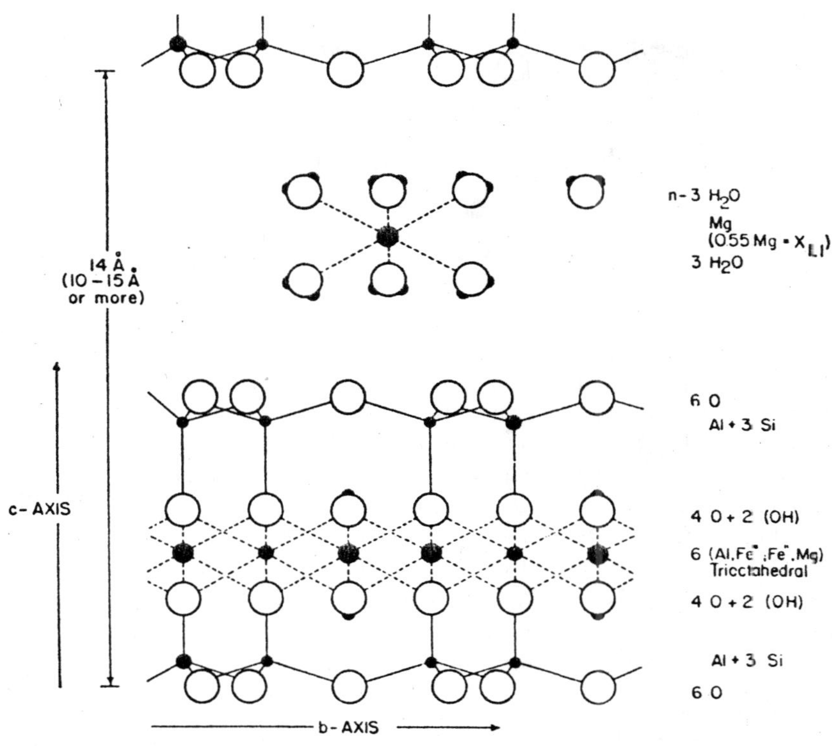

Figure 2.8. Some crystal structure features of vermiculite, a 2:1 layer silicate mineral with limited expansibility. The interlayer cations are exchangeable. The interlayer water can be replaced by some polar organic molecules. The trioctahedral variety is shown here.

tensive positive interlayers of chlorite. There is evidence that several or all of these possible types of substitution may occur in various samples.

Dioctahedral vermiculite occurs commonly in soils, and is almost exclusively the vermiculite variety found in acid soils. Its origin may be by the weathering of the dioctahedral types of mica such as muscovite or paragonite (51). The occurrence of some vermiculite layers as a high charge part of an expansible mineral complex (Figure 2.7) was proposed (103) to explain potassium fixation by montmorillonites. The development of some positively charged hydroxy interlayers when vermiculite is formed by mica weathering in soils has strong support in the extensive occurrence of vermiculite-chlorite intergrade minerals in soils, as discussed in a later sections on layer silicate intergrades.

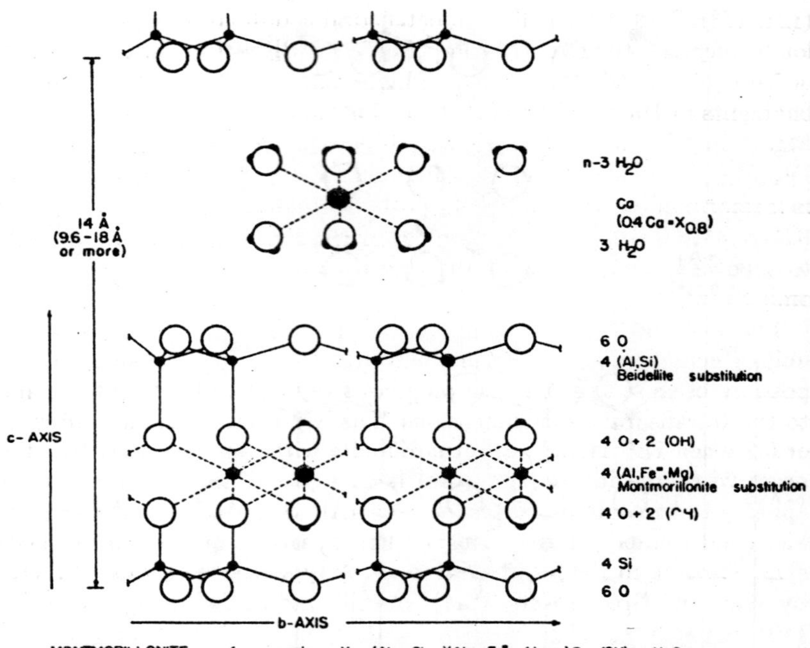

Figure 2.9. Some crystal structure features of the montmorillonite isomorphous series of freely expansible 2:1 layer silicate minerals. The breaks in the bond lines (111) indicate an omitted portion to bring the two types of substitution closer than they occur on the average. The interlayer cations are freely exchangeable. The *c*-spacing varies with water content, and the water can be replaced by polar organic molecules.

Montmorillonite

The minerals of the *montmorillonite* or *smectite* isomorphous series (Figure 2.9) are freely expansible layer silicates (94, 137, 171). The spacing of the layers ranges from 12 to 18 Å, and is variable with the exchangeable cation species and the degree of interlayer solvation (24). Complete drying yields a spacing of less than 10 Å. Full hydration can float the layers apart, independent of each other (155).

A model soil montmorillonite formula is $X_{0.8}(Al_{0.3}Si_{7.7})$ $(Al_{2.6}Fe^{3+}_{0.9}Mg_{0.5}\cdot$ $O_{20}(OH)_4\cdot nH_2O$, representative of that of Tama silt loam (Brunizem, Prairie or Mollisol) (74), Miami silt loam (Gray Brown Podzolic or Alfisol) formed in calcareous loess in Wisconsin, and Houston Black clay (Grumusol or Vertisol) formed in calcareous marl Coastal Plain sediments in Texas

(124, 173). The Mg and Fe^{2+} in octahedral coordination in soil montmorillonite increase to give $(Al_{2.3}Fe^{3+}_{1.1}Mg_{0.6})$ in Black Cotton soil (Grumusol or Vertisol) in India and a Dark Magnesium clay formed in brackish embayments in Hawaii. The octahedral Fe^{3+} increases to give $(Al_{0.9}, Fe^{3+}_{2.7}, Mg_{0.5})$ in Waipiata soil formed on basalt in New Zealand and to $(Al_{1.3} \cdot Fe_{2.6}Mg_{0.2})$ in another Dark Magnesium clay in Hawaii. These latter montmorillonites are grading toward *nontronite*—$X_{0.7}(Al_{0.7}Si_{7.3})\ Fe^{3+}_{4}O_{20} \cdot (OH)_4 \cdot nH_2O$, the ferric end-member of the series. Octahedral Al increases to give $(Al_{3.1}Fe^{3+}_{0.4}Mg_{0.5})$ in the clay fraction in the famous Upton, Wyoming bentonite (173).

The dioctahedral layer composition, involving approximately 0.4 to 0.6 unit of octahedral Mg is surprisingly widespread in soil montmorillonites, possibly because this composition gives a cell size which fits the octahedral to the tetrahedral layer dimensions. The Mg content tends to drop to 0.1 or 0.2 when Fe^{3+} is high. Montmorillonite with approximately half the net negative charge in the octahedral layer and half in the tetrahedral layer appears to be common. *Beidellite*—$X_{0.7}(Al_{0.7}Si_{7.3})Al_4O_{20}(OH)_4 \cdot nH_2O$ is the ideal end-member of the montmorillonite series with the entire negative charge arising in the tetrahedral layer. A type specimen was characterized by Weir and Greene-Kelly (211); in soils, the beidellite substitution (111, 140) in the tetrahedral layer is common but is usually supplemented by octahedral substitution as illustrated in the typical formulas given above. The trioctahedral montmorillonites of the *saponite* and *hectorite* series—$X_{0.7}(Al_{0.7}Si_{7.3})$ (LiAl, 2Mg)$_6O_{20}(OH, F)_4 \cdot nH_2O$ rarely occur in soils (171).

An abundance of the exchangeable cations Ca and Mg in silica-rich aqueous (moist) environments favors the formation of freely expansible layer silicates of the montmorillonite series. The most typical occurrences of montmorillonite are: bentonite clay formed from volcanic ash deposits in fresh water; clay formed in marl; clay of hydromorphic soils in many regions; clay in soils formed from basalts and limestones weathered to intermediate stages in humid climates; and as clays formed by weathering of micas in cool humid climates. Montmorillonite formed from micas occurs as frayed edges around yet unweathered mica cores (112) as represented in Figures 2.4 to 2.7.

Montmorillonite structures may include lattice "mistakes" with some tetrahedra pointed toward the interlayer space (62). A well-supported basis for the excess hydroxyls in montmorillonite lies in its ability to collect positive hydroxyaluminum or sesquioxidic interlayers on the layer surfaces. When the hydroxy units adhere to only one layer (112, 188, 192) the clay still can swell. The hydroxy units may attach to adjacent layers, as discussed in connection with intergradient 2:1 to 2:2 layer silicates below.

Attapulgite

Attapulgite, which is also widely known by an earlier name, *palygorskite* (42), occurs abundantly in the clay fraction of soils in a few areas, particularly in the areas around the Eastern Mediterranean Sea. Structurally, attapulgite consists of duochains of silica as in amphibole, joined through sharing the basal oxygen ions of the chains (25). The structure is essentially a modification of a 2:1 sheet structure of montmorillonite (Figure 2.9), with alternate duochains pointing in the opposite direction, giving an open channel beside each duochain instead of forming a continuous interlayer space. The two duochains coordinate octahedral cations.

The ideal end-member formula—$Si_8Mg_5(OH_2)_4O_{20}(OH)_2 \cdot 4H_2O$, as given by Bradley (25), has protons attached to OH groups along the channel edges. The $4H_2O$ group occurs in the channels. Substitution of $2Al^{3+}$ can be made for $3Mg^{2+}$ in the octahedral position. Usually some excess negative charge gives rise to exchangeable cations in the channels. Cleavage is through the channels along the (110) planes at the shared basal oxygens, giving sheets with optical properties which are similar to those of montmorillonite. The (110) spacing is 10.2 to 10.5 Å of a very strong diffraction intensity. This peak can be distinguished from a similar peak of a mixed-layer sequence high in mica by the absence of structural K in attapulgite. The (040) peak at 4.49 Å and the (440) peak at 2.62 Å are also strong and diagnostic. The fibrous habit of the crystals observed in the electron microscope (141) provides confirmation. A close mineralogical relative of attapulgite is *sepiolite*—a 3-chain silicate (41), the chains being joined through oxygen ions and alternating in orientation.

Attapulgite is formed (42) from amphiboles and pyroxenes through hydrothermal action, as veins in limestone and dolomite (69), in limestones (15), and in lagoonal deposits (Attapulgus, Ga.) or dry desert lakes (80). Redistribution in recent sediments gives rise to its occurrence in soil in Israel (15), Iraq, and Syria (152). There is evidence that attapulgite weathers to montmorillonite on exposure in the upper soil horizons.

Chlorite

Chlorite occurs extensively in soils, mainly inherited from mafic rocks, *serpentine*, and other rocks, but to some extent formed in soils. This mineral is a 2:1:1 or 2:2 layer silicate (Figure 2.10). A hydroxide interlayer (sometimes termed the "brucite" layer) of composition such as $Al_2Mg_4 \cdot (OH)_{12}^{2+}$ is sandwiched between negatively charged mica-like layers as a replacement for K_2 in the mica structure (Figure 2.3). A type formula for trioctahedral chlorite is $(Al_2Si_6)\ Al_2Mg_{10}O_{20}(OH)_{16}$ in which Fe^{2+} and other divalent cations may replace Mg, and Fe^{3+} and other cations may replace

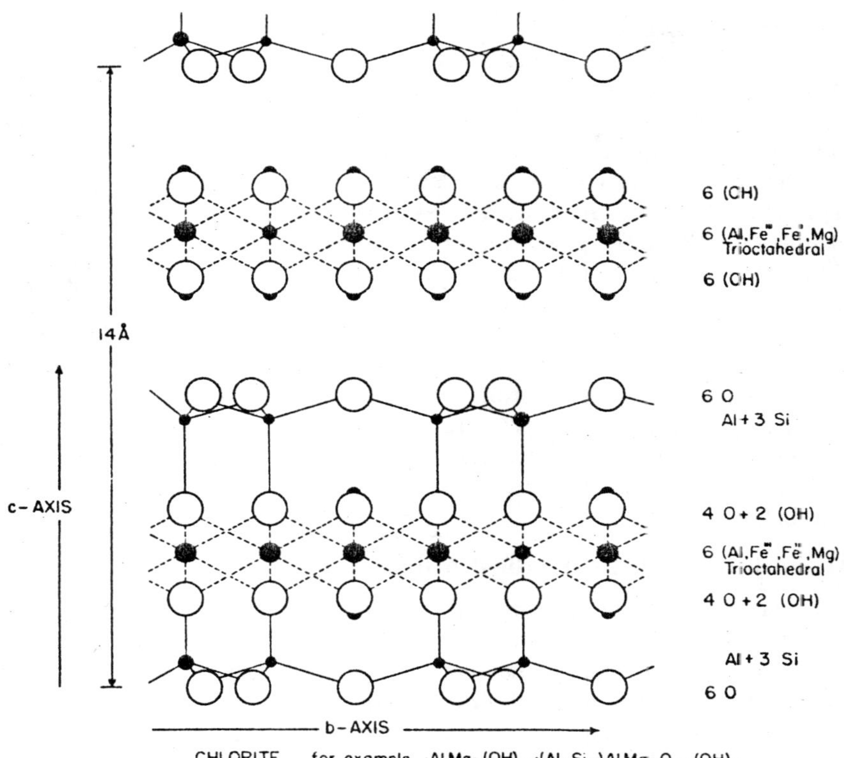

Figure 2.10. Some crystal structural features of the chlorite series of 2:2 layer silicate minerals. The trioctahedral variety is shown.

Al. This formula may be written $AlMg_5(OH)_{12}\cdot(Al_2Si_6)AlMg_5O_{20}(OH)_4$ to represent the positive charge symmetrically distributed between the hydroxide interlayer term (before the dot) and the 2:1 layer itself, balancing the 2(−) charge of the tetrahedral layer represented by (Al_2Si_6). The distribution of the (+) charge between the layer and the interlayer cannot be ascertained (154), but some charge of the interlayer is required for structural stability. Tetrahedral charge may range between $(AlSi_7)$ to (Al_4Si_4) or even outside this range (32).

The elemental composition of chlorites varies over an extremely wide range. Toxic elements, such as Cr and Ni, can occur in mafic chlorites. Serpentine-derived soils are sometimes outstanding for infertility and are known as "serpentine barrens," owing either to the presence of toxic elements or to a deficiency of certain essential elements such as Ca.

Chlorite commonly occurs in sedimentary rocks, and productive soils are derived therefrom. An outstanding example is the well-crystallized chlorite with illite, in the shale of Paleozoic age, which was pushed out of basins of the Great Lakes by glacial lobes which distributed these clays in parent materials of northcentral and northeastern United States.

Chlorite is usually trioctahedral, but *dioctahedral* chlorite, which may be represented by the formula $Al_{4.4}(OH)_{12}\cdot(Al_{1.2}Si_{6.8})Al_4O_{20}(OH)_4$, also occurs in soils (38, 73, 202); an aluminous dioctahedral chlorite has been synthesized (122). An intergrade series exists between dioctahedral chlorite and dioctahedral vermiculite. Weathering of chlorite to 2:2 to 2:1 intergrades has been suggested by the acid removal of Fe from the (Fe, Al, Mg) $(OH)_{12}$ interlayer (86) and by replacement (60) of OH by OH_2. A lacy residuum of aluminum (and possibly iron) hydroxide interlayers (59) would remain with internal edges or "holes" (111, 112) were Mg selectively removed by weathering. The resultant structures are similar to those produced by interlayer precipitation of sesquioxides in intergradient clays discussed in a section below.

Mixed-layer or Interstratified Minerals

Layer silicate clays of soils and sediments rather generally have some interstratification of different 2:1 and 2:2 layer silicate species along the *c*-axis. A given clay sample may consist predominantly of one layer silicate species, but within any crystallite a second phase is frequently mixed in the layering. For example, mica may be in binary mixture with vermiculite (Figure 2.11), chlorite with vermiculite, chlorite with mica, or montmorillonite with vermiculite. The mixtures may be binary in all combinations of 2:1 and/or 2:2 layer silicates. It may also be ternary (3 components) or quaternary (4 components). The mixing can be perfectly regular, as ABAB for components A and B, or a completely random sequence, or any intermediate degree between these limits. Combinations of two systems containing different proportions of components A and B can occur in one soil clay. Interstratification of layer silicates of soils is believed to be more complex (Figures 2.6 and 2.7) than for the idealized stacking.

Analytically, the determination of mixed layering is fairly easy for simple binary mixtures, that is mixtures of two components in a uniform type of statistical distribution, by the Hendricks and Teller (34, 91) intermediate spacings and diffraction intensities. The presence of ternary and quaternary mixtures can be detected by heating to 550°C to bring all of the 2:1 layer silicates (montmorillonite, vermiculite, and mica) to a 10 Å spacing (105), while the chlorite spacing remains at 14 Å (31). A binary interstratified system of 10 and 14 Å spacing is often thus obtained with peaks in the 10.3 to 12 Å range (105, 175).

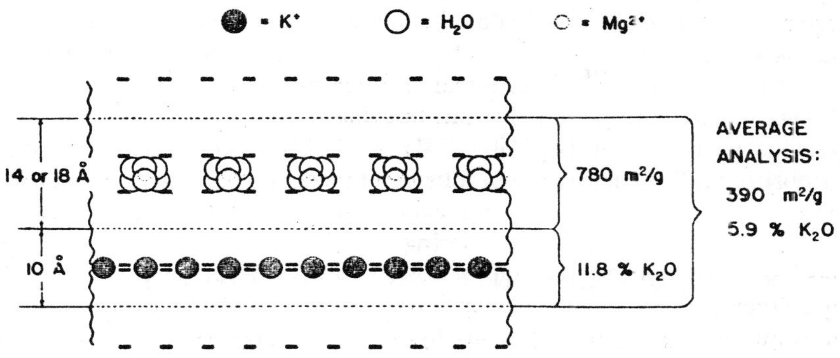

Figure 2.11. Interstratification type of interlayering caused by mica weathering, illustrated by an example of 50 per cent mica interlayers and 50 per cent expanded interlayers. The analytical data illustrate the tendency for interlayer specific surface constancy (112). (*Courtesy Pergamon Press, publishers of Clays and Clay Minerals.*)

The occurrence of interstratification in layer silicates in certain soil clays was recognized early by Alexander and others (2). A soil clay sample consisting predominantly of one species of 2:1 or 2:2 layer silicate generally contains some of one or more other layer silicate phases. A survey of soil clays (105) in representative soils from diverse sources and weathering conditions showed the rather general occurrence of interstratification in soils. The wide occurrence of interstratification in oceanic sediments was demonstrated through thousands of analyses (210).

Though creating analytical difficulties, the occurrence of mixed layering in soil clays is fortunate, because the resulting combination of properties enhances soil chemical characteristics important to plant nutrition. For example, expansible layers provide cation exchange, while mica layers can release K ions to the soil solution.

Montmorillonite-vermiculite-chlorite Intergrades

Precipitation of coatings of hydrous, positively charged sesquioxides on the negative surfaces of layer silicates has long been recognized as a phenomenon characteristic of soils, as has their fundamental influence on soil chemical properties such as phosphate fixation. In the past decade, the crystal chemistry and crystal structural relationships of this complex have been characterized. Precipitation of hydroxy-aluminum, hydroxy sesquioxides, and possibly magnesium hydroxide as gibbsite-like (or brucite-like) structures in the interlayer spaces of montmorillonite and vermiculite produces structures the properties of which are intergradient between those of the expansible mineral and those of chlorite.

Most natural clay intergrades so far described have shown 14 Å diffraction spacings and have been termed "dioctahedral vermiculite" (36, 164), "vermiculite" (130), "chlorite-like" (121), or chlorite (26), according to the end-member to which they are compared. The interlayers are heterogeneous with respect to *islands* of the "brucite-like" structures (81) or "gibbsite-like" structures (59, 112) as shown in Figure 2.12, which are

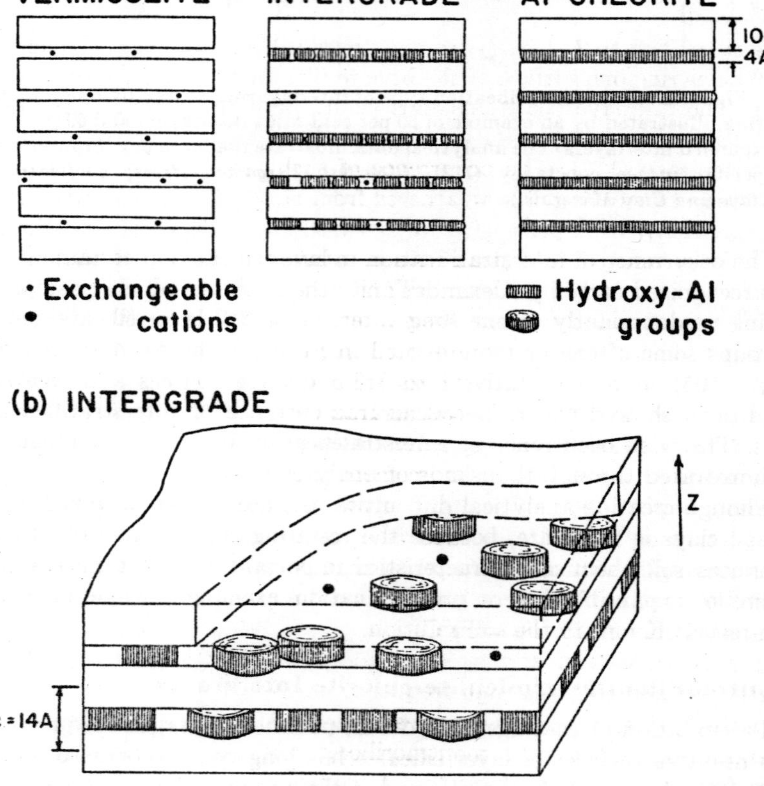

Figure 2.12. Some structural features of intergradient chlorite-expansible 2:1 layer-silicate structure, and its relation to end-member vermiculite (*left*) or chlorite (*right*) in edge view (*a*), and in cut-away of the intergrade (*b*), showing hydroxy aluminum islands as they are believed to occur more concentrated near the interlayer edges of the crystal (59).

distributed in interlayer spaces otherwise filled with water and exchangeable cations, as is characteristic of vermiculite or montmorillonite. The cation exchange capacity is consequently decreased to the extent of the positive charge of the nonreplaceable hydroxy cation interlayers. The interlayer specific surface is also decreased. The interlayer space becomes increasingly resistant to collapse by heating as hydroxy interlayers become more extensive. The products have appropriately been designated as intergradient 2:1 to 2:2 layer silicates or intergradient montmorillonite-vermiculite-chlorite (57, 110, 111, 112). When interlayer islands are extracted by citrate, fluoride, or NaOH solutions, the resultant layers of different samples show either vermiculite (165, 174) or montmorillonite spacings (57, 199).

Swelling 2:1 to 2:2 Intergrades. "Swelling chlorite" has been described (192) as having one surface of the brucite-like layer unattached to a silicate layer. The interlayer positions thus would be heterogeneous with respect to brucite and water-cation layers, and the complex is a swelling 2:1 to 2:2 intergrade (112). The occurrence of such interlayering in nature was indicated in a swelling clay weathered from basalt in Queensland, Australia (173). A sharp 18 Å peak with Mg saturation and glycerol solvation, and a greatly broadened 10 Å peak with K saturation and 500°C heating, characterize the expanding montmorillonite-chlorite intergrade type of clay. Extensive occurrences of swelling interlayered soil clays have now been observed, one of which is represented in Figure 2.13. A little true chlorite frequently occurs with the 2:1 to 2:2 intergrade clays as found, for example, in studies of Chestnut soils of Iraq (by A. B. Hanna and the writer).

Flexibility of the layers accommodates some interlayering without disturbing the main spacing. Reinforcement of the 14 Å peak occurs on heating of many expansible clay samples, showing chlorite-like characteristics in the portion of the sample which is more completely interlayered. When interlayering is still more complete, that is when hydroxy layers are attached to each surface of the swollen montmorillonite [Figure 2.13(e)], the collapse is only to 18 Å at 550°C.

In alkaline soils of arid regions, montmorillonite is relatively stable to weathering and so has a high frequency distribution. Characteristically, it is partially interlayered. The weathering mechanism by which swelling intergradient 2:1 to 2:2 montmorillonite-chlorite is formed appears to arise from the swollen character of montmorillonite when moist [Figure 2.13(a)]. The presence of abundant montmorillonite creates moist conditions and, as a result, the clay in the soil is kept swollen to 20 to 40 Å spacing (155). Intercalated hydroxyaluminum sheets can attach to only one layer of montmorillonite so long as the layers are widely separated. In a feldspar-

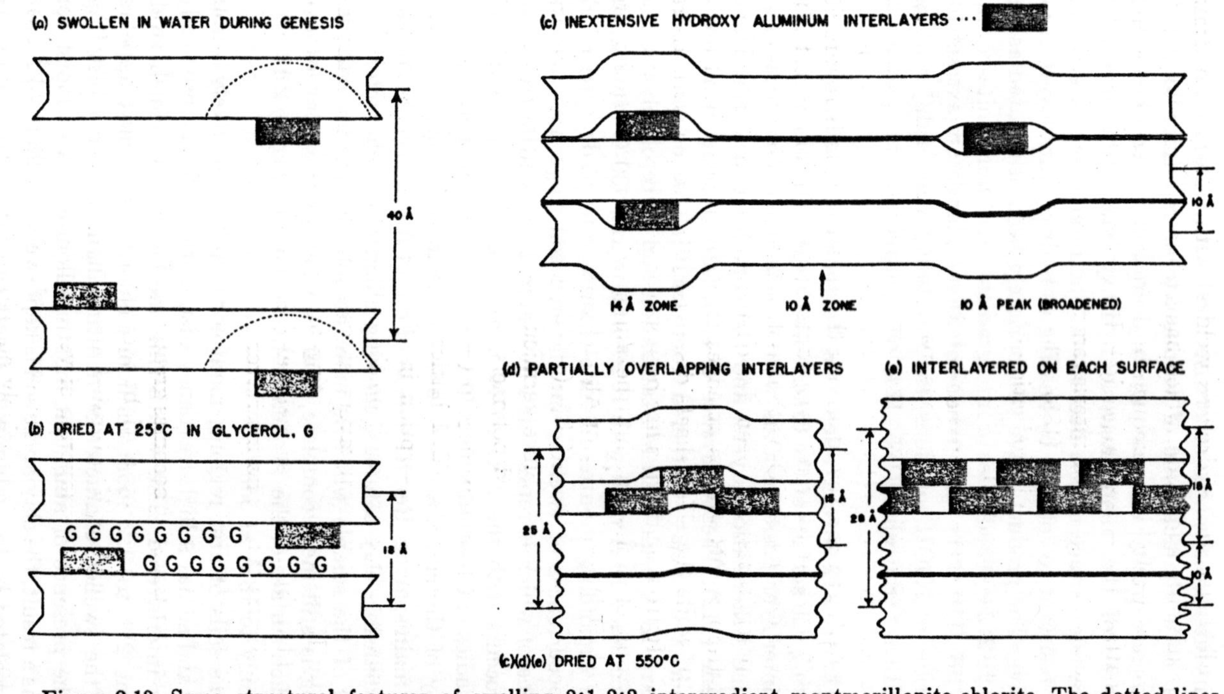

Figure 2.13. Some structural features of swelling 2:1–2:2 intergradient montmorillonite-chlorite. The dotted lines in (*a*) suggest possible spheres of (−) counter-charge affected by (+) charges of the hydroxy interlayers. Although most of the lattice can collapse to 10Å, as in (*c*) and (*d*), on heating, some of it cannot. The result is X-ray diffraction peak broadening and mixed-layer effects (112). (*Courtesy Pergamon Press, publishers of Clays and Clay Minerals.*)

rich mantle, alkalinity and hydrolysis yield a plentiful supply of freshly precipitated aluminum hydroxide, which has a positive charge when below pH 8.3 (111). The (+) valence charges on the hydroxy sesquioxide interlayer cations attach to (−) exchange sites. Laboratory syntheses also have involved attachment of gibbsitic interlayers to one surface while the montmorillonite is in a swollen state (188). The cation exchange capacity was considerably lowered by the introduction of hydroxyaluminum interlayers in synthetic preparations (182).

Weathering Processes Forming 2:1 to 2:2 Intergrades of 14 Å Spacing. Mobilization of layer cations, particularly aluminum from layer edges, occurs in acid soils through the bonding of protons in silicic acid edges, as outlined previously (109, 111), the silicic acid ($pk_1 = 9.5$) acting as a "proton sink." This is a key to driving the release of aluminohydronium ion, $Al(OH_2)_6^{3+}$, ($pk_1 = 5$) from layer edges by H_2CO_3 ($pk_1 = 6.4$). The decomposition of clay by H_2CO_3 with dissolution of silica in excess of alumina (which precipitates) has been demonstrated for clays (52). The exchange of H from Si-OH groups at edges of clays occurred in alcohol at pH 10.1 (212), while the silica sheets dissolved in aqueous alkali. Release of octahedral Mg from acid clays has been shown (16, 46).

Polymerization of aluminum into hexaluminohydroxyhydronium units (111) such as $Al_6(OH)_{12}(-OH_2)_{12}^{6+}$ or $Al_6(OH)_{12}(OH \cdots OH_2)_3(OH_2)_6^{3+}$, and larger units (112) such as $Al_{36}(OH)_{90}(-OH^{0.5-})_{14}(-OH_2^{0.5+})_{22}$ can occur in the vermiculite interlayer. Since vermiculite is characteristically spaced at 14 Å, the positively charged polymers can attach to both surfaces and the interlayer (+) charge proximity to (−) charge of layers is closer than for swelling intergrade. The "2:2 lattice-building" process is sterically assisted by cation exchange sorption in the vermiculite interlayer space because the charge concentration is on the order of 10 molar (112).

Flexibility of the structural layers of vermiculite permits collapse of vermiculite by K saturation and heating to a 10 Å spacing in areas not held open to 14 Å by interlayer material (180), as represented for montmorillonite in Figure 2.13. The interlayer $Al\text{-}OH_2$ functional group of interlayers appears to be the main inorganic site of the pH-dependent charge of soil clays (111, 179, 180).

As the interlayer material becomes greater in amount, the mixed layering phenomenon is recognized by a slight increase of the spacing to 10.4 or 10.6 Å when vermiculite is K-saturated and heated to 300°C. Progressive build-up of sesquioxide interlayering of vermiculite, through the operation of processes of soil acidity and chemical weathering of edges (109), gradually props the layers open extensively enough so that a noticeable mixed layering, giving a 12 Å bump on the side of the 14 Å peak (59) and, finally, as the interlayers become more complete, a stable 14 Å spacing, is recognized (73) as the intergradient structure approaches that of chlorite (Figure 2.10).

The development of hydroxy sesquioxide interlayering is so rapid in the weathering of micas according to equation (*1*) and so widespread in expansible layer silicates as to suggest it as an important form of the extra hydrogen found in hydrous micas or illite. Flexibility of the layers permits occlusion of appreciable amounts of hydroxy interlayers without precluding 10 Å spacings in most of the layers. The interlayers provide retention sites for OH (170) and OH_2 (209) having the observed properties.

When interlayered soil clays enter poorly oxidized positions, interlayering tends to decrease (100, 109), suggesting that the Fe^{3+} associated with the hydroxyaluminum interlayers may be reduced and leached from the interlayer structure. Aluminum and iron ions apparently become more stable in the form of true montmorillonite crystal layers rather than as hydroxy interlayers in a hydromorphic environment. The presence of silica and divalent cations in seepage waters entering hydromorphic situations may favor resilication (substitution of increased amounts of tetrahedral Si for Al relative to that in vermiculite), yielding an expansible mineral with a lower layer charge, that is yielding montmorillonite (Figure 2.9). Under alternating oxidation and reduction, swelling 2:1 to 2:2 intergrades (112, 173) tend to increase.

Kaolinite and Halloysite

Kaolinite—$Si_4Al_4O_{10}(OH)_8$, a 1:1 layer silicate, occurs commonly in soils. It has a triclinic symmetry and often occurs as crystals with hexagonal shape. The structure of the mineral (Figure 2.14) involves hydrogen bonding between adjacent layers spaced at intervals of 7.2 Å. The presence of hydrogen bonding prevents the expansion of kaolinite beyond its basal spacing in water or other organic liquids; however, grinding kaolinite with potassium acetate causes the layers to expand to 14 Å (208). After initial intersalation with KOAc, other salts, potassium and ammonium salts, may be introduced in between the layers of both kaolinite and dickite (8), giving varying x-ray spacings. This serves as a basis for the differentiation in mineralogical analysis (110) of kaolinite from chlorite (which does not expand in salt).

Kaolinite is a dioctahedral mineral, and in pure kaolinite, the Al atoms occupy the same positions in all the layers. There is, however, a continuous series from kaolinite to fire-clays, with increasing disorder in the arrangement of the two Al atoms in the three positions they can occupy, with fire-clays having completely random distribution (32). *Dickite* and *nacrite*

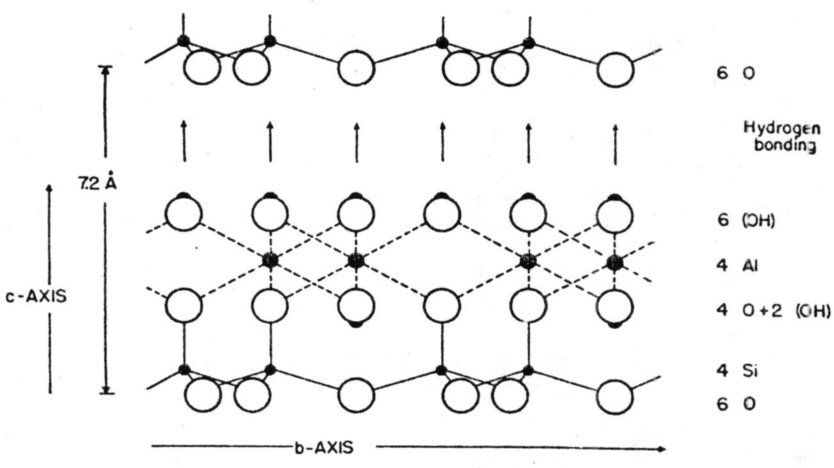

Figure 2.14. Some crystal structural features of kaolinite, a dioctahedral 1:1 layer silicate mineral. The hydrogen bonding can be broken by the reaction in a KOAc salt slurry, yielding a salt interlayer in the expanded structure.

are isomers of kaolinite, but differ from the latter in the position of the Al atoms in the adjacent layers. Dickite has a two-layer structure and nacrite a more complex six-layer structure. The available evidence suggests that there is little isomorphous substitution of ions in the kaolinite, dickite, and nacrite composition, at least in samples which can be obtained in a monomineralic state. The layers are therefore electrically neutral.

Halloysite has a structure somewhat similar to kaolinite except that a layer of water is hydrogen-bonded between the 1:1 silicate layers (Figure 2.15). The water molecules are attached to the adjacent silica and alumina sheets by hydrogen bonding. The fully hydrated halloysite—$Si_4Al_4O_{10}(OH)_8 \cdot 4H_2O$—is also known as *endellite* (3) or *hydrated halloysite* (89). The structure from which the interlayer water has been expelled is called *halloysite* or *metahalloysite* (133).

Halloysite readily dehydrates to give 7.2 to 7.6 Å basal spacing. The slightly larger basal spacing of halloysite, in contrast to that for kaolinite (7.2 Å), arises from the residual water trapped in between the collapsed layers (31). Bates *et al.* (18) found halloysite crystals to consist of tubes, some of which are collapsed, split, and partly or completely unrolled. The curling of the layers is attributed to the variations in the dimensions of the tetrahedral and octahedral sheets. The slightly larger tetrahedral sheets

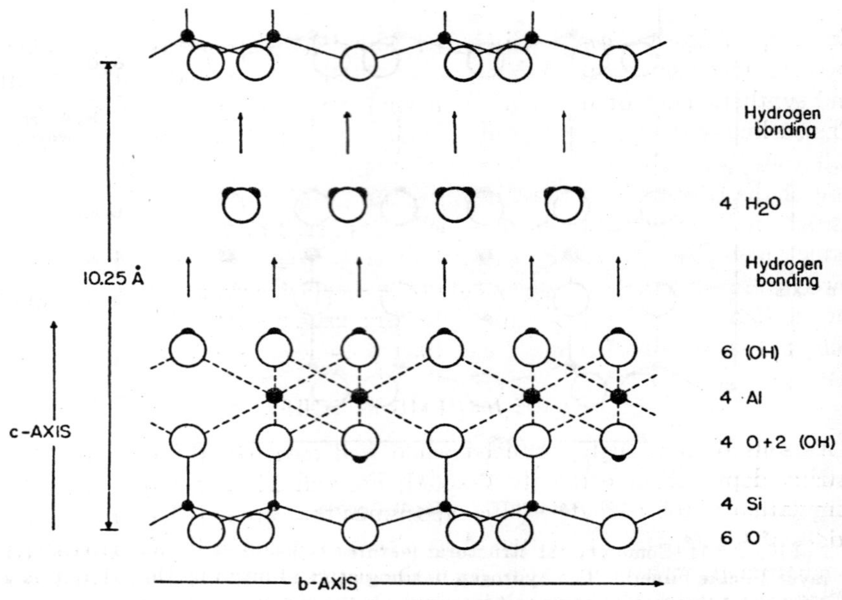

Figure 2.15. Some crystal structural features of hydrated halloysite, a 1:1 layer-slicate mineral which tends to curve into tubes. The water is replaceable by potassium and ammonium salts and by polar organic molecules.

occur on the outsides of the tubes. Study of the halloysite-organic complexes indicate that the mineral can take only one molecular layer of the liquids between the layers and therefore its expansibility is limited (134). Intersalation of halloysite (68, 207) results in a restoration of its interlayer solvation capacity.

Halloysite can be formed from extremely small particles of allophane by crystallization (193) and by the weathering of feldspathic rocks (172) and shales in association with alunite (195), and from kaolinite.

Halloysite occurs very commonly in soils. There are several reports of the presence of halloysite in the soil clays of well-weathered soils under acid conditions, as for example in the Humic Latosols of Hawaii.

Serpentines are a group of trioctahedral 1:1 layer silicates with a basal spacing of 7 Å. The octahedral cation is primarily magnesium, but other ferruginous and aluminous serpentines are known (215). The magnesium end-member is *chrysotile*—$Si_4Mg_6O_{10}(OH)_8$. There is considerable substitution of Al for Mg and there is a continuous series from chrysotile to *lizardite*, with *antigorite*—$Si_4(Mg_{4-5},Al,Fe^{2+})_6O_{10}(OH)_8$—as an intermediate member. Electron microscopic studies show that both natural and synthetic chrysotile crystallize in the form of hollow cylindrical tubes. The curling of the tube, however, is in a direction opposite to that found in halloysite, with the hydroxyl layer on the outside, because the *b* dimension of the trioctahedral sheet is larger than that of the tetrahedral sheet. Antigorite has a platy structure, which is supposed to arise partly from the reduction in the size of the octahedral layer when Al replaces Mg, and partly from a reversing arrangement of the sheets in a super-lattice. Chrysotile, antigorite, and other serpentines are usually not very common in soils, but there are a few reports of their presence in soil clays.

OXIDE MINERALS IN SOILS

As soils become highly leached while well oxidized, the element Si is usually depleted more rapidly than Al, Fe, and Ti. The result is an accumulation of soil colloids with compositions that are enriched in hydrous oxides of Al, Fe, and Ti relative to Si. These inorganic colloidal substances range from amorphous to crystalline in the degree of organization. The types of reaction products can be summarized (106) as follows:

Orderliness of Ion Stacking (↓) — Randomness in Ion Stacking (↑)

- Amorphous (ions randomly placed in gel structure; no electron diffraction)
  - Allophane (Si-Al-Fe-O-OH-OH_2 gel)
  - Hydrous oxide gels
    - Fe-Al-O-OH-OH_2 gel
    - Si-O-OH-OH_2 gel
- X-amorphous (little or no x-ray diffraction)
  - Poorly organized crystals emerging from gel state
  - Layer relics (some ions lost from layer structure)
  - Metamict crystals (some ions displaced in structure)
- Intergradient along *X* and *Y* axes (interlayers of OH-OH_2-R_2O_3 in montmorillonite-vermiculite-chlorite intergrades).
- Interstratification along *Z* axis (mixed layer stacking of 2:1 and 2:2 silicate layers).
- Defect structure—spaces in which electrons substitute for anions.
- Perfect crystals.

The intergradient and interstratified systems of disorder have been considered in connection with layer silicate structures (*above*). *Allophane* and the *hydrous oxides* of aluminum, iron, and titanium are important so-called "free" oxide minerals of soils.

Allophane

When it became known in the late 1920's that much of the soil colloidal material was crystalline, the content of amorphous mineral material present, such as allophane and free amorphous oxides, was for a time almost neglected.

Present evidence points to a continuous gradation in the degree of order-disorder from perfect order to complete disorder (amorphous state) in atomic structural arrangements in soil mineral colloidal material. Because amorphous materials may have great specific surface and high anion- and cation-exchange capacities, they greatly influence the chemistry of the soil.

Allophane is a general term for amorphous aluminosilicate gels of a wide range and composition, but containing Al_2O_3, SiO_2, H_2O and varying amounts of other constituents, including Fe_2O_3. Some hydroxyl as well as tightly bound H_2O may be present. Occasionally, in certain allophane deposits, large amounts of P_2O_5 are present (214), although allophanic soils are usually deficient in phosphorus. The SiO_2:Al_2O_3 mole ratio frequently falls in the range of 0.5 to 1.3, but can go to 2.42 (Table 2.5).

Allophane occurs characteristically in soils as a product of weathering of volcanic ash, and in surface soils is strongly associated with humus (10), giving dark-colored soils classified as the Ando great soil group (*Ando* in Japanese means "dark soil").

Allophane gives a stable porous structure to soils, predisposing them to high permeability, exhaustive leaching, and hence infertility. The porosity of freely drained sandy soils subjects the limited amount of clay present to intensive leaching in humid climates, and can result in the production of amorphous, allophane-like clay (108, 216). The colloid of Podzol soils resembles allophane in SiO_2/Al_2O_3 ratio (Table 2.5).

Allophane has the characteristic property of having a higher cation-exchange capacity after pretreatment with a mildly alkaline sodium carbonate solution and a much lower cation-exchange capacity when pretreated with pH 3.5 sodium acetate buffer, when the cation-exchange capacity is determined by washing with KOAc solution of pH 7 in each case (11). The cation-exchange capacity difference, or delta value, is approximately 100 meq per 100 g of allophane. Moreover, the cation exchange capacity of allophane varies greatly according to the pH of the washing solution employed, that is, allophane has a high "pH-dependent" charge. Allophane is spectacular in the degree to which exchangeable cations hydrolyze from the exchange positions during washing with aqueous alcohol solutions.

Allophane is often overlooked in soils containing an appreciable quantity of crystalline minerals, which show their characteristic x-ray diffraction and differential thermal peaks and other analytical characteristics. The large amount of water present is lost progressively with rise in temperature, coming off abundantly in the temperature range of 100 to 150°C and then being lost gradually as the temperature is raised to over 400°C. The cation-

TABLE 2.5. CHEMICAL COMPOSITION OF CLAY FROM THE B HORIZON

Soil type	Barnes loam*	Miami silt loam*	Cecil sandy clay loam*	Columbiana clay*	Au Train sand*	Choyo soil†	Whakamaru soil‡
Great soil group§	Chernozem	Gray-brown Podzolic	Red-yellow Podzolic	"Aluminous laterite"	Podzol	Ando	Ando
Order**	Mollisol	Alfisol	Ultisol	Oxisol	Spodosol	Inceptisol	Inceptisol
Percentage:							
SiO_2	49.21	41.07	34.27	26.59	12.38	25.46	52.81
Al_2O_3	19.44	23.38	33.03	36.51	29.95	38.61	37.04
Fe_2O_3	9.76	11.50	17.10	15.48	4.70	14.71	5.45
TiO_2	0.74	0.67	0.83	2.20	0.26	0.79	0.47
CaO	5.03	0.96	0.21	0.14	0.41	0.15	Tr
MgO	2.68	2.09	0.13	0.77	0.25	0.65	0.66
MnO_2	0.10	0.09	0.08	0.12	0.09	0.39††	ND
K_2O	1.44	4.37	0.26	0.08	0.48	0.49	ND
Na_2O	0.06	0.24	0.08	0.00	0.36	1.94	3.04
P_2O_5	0.31	0.33	0.27	0.61	0.39	1.11	ND
Ignition loss	11.76	9.45	12.97	17.67	57.47	16.52	15.02
Total	*100.53*	*100.27*	*99.23*	*100.17*	*100.74*	*100.82*	*99.56*
Organic matter‡‡	1.94	1.68	0.55	1.75	0.46		
CO_2 from carbonates‡‡	2.59	0.00	0.00	0.00	0.00		
Mole ratio:							
SiO_2/R_2O_3	3.25	2.60	1.32	0.97	0.78	0.90	2.22
SiO_2/Al_2O_3	4.29	3.41	1.76	1.24	0.88	1.12	2.42

Tr = trace; ND = not determined.

* Colloid, less than 0.3 μ [after Byers, Alexander, and Holmes (40)].

† Na-saturated, H_2O_2-treated, allophanic clay, less than 2 μ, dispersed in 0.002 *N* HCl [after Aomine and Yoshinaga (10)].

‡ Allophanic clay, less than 2 μ, which had been treated for removal of free iron oxides and dispersed in Na_2CO_3 ; depth of this sample not given [after Birrell and Fieldes (21)].

§ Older classification [Thorp and Smith (203)].

** 7th Approximation (189).

†† MnO.

‡‡ Also included in the ignition loss.

exchange capacity delta value method and selective dissolution in dilute NaOH are additional analytical means of determining allophane.

Amorphous aluminum and iron oxides become stabilized by silica only temporarily in the form of allophane. With time and weathering the two crystalline phases, halloysite and gibbsite, frequently separate out. When silica influx is maintained from freshly laid upper horizons of ash, the crystalline phase developed at depth with time is entirely halloysite.

Because intensive leaching has removed most of the primary minerals and most of the essential nutrient elements, allophanic soils are generally infertile. The high specific surface and high aluminum and iron activity of allophane cause a problem of high phosphate-fixation capacity in soils containing much allophane.

Aluminum Oxides

The crystalline mineral *gibbsite*—$Al(OH)_3$ (or $Al_2O_3 \cdot 3H_2O$)—is the most abundant free hydrous oxide of alumina in soils, occurring primarily in soils of tropical and subtropical regions which have undergone intensive leaching of silica. Intensive leaching of silica, as feldspars and other minerals weather, results in the formation of aluminum hydroxide. Gibbsite consists of paired sheets of hydroxyl held together dioctahedrally by aluminum ions. The series of paired sheets are held together by hydrogen bonding between adjacent hydroxyls, arranged directly above and below one another.

Amorphous hydrous oxides of aluminum are rare except as they occur in gels involving other ions such as iron. Drying of this type of gel from a Hydrol Humic Latosol results in the formation of separate complex iron amorphous residue as separate particulates and crystallization of visually separate granules of gibbsite (184).

Hydroxy alumina and possibly mixtures of iron and other ions may occur as positively charged hydroxy species held by the negatively charged layer silicate particles of soils, as mentioned in the section on intergradient 2:1 to 2:2 clays. The isoelectric pH of hydroxyalumina particles appears to be on the order of pH 4.8; above this pH the alumina may become negatively charged and separate as a free gibbsite phase in soils (111). The clay particles of soils thus may nucleate and promote the formation of gibbsite in soils (13).

The hydroxyaluminum oxide, *boehmite*—AlOOH, in which the aluminum ions are octahedrally coordinated by oxygen and hydroxyl ions (28), occurs in intensively leached, highly weathered soils, frequently in association with gibbsite. Soils that have been weathered for long time periods through unusual geomorphological circumstances that permit the weathering to extend for periods of several million years may develop appreciable amounts of boehmite along with the large amounts of gibbsite even from sediments that might have been fairly siliceous originally. Such deposits are known as bauxite. Boehmite with gibbsite has been noted in highly leached soils in Australia (129), Haiti (102), and other tropical areas. The isomer of boehmite, *diaspore*—AlOOH (which is a crystal structural analogue of goethite), ordinarily presumed to have involved low-temperature hydrothermal action, occurs in some aluminous fireclay deposits (116).

Iron Oxides

The most common iron oxides in soils are *hematite*—Fe_2O_3, which gives the pink to bright red color of soils, and *goethite*—FeOOH (or $Fe_2O_3 \cdot H_2O$), which gives the brown and dark reddish brown colors to soils. Iron oxides provide an extremely important reflection of the chemical properties of soils and the genetic processes that have governed soil formation (53, 176, 190). The iron oxides tend to occur as amorphous coatings (201) gradually transforming to crystalline forms as the amounts present increase.

Hematite of coarsely crystalline form may occur in silt and sand fractions of soils; it then has a deep purple color, and in certain well-oxidized-Ferruginous Humic Latosols (Oxisols) it gives a purplish tint to the dark red matrix. *Martite*—Fe_2O_3—has a structure similar to hematite. Dark iron oxide coatings are often associated (125) with coatings of *pyrolusite*—MnO_2.

When finely divided, the bright red color of hematite is intensified (as in the hematite streak), while the brownish color of goethite grades toward yellowish. Finely divided yellowish hydrous goethite is sometimes designated as *limonite*, although still giving a goethite x-ray pattern. Finely divided goethite (limonite) sorbs much water, and the mixture has the proportion of oxides represented in the approximate formula $2Fe_2O_3 \cdot 3H_2O$.

Goethite is abundant (20 to 80 per cent of the soil) in concretions in certain Latosols (Oxisols), ironstone (Ortstein), and B_{ir} horizons of Podzols (Spodosols). Films of goethite and hematite occur as small percentages on the surfaces of grains of most soils to give the reddish and brownish tints common in soils. In poorly drained horizons, the *mottled* yellowish to brownish streaks are the colors of iron oxide films. Aluminum has been found in certain Australian soils to substitute for ferric iron ions in goethite structures up to approximately 20 atomic per cent, resulting in slight decrease in the goethite unit cell size (156). In poorly drained organic soils, *lepidocrocite*—FeOOH, an isomer of goethite, frequently has been reported in Europe (176) as occurring as *bog ores* in poorly drained soils. It gives a bright orange soil color.

Iron oxides of soils are usually products of weathering of iron-bearing minerals. Hematite may also be inherited from the parent rocks, sedimentary, metamorphic, and certain igneous rocks.

The mineral *magnetite*—Fe_3O_4—is a magnetic iron oxide of spinel structure, inherited from parent rock. It occurs in soils usually as a sand-size mineral of specific gravity 5.17. Dark in color, it occasionally has been enriched by sorting, and is abundant in the "black sands" of certain beaches and in soils derived from such parent rock. Oxidation of the iron in Fe_3O_4 yields *maghemite*—Fe_2O_3, which is still magnetic (143).

Determination of free iron oxides is carried out by microscope, x-ray diffraction, and selective dissolution after reduction of the iron. *Dithionite*—$Na_2S_2O_4$ has been the most successful reagent used without (136) or with chelating agents (1) and $NaHCO_3$ buffer (147) to maintain pH 7.3 at which the optimal combination of oxidation potential, rate of oxide dissolution, and nondestruction of nontronitic (iron-bearing) clays is obtained. Amorphous iron oxides and hematite are dissolved most rapidly. Concretionary goethite is dissolved more slowly. Magnetite and ilmenite are little attacked by neutral dithionite reagents. Iron reduction with H_2S and with H_2 is also possible, but solutions of low pH are required and iron-bearing colloidal minerals are severely attacked. Acid oxalate solutions have been used for chelation and extraction of free iron oxides from soils.

Titanium Oxides

Rutile—(TiO_2) and *anatase*—(TiO_2) commonly occur in soils and clay sediments. Rutile in soils may be inherited from rocks, while anatase may be pedogenic—formed during soil weathering. Although the Ti content of layer silicate samples is frequently placed in the octahedral layer in the calculation of formulas for soils and clay deposits, it has gradually become recognized that Ti almost always can be concentrated as euhedral rutile and anatase crystals which had been occluded in the layer silicate specimens. Amorphous hydrous titanium oxide is known as *leucoxene*—$TiO_2 \cdot nH_2O$.

Ilmenite—$TiFeO_3$ (or $TiO_2 \cdot FeO$)—is common in Hawaiian soils inherited from mafic rocks. Substitution of Ti for Fe in magnetite gives the *titanomagnetite* ($TiFe_2O_4$) to *titanomaghemite* ($TiFe_2O_5$) series, which occurs in Hawaiian soils (196).

FREQUENCY DISTRIBUTION OF MINERALS IN SOILS

The frequency distribution of minerals present in soils is controlled (*a*) by the minerals originally inherited from the parent rock, and (*b*) by soil-forming processes which destroy some minerals, selectively removing some elements while accumulating other elements as a residue. Some of the mobilized elements are deposited in deeper horizons of the soil profile and in positions further downslope.

Soil-forming Processes

Eluviation and Illuviation. Removal of elements from any horizon, but especially from the surface soil horizon, is known as *eluviation*. The chief processes are: (*a*) *soluviation*—ordinary dissolution in water of element from the minerals, followed by leaching (194) into lower horizons,

TABLE 2.6. CHEMICAL ANALYSIS OF THE SOIL, BARNES LOAM

Constituent	% of Soil* for 4 horizons			
	A, 0-9 in.	B_1, 9-17 in.	B_2, 17-33 in.	C, 33-50 in.
SiO_2	69.32	73.56	53.43	57.58
Al_2O_3	11.39	11.31	10.18	10.39
Fe_2O_3	3.84	3.96	3.70	3.93
TiO_2	0.49	0.49	0.46	0.48
CaO	1.61	1.53	13.42	10.52
MgO	0.92	0.72	2.00	2.37
MnO	0.16	0.12	0.15	0.18
K_2O	1.80	1.76	1.42	1.62
Na_2O	1.14	1.16	0.84	0.89
SO_3	0.14	0.11	0.07	0.06
P_2O_5	0.17	0.01	0.18	0.15
Ignition loss	9.53	5.26	14.48	12.02
Total	*100.51*	*99.99*	*100.33*	*100.19*
Organic matter†	5.98	2.36	1.12	0.59
CO_2 from carbonates	0.0	0.0	10.50	8.66

* Barnes loam is a member of the Chernozem great soil group. This soil was sampled 2 miles east of Le Bolt, South Dakota [after Byers, Alexander, and Holmes (40)].

† Included also with ignition loss.

as with $CaCO_3$ (Table 2.6); (*b*) *cheluviation*—chelation or complexing by organic molecules (22, 194), particularly of Al or Fe, followed by leaching or translocation to a lower position in the profile; (*c*) *reduction*, particularly of Fe^{3+} and Mn^{4+}, followed by translocation through leaching; and (*d*) *suspension* of colloidal clay (and other mineral and organic substances) in water and its subsequent removal by erosion, or its movement through soil pores into a lower position in the soil profile. The first three of these processes operate as part of *podzolization* and result in the bleaching of the A_2 horizon of podzols. The horizon of accumulation of the material moved from the upper part of a soil profile is the *illuvial* horizon, and the process of receiving material is *illuviation*. Aluminosilicates and sesquioxides thus accumulated characterize the B horizon of soils. Soluviation under good oxidation selectively removes silica from minerals and causes the accumulation of residual sesquioxides in the soil, a soil-forming process known as *laterization*. Soils high in sesquioxide accumulations frequently, but not always, are high in layer silicate clay accumulations as well.

Alluviation. When material is moved downslope in suspension, the accumulation is termed *colluvium* at the footslope and *alluvium* when occurring along streams or in basins. When materials in solution move downslope by seepage and are redeposited by evaporation in a lower basin,

TABLE 2.7. CHEMICAL COMPOSITION OF ASH OF VARIOUS PEATS

Constituent	% of Ash for various depths of 5 peats*									
	(1)		(2)		(3)		(4)		(5)	
	2–4	8–12	0–3	3–9	0–4	32	15–30	42–48	0–6	10–16
Ash	4.5	2.3	7.2	2.3	8.5	10.7	47.6	10.3	10.8	6.5
SiO_2	51.8	31.8	65.3	46.1	26.0	3.5	59.6	35.7	34.3	17.7
Al_2O_3	14.9	11.6	10.6	10.0	1.7	2.1	12.1	7.4	11.4	15.3
Fe_2O_3	6.8	4.2	4.2	4.6	2.1	1.6	10.0	7.3	28.2	24.9
TiO_2	0.6	0.4	0.6	0.6	0.2	0.2	1.6	0.7	0.5	0.5
CaO	10.6	24.3	8.1	10.2	34.8	48.9	5.5	22.4	11.9	21.9
MgO	5.2	9.9	4.5	13.5	4.2	6.0	7.3	5.9	2.3	2.0
MnO	0.4	—	0.1	0.1	0.2	0.1	0.1	0.1	0.8	0.8
K_2O	0.7	1.4	0.6	1.3	0.8	0.2	0.7	0.4	0.9	0.4
Na_2O	1.9	1.0	0.7	1.3	0.7	0.5	0.2	0.6	—	0.8
SO_3	4.5	5.9	3.2	8.8	6.4	6.5	2.3	11.6	4.5	7.6
P_2O_5	2.2	4.5	1.8	3.1	2.6	0.4	0.4	1.7	4.0	3.8
CO_2	0.8	4.9	0.1	0.8	20.1	30.4	0.1	6.1	1.2	4.4

* (1) Highmoor sphagnum, Orono, Me.; (2) Heath-forest peat, Beaufort, N.C.; (3) Saw-grass peat, Belle Glade, Fla.; (4) Sedimentary peat, Miami Canal Lock, Fla.; (5) Woody sedge peat, Monroe, Wash. Depths in inches. [data from Feustal and Byers (65).]

the process is an alluviation process, and the deposit may be designated an *evaporite* in the case of fairly soluble salts—*marl* (soft $CaCO_3$) in lagoons (Table 2.7)—or *caliche* in the case of hardened $CaCO_3$-rich deposits (gypsum, and some silica may be included). When the accumulation is sesquioxidic (high in hydrous oxides of Fe and Al), the dense accumulation may be recognized as *orterde, ortstein, ironpan, bog iron*, or *laterite* crust. Since sesquioxidic deposits may harden and become resistant to erosion, geomorphic erosional processes sometimes elevate the sesquioxidic formation to form plateaus, ridges, or hilltops in subsequent landscapes on which the soils are designated as *laterites* (5, 187) or, if highly aluminous, as *bauxites*. *Laterization* thus may involve alluviation, as well as simple residual accumulation.

Biocycling. Elements brought into plants through roots extending into deep-lying soil horizons are returned as plant parts fall to the soil surface. Reincorporation, often by small animals, results in significant accumulations in surface soil horizons of Ca, Mg, K, Na, Mn, Cu, P, S, B, and N. The element Si is accumulated in grass cells and later found in the surface soil as plant opal silt (19). Calcium accumulation in this way is sufficient in many places to make the surface horizon neutral even though the B horizon is as acid as pH 4.5 to 5.5. Occurrence of a calcareous horizon within reach of the roots aids phytocycling of calcium, a process known as *calcification*, but calcification also operates through weathering of other calcium-bearing minerals if they are present in the subsoil. Removal of water from the soil by plants aids in the formation of $CaCO_3$ in the C_{ca} horizon of soils of subhumid regions at the average depth to which the soil solution carries $Ca(HCO_3)_2$, as illustrated in the B_2 horizon of the Barnes soil (Table 2.6).

Pedoturbation. Most of the soil-forming processes described above cause horizon differentiation or vertical differences in the chemical and mineralogical composition of the soil. In contrast to the processes of horizon differentiation, other processes operate which tend to make a soil profile more uniform vertically. These mixing processes have been grouped (95) together under the term *pedoturbation* and subdivided according to the agent causing the mixing. *Faunal pedoturbation* is mixing by animals; *floral pedoturbation* is mixing by plants, as by treefall; and *argillipedoturbation* is mixing by the shrinking and swelling of clay. The Vertisol order of soils of the Seventh Approximation (189) is an example of an extreme case of argillipedoturbation. The chemical and mineralogical composition of a Vertisol is essentially uniform with depth.

Factors Affecting Frequency Distribution of Soil Minerals

The frequency distribution or relative abundance of minerals in soils varies (102, 108) with Dokuchaiev's five principal classes of factors that govern soil formation (189). The minerals of the soil *parent material* undergo alteration over varying periods of *time* in response to *climatic* factors, *relief* factors, and *biotic* factors. Each class of factors can be shown to have important independent effects on mineralogy of soils under certain circumstances.

The soil parent material provides the initial suite of minerals present in soil. These initial minerals in turn influence soil mineralogy by their relative susceptibility to chemical weathering. Permeability of parent material influences the rate of weathering through controlling the effectiveness of leaching, which in turn depends on the rainfall and temperature, the slope, the effects of vegetation, and the time available since deposition of the parent material. Porous soils, for example, very sandy soils (33, 216) (Table 2.8), and volcanic ash soils of uplands tend to have more highly weathered clay fractions (Table 2.5) than soils of less porosity in similar sites.

Inherited minerals, such as quartz, feldspars, micas, mafic minerals, carbonates, and gypsum, are most abundant in less-weathered zonal soils of the Desert, Brown, Chestnut, and Tundra groups (Table 2.9) as well as

TABLE 2.8. CHEMICAL ANALYSIS OF THE SOIL AU TRAIN SAND

Constituent	% of Soil* for 5 horizons				
	A_1 0–2 in.	A_2 2–10 in.	B_1 10–12 in.	B_2 12–40 in.	C 40–60 in.
SiO_2	51.84	95.49	87.40	90.84	92.99
Al_2O_3	2.39	2.12	4.92	4.29	3.76
Fe_2O_3	0.47	0.16	0.79	0.50	0.50
TiO_2	0.14	0.09	0.17	0.11	0.12
CaO	0.86	0.15	0.26	0.26	0.24
MgO	0.07	Tr	0.01	Tr	0.02
MnO	0.24	0.01	0.02	0.01	0.01
K_2O	0.86	0.98	1.96	1.73	1.66
Na_2O	0.13	0.18	0.24	0.31	0.36
P_2O_5	0.12	0.01	0.06	0.03	0.02
SO_3	0.28	0.01	Tr	Tr	Tr
Ignition loss	42.49	0.68	3.70	1.41	0.24
Total	*99.89*	*99.88*	*99.53*	*99.49*	*99.92*
Organic matter†	—	0.63	2.11	1.02	0.10

Tr = trace.

* Au Train sand is a member of the Podzol great soil group. The sample was collected from Luce Co. Mich. [after Byers, Alexander, and Holmes (40)].

† Included also with ignition loss.

the intrazonal Mountain soils and azonal Regosols and Lithosols. Kaolinite and sesquioxides of Fe and Al in ancient soils can persist through a cycle of sedimentary rock formation and appear in a young soil subsequently developed from the rock.

Secondary layer silicate minerals, such as montmorillonite, vermiculite, and 2:1 to 2:2 intergrades, occur mixed with inherited minerals in moderately weathered soils of zonal Prairie, Gray-Brown Podzolic, and Podzolic soils (Table 2.9). Kaolinite, halloysite, and hydrous sesquioxides as well as 2:1 to 2:2 intergrades (intergradient vermiculite chlorite) increase in the more highly weathered Red-Yellow Podzolic soils (Ultisols). The intrazonal hydromorphic soils, including Humic Gley and Grumusols (Vertisols) tend to be high in montmorillonite (96, 173).

Secondary sesquioxide minerals, such as hematite, goethite, allophane, and gibbsite, are formed in highly weathered laterites, Latosols (Oxisols) such as the Ferruginous Humic Latosols, Hydrol Humic Latosols, and Latosolic Brown soils (50, 197, 198). These highly weathered soils tend to have enhanced amounts of resistant assessory minerals such as ilmenite, magnetite, and anatase.

Weathering Sequence of Clay-size Minerals

The reaction rates of chemical weathering, and therefore the persistence of a clay-size mineral species in soils, are controlled by various intensity

TABLE 2.9. DOMINANT CLAY MINERAL TYPES IN SOIL GROUPS

Order*	Great soil group†	Dominant clay minerals‡
Oxisol	Laterite soils and Latosols	Sesquioxides, gibbsite, kaolinite, 2:1 to 2:2 intergrades
Ultisol	Red-Yellow Podzolic	Kaolinite, halloysite, vermiculite, 2:1 to 2:2 intergrades, sesquioxides, gibbsite
Spodosol	Podzol	Sesquioxides, interstratified layer silicates, 2:1 to 2:2 intergrades, mica
Vertisol	Grumusols	Montmorillonite
Alfisol	Gray-Brown Podzolic, Planosol	Mica, montmorillonite, 2:1 to 2:2 intergrades, chlorite, kaolinite
Mollisol	Prairie, Chernozem, Chestnut, Rendzinas	Montmorillonite, mica, vermiculite, chlorite
Inceptisol	Ando	Allophane
	Brown Forest, Sol Brun Acide	Mica, interstratified layer silicates
Aridisol	Desert, Sierozem	Mica, vermiculite, interstratified layer silicates, chlorite
Entisol	Azonal soils	Highly variable
—	Tundra soils	Mica
—	Humic Gley	Montmorillonite

* 7th Approximation (189).
† Older soil classification.
‡ Each mineral given has been reported dominant in the clay of one or more soil profiles.

and capacity factors operating as a function of time. The capacity factor includes the relative ease with which the mineral can weather. Different combinations of intensities, capacities, and time of weathering may produce a given degree or stage of weathering. The weathering sequence (Table 2.10) of Jackson and associates (102, 103, 104), as slightly modified, gives weathering stability index numbers to clay-size minerals. Increasing index numbers indicate increasing relative resistance to weathering. Ordinarily three to five minerals dominate the clay composition of a given soil horizon. When the percentage content of clay-size minerals of a soil horizon is plotted against the weathering index or "stage of weathering," a distribution curve is obtained, as represented in Figure 2.16. The weighted average of the index numbers calculated for the minerals present in the clay fraction is the weathering mean (98), and this mean increases as a function of the time and intensity of weathering of the soil. This is to say, the distribution curve moves to a position further to the right in Figure 2.16 as the soil reflects increased chemical weathering in its development.

Weathering of ferromagnesian chlorite or the incomplete reconstitution of ferromagnesian chlorite through diagenesis yields in soil clays a second-

TABLE 2.10. WEATHERING INDEXES OF CLAY-SIZE MINERALS IN SOILS AND SEDIMENTARY DEPOSITS†

Weathering index and symbol	Typical clay-size minerals*	Colloidal minerals**	
		Of soils	Of sedimentary deposits
1 Gp	Gypsum (also halite, sodium nitrate, ammonium chloride, sodium sulfate)	Pierre clay, C horizon (South Dakota)	Polders clay (Holland)
2 Ct	Calcite (also dolomite, aragonite, apatite)	Minatare, B horizon (Nebraska)	Calcite limestones (Michigan)
3 Hr	Olivine-hornblende (also pyroxenes, diopside)	Abitibi, C horizon‡ (James Bay, Canada)	Fresh rock flour
4 Bt	Biotite§ (also glauconite, ferromagnesian chlorite, antigorite, nontronite)	Abitibi, C horizon*** (James Bay, Canada)	Glauconite in Cambrian sandstone (Wisconsin)
5 Ab	Albite (also microcline)	Rideau clay, C_1 horizon (So. Ontario)	Authigenic feldspar
6 Qr	Quartz (also cristobalite)	Rideau clay, B horizon (So. Ontario)	Authigenic quartz
7 Il	The 10 Å zones of illite and sericite (also muscovite and other dioctahedral micas)	Elliott silty clay loam B horizon (Wisconsin)	Pennsylvanian underclays (Illinois)
8 Vr	Vermiculite (also expanded zones of interstratified 2:1 layer silicates, and intergradient 2:1 to 2:2 layer silicates where the brucite component is primarily ferromagnesian)	Kilcolgan loam†† (Ireland)	Ordovician bentonite (Kentucky)
9 Mt	Montmorillonite (also beidellite)	Barnes silt loam, A horizon (South Dakota)	Bentonite (Wyoming)
9 Ig	Pedogenic 2:1 to 2:2 intergradient montmorillonite-vermiculite-chlorite	Loring silt loam, A_2 horizon (Mississippi)‡‡	
10 Kl	Kaolinite (also halloysite)	Cecil clay, B horizon (Alabama)	China clay deposits (Georgia)
11 Gb	Gibbsite (also boehmite)	Fannin sandy loam, C horizon (N. Carolina)	Bauxite (Arkansas)
11 Allo	Allophane	Ando soils (Japan)	Allophane (Indiana)
12 Hm	Hematite (also geothite, limonite)	Nipe clay, B horizon (Puerto Rico)	Bog iron (Minnesota)
13 An	Anatase (also zircon, rutile, ilmenite, leucoxene, corundum)	Naiwa A_2 or B horizon (Kauai Island, Hawaii)	Metamorphosed bauxite

ary chlorite having less thermal stability but more weathering stability (index 8) than the typical ferromagnesian chlorite (index 4). It appears that the ferromagnesian-related secondary chlorites, given (108) weathering index 8 (with vermiculite, including both trioctahedral and dioctahedral), have a lower stability index than the pedogenic 2:1 to 2:2 intergrades of both 18 Å swelling and 14 Å type. The latter have aluminous interlayer precipitates sufficient to prevent complete collapse to 10 Å at 300°C on K saturation and are given weathering index 9 (112). Thus both montmorillonite and 2:1 to 2:2 intergrades have a weathering stability index of 9, and these minerals are superseded by kaolinite (6, 102) of weathering stability index 10. The index number 11 for allophane (197) reflects its approximate equality of resistance to that of gibbsite (index 11). Iron oxides (index 12) accumulate under good oxidation combined with high intensity of weathering. Titanium oxides (index 13) are most resistant to weathering.

Weathering Reactions

The formation of vermiculite-chlorite intergrades by weathering (36, 164) and their survival even in warm humid climates (1, 130) are explained on the basis of the continual formation of vermiculites by weathering of the sand and silt fractions of mica particles. As shown by equation (*1*), vermiculite can weather progressively to montmorillonite in an acid environment while (and after) being interlayered. Thus when some of the hydroxy Al or Fe interlayers (57, 199) are extracted, montmorillonite sometimes appears. Aggregation of the fine montmorillonite layers into coarse clay aggregates showing a 14 Å spacing because of the bridging by hydroxy sesquioxides has been proposed (157).

Montmorillonite, which characteristically has smaller crystal size and hence greater specific edge surface (Figure 2.17), may be attacked by weathering acids faster than are large particles of vermiculite freshly formed by mica cleavage and of lower specific edge surface (109, 112). An amorphous relic of fine-grained layer silicate clay which appears to reflect layer lattice decomposition was found in the less than 0.08 micron fraction

---

FOOTNOTES TO TABLE 2.10.

* Occurring with various degrees of weathering.

† After, Jackson, Tyler, *et al.* (102), Jackson and Sherman (104), Jackson (108, 112).

‡ Slightly altered "rock flour," bearing amphibole. Other members of indexes 3 and 4 are postulated as occurring in soils on the basis of petrographic microscopic observations of occurrence in sand and silt of youthful soils.

§ Layer silicates containing much Fe^{+2} or Fe^{+3} and Mg in the octahedral layer.

** Examples of occurrence of colloidal minerals of various weathering indexes.

†† After Sawhney and Jackson (unpublished).

‡‡ After Glenn (73).

*** Slightly altered "rock flour" bearing chlorite. Biotite may also occur in the clay of various little-weathered soils in association with illite, for example in the mica of Rideau clay, C_1 horizon.

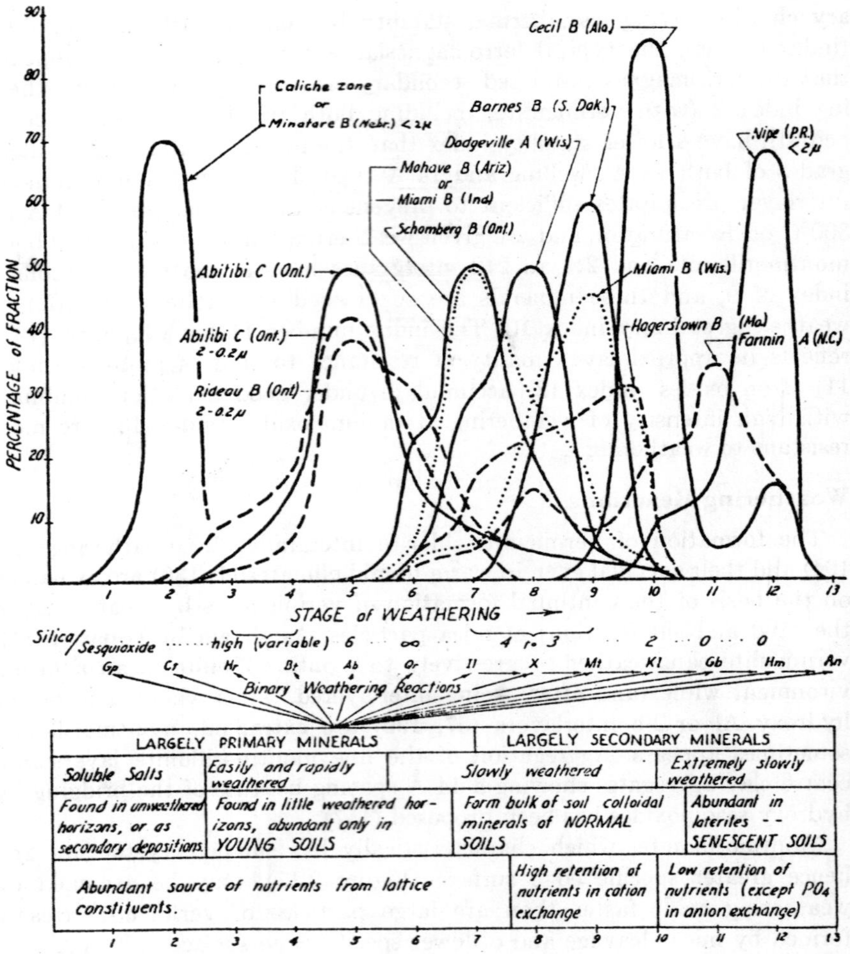

Figure 2.16. Frequency distribution curves for mineral colloids of various stages of weathering, and their significance. Particle sizes are less than 0.2 micron unless otherwise stated (102). (*Courtesy Williams and Wilkins Co., publishers of the Journal of Physical and Colloid Chemistry*)

of the Chester soil derived from schist in Maryland (158). Although montmorillonite can persist in highly montmorillonitic soils long after the soils become acid (108), weathering of such soils finally produces kaolinite (6, 64, 102), possibly through interlayered vermiculite-like and chlorite stages

Extensive weathering and soil development under good drainage ap-

WEATHERING RATES OF 2:1-2:2 INTERGRADES

VERMICULITE-CHLORITE INTERGRADE
(larger particle)

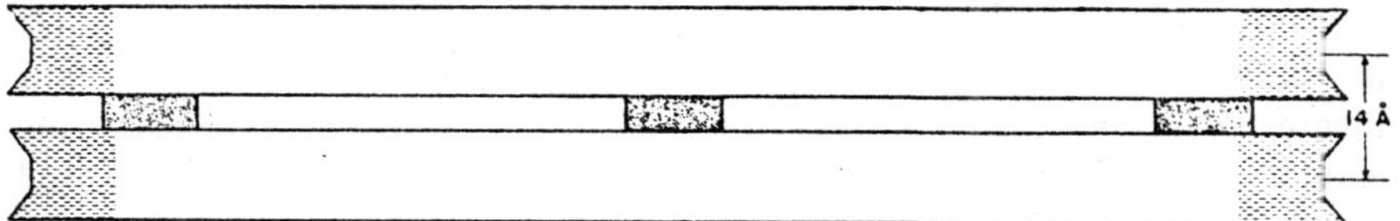

MONTMORILLONITE-CHLORITE INTERGRADE
(smaller particles; more edge surface per unit weight)

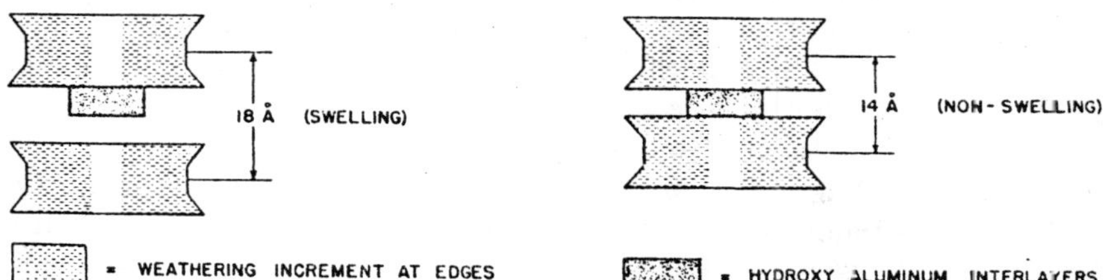

Figure 2.17. Selective weathering away of montmorillonite-chlorite intergrade faster than vermiculite-chlorite intergrade in an acid environment, owing to higher specific edge surface of montmorillonite particles, which are mainly of finer particle size. Concurrently, vermiculite is formed from cleavage of larger mica particles, and becomes interlayered at 14Å by hydroxy aluminum. The resulting vermiculite-chlorite intergrade accumulates as weathering stage or index 9, just ahead of kaolinite of index 10 (112). (*Courtesy Pergamon Press, publishers of Clays and Clay Minerals*)

parently can rather completely develop interlayer precipitates and approach the 2:2 end-member of the series, that is, dioctahedral chlorite (38, 73, 202). The 2:2 to 2:1 intergrade clays appear to accumulate alumina in acid soils nearly to the 2:2 level during weathering; then a subsequent weathering step, 2:2 → 1:1, involving silica tetrahedra inversion, could give kaolinite, as proposed by Glenn and others (74), by translation of silicon atoms, as proposed by Brindley and Gillery (30) for daphnite (interstratified chlorite-kaolinite). Some inversion of silica tetrahedra has been proposed for montmorillonite (62) and for chlorite (26). Electron microscopic evidence for the transformation of montmorillonite to kaolinite by epitaxy has been presented by Altschuler and others (6) under low-temperature supergene weathering conditions. It has also been observed that micas weather to kaolinite (172) under fairly intensive weathering environments. Kaolinite formation thus may be represented as a 2:1 → 1:1

weathering transformation and apparently may involve the 2:1 → 2:2 → 1:1 pedogeochemical reaction (74).

The silicate layers may serve sterically (109) as a template for nucleating gibbsite-like crystal units during the interlayer polymerization of positive hydroxy ion units discussed above. The net result has been termed (112) the "antigibbsite effect," namely, as long as there are expansible layer silicates in the weathering mantle, the aluminum released by weathering tends to be deposited in soils as aluminohydroxyhydronium interlayers of 2:1 to 2:2 intergrades. After intensive leaching of the soil, the pH of the soil solution system tends to rise above the isoelectric pH of hydroxy-aluminum, which appears to be near pH 4.8. A separate crystalline gibbsite phase should be nucleated (111), as the templated (+) hydroxyaluminum sorbed on clay below pH 4.8 becomes negative and thus repelled and released above pH 4.8. This effect has been demonstrated with synthetic systems (13). Intense pedogeochemical leaching removes soluble anions and accelerates crystal growth of the released gibbsite phase.

The replication (nucleation) of montmorillonite crystals appears to be an extension of the foregoing mechanism; templated hydroxy (Al, Fe, Mg) units formed on pre-existing montmorillonite layers, or volcanic glass feldspar cleavages, become silicated to produce new montmorillonite layers, a system observed for example in basalt weathering (173). Under more severe leaching, purer hydroxyaluminum units may be produced which silicate epitaxially to kaolinite, thus serving as the mechanism for the observed (6) montmorillonite to kaolinite transformation. The almost universal occurrence (112) of hydroxy cation units precipitated on the surfaces of layer silicates exposed to weathering is thus extremely fundamental to clay mineralogy and clay genesis.

The various soil clay mineral weathering reactions and indexes may be summarized (108, 111, 112) by equation (*2*) in which the numbers in parenthesis represent weathering indexes:

Mica Biotite (4) Muscovite (7) Illite (7)	→	Vermiculite (8) ⇘	→	Montmorillonite (9) ⇗⇙	⇌	Pedogenic 2:1 to 2:2 swelling 18 Å intergrade (9) ↙					(2)
(Fe, Mg, Al) chlorite (4)	→	Secondary chlorite (8)	→	Pedogenic 2:1 to 2:2 14 Å intergrade (9)	→	Al chlorite (9)	→	Kaolinite and halloy- site (10)	→	Gibbsite (11)	

CONTENT OF CHEMICAL ELEMENTS IN SOILS

The bulk of the inorganic material of soils consists of the four elements oxygen, silicon, aluminum, and iron, much as in the earth's crust (Table 2.1). At least 90 per cent of the mineral matter of most soils consists of the

TABLE 2.11. CHEMICAL ANALYSIS OF MAUI SOIL

Constituent	% of soil* for 4 depths			
	0–7 in.	7–19 in.	19–31 in.	31+ in.
SiO_2	20.33	17.17	14.72	18.30
Al_2O_3	11.68	18.49	21.46	24.60
Fe_2O_3	41.60	38.94	36.96	32.00
TiO_2	8.92	6.25	3.90	3.90
CaO	0.59	0.62	0.66	0.76
MgO	0.88	0.57	0.50	0.58
MnO	0.19	0.10	0.07	0.06
K_2O	0.90	0.62	0.48	0.19
Na_2O	0.29	0.44	0.44	0.39
SO_3	0.24	0.21	0.24	0.23
P_2O_5	0.21	0.20	0.26	0.25
Ignition loss	14.35	16.16	20.70	19.00
Total	*100.18*	*99.77*	*100.39*	*100.26*
Organic matter†	6.88	5.67	5.58	4.51

* This profile was given as being representative of a Humic Latosol by Sherman [after Hough and Byers (97)].

† Included also with ignition loss.

combined oxides of silicon, aluminum and iron (Tables 2.6 to 2.8). Calcareous soils contain as high as 20 to 50 per cent $CaCO_3$ at certain depths (Tables 2.6 and 2.7). The proportions of these and other constituents are greatly changed by chemical weathering and leaching, and this is markedly reflected in the composition of the fine fractions of the illuvial or B horizons (Table 2.5) even before much change is noted in the composition of the whole soil mass. Oxides of iron, aluminum, and titanium increase markedly even on a whole soil basis (Tables 2.11 to 2.13) when nearly all of the minerals have undergone pedogeochemical transformations.

The oxides of calcium, magnesium, sodium, and potassium each make up about 1 to 2 per cent, and the total of these oxides constitute about 5 to 7 per cent of many soils of the humid region. The oxides of manganese, phosphorus, and sulfur generally make up fractions of a per cent of the mineral soil components.

Mineral elements found in the ash of organic soils (Table 2.7) include quantities of Si, Al, Fe, Ti, Na, and K, which are not too far from those in mineral soils. The Ca and Mg increase considerably, as do SO_3, P_2O_5, and CO_2. The latter constituents are relatively higher in amounts because they are the ignition products of organic compounds in humus and in some horizons are calcareous constituents of sedimentary marl (for example, soil numbers 3 and 4 of Table 2.7).

TABLE 2.12. CHEMICAL COMPOSITION OF SELECTED LATERITES

Constituent	% of Laterites* from 5 sites				
	(1)	(2)	(3)	(4)	(5)
SiO_2	1.77	0.37	1.93	17.08	31.37
Al_2O_3	4.32	43.83	62.32	20.83	19.22
Fe_2O_3	80.02	26.61	1.88	40.18	38.51
TiO_2	6.06	4.45	11.87	1.72	1.12
CaO	—	0.86	—	—	0.10
Quartz	0.76	ND	ND	4.32	ND
Feldspar	ND	ND	ND	2.35	ND
H_2O (loss on ignition)	7.06	23.88	21.54	11.05	9.10
Total	*99.99*	*100.00*	*99.54*	*97.53*	*99.42*

ND = not determined.

* After Sivarajasingham, Alexander, Cady, and Cline (187); site and source: (1) Coolgardie, Australia (Simpson, 1912); (2) Satara, Bombay, India (Warth and Warth, 1903); (3) Bagru Hill, Bihar, India (Fox, 1936); (4) Cheruvannur, India, Buchanan's original site (Fox, 1936); (5) Djougou, Dahomey, laterite on granite (Alexander and Cady, 1962).

TABLE 2.13. CHEMICAL ANALYSES OF SELECTED WEATHERED SHELLS, OF HARD LATERITE, AND OF PARENT DOLERITE AT MAMOU, GUINEA*

Constituent	Percentages				
	Rock	Scrapings of rock margin	Shell adjacent to rock	Outer weathered shell	Hard laterite knobs
SiO_2	54.0	45.3	10.8	9.5	16.4
Al_2O_3	15.0	9.4	37.4	37.5	23.5
Fe_2O_3	1.8	23.3	29.4	30.3	45.0
FeO	8.9	<0.5	<0.5	<0.5	<0.5
TiO_2	1.2	2.1	2.0	2.0	1.8
CaO	9.3	5.6	<0.1	<0.1	<0.1
MgO	5.7	7.7	0.1	<0.1	<0.1
MnO	0.18	0.3	<0.1	0.05	0.05
K_2O	1.0	1.0	0.2	—	<0.1
Na_2O	2.4	0.5	<0.1	—	<0.1
P_2O_5	0.05	0.1	0.3	0.17	0.17
H_2O†	0.2	5.1	20.4	20.8	13.3
Total	*99.7*	*100.4*	*100.4*	*100.3*	*100.2*

* After Alexander and Cady (5).
† Ignition loss.

Silicon

The silicon content, expressed as percentage of SiO_2, makes up the bulk of mineral soils that are not greatly modified from the mantle rock sediments from which they are derived (Tables 2.6 to 2.8). The SiO_2 content ranges from 50 to 70 per cent of many soils, which averages about the same as for the earth's crust (Table 2.1). In sandy soils (Table 2.8) the SiO_2 content rises as in sandstone (Table 2.2). The silicon percentage is decreased by dilution with organic matter in organic soils (Table 2.7) and in the A_1 horizon of others (Table 2.6) and is sometimes diluted by $CaCO_3$ as in marl (Table 2.7, soils 3 and 4).

Silicon is depleted from many soils in humid tropical soils which have been subjected to intensive weathering and leaching (Tables 2.11 to 2.13).

Aluminum

After oxygen and silicon, aluminum is the most abundant element in the earth's crust and in the majority of rocks and soils. The aluminum content of soils, expressed on the basis of Al_2O_3, frequently is in the range of 2 to 12 per cent (Tables 2.6 and 2.8). The content ranges up to 20 to 60 per cent in highly weathered soils and laterites (Tables 2.11–2.13), compositions qualifying such soils as bauxite ores. These examples are extreme and should not be taken as a measure of the majority of the soils of the humid tropics. Aluminum occurs mainly in the aluminosilicate minerals, feldspars, amphiboles, pyroxenes, and layer silicates.

In igneous rocks formed at highest temperatures, aluminum is largely bonded to oxygen ions in tetrahedral or fourfold coordination as in feldspars, but as pedogeochemical weathering proceeds, aluminum progressively acquires more octahedral coordination with hydroxyls, a relationship (111) shown in equation (*3*):

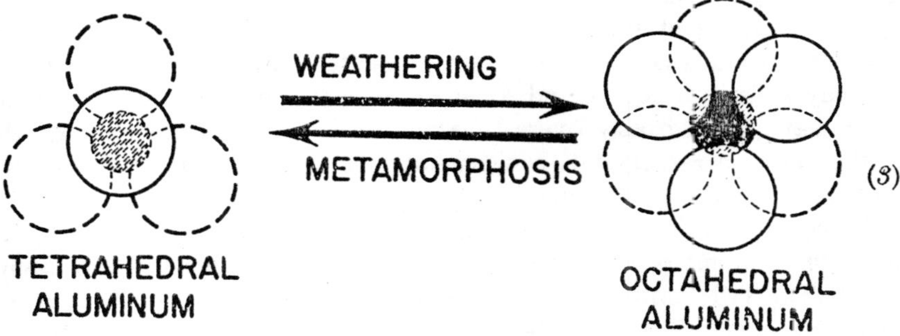

a most fundamental equation, representing the aluminum bonding cycle in pedogeochemical weathering and metamorphic and igneous processes. Layer silicates and hydrous oxides contain the octahedral aluminum represented on the right-hand side of equation (*3*).

As silicon is depleted and aluminum is enriched, the molar ratio, SiO_2/Al_2O_3 in soil colloids falls from over 4 in colloids high in layer silicate to less than 1 in colloids high in allophane (Table 2.5).

Iron

The iron content, expressed as percentage of Fe_2O_3, makes up 1 to 6 per cent of many soils (Tables 2.6 to 2.8), which is comparable to 7 per cent in the earth's crust and 1 to 7 per cent in various rocks (Table 2.2). Iron is subject to increase in concentration through soil development processes, as reflected by contents of 10 to 15 per cent in many soil colloids (Table 2.5). In Latosols and Laterites, the Fe_2O_3 content is frequently 20 to 80 per cent (Tables 2.11 to 2.13). Soils having the higher percentages are approaching the iron content of low-grade iron ore.

High stability of soil aggregates and high soil porosity are usually associated with high iron oxide contents. Under poor drainage the iron becomes reduced and in the presence of organic matter is frequently mobilized. It can be removed from the soil profile to a lower position in the landscape (soil 5, Table 2.7), but frequently becomes concentrated as an iron-rich layer in the lower portion of the soil profile.

The principal form of iron in soils is generally as hydrous oxide, but iron freely enters the 2:1 and 2:2 layer silicate structures of soils, including biotite, vermiculite, and montmorillonite. Smaller percentages of iron occur in soils as pyroxenes and amphiboles.

Titanium

The titanium content, expressed as TiO_2, characteristically makes up 0.2 to 1 per cent in many soils of temperate regions. This value is similar to the range for sedimentary and igneous rocks (Table 2.2).

The titanium of soils occurs primarily as fine-gained crystals of free TiO_2 (rutile and anatase) and $FeTiO_3$ (ilmenite). Although small amounts of Ti may substitute for Fe and Al in silicates, the reported amounts of this form have been grossly overestimated. Recent studies have shown that rutile and anatase crystals can generally be isolated from clays and silts in which they occur as intimate mixtures.

Since titanium oxide minerals are relatively resistant to weathering, the titanium content of soils tends to increase as other elements are leached away. The TiO_2 content of highly weathered soils tends to run 2 per cent or more on a dry-soil basis (Table 2.11). That of laterites in various localities (Table 2.12) and Humic Ferruginous Latosols of Hawaii (198) increase into the range of 5 to 20 per cent or more, and approach the range of titanium ores. Not all latosols and laterites are extremely high in TiO_2 (Tables 2.12 and 2.13).

Calcium

The calcium content, expressed as CaO, is generally in the low range of about 1 per cent in soils (Tables 2.6, 2.8, and 2.11) except when calcium occurs in carbonate or sulfate form (Table 2.6, B_2 horizon). The content of calcium in soils is thus low, compared to that in igneous and sedimentary rocks, the CaO content of which averages about 5 per cent (Table 2.1). Limestones average 43 per cent CaO. The decrease reflects the fact that carbonates and sulfates of calcium leach readily from the upper horizons of well-drained soils except under arid climates.

The calcium of igneous and metamorphic rocks occurs mainly in the plagioclase series of feldspars. In soils, the high calcium end of the plagioclase series weathers fairly rapidly, as explained in the discussion of feldspars above. Noncalcareous soils contain calcium in the form of high sodium plagioclase, augite, hornblende, and epidote.

Calcium carbonate occurs in many soils in lower horizons. This form of calcium is found in soil horizons at lesser depths as one traverses from more humid to more arid regions. The black earth soils of the world, including such broad groups as Chernozems and Rendzinas in North and South Dakota, Texas, Alabama, South Africa, Central India, Russia, and Australia frequently contain 40 to 50 per cent $CaCO_3$ in the subsoil horizons. Younger soils in the humid region may contain calcium carbonate, as calcite ($CaCO_3$) or dolomite ($MgCO_3 \cdot CaCO_3$) inherited from the parent rock. In calcareous drift areas of the north central United States, calcium carbonate normally occurs closer to the soil surface as one traverses from well-drained upland soils toward the hydromorphic end of the catena.

Pedogenic $CaCO_3$ is mainly the mineral calcite, precipitated when leaching water carries $Ca(HCO_3)_2$ downward to a certain depth from which the water is removed by roots or is evaporated. Crystalline gypsum—$CaSO_4 \cdot 2H_2O$ also is accumulated in soils of semiarid to arid regions. A few hundred parts of calcium phosphate per million of soil occur in nearly neutral to calcareous soils, as discussed below under phosphorus.

Magnesium

In well-leached soils, magnesium is found chiefly in mafic minerals, such as biotite, augite, hornblende, and montmorillonite. The magnesium content of soils expressed as MgO frequently is less than 1 per cent (Tables 2.6, 2.8, 2.11, 2.13) in noncalcareous soils. Dolomite—$CaCO_3 \cdot MgCO_3$ and Mg substituted calcite $(Ca,Mg)CO_3$ occur in substantial quantities in some soils, inherited from parent rocks and formed pedogenically (185). Soil contents of 2 per cent or more of MgO (Table 2.6) in carbonate form occur in Brown, Chestnut, and Chernozem soils of the semiarid parts of the world.

Potassium

The potassium content, expressed as K_2O, ranges between 0.05 to 3.5 per cent for mineral soils (Tables 2.6–2.13). Most of the agricultural soils of the United States contain amounts ranging from 1 to 2 per cent or 20,000 to 40,000 parts per 2 million (pounds per acre in a 6-inch depth). On a percentage basis, this is somewhat lower than the average content of potassium in igneous rocks (Table 2.2) but somewhat higher than that in some sedimentary rocks. The distribution of potassium in soils on a world-wide basis follows a definite geomorphologic pattern, and is related more to the conditions of weathering of the potash feldspars and micas than to the composition of the parent rocks themselves. Sandy soils, such as those derived from coarse-textured sediments in the Coastal Plain of southeastern United States, the Cambrian sandstone-derived soils of central Wisconsin, and the Oxisols and Ultisols of humid tropic areas, are examples of soils low in potassium formed during two or more cycles of weathering from rocks relatively rich in potash-bearing minerals. On the other hand, soils containing large quantities of micas and feldspars, and thus large percentages of potassium, have developed from shale-derived till in north central and northeastern United States, from glacial outwash containing small particles of glacial rock flour in Ontario (39, 102) and in northeastern Wisconsin, and loessial wind deposits of the central United States, China, and elsewhere.

The proportion of the total potassium in soils held in soluble and exchangeable forms is usually relatively small. The majority of it resides in potassium-bearing feldspars and micas. The high potassium primary silicate minerals are muscovite, biotite, orthoclase, and microcline; but other micas, feldspars, and other minerals may contain substantial amounts of potassium.

The chemical relationships between the various forms of soil potassium can be expressed by equation (*4*):

$$
\begin{array}{lcccc}
\text{Micas and feldspars} & & & & \\
\downarrow \text{Increasing specific surface of mineral resulting from weathering} & & & & \\
\text{Moderately available K} & & \xrightleftharpoons[\text{Fixation}]{\text{Release}} & & \text{Exchangeable K} \\
 & \updownarrows & & \updownarrows & \\
\text{Applied K} \rightarrow & & \text{Water-soluble K} & \rightleftharpoons & \text{Root K}
\end{array}
\qquad (4)
$$

Because of equilibrium (12, 218) between primary mineral forms and exchangeable and water-soluble potassium, plants can sometimes survive on soils having primary minerals in fine particle sizes as the sole source of potassium. Of the layer-silicate clays, only mica or illite has a substantial potassium content. A very close correlation has been found between the specific surface of mica in soils and potassium released for plant growth (149). In silt loam and clay loam soils of the subhumid region, such as of Nebraska, North and South Dakota, and western Oklahoma and Kansas, the exchangeable potassium is kept at adequate levels by weathering release of potassium. Intensive cropping and removal of potassium in the humid region, and on sandy soils generally, results in the development of stunted growth of plants exhibiting potassium deficiency symptoms. Application of water-soluble forms of fertilizer such as KCl and K_2SO_4 is necessary for intensive agriculture on such areas.

Sodium

The sodium content, expressed as Na_2O, ranges from 0.1 to 1 per cent of many soils (Tables 2.6 to 2.13). This is much smaller than the average of 3.7 per cent in the earth's crust (Table 2.1). The Na_2O content in sedimentary rocks ranges from 1.3 per cent in shales (Table 2.2) to 0.4 per cent in sandstone and 0.05 per cent in limestone. The decreases in Na_2O content in soils and sedimentary rocks reflect weathering away of sodium-bearing minerals, mainly plagioclase feldspars. The high sodium content of ocean waters reflects the leaching of sodium from soil minerals.

The sodium in well-leached soils occurs in high-albite plagioclases and as small amounts in micas, pyroxenes, and amphiboles, chiefly in the fine sand and silt fractions. Sodium occurs as $NaCl$, Na_2SO_4, and sometimes as Na_2CO_3 and other soluble salts in saline and alkaline saline soils, soils in which leaching has not been intensive enough to remove the soluble salts. Sodic soils, containing exchangeable sodium on the order of 15 per cent or more of the cation-exchange capacity, tend to disperse and develop adverse tillage properties.

Phosphorus

The phosphorus content of most mineral soils falls between 0.02 and 0.5 per cent P, and a general average of 0.05 per cent (0.12 per cent P_2O_5) frequently is representative of soils [compared to an average of 0.12 per cent P in the earth's crust (Table 2.1)]. About half the soil phosphorus occurs in combination with organic matter of surface soils, and the remainder occurs in mineral or inorganic combination.

Inorganic phosphorus occurs mainly as calcium phosphate in alkaline and calcareous soils and many parent rocks from which soils are formed. The main calcium phosphate in soils (191) is *fluorapatite*—$Ca_{10}(PO_4)_6F_2$. As weathering proceeds and acidity develops in soils, the phosphate becomes increasingly bonded to aluminum and iron ions released from silicate minerals by weathering (63, 87, 119, 120, 131). Reactions of this type are complex ion-exchange reactions in which PO_4 replaces OH from hydroxy minerals or oxygen from oxide minerals from which the trivalent ions are derived. The chemical properties of phosphate in this form resemble those of the minerals *variscite*—$Al(OH)_2H_2PO_4$ or $AlPO_4 \cdot 2H_2O$ (49, 126), *strengite*—$Fe(OH)_2H_2PO_4$ or $FePO_4 \cdot 2H_2O$ (44), and the isomorphous intermediate *barrandite*—$(Al,Fe)(OH)_2H_2PO_4$ or $(Al,Fe)PO_4 \cdot 2H_2O$ (49, 219). When K or NH_4 is high, as in fertilizer bands, these ions are incorporated as in *taranakite*—$K_3Al_5H_6(PO_4)_8 \cdot 18H_2O$ or its ammonium analogues (206), or other (K,NH_4) salt such as $(K,NH_4)(Al,Fe)_3H_8(PO_4)_6 \cdot 6H_2O$ (87, 127). *Wavellite*—$Al_6(F,OH)_6(PO_4)_4 \cdot 9H_2O$ has been identified (61) in soils.

As weathering in soils proceeds, some of the phosphate becomes occluded (107) with hydrous iron oxides (43), and acquires properties similar to those of *dufrenite*—$Fe_2(OH)_3PO_4$ in solid solution with FeOOH. Occluded iron phosphate is abundant in Ferruginous Latosols.

Under reducing conditions, *vivianite*—$Fe_3(PO_4)_2 \cdot 8H_2O$ forms in soils and, being fairly soluble, may be deposited in lower levels in the soil profile. Reduction of ferric iron renders the oxides soluble and the phosphate becomes more available to plants, as to rice in paddy culture. The phosphate which had been associated with iron may leach out of highly reduced soils, while the relative amount of aluminum phosphate form increases in acid gley soils (45). Decrease of the phosphate content in subsoils is effected also by phytocycling, accompanied by accumulation in the organic matter of surface soils, and depletion from the surface by fire, runoff, and erosion.

The degree of chemical weathering of soils is reflected in the transformation of the discrete inorganic phosphate species. The degree of weathering is related to such soil factors as pH, activities of various cations (particularly of calcium, aluminum, and iron), the solubility products of the various phosphates, and the fertilizer and liming practice, pH especially being a controlling factor (99). The solubilities of phosphate from calcium, aluminum, and iron compounds are close to each other in the soil pH range between pH 6 and 7. Below this pH range, the aluminum and iron phosphates are more stable than calcium phosphate; above this range, various calcium phosphates are formed. The calcium phosphates formed include *hydroxyapatite*—$Ca_{10}(PO_4)_6(OH)_2$, *dicalcium phosphate*—$CaHPO_4$, and complex calcium phosphates intermediate between these two, one of which is known as *octocalcium phosphate* (128). Finely ground phosphate rock containing much *fluorapatite*—$Ca_{10}(PO_4)_6F_2$ is sometimes added to soils as an amendment. This phosphate in acid soils is rapidly transformed into combination with aluminum and iron phosphate (115).

Some phosphate is dissolved from calcium phosphate and precipitated as aluminum and iron phosphates even in soil horizons with an average pH above 7, probably in local acid root and leaching channels. Slow back-transformation to calcium phosphate, perhaps as hydroxyapatite, is indicated in acid soil layers that have been buried by calcareous horizons with the accompanying rise in pH, but considerable aluminum and iron phosphates persist (99). Since the soil is a heterogeneous system, the transformation of one phosphate form to another in soils requires a long time, and the transformation rates are controlled by the slow rate of diffusion of ions along moisture films joining particles of different species. The back-transformation rate is further slowed down by iron oxide coatings.

Although the rates of transformations of soil phosphates are slow, the availability of the different phosphates to plants is greatly affected by the ionic activities in the soil solution. Calcium phosphate is highly unavailable at high soil pH values, usually accompanied by high calcium activity. Iron and aluminum phosphates of acid soils become more available when the soil pH is raised by liming, whereby the activity of iron and aluminum ions is lowered. The extent to which the inorganic phosphate is exchangeable or occurs in forms with a high specific surface (43, 200) appears to be closely correlated with plant growth and crop response to phosphate fertilizer.

Molybdenum

The molybdenum content of soils generally ranges from 1 to 10 ppm but in certain soils it may rise to 20 or 30 ppm or more (107). The element Mo occurs in soil mainly as MoO_4 ion and undergoes fixation as basic iron and aluminum molybdates, much as phosphate does. Likewise, the availability of soil molybdenum rises with the rise in soil pH on liming. Plant deficiencies in molybdenum occur in soils low in this element. Plant contents of Mo may become sufficiently high as to become toxic to animals when the level of available Mo in soil is high.

Nitrogen

The quantity of nitrogen in surface soils generally ranges from 0.02 to 0.25 per cent and is closely related to the amount of soil organic matter of which N makes up approximately 5 per cent. The nitrates, nitrites, and exchangeable ammonium, which are the forms available for plant nutrition and which can be extracted by neutral salt solution, usually make up less than 1 per cent of the total soil nitrogen content of mineral soils. An appreciable fraction of the nitrogen content of subsoils and rocks occurs as NH_4^+ ions substituting for K^+ ions in micas, as revealed by NH_4^+ released by HF treatment of mineral fractions. Mineral-humus complexes also account for much additional nitrogen of soils that can be released by HF treatment.

Sulfur

Sulfur is present in soils in both inorganic and organic forms. In well-leached surface soils, much of the sulfur is combined with organic matter. Field soils of the humid temperate region frequently have 50 to 500 ppm of water-soluble sulfates and 100 to 1500 ppm of total sulfate. Soils high in free iron oxides tend to contain SO_4 ions substituting for hydroxyl in basic sulfate complexes. This form of sulfate is gradually extractable in NaOH solutions and forms a sulfur reservoir of slowly available sulfate which supplements that released by decomposing organic matter.

Crops remove 10 to 30 pounds of sulfur per acre per year, and 40 to 60 pounds of sulfur per acre appear in the leachate of soils in lysimeters. Only a few areas show sulfur deficiency. Sulfur is added to soils in rain and snowfall, amounting to as much as 30 pounds of sulfur per acre in industrial and urban areas where coal, oil, and gas are burned, and as little as 5 pounds in thinly populated areas. Sulfur is added to soils in many commercial fertilizers and in irrigation waters used for croplands and greenhouse culture.

Selenium

Selenium, like other rarer elements, is found in trace amounts in most soils and in most rocks. But crops grown on soils developed on certain geological formations which are exceptionally high in selenium have an excessive selenium content. For example, soils developed from Cretaceous shale of South Dakota, Montana, Wyoming, Nebraska, Kansas, Utah, Colorado, and New Mexico tend to be high in Se, ranging from 2 to 10 ppm. Under conditions of limited rainfall enough Se can be present in the crops grown on these soils to produce vegetation toxic to livestock. Certain native plants are "accumulators" of Se and analysis of them may be used as a guide to soil areas of excess selenium supply.

The element Se behaves much like S, oxidizing to selenate on weathering, being subject to leaching, and combining with free iron oxides to form basic selenates.

Boron

The total boron content of soils generally ranges from 4 to 98 ppm and may average about 30 ppm (107). This compares to 10 ppm in igneous rocks and 4.5 ppm in sea water. Sandy soils may run as low as 2 to 6 ppm. The boron soluble in 85 per cent H_3PO_4 at 100°C, averaging 17 ppm, consists of that present in organic matter and associated with layer silicates. This acid extraction excludes that present as the mineral tourmaline (3.5 per cent B). Acid-soluble boron is associated with the mica fraction of marine sediments and mica-derived clays of soils.

Plant availability of soil boron is most closely correlated with boiling water-soluble boron, which frequently ranges from 0.2 to 1.5 ppm in soils of humid temperate regions; soils of arid and semiarid areas may contain 10 to 40 ppm or more. Extensive leaching of fine-textured desert soils may be required to decrease the boron below toxic levels before crop production under irrigation can be carried out.

REFERENCES

1. Aguilera, N. H., and Jackson, M. L., *Soil Sci. Soc. Am. Proc.* **17**, 359–364; **18**, 223 and 350 (1953).
2. Alexander, L. T., Hendricks, S. B., and Nelson, R. A., *Soil Sci.* **48**, 273–279 (1939).
3. Alexander, L. T., Hendricks, S. B., and Faust, G. T., *Soil Sci. Soc. Am. Proc.* **6**, 52–57 (1942).
4. Alexander, L. T., Faust, G. T., Hendricks, S. B., Insley, H., and McMurdie, H. F., *Am. Mineralogist* **28**, 1–18 (1943).
5. Alexander, L. T., and Cady, J. G., *U. S. D. A. Tech. Bull.* **1282**, 1962.
6. Altschuler, Z. S., Dwornik, E. J., and Kramer, H., *Science* **141**, 148–152 (1963).
7. Anderson, M. S., and Byers, H. G., *U. S. D. A. Tech. Bull.* **228** (1931).
8. Andrew, R. W., Jackson, M. L., and Wada, K., *Soil Sci. Soc. Am. Proc.* **24**, 422–424 (1960).
9. Antipov-Karataev, I. N., "Reclamation of solonetz soils in U.S.S.R.," Academy of Sciences, U.S.S.R., 1953.
10. Aomine, S., and Yoshinaga, N., *Soil Sci.* **79**, 349–358 (1955).
11. Aomine, S., and Jackson, M. L., *Soil Sci. Soc. Am. Proc.* **23**, 210–214 (1959).
12. Attoe, O. J., and Truog, E., *Soil Sci. Soc. Am. Proc.* **10**, 81–86 (1946).
13. Barnhisel, R. I., and Rich, C. I., *Soil Sci. Soc. Am. Proc.* **27**, 632–635 (1963).
14. Barshad, I., *Am. Mineralogist* **33**, 655–678 (1948).
15. Barshad, I., Halevy, E., Gold, H. A., and Hagin, J. *Soil Sci.* **81**, 423–433 (1956).
16. Barshad, I., *Science* **131**, 988–990 (1960).
17. Barth, T. F. W., "Theoretical Petrology," New York, John Wiley & Sons, Inc., 1952.
18. Bates, T. F., Hildebrand, F. A., and Swineford, A., *Am. Mineralogist* **35**, 463–484 (1950).
19. Beavers, A. H., *Soil Sci.* **86**, 1–5 (1958).
20. Berry, L. G., and Mason, B., "Mineralogy," San Francisco, W. H. Freeman and Co., 1959.
21. Birrell, K. S., and Fieldes, M., *J. Soil Sci.* **3**, 156–166 (1952).
22. Bloomfield, C., *J. Soil Sci.* **4**, 5–23 (1953).
23. Bowan, N. L., *Am. J. Sci.* **40**, 161–185 (1915).
24. Bradley, W. F., Grim, R. E., and Clark, G. L., *Z. Krist.* **97**, 216–222 (1937).
25. Bradley, W. F., *Am. Mineralogist* **25**, 405–410 (1940).
26. Bradley, W. F., *Clays and Clay Minerals*, Washington, D.C., National Academy of Sciences–National Research Council, **395**, 94–102 (1955).
27. Bradley, W. F., and Serratosa, J. M., *Clays and Clay Minerals*, London, Pergamon Press, 7th Conf., pp. 260–270 (1960).
28. Bragg, W. L., "Atomic Structure of Minerals," Ithaca, N. Y., Cornell University Press, 1937.

29. Bray, R. H., *Soil Sci.* **43**, 1–14 (1937).
30. Brindley, G. W., and Gillery, F. H., *Clays and Clay Minerals*, Washington, D.C., National Academy of Sciences–National Research Council, **327**, 349–353 (1954).
31. Brindley, G. W., "Clays and Clay Technology, Bull. 169," pp. 33–43, San Francisco, California Division of Mines, 1955.
32. Brindley, G. W., in Brown, G., ed., "The X-ray Identification and Crystal Structures of Clay Minerals," pp. 51–131, 242–296, London, Mineralogical Society, 1961.
33. Brown, B. E., and Jackson, M. L., *Clays and Clay Minerals*, Washington, D.C., National Academy of Sciences–National Research Council **566**, 213–226 (1958).
34. Brown, G., and MacEwan, D. M. C., in Brindley, G. W., ed., "X-ray Identification and Crystal Structures of Clay Minerals," pp. 266–284, London, Mineralogical Society, 1951.
35. Brown, G., and Norrish, K., *Mineral. Mag.* **29**, 929–932 (1952).
36. Brown, G., *Clay Minerals Bull.* **2**, 64–69 (1953).
37. Brown, G., ed., "The X-ray Identification and Crystal Structures of Clay Minerals," London, Mineralogical Society, 1961.
38. Brydon, J. E., Clark, J. S., and Osborne, V., *Can. Mineralogist* **6**, 595–609 (1961).
39. Brydon, J. E., and Patry, L. M., *Can. J. Soil Sci.* **40**, 169–191 (1961).
40. Byers, H. G., Alexander, L. T., and Holmes, R. S., *U. S. D. A. Tech. Bull.* **484** (1935).
41. Caillere, S., and Henin, S., in Brown, G., ed., "The X-ray Identification and Crystal Structures of Clay Minerals," pp. 325–342, London, Mineralogical Society, 1961.
42. Caillere, S., and Henin, S., in Brown, G., ed., "The X-ray Identification and Crystal Structures of Clay Minerals," pp. 343–353, London, Mineralogical Society, 1961.
43. Chang, S. C., and Jackson, M. L., *Soil Sci.* **84**, 133–144 (1957).
44. Chang, S. C., and Jackson, M. L., *Soil Sci. Soc. Am. Proc.* **21**, 265–269 (1957).
45. Chang, S. C., and Jackson, M. L., *J. Soil Sci.* **9**, 109–119 (1958).
46. Chernov, V. A., *Soviet Soil Sci.* **10**, 1150–1156 (1959).
47. Clarke, F. W., "U.S. Geol. Survey Bull. 770," 1924.
48. Clarke, F. W., and Washington, H. S., "U.S. Geol. Survey Prof. Paper 127," 1924.
49. Cole, C. V., and Jackson, M. L., *Soil Sci. Soc. Am. Proc.* **15**, 84–89 (1951).
50. Coleman, N. T., Jackson, M. L., and Mehlich, A., *Soil Sci. Soc. Am. Proc.* **14**, 81–85 (1949).
51. Cook, M. G., and Rich, C. I., *Soil Sci. Soc. Am. Proc.* **26**, 591–595 (1962).
52. Correns, C. W., *Clay Minerals Bull.* **4**, 249–265 (1961).
53. Daniels, R. B., Simonson, G. H., and Handy, R. L., *Soil Sci.* **91**, 378–382 (1961).
54. DeMumbrum, L. E., *Soil Sci. Soc. Am. Proc.* **23**, 192–194 (1959).
55. Denison, I. A., Fry, W. H., and Gile, P. L., *U. S. D. A. Tech. Bull.* **128** (1929).
56. DeSigmond, A. A. J., "The Principles of Soil Science," London, Thomas Murby and Sons, 1938.
57. Dixon, J. B., and Jackson, M. L., *Science* **129**, 1616–1617 (1959).
58. Dixon, J. B., and Jackson, M. L., *Clays and Clay Minerals*, London, Pergamon Press, 8th Conf., pp. 274–286 (1960).
59. Dixon, J. B., and Jackson, M. L., *Soil Sci. Soc. Am. Proc.* **26**, 358–362 (1962).
60. Droste, J. B., *Geol. Soc. Am. Bull.* **67**, 911–918 (1956).
61. Dyal, R. S., *Soil Sci. Soc. Am. Proc.* **17**, 55–58 (1953).

62. Edelman, C. H., and Favejee, J. C., *Z. Krist.* **102**, 417–431 (1940).
63. Ensminger, L. E., *Soil Sci. Soc. Am. Proc.* **13**, 170–174 (1949).
64. Ferguson, J. A., *Australian J. Agric. Research* **5**, 98–108 (1954).
65. Feustal, I. C., and Byers, H. G., *U. S. D. A. Tech. Bull.* **214** (1930).
66. Fine, L. O., Bailey, T. A., and Truog, E., *Soil Sci. Soc. Am. Proc.* **21**, 52–58 (1941).
67. Frederickson, A. F., and Reynolds, R. C., Jr., "*Clays and Clay Minerals*," London, Pergamon Press, 8th Conf., pp. 203–213 (1960).
68. Garret, W. G., and Walker, G. F., *Clay Min. Bull.* **4**, 75–80 (1959).
69. Garasimov, N. P., Grushko, T. E., and Chirvinsky, P. N., *Vsesoyur Miner. Obshch., Zap.*, **78**, 95–100 (1949). Quoted in Mumpton, F. A., and Roy, R., *Clays and Clay Minerals*, National Academy of Sciences–National Research Council, **566**, 136–143 (1958).
70. Gieseking, J. E., *Advances in Agron.* **1**, 159–204 (1949).
71. Gilluly, J., Waters, A. C., and Woodford, A. O., "Principles of Geology," 2nd ed., San Francisco, W. H. Freeman and Co., 1959.
72. Glass, H. D., *Clays and Clay Minerals*, Washington, D.C., National Academy of Sciences–National Research Council, **566**, 227–241 (1958).
73. Glenn, R. C., *Trans. Intern. Congr. Soil Sci. 7th Congr.* **4**, 523–531 (1960).
74. Glenn, R. C., Jackson, M. L., Hole, F. D., and Lee, G. B., *Clays and Clay Minerals*, London, Pergamon Press, 8th Conf., pp. 63–83 (1960).
75. Goldich, S. S., *J. Geol.* **46**, 17–58 (1938).
76. Goldman, M. I., and Tracey, J. I., Jr., *Econ. Geol.* **41**, 567–575 (1946).
77. Goldschmidt, V. M., *J. Chem. Soc.*, **59**, 655–673 (1937).
78. Goldschmidt, V. M., *Soil Sci.* **60**, 1–8 (1945).
79. Grim, R. E., Bray, R. H., and Bradley, W. F., *Am. Mineralogist* **22**, 813–829 (1937).
80. Grim, R. E., "Clay Mineralogy," New York, McGraw-Hill Book Co., 1953.
81. Grim, R. E., and Johns, W. D., *Clays and Clay Minerals*, Washington, D.C., National Academy of Sciences–National Research Couneil, **327**, 81–103 (1954).
82. Gruner, J. W., *Z. Krist.* **83**, 75–88 (1932).
83. Gruner, J. W., *Am. Mineralogist* **19**, 557–575 (1934).
84. Hanway, J. J., Scott, A. D., and Stanford, G., *Soil Sci. Soc. Am. Proc.* **21**, 29–34 (1957).
85. Harrassowitz, H., "Boden der tropischen Regionen," Berlin, Julius Springer, 1930.
86. Harrison, J. L., and Murray, H. H., *Clays and Clay Minerals*, London, Pergamon Press, 6th Conf., pp. 203–213 (1959).
87. Haseman, J. F., Brown, E. H., and Whitt, C. D., *Soil Sci.* **70**, 257–271 (1950).
88. Hellman, N. N., Aldrich, D. G., and Jackson, M. L., *Soil Sci. Soc. Am. Proc.* **7**, 194–200 (1943).
89. Hendricks, S. B., *Am. Mineralogist* **23**, 295–301 (1938).
90. Hendricks, S. B., *J. Geol.* **50**, 276–290 (1942).
91. Hendricks, S. B., and Teller, E., *J. Chem. Phys.* **10**, 147–167 (1942).
92. Hensel, D. R., and White, J. L., *Clays and Clay Minerals*, London, Pergamon Press, 7th Conf., pp. 200–215 (1960).
93. Hilgard, E. W., "Soils, Their Formation, Properties, and Composition," New York, The Macmillan Company, 1906.
94. Hofmann, U., Endell, K., and Wilm, D., *Z. Krist.* **86**, 340–348 (1933).
95. Hole, F. D., *Soil Sci.* **91**, 375–377 (1961).

96. Hosking, J. S., *Trans. Roy. Soc. S. Australia* 59, 168–200 (1935).
97. Hough, G. H., and Byers, H. G., *U. S. D. A. Tech. Bull.* 584 (1937).
98. Hseung, Y., and Jackson, M. L., *Soil Sci. Soc. Am. Proc.* 16, 294–297 (1952).
99. Hsu, P. H., and Jackson, M. L., *Soil Sci.* 90, 16–24 (1960).
100. Hutcheson, T. B., Jr., Lewis, R. J., and Seay, W. A., *Soil Sci. Soc. Am. Proc.* 23, 474–478 (1959).
101. Jackson, M. L., and Hellmann N. N., *Soil Sci. Soc. Am. Proc.* 6, 133–145 (1942).
102. Jackson, M. L., Tyler, S. A., Willis, A. L., Bourbeau, G. A., and Pennington, R. P., *J. Phys. Colloid Chem.* 52, 1237–1260 (1948).
103. Jackson, M. L., Hseung, Y., Corey, R. B., Evans, E. J., and Vanden Heuvel, R. C., *Soil Sci. Soc. Am. Proc.* 16, 3–6 (1952).
104. Jackson, M. L., and Sherman, G. D., *Advances in Agron.* 5, 219–318 (1953).
105. Jackson, M. L., Whittig, L. D., Vanden Heuvel, R. C., Kaufman, A., and Brown, B. E., *Clays and Clay Minerals*, Washington, D.C., National Academy of Sciences–National Research Council, 327, 218–240 (1954).
106. Jackson, M. L., "Soil Chemical Analysis—Advanced Course," Mimeo. published by the author, Dept. of Soil Science, University of Wisconsin, Madison, 1956.
107. Jackson, M. L., "Soil Chemical Analysis," Englewood Cliffs, New Jersey, Prentice-Hall, Inc., 1958.
108. Jackson, M. L., *Clays and Clay Minerals*, London, Pergamon Press, 6th Conf., pp. 133–143 (1959).
109. Jackson, M. L., *Trans. Intern. Soc. Soil Sci., 7th Congr.*, Madison, 2, 445–455 (1960).
110. Jackson, M. L., *Clays and Clay Minerals*, London, Pergamon Press, 9th Conf., pp. 424–430 (1962).
111. Jackson, M. L., *Soil Sci. Soc. Am. Proc.* 27, 1–10 (1963).
112. Jackson, M. L., *Clays and Clay Minerals*, London, Pergamon Press, 11th Conf., pp. 29–46 (1963).
113. Jackson, M. L., Chapter in *Methods of Soil Analysis*, Agronomy Monograph, C. A. Black, ed., (in press).
114. Joffe, J. S., "Pedology," New Brunswick, New Jersey, Pedology Publications, 2nd ed., 1949.
115. Joos, L. L., and Black, C. A., *Soil Sci. Soc. Am. Proc.* 15, 69–75 (1950).
116. Keller, W. D., Westcott, J. F., and Bledsoe, A. O., *Clays and Clay Minerals*, Washington, D.C., National Academy of Sciences–National Research Council, 327, 7–46 (1954).
117. Kelley, W. P., "Cation Exchange in Soils," New York, Reinhold Publishing Corp., 1948.
118. Kellogg, C. E., "The Soils that Support Us," New York, The Macmillan Co., 1941.
119. Kittrick, J. A., and Jackson, M. L., *Science* 120, 508–509 (1954).
120. Kittrick, J. A., and Jackson, M. L., *Soil Sci. Soc. Am. Proc.* 19, 292–295 (1955).
121. Klages, M. G., and White, J. L., *Soil Sci. Soc. Am. Proc.* 21, 16–20 (1957).
122. Koizumi, M., and Roy, R., *Am. Mineralogist* 44, 788–805 (1959).
123. Kunishi, H. M., and Corey, R. B., "Agronomy Abstracts," American Society of Agronomy, Madison, Wisconsin, 1962.
124. Kunze, G. W., Templin, E. H., and Page, J. B., *Clays and Clay Minerals*, Washington, D.C., National Academy of Sciences–National Research Council, 395, 373–383 (1955).
125. Leeper, G. W., *Soil Sci.* 63, 79–94 (1947).

126. Lindsay, W. L., Peech, M., and Clark, J. S., *Soil Sci. Soc. Am. Proc.* **23**, 357–360 (1959).
127. Lindsay, W. L., and Stephenson, H. F., *Soil Sci. Soc. Am. Proc* **23**, 440–445 (1959).
128. Lindsay, W. L., and Moreno, E. C., *Soil Sci. Soc. Am. Proc.* **24**, 177–182 (1960).
129. Loughnan, F. C., and Bayliss, P., *Am. Mineralogist* **46**, 209–217 (1961).
130. Loughnan, F. C., Grim, R. E., and Vernet, J., *J. Geol. Soc. Australia* **8**, 245–257 (1962).
131. Low, P. F., and Black, C. A., *Soil Sci. Soc. Am. Proc.* **12**, 180–184 (1947).
132. Lutz, H. J., and Chandler, R. F., "Forest Soils," New York, John Wiley & Sons, Inc., 1946.
133. MacEwan, D. M. C., *Mineral. Mag.* **28**, 36–44 (1947).
134. MacEwan D. M. C., *Trans. Faraday Soc.* **44**, 349–368 (1948).
135. MacEwan, D. M. C., *J. Soil Sci.* **1**, 90–103 (1949).
136. Mackenzie, R. C., *J. Soil Sci.* **5**, 167–172 (1954).
137. Maegdefrau, E., and Hofmann, U., *Z. Krist.* **98**, 299–323 (1937).
138. Mankin, C. J., and Dodd, C. G., *Clays and Clay Minerals*, London, Pergamon Press, 10th Conf., pp. 372–379, 1963.
139. Marel, H. W. van der, *Soil Sci.* **78**, 163–179 (1954).
140. Marshall, C. E., *Z. Krist.* **91**, 433–449 (1935).
141. Marshall, C. E., Humbert, R. P., Shaw, B. T., and Caldwell, O. G., *Soil Sci.* **54**, 149–158 (1942).
142. Marshall, C. E., "The Colloid Chemistry of the Silicate Minerals," New York, Academic Press, Inc., 1948.
143. Matsusaka, Y., and Sherman, G. D., *Soil Sci.* **91**, 239–245 (1961).
144. McConnell, D., *Am. Mineralogist*, **35**, 166–172 (1950).
145. Mehmel, M., *Z. Krist.* **90**, 35–43 (1935).
146. Mehra, O. P., and Jackson, M. L., *Soil Sci. Soc. Am. Proc.* **23**, 101–105 (1959).
147. Mehra, O. P., and Jackson, M. L., *Clays and Clay Minerals*, London, Pergamon Press, 7th Conf., pp. 317–327 (1960).
148. Merrill, G. P., "A Treatise on Rocks, Rock Weathering, and Soils," New York, The Macmillan Company, 1921.
149. Milford, M. H., and Jackson, M. L., "Agronomy Abstracts," American Society of Agronomy, Madison, Wisconsin, p. 21, 1962.
150. Mohr, E. C. J., "The Soils of Equatorial Regions with Special Reference to the Netherland East Indies," Ann Arbor, Michigan, J. W. Edwards, 1944.
151. Mortland, M. M., and Gieseking, J. E., *Soil Sci.* **71**, 381–385 (1951).
152. Muir, A., *J. Soil Sci.* **2**, 163–183 (1951).
153. Murray, H. H., and Leininger, R. K., *Clays and Clay Minerals*, Washington, D.C., National Academy of Sciences–National Research Council, **456**, 340–347 (1956).
154. Nelson, B. W., and Roy, R., *Am. Mineralogist* **43**, 707–725 (1958).
155. Norrish, K., *Discussions Faraday Soc.* **18**, 120–134 (1954).
156. Norrish, K., and Taylor, R. M., *J. Soil Sci.* **12**, 294–306 (1961).
157. Pawluk, S., *Clays and Clay Minerals*, London, Pergamon Press, 11th Conf., pp. 74–82 (1963).
158. Pennington, R. P., and Jackson, M. L., *Soil Sci. Soc. Am. Proc.* **12**, 452–457 (1948).
159. Pierre, W. H., "Soil and Fertilizer Phosphorus in Crop Nutrition," Agronomy Monograph Series IV, New York, Academic Press, Inc., 1954.

160. Polynov, B. B., "Cycle of Weathering" (trans. by A. Muir), London, Thomas Murby and Co., 1937.
161. Radoslovich, E. W., *Acta Cryst.* 13, 919–932 (1960).
162. Rankama, K., and Sahama, T. C., "Geochemistry," Chicago, University of Chicago Press, 1950.
163. Reitemeir, R. F., *Advances in Agron.* 3, 113–159 (1951).
164. Rich, C. I., and Obenshain, S. S., *Soil Sci. Soc. Am. Proc.* **19**, 334–339 (1955).
165. Rich, C. I., *Soil Sci. Soc. Am. Proc.* 24, 26–32 (1960).
166. Rich, C. I., and Thomas, G. W., *Advances in Agron.* 12, 1–39 (1960).
167. Rich, C. I., and Black, W. R., *Soil Sci.* (in press).
168. Robinson, G. W., "Soils, Their Origin, Constitution and Classification," 2nd ed., London, Thomas Murby and Co., 1949.
169. Robinson, W. O., *U. S. D. A. Bull.* 122 (1914).
170. Rosenquist, I. Th., *Clays and Clay Minerals*, London, Pergamon Press, 11th Conf., pp. 117–135 (1963).
171. Ross, C. S., and Hendricks, S. B., "U.S. Geol. Survey Prof. Paper 205-B," pp. 23–77, 1945.
172. Sand, L. B., "On the genesis of residual kaolins," *Am. Mineralogist* **41**, 28–40 (1956).
173. Sawhney, B. L., and Jackson, M. L., *Soil Sci. Soc. Am. Proc.* **22**, 115–118 (1958).
174. Sawhney, B. L., *Soil Sci. Soc. Am. Proc.* 24, 221–226 (1960).
175. Schmehl, W. R., and Jackson, M. L., *Clays and Clay Minerals*, Washington, D.C., National Academy of Sciences–National Research Council, **456**, 423–428 (1956).
176. Schwertmann, U., *Neues Jb. Miner.* **93**, 67–86 (1959).
177. Schwertmann, U., *Die Naturwissenschaften* **6**, 159–160 (1961).
178. Schwertmann, U., *Beitrage zur Mineralogie und Petrographie* **8**, 199–209 (1962).
179. Schwertmann, U., and Jackson, M. L., *Science* **139**, 1052–1054 (1963).
180. Schwertmann, U., and Jackson, M. L., *Soil Sci. Soc. Am. Proc.* **28**, 179–183 (1964).
181. Scott, A. D., Hunziker, R. R., and Hanway, J. J., *Soil Sci. Soc. Am. Proc.* **24**, 191–194 (1960).
182. Shen, M. J., and Rich, C. I., *Soil Sci. Soc. Am. Proc.* **26**, 33–36 (1962).
183. Sherman, G. D., "Soil Survey of the Territory of Hawaii, U.S.D.A. and Hawaiian Agr. Exp. Sta.," pp. 110–124, 1955.
184. Sherman, G. D., "Hawaii Agr. Exp. Sta. Bull. 116," 1958.
185. Sherman, G. D., Schultz, F., and Allway, F. J., *Soil Sci.* **94**, 304–313 (1962).
186. Simon-Sylvestre, G., *Ann. Agron. Paris* **11**, 309–330 (1960).
187. Sivarajasingham, S., Alexander, L. T., Cady, J. G., and Cline, M. G., *Advances in Agron.* **14**, 1–60 (1962).
188. Slaughter, M., and Milne, I. H., *Clays and Clay Minerals*, London, Pergamon Press, 7th Conf., pp. 114–124 (1960).
189. Soil Survey Staff, "Soil Classification, A Comprehensive System, 7th Approximation," Washington, D.C., U.S.D.A., Soil Conservation Service, 1960.
190. Soileau, J. M., and McCracken, R. J., "Agronomy Abstracts," American Society of Agronomy, Madison, Wisconsin, 1962.
191. Stelly, M., and Pierre, W. H., *Soil Sci. Soc. Am. Proc.* **7**, 139–147 (1943).
192. Stephen, I., and MacEwan, D. M. C., *Clay Min. Bull.* **1**, 157–162 (1951).
193. Sudo, T., *Clay Min. Bull.* **2**, 96–106 (1954).

194. Swindale, L. D., and Jackson, M. L., *Trans. Intern. Soc. Soil Sci., 6th Congr., Paris* **5**, 233–239 (1956).
195. Swineford, A., McNeal, J. D., and Crumpton, C. F., *Clays and Clay Minerals*, Washington, D.C., National Academy of Sciences–National Research Council, **327**, 158–170 (1954).
196. Takashi, K., Kushiro, I., Akimoto, S., Walker, J. L., and Sherman, G. D., *J. Sed. Petrology* **32**, 299–308 (1962).
197. Tamura, T., Jackson, M. L., and Sherman, G. D., *Soil Sci. Soc. Am. Proc.* **17**, 343–346 (1953).
198. Tamura, T., Jackson, M. L., and Sherman, G. D., *Soil Sci. Soc. Am. Proc.* **19**, 435–439 (1955).
199. Tamura, T., *Am. Mineralogist* **42**, 107–110 (1957).
200. Taylor, A. W., Gurney, E. L., and Lindsay, W. L., *Soil Sci.* **90**, 25–31 (1960).
201. Taylor, R. M., *J. Soil Sci.* **10**, 309–315 (1959).
202. Theisen, A. A., Webster, G. R., and Harward, M. E., *Can. J. Soil Sci.* **39**, 244–251 (1959).
203. Thorp, J., and Smith, G. D., *Soil Sci.* **67**, 117–129 (1948).
204. Venkata-Raman, K. V., and Jackson, M. L., *Clays and Clay Minerals*, London, Pergamon Press, 12th Conf., pp. 423–429 (1964).
205. Vilenskii, D. G., "Soil Science," Moscow, Russian Socialist Federated Soviet Republic, State Teachers College Publishing House, Ministry of Culture, 1957. Translated by Dr. A. Birron and Z. S. Cole, available from Office of Technical Services, U.S. Dept. of Commerce, Washington, 25, D.C., 1960.
206. Wada, K., *Soil Sci.* **87**, 325–330 (1959).
207. Wada, K., *Am. Mineralogist* **44**, 153–165 (1959).
208. Wada, K., *Am. Mineralogist* **46**, 78–91 (1961).
209. Walker, G. F., *Mineral. Mag.* **28**, 693–703 (1949).
210. Weaver, C. E., *Am. Mineralogist* **41**, 202–221 (1956).
211. Weir, A. H., and Greene-Kelly, R., *Am. Mineralogist* **47**, 137–146 (1962).
212. Weiss, A., *Z. anorg. allg. Chemie* **297**, 258–286 (1958).
213. White, J. L., *Soil Sci. Soc. Am. Proc.* **15**, 129–133 (1951).
214. White, W. A., *Am. Mineralogist* **38**, 634–642 (1953).
215. Whittaker, E. J. W., and Zussman, J., *Mineral. Mag.* **31**, 107–126 (1956).
216. Whittig, L. D., and Jackson, M. L., *Clays and Clay Minerals*, Washington, D.C., National Academy of Sciences–National Research Council, **456**, 362–371 (1956).
217. Whittig, L. D., and Janizky, P., *J. Soil Sci.*, **14**, 322–333 (1963).
218. Wood, L. K., and DeTurk, E. E., *Soil Sci. Soc. Am. Proc.* **5**, 152–161 (1941).
219. Wright, B. C., and Peech, M., *Soil Sci.* **90**, 32–43 (1960).
220. Yoder, H. S., *Clays and Clay Minerals*, London, Pergamon Press, 6th Conf., pp. 42–60, (1959).

CHAPTER 3

The Physical Chemistry of Soils

S. J. Toth

Rutgers—The State University, New Brunswick, New Jersey

The physical chemistry of soils is essentially a study of the most reactive fractions of soils, or the so-called colloidal phase. The colloid-sized particles play an important role in determining the physical and physiochemical properties and reactions that occur in soils. From a broad viewpoint, the colloidal particles consist of mineral and organic matter intimately associated and held together by various forces. Under certain circumstances the mineral and organic phases may exist separately. In general, as the colloidal content increases, soils tend to lose friability, to retain more water against loss by drainage and evaporation, and to increase their storage capacity for plant nutrient elements in forms available for plant use. The effects of the colloidal phase can be modified to some extent by varying the system of soil management. The purpose of this chapter is to acquaint the reader with the nature of these particles, their properties, both natural and as altered by man, and their relationship to plant growth and crop production.

NATURE OF THE SOIL COLLOIDAL FRACTION

Soil colloids can be roughly grouped into organic and inorganic phases. The inorganic phase may consist either of residues of the original minerals present in the parent material or of new formations produced during the weathering of these minerals. To some degree, the nature of the inorganic phase is dependent upon the forces in the biosphere in any specific area. The organic phase consists of fresh or decomposed residues of plant, animal, and microbial remains that may or may not be combined with the inorganic phase. Considerably more research has been devoted to the study of the nature of the inorganic phase than the organic, even though the cation-exchange capacity of the former far exceeds that of the latter. Many oxidation-reduction phenomena are often associated with the decomposition of the organic phase.

Earlier concepts concerned with the nature of the soil colloidal fraction were, in general, the results of attempts to apply purely colloidal postulations to soils. On this basis, soil colloids did not differ, except in size, from the primary soil minerals, and supposedly were identical in composition to them. This concept did not, however, adequately explain the greater reactivity of the soil colloidal fraction. To explain this behavior, the soil colloidal particles were considered to consist chiefly of zeolitic minerals, which are highly reactive.

Classical studies along these lines include the research of Schloessing, Van Bemmeln, Blanch, Oden, Hall and Russel, Byers, and many others. Perhaps the most important of these was the work of Van Bemmeln, who studied the behavior of the soil colloidal fraction when subjected to different concentrations of various acids and of sodium hydroxide. Although the terms "silicate A" and "silicate B" used by Van Bemmeln have lost their significance in the light of our present knowledge of the nature of the colloidal fraction, his studies are of special importance in that they were probably the first in which a differentiation was made between amorphous gels and lattice fragments of clay minerals. The development of the concept of the so-called adsorption compounds of indefinite chemical composition, which are the reaction products of silica with alumina, iron and water, and small amounts of alkali and alkaline earth cations, may well be considered as the basis for later work that deals with colloidal studies of soil clays. In general, the views of the earlier investigators on the nature of the soil colloidal fraction may be summarized into three categories: the soil colloidal fraction was thought (*a*) to consist of hydrous aluminum silicates of variable composition, and (*b*) to be heterogenous in nature and to contain hydrous aluminum silicate, free silicic acid, hydrated oxides of Fe and Al, quartz fragments, and colloidal-sized particles of unaltered primary soil minerals, and (*c*) the soil colloids were believed to be adsorption complexes of indefinite composition.

Prior to the development of physical methods for identifying the constituents of soil colloids, attempts were made to separate the colloidal fraction into various groups by colloidal-separation techniques which were then in use and which had been successfully applied to the separation of purely hydrophobic organic colloids. The procedures attempted included fractional precipitation and the use of freezing and thawing cycles. These attempts were, however, unsuccessful when applied to the inorganic soil colloids.

Development of physical methods for identifying the structure of clay minerals, including x-ray diffraction and differential thermal analysis (DTA), has led to the formation of the so-called clay mineral concept. According to this theory, soil clays are composed primarily of colloidal-sized particles of crystalline clay minerals, and usually one or more members of a relatively small group of minerals are present in them. The important properties of the soil colloidal fraction are related to these minerals, and the role of the amorphous gels in determining certain electrokinetic properties has largely been neglected. Later in this chapter it will be shown that the amorphous gel portions of soil colloids are equally as important as the crystalline minerals.

For approximately 30 years, the clay mineral concept has been intensively developed by both soil and clay-mineral investigators, and the literature dealing with it has been voluminous. Important contributions to this concept in the field of soils include Anderson (1), Hendriks (8), Kelly (12), Jenny (10), Jackson (9), and many others.

Identification of the crystalline minerals present in the clay fraction of soils has proceeded along the lines of evidence obtained by the use of x-ray diffraction, dehydration, and differential thermal, optical, electron microscopic, and chemical methods. Usually no one of these methods enables absolute identification or accurate delineation of the various percentages of the clays present, but by combining two or more techniques it is usually possible to determine the dominant clay mineral types and to arrive at a reasonable estimation of the crystalline mineral or minerals present in the soil colloidal fraction.

Attempts to modify the clay mineral concept by considering the role of hydrated oxides of Al and Fe in electrokinetic phenomena exhibited by soil colloids has led to our modern concepts of soil colloids, which will be discussed later.

CLAY MINERAL TYPES IDENTIFIED IN SOIL GROUPS

The structures of the clay minerals are adequately described elsewhere in this book. Although the evidence that has accumulated to date is sufficient to attempt to list (Table 3.1) the dominant clay mineral types that occur in the various soil groups of the world, these data are, at best, approximations and are subject to change as further data are accumulated.

Although clay minerals of the chlorite, vermiculite, and mixed layer types have also been identified in clay fractions of soils from the broad zonal groups, no attempt has been made to list their occurrence in Table 3.1 except in the case of the desert soils. The reason for this is that their rela-

TABLE 3.1. DOMINANT TYPES OF CLAY MINERALS IN SOIL GROUPS

Soil group	Dominant clay mineral type	Soil group	Dominant clay mineral type
Tundra	Illite	Red-Yellow Podzolic	Kaolinite
Desert	Mixed lattice (montmorillonite +?)	Tropical and subtropical latitudes	Kaolinite (halloysite?)
Red Desert	Illite	Solonchak	Silimar to surrounding zonal type
Chestnut	Montmorillonite		
Chernozem	Illite or montmorillonite	Solonitz	Similar to surrounding zonal type
Prairie	Illite or montmorillonite	Solodi	Variable
		Humic-glei and Weisenboden	Variable
Noncalcic Brown or Shantung	Variable	Plansools	Variable
Podzols	Illite	Rendzina	Montmorillonite and kaolinite
Gray-Brown Podzolic	Illite or kaolinite	Alluvial	Similar to surrounding zonal types

tive importance in soil clays needs additional elucidation. The most common of the mixed-layer minerals are combinations of illite and montmorillonite and of chlorite and vermiculite.

The most important property of the clay minerals from the viewpoint of the soils chemist is that of cation adsorption and exchange, since it is this property that largely determines the storage capacity of a soil for plant nutrient elements. Although it was known that soils differ widely in this property, there was no satisfactory explanation for the variations noted until determinations of the cation-exchange capacities of the clay minerals encountered in soils were determined. Values for the cation-exchange capacities of clay minerals identified as being found in soil clays are listed in Table 3.2. As is the usual practice in soil studies, the values reported were determined at pH 7.0.

The factors that determined the cation-exchange capacity of the clay minerals, such as particle size and degree of isomorphous replacement of ions in the lattice, are responsible for the fact that the values are reported in ranges rather than as set values.

TABLE 3.2. CATION-EXCHANGE CAPACITIES OF SOME CLAY MINERALS*

Minerals	Cation-exchange capacity (meq/100 g)
Kaolinite	3–15
Halloysite (2 to 4 H_2O)	5–50
Montmorillonite	80–150
Illite	10–40
Chlorite	10–40
Vermiculite	100–150

* By permission from "Clay Mineralogy" by R. Grim, Copyright 1953, McGraw-Hill Book Co., Inc.

COLLOIDAL PROPERTIES OF SOIL COLLOIDS AND CLAY MINERALS

Colloidal-sized particles of soil exhibit characteristics of the colloidal state, such as light scattering, osmotic effects, and charge. With soil colloids, however, it is possible to reverse the sign of the electrokinetic charge, but this is not the case with colloidal-sized particles of clay minerals. Table 3.3 lists the isoelectric point of some soil colloids and the point of minimum migration rates of three clay minerals.

These data indicate that colloidal-sized particles of clay minerals are never isoelectric, and must be considered as strongly electronegative colloids. Soil colloids, depending upon their SiO_2/R_2O_3 ratios, have various isoelectric points, which indicates that they are in most cases amphoteric (14). The amphoteric nature of the soil colloids is associated with the presence of gels of iron, aluminum, and manganese, which coat the crystalline nucleus.

Whether the soil colloidal clay fraction is considered as almost wholly crystalline, or as a central crystalline nucleus surrounded by amorphous precipitates of indefinite composition, does not matter. There are strong reasons for believing that the importance of the crystalline fraction may

TABLE 3.3. ISOELECTRIC pH VALUES OF SOME SOIL COLLOIDS AND POINT OF MINIMUM MIGRATION OF THREE CLAY MINERALS

Soil colloid or clay mineral	Isoelectric pH*	pH of minimum migration†
Bentonite	—	3.10
Kaolinite	—	3.05
Illite	—	3.45
Aragon	6.45	—
Colts Neck	3.55	—
Penn	2.75	—

* Na-saturated, flocculated with HCl.
† Unpublished data.

TABLE 3.4. EFFECTS OF REMOVAL OF AMORPHOUS MATERIAL FROM SOIL COLLOIDS ON CATION-EXCHANGE CAPACITY

Soil series*	Cation-exchange capacity (meq/100 g)	
	Isolated colloid	Cleaned residue
Cecil	10.7	10.6
Colts Neck	28.8	20.5
Norton	22.5	20.2
Sassafras	20.6	20.8

* Except for Cecil, illite is the dominant clay mineral type of the series listed.

in some instances overshadow the role of the precipitates with respect to such properties as adsorption, fixation, and release of cations.

The major importance of the crystalline clay minerals and the minor role of the amorphous isoelectric precipitates in determining the cation-exchange capacity of isolated soil clay fractions are presented in Table 3.4. These data, although rather limited in scope, tend to show that the bulk of the cation-exchange powers is concentrated in the crystalline clay minerals of the soil colloids examined.

Although in most instances the cation-exchange capacities of soil colloids, and presumably those of soil clay minerals, are not greatly affected by removal of amorphous material from the colloids, marked changes are encountered in the retention of anions. This is especially notable when anions that are retained by chemisorption are employed in studies. The retentions of PO_4 ions by two soil colloids are listed in Table 3.5.

Flocculation studies and electrokinetic measurements of soil colloids from which the amorphous materials have been largely removed, in systems containing Cl and SO_4 ions, also seem to indicate marked reductions in the adsorption of these ions (19, 20).

In most parts of the world, establishment of proper media for growth of plants in the field requires the addition of liming materials or fertilizers, or both. Addition of these substances modifies certain colloidal properties of the soil clay fraction because of the operation of exchange adsorption phenomenon. The various cations, because of their size, charge, and hydration, are differentially adsorbed by soil clays. The adsorption of cations by soil clay minerals is related to the type of mineral involved and the forces that are responsible for exchange. A consideration of the causes of adsorption shows that, in the kaolinite and halloysite minerals, broken bonds around the edges of the silica-alumina units are largely involved, as well as the replacement of hydrogen of exposed hydroxyl groups; in illite and chlorite, broken bonds are again of major importance, although lattice

TABLE 3.5. RETENTION OF PHOSPHATE IONS BY TWO SOIL COLLOIDS

Soil series and treatment	P_4 adsorbed (meq/10 g untreated colloid)
Colts Neck	
Untreated	3.9
Residue	1.3
Norton	
Untreated	2.3
Residue	0.9

substitutions are also involved; and in the montmorillonites, vermiculites and, presumably, the mixed-layer minerals, broken bonds are of only minor importance, and it is the lattice substitutions that contribute the major part (7).

ADSORPTION AND EXCHANGE OF CATIONS

Colloidal particles of clay minerals and of soil colloids are usually charged electronegatively (Zeta potential). The charge may be the result of the adsorption of an excess of anions or the result of an unbalance in the atomic charges in the crystal lattices. To balance the negative charge, the particle tends to adsorb counterions (cations). These counterions serve as the main source of plant nutrient elements.

The adsorption of counterions is governed by many factors, the most important being type of cations, ion concentration, nature of the anion associated with the cation, and nature of the colloidal particle.

The effect of type of cation has usually been studied from the viewpoint of cation replaceability in clays. Systems employed have varied from homoionic to multi-ionic with respect to ion populations on the clay surfaces and in the replacing solutions. It has been shown that the higher the valence of the ion, the greater its replacing power and the more difficult its replacement from the clay surface. Hydrogen ions tend to behave as divalent or trivalent ions. Further, with ions of the same valence the replacing power tends to increase with the size of the ion. The valence effect is not new, since it is only a modification of the well-known Schulze-Hardy rule. Similarly, the influence of the hydration sphere of ions on cation replaceability has also received considerable attention by many investigators. With ions of equal valence, those which are least hydrated have the greatest replacing power and are the most difficult to exchange when present on the clay. Many orders for cation replaceability are listed in the literature (2, 7, 10, 13). A typical order is Li > Na > K. These orders tend to vary according to the nature of the clay. It is of interest to note that in an "ideal soil," the ion population consists of Ca, 65 per cent; Mg, 10 per cent; K, 5 per cent; and H, 20 per cent.

The effect of ion concentration on the replacement of cations can be predicted from the laws of mass action, since exchange adsorption is considered to be stoichiometric in nature. It is to be expected that as the concentration of an ion increases in the replacing solution, it will result in greater cationic exchange by the cation involved. This has been confirmed by the studies of Kelley (12) and others (3, 17). It has also been demonstrated that with soil clays, other factors such as the nature of the cation and its valence, as well as the ion concentration, play important roles in cation replaceability.

The nature of the anion associated with the cation also exerts marked effects. It is to be expected from the Paneth-Fajans-Hahn rule of adsorption by crystal lattices that, when an anion which is strongly adsorbed by clays is used, cation adsorption will increase. This has been demonstrated to occur in soils and is indicated by the variations in cation-exchange capacity values obtained when clays are leached with different neutral salts of a given cation.

The type of clay or mineral and its effects on cation adsorption and replaceability have been the subjects of many investigations (7, 13), the principal result of which is that no one order of replaceability of cations tends to hold or is characteristic of all soil clays or minerals. In light of the knowledge of the causes of exchange in the various clay minerals, such a result is to be expected.

A discussion of cation exchange and adsorption would not be complete without a brief discussion of the fixation of certain plant-nutrient cations by soil clays and minerals and the factors that determine fixation. Fixation of cations may be defined as conversion of exchangeable forms of cations to nonexchangeable forms. Potassium fixation in soils is principally related to the action of the illitic minerals, although there is some evidence to indicate that montmorillonite fixes lesser amounts than does illite. Ammonium fixation has also been demonstrated, and the mechanism involved is similar to that noted for K. Indications that Mg fixation leads to the production of a chlorite-type structure have also been noted.

The effect of cations on the flocculation and the electrokinetic behavior of soil clays has been rather intensively investigated by many workers. Studies of the electrokinetic behavior of clay minerals are influenced by the dissociation of ions on the flat surface of the crystalline plates or flakes and by the unsaturation of free valencies created by the breakage of bonds on the cleavage flake edge. This concept is at variance with that proposed by Mattson, who attributes variations in Zeta potentials to differences

TABLE 3.6. RELATION BETWEEN NATURE OF CATION AND ZETA POTENTIAL*

Cation	Zeta potential (mv)	Cation	Zeta potential (mv)
Li	58.8	H	48.4
Na	57.6	Mg	53.9
K	56.4	Ca	52.6
NH_4	56.0	Sr	51.8
Rb	54.9	Ba	50.8
Cs	51.2		

* Courtesy, H. Jenny and R. F. Reitemeier and the Williams & Wilkins Company, Baltimore, Md., publishers, *J. Phys. Chem.*

in the composition of clays. The effect of ions on the Zeta potential is apparently related to charge of the ion and its hydration size. This is illustrated by Table 3.6. The clay used in the study was a Putnam, and was equally saturated with the various cations (60 meq per 100 g). The results show that for ions of equal size the value of the Zeta potential tended to decrease as the charge associated with the cation increased. Flocculation of these homoionic clays with KCl showed that more of the electrolyte was required as the Zeta potential increased. The factors which determine the effect of a cation on the Zeta potential are the charge associated with the ion and its hydration and concentration.

The possibility of differential and even distinctive values of the Zeta potential for the different clay minerals has been proposed by Bergna (4). Standard clays, kaolinite, illite, and montmorillonite, were prepared in H forms, and their Zeta potentials were measured over a range of increasing pH values (Table 3.7). The significance of these values at the present time is not well known; in fact, there is a question as to whether the clays should be considered as Al rather than H-saturated.

ADSORPTION AND EXCHANGE OF ANIONS

Early studies of the adsorption of anions by soil clays indicated a relation between the SiO_2/R_2O_3 ratio and the magnitude of adsorption (14). As the ratio decreased, the adsorption of anions increased. Typical data illustrating this relation are presented in Table 3.8.

Mechanisms involved in the adsorption of anions by soil clays are not clear. Considerably more information has been accumulated on the adsorption of PO_4 than any other anion. There are indications that the anion adsorptive powers of soil clays are associated with the presence of Fe and Al compounds, since removal of these substances by various methods tends either completely to destroy or markedly to reduce anion ad-

TABLE 3.7. ZETA POTENTIAL VALUES OF THREE CLAY MINERALS AT VARIOUS pH VALUES*

pH	Zeta potential (mv)		
	Kaolinite	Illite	Montmorillonite
4	23	30	—
5	29	32	40
6	34	35	45
7	38	44	49
8	42	52	51
9	46	59	53
10	49	51	43
11	47	46	33
12	44	42	21

* Courtesy, H. E. Bergna and the International Society of Soil Science, publisher *Trans. 4th Intern. Congr. Soil Sci.*

TABLE 3.8. ADSORPTION OF PO_4, SO_4, AND Cl IONS BY SOIL COLLOIDS OF VARYING SiO_2/R_2O_3 RATIOS*

Colloid	SiO_2/R_2O_3	PO_4	SO_4	Cl
Fallon	3.82	0.52	—	—
Marshall	2.82	0.93	0.04	—
Sassafras	1.89	1.15	0.15	0.03
Aragon	0.55	1.60	0.27	0.04

* Courtesy, S. Mattson and the Williams & Wilkins Company, Baltimore, Md., publishers *Soil Sci.*

sorption (19). Proposed mechanisms for anion adsorption by pure clay minerals include replacement of OH from the clay-mineral surfaces, especially around the flake edges, and adsorption via anion-exchange sites existing along the basal plane surfaces. Replacement of adsorbed anions which have not been adsorbed via chemisorption is easily brought about by varying the pH. This reaction favors the theory of deactivation of the Fe and Al compounds present. Definite proof of replacement of adsorbed PO_4 on kaolinite by fluoride ions has also been presented (Table 3.8).

Studies of phosphate fixation by soils and clay minerals indicate that the retention may be the result of reaction between free oxides of Fe and Al; of formation of insoluble salts of Fe, Al, or Ca; as well as of fixation by clay minerals. Fixation by the clay minerals is, however, rather slight, ranging from about 0.03 to 0.07 millimole PO_4 per gram of clay. Fixed PO_4 ions can be replaced or released from soil clays by divalent arsenate, silicate, citrate, oxalate, hydroxyl, and fluoride ions. It is believed that the citrate and oxalate ions operate by dissolving iron compounds, hydroxyl by decomposing clays, and only the arsenate and silicate by typical exchange reactions (7, 10).

CATIONIC ACTIVITY IN CLAY MINERALS AND SOILS

The ionic environment in which a plant root exists in soils is determined largely by the ions dissociated from the soil colloids. The dissociation of the ions is related to the cation status of the exchange complex and to the nature of the clay minerals. Prior to development of the use of mineral membranes to determine ion activities, the only procedure available was to determine the cation status of the soil before and after the growth of plants and to analyze the plant tissue. It was extremely difficult to relate the high absorption values of monovalent cations (such as K) by plants to their relatively low degree of saturation in the exchange complex. The use of mineral membranes, as developed by Marshall (13), has been responsible to some degree for the understanding of cation activities and their relation to the absorption of cations by plants.

Descriptions of the techniques employed in these studies have been adequately discussed elsewhere. Only pertinent information concerning the activities of various ions, employing different mineral membranes, will be presented here. With homoionic-clay suspensions, at equivalent cation contents, the activities of the alkali metals decrease in the following order with respect to the clay mineral types: kaolinite > montmorillonite > illite. With the alkaline earth metals the decrease in ion activities under similar conditions is kaolinite > illite > montmorillonite. These facts hold only when the clay minerals are at the equivalence point. With any specific clay mineral, the difference in activity of Na, K, and NH_4 ions at the equivalent point was observed to be for kaolinite and montmorillonite, Na > K > NH_4 ; and for illite, K > NH_4 > Na. The ionization of Ca is greater than that of Mg from the clay types investigated when the degree of saturation with the specific cation is greater than 70 per cent. At less than 70 per cent saturation, Ca ionizes to a greater degree from kaolinite and montmorillonite than does Mg, whereas approximately equal ionization of the two alkaline earth cations occurs with illite (17).

In considering the ionization of cations from clays containing two ions, it has been noted with montmorillonite that K activity increases markedly when Ca replaces H. The substitution of K for H, on the other hand, decreases Ca activity, especially when the degree of Ca saturation is low. With illite, Ca activity is increased when K replaces H. The findings for kaolinite are similar to those for montmorillonite. Other proposals for determining ion activities include the application of the theory of the Donnan equilibrium to the distribution of ions (14).

The influence of the nature of the colloid (clay mineral) on the availability of ions to plants has been carefully considered by Mehlich and his co-workers (17). In general, it has been observed that Ca availability is higher in soils containing clay minerals of the 1:1 type than in soils containing the 2:1 types. The inference from these studies is that soils containing clay minerals of the 2:1 lattice types must be limed to a higher degree of Ca saturation than those containing 1:1 lattice-type minerals. The problem of the availability of K from soils containing the various lattice types of minerals is not quite clear.

ADSORPTION OF ORGANIC COLLOIDAL MATERIALS AND PROTEINS

The chemical, physical, and biological properties of soils are affected by soil organic matter. For this reason the relationships between the organic and inorganic soil colloidal fractions are of importance. Soil organic matter has several characteristics which suggest that it occurs in soils in combination with the inorganic colloids. In clay soils it is immobile, resists oxidation, and is removed with difficulty by chemical extractants. Since proteins constitute one of the most important groups of organic compounds added to the soil in the form of manures and plant residues, most of the studies have dealt with this group of materials.

Demolon and co-workers (5) have demonstrated that humus and clay form definite compounds, the nature of which depends upon the concentrations of the interacting complexes and the prevailing pH of the media. Changes in colloidal properties induced by mixing organic and inorganic colloidal systems have been studied in detail by Meyers, Mattson, and Waksman and include variations in viscosity, in cation-exchange capacity, and in isoelectric and ultimate pH values (18, 14, 22). The presence of organomineral gel complexes in soil has been indicated by the studies of Tyulin (21). Mechanisms for the reaction between organic matter and clays have been ascribed to polar adsorption and to the formation of amphoteric isoelectric compounds (14).

Studies of the clay minerals with various organic compounds have been rather extensive. Gieseking (6) has shown that replacement of Ca and H ions by large substituted ammonium ions, such as piperidene, methylene blue, pyridine, aniline, brucine, napthalamine, and gelatin, gives rise to greater (001) spacing in the montmorillonite-biedellite-nontronite type of clay minerals, and that the intensity of the diffracted radiation from the (001) planes increases with an increase in the amount of complex ion added. This indicates that a portion of the cations is being adsorbed within the variable spacing. The spacings obtained, however. are not always dependent upon the dimensions of the added cation in the free state. Adsorption of proteins by montmorillonite has been found to decrease as the pH values of the suspensions decrease. This is to be expected, because of the amphoteric behavior of proteins. The addition of acid increases the ionization of the amino groups and decreases the ionization of the carboxyl group, and the protein behaves as a cation. Removal of the amino groups by nitrous acid treatments produces materials which do not affect the cation-exchange capacity. The forces involved in the retention of proteins by clay minerals have been described as both Van der Waals and Coulombic.

Evidence has also been presented that combination of organic substances, especially proteins or protein-derived materials, with the inorganic portions of soils may be a factor in conserving organic matter supplies in soils (18). This seems to be especially true in soils containing the 2:1-type clay minerals. This behavior leads to the inference that in soils containing 1:1-type clay minerals, organic matter additions containing large amounts of lignin or other resistant types might be more desirable than high-protein plant residues.

COLLOIDAL PROPERTIES OF SOIL ORGANIC MATTER

The importance of the organic phase in soils is shown by the fact that the cation-exchange capacity can be reduced by as much as 20 to 50 per cent by removal of this phase, even though it constitutes only 3 to 5 per cent of the total soil mass. The organic phase is referred to as humus and consists of a large group of compounds which are formed by decomposition of the plant and animal residues added to soils. Evidence indicates that the chemical composition of humus depends largely upon the type of organic residues and the environment under which the decomposition is occurring.

Although this phase of the soil has been under continuous study for more than 75 years, very little definite information has been accumulated with respect to its composition and colloidal behavior. It is generally accepted at present that lignin is the main constituent, and that nitrogenous complexes of microbial origin as well as polyuronides constitute the major part of the complex.

Three general methods have been rather widely employed to study the effect of humus on the colloidal properties of soil: (*a*) determination of change in colloidal behavior after removal of the organic phase by suitable procedures; (*b*) removal of the organic phase by chemical extractions followed by an examination of the humus; and (*c*) preparation of synthetic types of humus.

One of the most important surface phenomena associated with colloidal humic compounds is its cation-exchange capacity (15, 16). At this point it is necessary to emphasize again that values reported for this property

TABLE 3.9. SPECIFIC CONDUCTIVITY AND MIGRATION VELOCITY OF COLLOIDAL HUMUS SYSTEMS*

Cation	Specific conductivity (10^{-4} mhos)	Migration velocity (μ/sec/volt/cm)
H	2.35	10.4
Li	8.14	14.2
Na	10.74	14.2
K	17.33	13.3
Ca	3.19	6.6
Ba	3.14	4.4

* Courtesy, L. D. Baver, N. S. Hall, and the Missouri Agricultural Experiment Station.

are those of the separated compounds. Values have been reported that range between 250 and 450 meq per 100 g. The fact that a definite value cannot be assigned to this property is in part related to the source from which the "humic compounds" have been isolated, and its stage of decomposition. Evidence indicates that the bulk of the exchange sites are associated with the lignin fraction and are probably due to the carboxyl, phenolic, and enolic hydroxyl groups. Lignin reacts with amino acids to yield "ligno-protein complexes," and the exchange sites in these complexes may also be uronic carboxyl groups. The pH values of the "humic compounds" in their standard state (H-OH) range from 2.3 to 3.7.

Specific conductivities and the migration velocities of various "humic compound" cation systems have been examined in detail by many researchers. Baver (3) studied the colloidal "humic compounds" extracted from soil and found that changes in migration velocities and Zeta potential occur as the nature of the cation associated with these compounds is varied (Table 3.9).

The cation order for decreasing conductivities was found to be:

$$K > Na > Li > Ca > Ba > H$$

For "ligno-humate" systems investigated by McGeorge (16) the cation order was:

$$K > Na > H > Ca > Ba$$

The order for decreasing migration velocities in Baver's system is:

$$Li = Na > K > H > Ca > Ba$$

These data (Table 3.9) indicate that cations of the alkali metal salts of the isolated "humic compounds" of Baver are dissociated more than those of the alkaline earth metals.

The flocculation and viscosity of "humic compounds" sols have also received detailed examination. Sols of these compounds can be flocculated in the same manner as inorganic sols, except that, being more hydrophyllic than the latter, they will require higher concentrations of electrolytes for flocculation. Oden used "humic compounds" isolated from peat and noted that monovalent cations were less effective than divalent cations, and the polyvalent cations were the most active in flocculation of these sols. The cation series, listed in the order of decreasing abilities to flocculate the sol, was as follows:

$$La^{+++}, H^{+}, Ca^{++}, Sr^{++}, Mg^{++}, K^{+}, Na^{+} = Li^{+}$$

It is interesting to note that the Mg ion acts more like a monovalent cation with respect to flocculation and that the H^+ behaves like a divalent cation. The viscosities of "humic compound" sols do not differ greatly from that of water and usually range between 1.10 and 1.18 poise units, depending upon the nature of the exchangeable cation. At symmetry concentrations of cations, Baver (3) found that the cation order for decreasing viscosities was:

$$Ba > Ca > H > Li > K$$

At cation contents greater than the symmetry concentration, viscosities of divalent salts of the "humic compounds" increased, being associated with flocculation and the inclusion of occluded water in the floc. With the monovalent cations, viscosities decreased at cation contents greater than the symmetry concentration, due to a decrease in the ionization of the cation and its subsequent decrease in the water bulk.

The hydration of "humic compounds" and their salts can be calculated from viscosity measurements by use of the Mark and Meyer equation:

$$\frac{N_s}{N_o} = 1 + \left(2.5 \times \frac{\phi}{V - \phi}\right)$$

where: ϕ = effective volume of the dispersed phase; V = total volume of the sol; and N_s/N_o = relative viscosity of the sol. The calculated values are expressed in terms of ml of water per gram of sol, and in flocculated systems it will also include occluded water in the floc. Values calculated from viscosity measurements indicated that the hydration of various-cation-saturated "humic compound" sol is as follows:

$$Ba > Ca > H > Li > Na > K$$

The relative strength of adsorption and release of cations for "humic compound" sol vary as indicated by the data in Table 3.10. In general, the order of release of H^+ was: polyvalent > divalent > monovalent

TABLE 3.10. RELEASE OF CATIONS FROM HUMIC ACID SYSTEMS*

System	Added electrolyte	Ionic radius of cation (Å)	% Cation released by symmetry concentration of added cation
H	LiCl	0.78	11.0
H	NaCl	0.98	11.3
H	KCl	1.33	13.3
H	$MgCl_2$	0.78	21.4
H	$CaCl_2$	1.02	22.1
H	$SrCl_2$	1.28	19.0
H	$BaCl_2$	1.43	23.6
H	$La(NO_3)_3$	1.22	65.4
H	$ThCl_4$	—	98.2
Li	HCl	—	84.7
Na	HCl	—	79.1
K	HCl	—	79.4
Ca	HCl	—	80.4
B	HCl	—	76.0

* Courtesy, L. D. Baver, N. S. Hall, and the Missouri Agricultural Experiment Station.

cations. The reverse release, namely cations by H^+, was rather erratic, but generally more monovalent cations tended to be released by this exchange.

Another extremely interesting property of the "humic compounds" is the exhibition of a "hysteresis effect" upon dehydration. When water is removed from the Ca salt of the "humic compounds," a point is reached at which a black, nonelastic gel is formed. Continued drying beyond this point produces a black brittle solid which does not again pass back completely into the sol state upon the addition of water. Only about 23 per cent of the original material will redisperse when the calcium salt is dried at 50°C. The explanation of this behavior is that orientation of the "humic compounds" with respect to one another occurs, and it is this orientation which is responsible for the hysteresis effect.

AMPHOTERIC NATURE OF SOIL COLLOIDS

Up to this point discussion of the colloidal phase of soils has been primarily limited to the nature and properties of clay minerals and of organic matter. Mention has been made of hydrous gels and their role in the adsorption and retention of anions. Unfortunately, however, it soon becomes apparent to a researcher in the field of colloidal chemistry of soils that it is extremely difficult to believe that studies of pure clay mineral systems will yield all the information needed to explain the migration and translocation of sols within a profile, or that adsorption and retention of fertilizer anions can be entirely due to the role of the pure clay mineral systems in this process. Further, studies (20, 21) have shown that removal of free oxides of Fe and Al from soil colloids usually changes the colloidal nature of the naturally occurring soil-clay sols.

All this information leads to a consideration of the role of hydrated complexes of iron and aluminum, and possibly manganese, in the colloidal behavior and other properties of soil colloids. The study of hydrous gels in situ is, however, extremely difficult, although researches along these lines have been attempted (21). An alternate procedure for a study of this type is to prepare in the laboratory synthetic gel and sol complexes, evaluate their properties, and attempt to apply the findings so obtained to soil colloidal systems (14).

In a series of publications Mattson (14) reported studies of the sols of Fe and Al in combination with silicates, phosphates, and humates. Dilute solutions were used in preparation of the sols to simulate the conditions that might occur in the soil solution. Iron and aluminum were considered to be "basoids" or "amphylotoids," and silicates, phosphates, and humates to be "acidoids." Reactions between the amphylotoids and acidoids yielded sols of various chemical composition. The electrokinetic behavior of these sols, changes in cation and anion adsorption in relation to composition, and other physico-chemical properties were found to be related to the acidoid/amphylotoid ratios. Mattson (14) points out that it is possible to vary the amphylotoid/acidoid ratio by modifying the method of preparing the sol, by varying the nature of the anion associated with the amphylotoid, or by altering the state of dispersion of the acidoid (silicate).

From these studies certain general rules soon were apparent. These were:

1. Aluminum sols possessed higher isoelectric points than iron sols. This was true even when the metals had reacted with silicate, phosphate, or humate acidoids.

2. In terms of decreasing isoelectric points (pH), the order for the three acidoids was silicate, phosphate, and humate.

3. Cation and anion adsorption by the sols (precipitated and dried) was related to the acidoid-amphylotoid ratio. As this ratio decreased, cation adsorption decreased and anion adsorption increased. In all sols the relation was not, however, linear.

4. Iron sols possessed higher cation-exchange capacities per mole of combined acidoid than did the aluminum sols.

Application of these facts to soil colloidal systems indicated that if all the silica, iron, and aluminum present in soil colloids were considered to be in combination, the cation-exchange capacities of the colloids were related to the SiO_2/R_2O_3 ratio. This is indicated by the data presented in Table 3.11.

TABLE 3.11. RELATION BETWEEN SiO_2/R_2O_3 AND CATION-EXCHANGE CAPACITIES OF SOIL COLLOIDS*

Colloid	SiO_2/R_2O_3	Cation-exchange (meq/g) capacity
Fallon	3.82	0.947
Sharkey	3.18	0.796
Marshall	2.82	0.671
Sassafras	1.89	0.331
Norfolk	1.63	0.207
Aragon	0.55	0.164

* Courtesy, S. Mattson and the Williams & Wilkins Co., Baltimore, Md., publishers *Soil Sci.*

TABLE 3.12. EFFECT OF ASSOCIATING ANIONS ON THE COLLOIDAL PROPERTIES OF A COLTS NECK COLLOID

Anion	Anion adsorbed	Cation-exchange capacity (meq/g)	Isoelectric pH
None	None	0.33	4.90
Phosphate	0.58 meq/g	0.73	3.90
Silicate	0.40 meq/g	0.51	4.75
Humate	297 mg/g	1.23	4.04

Additional studies by Mattson (14) have shown that soil colloidal systems can be degraded or regenerated. For example, when soil colloids react with easily replaceable ions, the reaction is reversible. On the other hand, if the system reacts with ions which are hard to displace or which "associate" (chemosorption) with the soil colloids, the reaction often is irreversible. Under the latter condition, the properties of the soil colloids, such as pH, cation-exchange capacity, and others, are markedly altered. The strongly associating anions are phosphates, silicates, and humates, and the difficultly displaceable cations are iron, aluminum, manganese, and magnesium.

To illustrate the effect of associating anions on colloidal properties, data are presented in Table 3.12, obtained with the colloid extracted from the Colts Neck soil. The SiO_2/R_2O_3 ratio of the colloid was 0.81.

These data (Table 3.12) show that the adsorption of strongly associated anions by a soil colloid markedly alters the cation-exchange capacity and isoelectric pH values. Actually, what is occurring is that the acidoid/amphylotoid ratio is being increased by the adsorption of acidoids.

In a similar fashion, it is possible to reduce the acidoid/amphylotoid ratio by introducing new amphylotoids into soil colloids. This effect is illustrated by the data presented in Table 3.13. The soil colloids were the Sharkey and Sassafras.

The amphoteric theory of soil colloids emphasize the fact that at pH

TABLE 3.13. EFFECT OF ADDITION OF AMPHYLOTOIDS ON COLLOIDAL PROPERTIES OF COLLOIDS

Colloid	Amphylotoid	Cation-exchange capacity (meq/g)	pH_{iso}
Sharkey	Al	0.37	5.8
Sassafras	Al	0.14	6.6
Sharkey	None	0.74	3.2
Sassafras	None	0.30	4.0
Sharkey	Mn	0.61	ND
Sassafras	Mn	0.18	ND

ranges above and below the isoelectric point, degradation of the complex results. On the acid side of the isoelectric point the amphylotoid groups are first dispersed, then rendered soluble, and finally transported as ions or inorganic complexes. Silica may become inert due to the formation of H_2SiO_3. On the alkaline side of the isoelectric point, the acidoids become mobile. The humate becomes highly dispersed, the silicate is ionized, and the amphylotoids (Fe, Al, Mn) accumulate. Both of these conditions are favorable for reduction in cation-exchange capacity, changes in isoelectric pH, and other properties. It should be also stated that the amphoteric soil colloids progress through a series of states of aggregation as one proceeds on either side of the isoelectric point. This can be represented as follows:

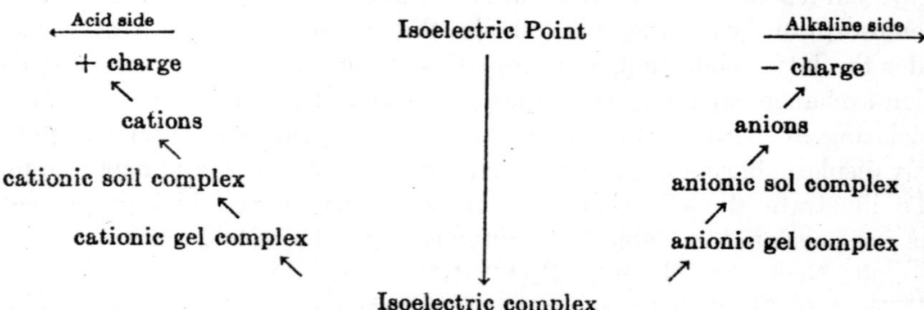

An extremely interesting development from the amphoteric theory of soil colloids and their behavior is the use of what might be called "soil points" or "soil colloidal constants." These are useful for evaluating the properties of the soil colloidal complex in various parts of the soil profile without separating the colloidal fraction or subjecting samples from various horizons in the profile to detailed chemical analysis. This so-called qualitative colloid chemistry of the soil, if the latter term is permissible, is based primarily on the reactions of electrodialyzed soils or H-soils in salt solutions under a definite set of conditions.

Useful data may of course be obtained with reference to the diffusible acid-base balance by titration of electrodialyzates, and it is possible to further differentiate between mineral and organic acids. The ultimate pH of the electrodialyzed soil yields additional information with reference to the base-holding powers of the complex. The combining capacity or neutralization of H-soils with bases yields information on the total cation-exchange capacity and a general idea with reference to the nature of the soil acidoids.

The reactions of the H-soils in salt solution of varying pH will, however, yield much more basic information on the nature of the soil complex than all the other procedures that have been discussed. Depending upon the pH, exchange acidity or alkalinity will develop. Suppose, however, that the pH of the salt solution is adjusted by adding acid to the salt solution so that the displacement of OH^- from the complex by anions of the salt solution just balances the displacement of H^+ by the cation of the salt solution. When this condition is fulfilled, the pH of the salt solution is not affected by the colloid; this soil point was formerly designated as "pH of exchange neutrality," but later was called the "equi-ionic point" (14). In general, the equi-ionic point is independent of the colloid content of the soil within a rather wide range. It is to be expected that the stronger the acidoid residue the lower the equi-ionic, ultimate, and isoelectric points. Some use has been made of these soil constants by pedologists (11) with varying results and opinions.

DYNAMIC CONCEPT OF SOIL COLLOIDS

The presentation of information concerning the composition, nature, and behavior of soil colloids up to this point has principally stressed the importance of the clay-mineral and amphoteric theory of soil colloids. From the viewpoint of colloidal behavior of "cleaned residues" of soil colloids and a knowledge of the transport of sols, it becomes somewhat difficult to accept the clay-mineral concept completely. The changes in ion adsorption, flocculation, and electrokinetic behavior, as well as other properties of the "cleaned residues" of isolated soil clays, emphasize these differences. Further, the rapid changes that occur in the colloidal behavior of soils upon addition of organic residues raise some doubt. It is also unlikely that the amphoteric concept will adequately explain soil colloidal behavior, especially since crystalline clay minerals have been identified in them.

The question arises as to how either the clay mineral or amphoteric theory can be modified or altered to describe adequately the colloidal behavior of soil colloids in situ. Several attempts have been made in recent years to do so, and in general these approaches picture the soil colloidal particle as consisting of a crystalline core (clay minerals), which forms the main mass of the clay particle; gel-like silicates of indefinite composition; and humus matter which may or may not be in combination with Fe and Al. The last two phases are attached to the surface of the silicate core by both physical and chemical means, and it is the combination of all three phases which can be designated as a "surface complex," that is, responsible for the dynamic nature of the clay particle. Depending upon changes in the environment, the composition of the surface complex will be modified to fit the prevailing pH. As was pointed out earlier, these modifications will alter the colloidal properties of the particles.

REFERENCES

1. Alexander, L. T., Hendriks, S. B., and Nelson, R. A., *Soil Sci.* **48**, 173–279 (1939).
2. Anderson, M. S., and Byers, H. G., *U. S. Dept. Agr. Tech. Bull.*, 377 (1933).
3. Baver, L. D., and Hall, N. S., *Missouri Agr. Expt. Sta. Research Bull.* 267 (1937).
4. Bergna, H. E., *Trans. Intern. Congr. Soil Sci. 4th Congr.* **3**, 75–80 (1950).
5. Demolon, A., and Barbier, G., *Compt. rend.* **188**, 654–656 (1939).
6. Ensminger, L. E., and Gieseking, J. E., *Soil Sci.* **48**, 467–473 (1939).
7. Grim, R., "Clay Mineralogy," New York, McGraw-Hill Book Company, Inc., 1953.
8. Hendriks, S. B., *J. Geol.* **50**, 276–290 (1942).
9. Jackson, M. L., and Hellman, N. N., *Soil Sci. Soc. Am. Proc.* **6**, 133–145 (1942).
10. Jenny, H., and Reitemeier, R. F., *J. Phys. Chem.* **39**, 593–601 (1934).
11. Kardos, L. T., and Bowlsby, C. C., *Soil Sci.* **52**, 335–350 (1941).
12. Kelley, W. P., "Cation Exchange in Soils," New York, Reinhold Publishing Corp., 1948.
13. Marshall, C. E. "The Colloid Chemistry of the Silicate Minerals," New York, Academic Press Inc., 1948.
14. Mattson, S., *Soil Sci.* **28–44** (1929–1937).
15. McGeorge, W. T., *Arizona Agr. Expt. Sta. Tech. Bull.* **30** (1930).
16. McGeorge, W. T., *Arizona Agr. Expt. Sta. Tech. Bull.* **31** (1931).
17. Mehlich, A., *Soil Sci.* **62**, 373–405 (1946).
18. Meyers, H. E., *Soil Sci.* **44**, 331–359 (1937).
19. Toth, S. J., *Soil Sci.* **44**, 1–10 (1937).
20. Toth, S. J., *Soil Sci.* **53**, 265–272 (1942).
21. Tyulin, A. T., *Soil Sci.* **34**, 1–10 (1938).
22. Waksman, S. A. "Humus," 2nd ed., Baltimore, The Williams and Wilkins Co., 1938.

CHAPTER 4

Cation and Anion Exchange Phenomena

LAMBERT WIKLANDER

Department of Pedology
Agricultural College of Sweden, Uppsala

By *ion exchange* is meant the reversible process by which cations and anions are exchanged between solid and liquid phases, and between solid phases if in close contact with each other. For the exchange of cations, the term *base exchange* has been widely used. Since we know by now that the hydrogen ion always takes part in cation exchange in soils and, principally, in a way similar to that of other cations, the process is more adequately expressed by the term *cation exchange* as distinguished from *anion exchange*.

Other terms to be used in this survey are ion *adsorption*, meaning an increase in concentration or an accumulation of an ion species on a solid, caused by ion exchange or other reactions; and *desorption*, which is the opposite process, that is, the replacement or release of an adsorbed ion species.

The soil is a heterogenous, polydisperse system of solid, liquid, and gaseous components in various proportions. The solid component of the soil is made up of primary minerals, clay minerals, and hydrous oxides, together with organic matter and living organisms, forming a polyphase system of more or less discrete particles or aggregates. In this heterogenous system, the soil solution acts as the medium by which chemical reactions between members of the different phases and of the same phase are made possible even when the reactants are not in direct contact. Because of the property of the soil material to bind and exchange cations and anions, many of these reactions are facilitated or made feasible. Chemical and physical processes more or less intimately connected with ion exchange include weathering of minerals, nutrient absorption by plants, swelling and shrinkage of clay, and leaching of electrolytes. Ion exchange may therefore be considered as the most important of all the processes occurring in a soil.

In this short survey only the most important aspects of ion exchange can be dealt with. Experimental data are included only to the extent necessary for the illustration of exchange processes and theories.

ORIGIN OF ION-EXCHANGE PROPERTIES OF SOIL

From numerous investigations on soils and clays it is well known that the ion-exchange property of a soil is due almost entirely to the clay and silt fractions ($<20\mu$) and the organic matter, the colloidal material of the soil being most important. The composition and general properties of the clay fraction and the organic matter are dealt with in Chapters 2, 3 and 5 of this monograph.

The soil particles have an amphoteric character as evidenced by their power to bind both cations and anions. In general, they carry a net negative charge, which can be directly demonstrated by electrophoresis. This charge arises essentially in two ways: first, by isomorphous ion substitutions and, second, by ionization of hydroxyl groups attached to silicon of broken tetrahedron planes, in the same way as for ordinary silicic acid, that is, $Si—OH + H_2O = SiO^- + H_3O^+$. The negative charges created in the former way are more uniformly distributed in the plate- or lath-shaped clay particles, whereas the latter are at corners and along edges. In addition to these two ways of formation, negative charges may also originate from humic (—COOH, —OH), phosphoric, and silicic acids, constituting a more or less integral part of the clay particle surface. The relative part played by these different ways of negative charge formation varies in the soil with the mineral composition, particle size, organic matter status, and, to some extent, phosphate fertilization.

The amphoteric nature of the clay fraction requires, however, the existence of positive charges. These may originate from hydrous oxides of iron, aluminum, and manganese, and from exposed octahedral groups, which react as bases by accepting protons from the surrounding soil solution, thus acquiring a positive electric charge. The basic groups of humus are due to nitrogen. The existence of positive and negative charges on the same mineral has been actually demonstrated by Thiessen, working with the adsorption of negative and positive gold sols on kaolinite. Electron microscope pictures showed adsorption of the colloidal gold particles on the edges of the kaolinite plates.

The electric charge and, at the same time, the surface charge density are not constant but vary more or less with the pH of the solution. The

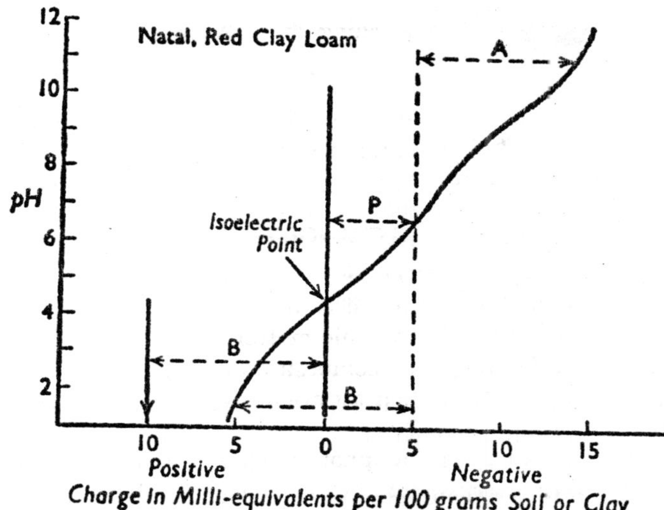

Figure 4.1. Change of the net negative charge on a clay soil caused by variation of the pH (titration curve). *P* is the permanent negative charge on the clay; *A*, the additional negative charge developed at high pH; *B*, the positive charge developed at low pH (32).

negative charge grows and the positive charge decreases with rising pH as a result of increasing ionization of the acid groups and decreasing proton addition to the basic groups. The change goes in the opposite direction if the pH decreases. In most cases the negative charge of the clay particle exceeds the positive charge, resulting in a net negative charge as manifested by electrokinetic phenomena.

Of interest in this connection are the findings of Schofield (35), who showed that in the pH range of 2.5 to 5 the permanent negative charge of a Rothamsted subsoil clay is constant, probably because of isomorphous ion substitutions. Ionization of hydroxyl groups is slight at pH 6 but considerable at pH 7. Figure 4.1 shows the relationship between the permanent negative charge and that formed by ionization on Natal red clay loam.

The decrease of the net negative charge below pH 6 is supposed to be caused not by an actual decrease of the negative charge itself but by an increase in the positive charge, which at a certain pH balances the negative charge and renders the clay particle isoelectric.

The carboxyl groups of humus ionize in the acid part of the pH scale, and the phenolic hydroxyls mainly above pH 6. The basic groups are weak, and since the acid properties predominate, humus never becomes isoelectric.

Because of the irregular shape of the clay particles and the nonuniform distribution of charges throughout the particles, the surface charge density varies not only with the kind of mineral but also on the same clay particle. Thus, the charge density and the potential are higher on edges and corners and in furrows and cavities than on plane surfaces; for soils in general, they vary not only with the nature of the soil colloids but also with pH, and hence, with liming, because of varying degree of ionization. As discussed later, this factor may exercise a considerable influence upon the ion-exchange properties of the soil material.

Kind of Exchangeable Ions and Exchange Capacity

The electric charge on the soil particles is neutralized by an equivalent amount of oppositely charged ions, the so-called *exchangeable*, *counter-* or *gegen* ions, held to the surface by mainly Coulomb forces. In addition to the Coulomb forces there seem to exist specific forces, Van der Waals-London forces, increasing the strength of bonding of certain ions.

In soils the most common exchangeable cations are Ca^{++}, Mg^{++}, H^+, K^+, Na^+, and NH_4^+, Ca generally being the dominant ion, with the relative abundance of the other ions varying greatly. In very acid soils $Al(OH)_2^+$ may constitute a considerable part of the counterions, the proportion increasing with falling pH. In alkali soils the content of Na is exceptionally high. Common anions are SO_4^{2-}, Cl^-, NO_3^-, $H_2PO_4^-$, HPO_4^{2-}, HCO_3^-, and anions of humic acids, though some of these anions do not always function as exchangeable ions but are merely present in the soil solution. In acid saline or sulfate soils, the content of SO_4^{2-} is very high.

The capacity of soils to adsorb and exchange cations and anions varies greatly with the content of clay and organic matter and the mineralogical composition. The cation-exchange capacity (CEC) is defined as the amount of a cation species bound at pH 7 or another suitable pH depending on the method used for its measurement. It varies slightly with the bonding strength of the ion and increases with the content of clay and organic matter. The CEC of mineral soils may range, in accordance with the clay content, from a few to 50 or 60 meq per 100 g, whereas the CEC of organic soils may exceed 200 meq per 100 g. The anion-exchange capacity is discussed in connection with the exchange of anions.

Table 4.1 gives cation-exchange capacities and major exchangeable ion contents of various soils.

STRUCTURE AND PROPERTIES OF THE DIFFUSE DOUBLE LAYER

Because of thermal motion, the counterions are distributed within a certain space, forming a diffuse layer or ion swarm, the structure of which

TABLE 4.1. CATION-EXCHANGE CAPACITIES (CEC) AND MAJOR EXCHANGEABLE IONS OF VARIOUS SOILS

Soils	pH	CEC (meq/100 g)	% of Total				
			Ca	Mg	K	Na	H
Chernozem soil*	7	56.1	84.3	11.0	1.6	3.0	—
Holland soils	7	38.3	79.0	13.0	2.0	6.0	—
California soils	7	20.3	65.6	26.3	5.5	2.6	—
Merced clay loam†	10.0	18.9	0	0	5.0	95.0	0
Lanna, unlimed‡	4.6§	17.3	48.0	15.7	1.8	0.9	33.6
Lanna, limed‡	5.9§	20.0	69.6	11.1	1.5	0.5	17.3

* Russia.
† U. S. A.
‡ Sweden.
§ In 1 *N* KCl.

for a given particle is determined by the surface charge density, kind of counterions, temperature, and concentration of electrolytes in the solution. The exchangeable ions are surrounded by water molecules and may thus be considered as forming a solution which is often called a *micellar solution* or *inner solution* in distinction to the outer solution of free electrolytes, the so-called *intermicellar* or *outer solution*.

The structure of the double layer is of great interest for an understanding of the colloidal behavior of such systems. Several physicists and chemists have focused attention on this problem. In 1879, Helmholtz advanced an electrokinetic theory of the double layer at a charged plane surface. The particle surface with counterions was presumed to behave like a condenser, the charge of the particle forming the inner layer and the counterions the outer layer close to the former (Figure 4.2).

Gouy, in 1910, and Chapman, later but independently, developed the theory of the diffuse double layer at a plane surface (Figure 4.2). The concentration of the counterions is highest in the immediate vicinity of the surface and decreases at first rapidly and then asymptotically to the intermicellar solution of uniform composition. Because of its fundamental importance to an understanding of the nature of the diffuse double layer, the main features of the Gouy-Chapman theory are given here. A detailed discussion is presented by Verwey and Overbeek (42).

The change in concentration of counterions follows Boltzmann's distribution law

$$n_j = n_{jo} \exp\left(-\frac{z_j e\psi}{kT}\right), \qquad (1)$$

where n_j means number of counterions of kind j per cubic centimeter at an

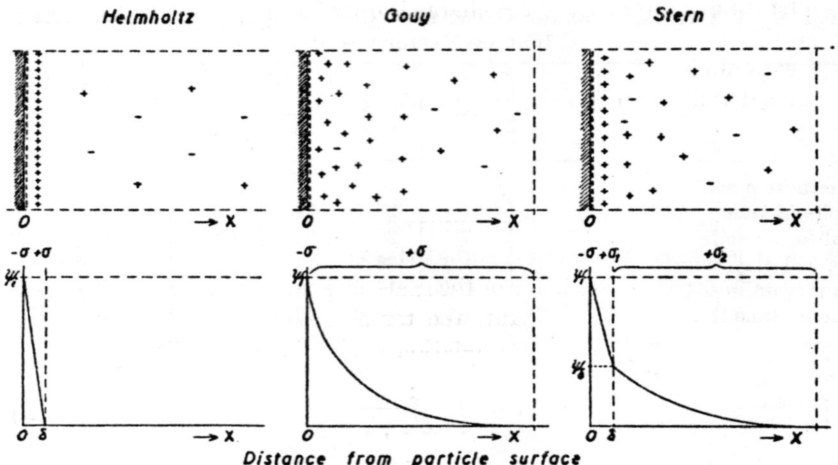

Figure 4.2. Schematic representation of ion and potential distribution in the double layer according to the theories of Helmholtz, Gouy, and Stern. ψ_i denotes total potential; ψ_δ, zeta potential; x, distance from particle surface; σ, surface charge density; δ thickness of Stern layer.

arbitrary point in the diffuse double layer, where the potential is ψ; n_{jo} is the corresponding concentration in the outer solution; z_j is the valence of ion j; e, the charge of proton; k, the Boltzmann constant; and T, the temperature.

The relation between the potential ψ, which changes from ψ_i at the interface to $\psi = 0$ in the outer solution, and the space net charge density ρ is given by the Poisson equation, which for a direction perpendicular to the surface reads

$$\frac{d^2\psi}{dx^2} = -\frac{4\pi\rho}{\epsilon}, \tag{2}$$

where x means distance from the surface and ϵ the dielectric constant. This formula also expresses the Coulomb interionic forces.

The space net charge density is equal to the algebraic sum of positive and negative charges per cubic centimeter, thus

$$\rho = \sum z_j e n_j \tag{3}$$

Consequently, in the intermicellar solution $\rho = 0$. Further, the space net charge is related to the surface charge density by

$$\sigma = -\int_0 \rho \, dx \tag{4}$$

meaning that the total space net charge is opposite and equal to the surface charge.

By a combination of equations (*1*), (*2*), and (*3*), a differential equation is obtained which can be integrated under the condition

$$\text{if } x \to \infty, \quad \psi \to 0 \text{ and } \frac{d\psi}{dx} \to 0,$$

expressing that $\psi = 0$ outside the diffuse double layer. Such a mathematical treatment leads to the conclusion that the "thickness" of the double layer (l), that is, the center of gravity of the space charge, is a function of valency, concentration, dielectric constant, and temperature. At room temperature and molar concentration (c), the relation reads

$$l \sim \frac{1}{3 \cdot 10^7 z \sqrt{c}} \tag{5}$$

It appears from equation (*5*) that the thickness of Gouy's diffuse double layer decreases with increasing valence and concentration of the electrolyte in solution. Owing to the neglect of the ionic dimensions, the Gouy theory is valid only for colloids having a low surface-charge density, as in the plane surfaces of the clay minerals, and at very low concentrations of the solution.

In a theory of Stern the picture of the double layer (Figure 4.2) has been further advanced by considering the ionic dimensions. Stern's double layer consists of an inner part, similar to Helmholtz's layer, and acting like a flat molecular condenser, the thickness of which (δ) depends on the volume of the adsorbed ions present in a monolayer. The outer part of the ion swarm is supposed to have a structure as pictured by Gouy. By the adsorption of part of the counterions in an inner layer, the space charge density and the potential are decreased to the extent that the application of the Gouy theory to the diffuse part of the double layer gives more reasonable results at higher concentration than does the pure Gouy theory. The charge density of the particle surface (σ) is equal to the sum of charges in the inner layer (σ_1) and in the outer layer (σ_2) or

$$\sigma = \sigma_1 - \int_0^\infty \rho d(x - \delta) \tag{6}$$

In the same way the total potential drop (ψ_i) is composed of an inner potential drop ($\psi_i - \psi_\delta$), and an outer one (ψ_δ). ψ_δ is the same as the electrokinetic or Zeta potential.

The ratio of ψ_δ to ψ_i varies with the concentration of the solution and with the valence and bonding strength of the counterion. Increasing concentration decreases this ratio, and so do higher valence and bonding strength and vice versa. Thus, it is likely that increasing concentration tends to change the Stern double layer in the direction of Helmholtz's layer, and dilution in the direction of the Gouy diffuse double layer. In the former case the ion swarm shrinks, and in the latter it expands.

As schematically shown in Figure 4.2 the double layer of Gouy and of Stern contains not only ions of opposite charge but also ions of the same charge emanating from the diffusible free electrolytes. Because of the thermal motion, electrolytes migrate into the diffuse layer and distribute themselves so that the concentration of counterions decreases from the particle surface outward, while the concentration of ions of the same charge as the particle increases in the same direction. The total concentration of ions decreases in the same direction, as does the osmotic pressure p. The electric attraction on the counterions and the osmotic pressure are counteracting forces which balance each other at equilibrium.

A drawback of this picture of the diffuse double layer is the fact that it is not possible to state a distinct transition line between the micellar and the intermicellar solutions. Mathematically the limit may be defined by

$$\frac{d\psi}{dx} = 0, \qquad \frac{d\rho}{dx} = 0, \qquad \frac{dc}{dx} = 0, \qquad \text{and} \qquad \frac{dp}{dx} = 0.$$

Experimentally, the volume of the micellar solution may be estimated from the anion distribution and the increase in the osmotic pressure when passing from the intermicellar to the micellar solution.

It may be objected that the foregoing discussion of the structure and properties of the double layer applies only to the large flat surfaces of the clay particles, but even though these theories are not directly applicable to edges and corners, the ion arrangement may be assumed to show about the same pattern as for a flat surface. This is strongly supported by the ion distribution as deduced from the Donnan equilibrium. For the pure soil organic matter, however, the theories of the diffuse double layer are probably not valid, partly because of the humus structure and partly because of the operation of forces other than Coulombic, leading to complex bondings.

ION-EXCHANGE FORMULAS

Because of the thermal motion of the adsorbed ions and the diffusible free electrolytes, there must be a continuous exchange of ions between the solid phase (the ion exchanger) with its surrounding micellar solution and the intermicellar solution. At equilibrium the same number of counterions migrate, per unit time, in and out of the diffuse double layer, that is, adsorption and desorption balance each other. That this occurs is easily demonstrated by isotopic exchange. By addition of a minute amount of a carrier-free salt of the ion, the equilibrium will not be disturbed and the exchange between soil and solution can be determined. The term "ion exchange," however, does not generally imply this permanently occurring process, but implies reactions involving displacement of the equilibrium state, caused either by addition of foreign ions or simply by altering the concentration of the outer solution.

To describe ion-exchange processes and predict ion distributions, several formulas have been proposed in the course of ion-exchange studies. One group is essentially empirical and mainly intended to provide mathematical expressions best fitting the experimental data. Among these formulas, Freundlich's (11) well-known adsorption isotherm may be mentioned. The Freundlich formula is of parabolic type and cannot therefore give a maximal adsorption value. The equation derived by Vageler (40) for calculation of the fertilizer requirements of soils is, on the other hand, of hyperbolic type. A theoretical formula of the hyperbolic type for ion exchange was deduced by Pauli and Valko (30).

The principle of the law of mass action has been utilized to a great extent for deriving ion-exchange formulas. It may be objected that ion exchange is not a real chemical reaction forming new compounds but merely a redistribution within the micellar solution and between it and the intermicellar solution. With our present opinion about the structure of the diffuse double layer and its relationship to the surroundings, unquestionably dependent on the active mass (the activity) of the diffusible ions and also connected with changes in free energy, we are certainly fully entitled to apply the law of mass action to the study of ion exchange.

In homovalent exchange such as

$$RK + Na^+ \rightarrow RNa + K^+$$

the equilibrium equation may be written as

$$\frac{[Na^+]_i(K^+)_o}{[K^+]_i(Na^+)_o} = k_{K,Na} \qquad (7)$$

and for monovalent and divalent ions

$$RCa + 2K^+ \rightarrow 2RK + Ca^{2+}$$

$$\frac{[K^+]_i^2(Ca^{2+})_o}{[Ca^{2+}]_i(K^+)_o^2} = k_{Ca,K} \qquad (8)$$

Brackets denote concentration; parentheses, activity; i, exchanger phase or inner solution; o, outer solution; and k, the equilibrium constant, which is no real constant but varies more or less with the mole fractions of the two ions. Because of the difficulties encountered in the determination of the activities of the ions present in the exchanger phase, concentration has generally been used for these.

As long ago as 1913 Gans and later Kerr (16) utilized the mass-action law in ion-exchange studies. Gapon's (12) and Jenny's (15) kinetically derived formulas are of mass-action type, as they are easily transformed into the same form as equations (7) and (8). Also, the Donnan equilibrium belongs to the mass-action equations. The variation of k of the mass-action equation in relation to the mole fraction and nature of the exchanging ions was particularly investigated by Vanselow (41), Møller (28), Damsgaard-Sørensen (6), and Wiklander (44). The equations derived by Vanselow and by Krishnamoorthy and Overstreet (17) have yielded satisfactory equilibrium constants for exchange of metal cations on selected bentonite clay, soil colloids, and synthetic resins, whereas Gapon's formula was found to be less suitable. But in the experiments concerned, the mole fractions of the exchanging ions were not varied much. Eriksson (10) has shown that the equations of Vanselow and of Krishnamoorthy *et al.* are formally equivalent to the Donnan equilibrium. Because of the growing attention paid to the Donnan equilibrium as a basis for elucidating various exchange phenomena, it is more fully treated in the next section.

APPLICATION OF THE DONNAN EQUILIBRIUM

The state of equilibrium of a system containing a membrane impermeable to one ion species was formulated by Donnan in 1911. For a system containing the nondiffusible electrolyte RM and the diffusible electrolyte MA, Donnan showed that the following relation holds:

$$(M^+)_i(A^-)_i = (M^+)_o(A^-)_o. \tag{9}$$

This so-called constant ion product is usually recognized as the Donnan membrane equilibrium.

Further studies have indicated that any system in which at least one ion species is restrained from freely diffusing throughout the whole system is characterized by a Donnan distribution. A colloid particle, for example, a clay particle, with its surrounding diffuse double layer may be looked upon as a micro-Donnan system, where the attractive electric forces between particle surface and counterions act as a restraint, causing a nonuniform distribution of the counterions in the micellar solution.

Derivations of Ionic Distributions

Thermodynamically, the equilibrium condition of a Donnan system is characterized by the fact that the chemical potential (μ) of every diffusible electrolyte, and the electrochemical potential of every diffusible ion species, are constant throughout the system. Thus, for a clay colloid in equilibrium with the intermicellar solution containing a strong electrolyte MA, temperature and pressure being constant, the state of equilibrium is characterized by

$$\mu_{MA_i} = \mu_{MA_o} \tag{10}$$

By use of the relation between chemical potential and activity,

$$\mu_{MA} = \overset{0}{\mu}_{MA} + RT \ln(MA) \tag{11}$$

equation (*10*) can be written

$$\overset{0}{\mu}_{MA} + RT \ln(MA)_i = \overset{0}{\mu}_{MA} + RT \ln(MA)_o$$

or

$$(MA)_i = (MA)_o \tag{12}$$

which is the general Donnan equilibrium. According to definition, (MA) = (M)(A), equation (*12*) can be written as the well-known ion product formula

$$(M)_i(A)_i = (M)_o(A)_o \tag{13}$$

Equations (*12*) and (*13*) infer that the activity of cation and anion varies in opposite directions, so that the activity of the electrolyte is constant at any point of the liquid phase.

If the electrolyte is of the general type M_nA_m, the ion product formula reads,

$$(M)_i^n(A)_i^m = (M)_o^n(A)_o^m \tag{14}$$

or

$$\left[\frac{(M)_i}{(M)_o}\right]^{1/m} = \left[\frac{(A)_o}{(A)_i}\right]^{1/n} \tag{15}$$

If distinction is made between ions originating from the colloid and the free salt, the corresponding formulas read:

$$[Z_M + Y_M]^n \cdot Y_A^m = X_M^n \cdot X_A^m \tag{16}$$

and

$$\left[\frac{Z_M + Y_M}{X_M}\right]^{1/m} = \left[\frac{X_A}{Y_A}\right]^{1/n} \tag{17}$$

where Z equals the activity of M ionized from the exchanger: Y, the activity of the ions of the free salt in the micellar solution; and X, the corresponding activity in the intermicellar solution.

For a system with more than one diffusible electrolyte, the activity of each must be constant throughout the aqueous phase. For an aqueous system containing the ions K^+, Ca^{2+}, La^{3+}, Cl^-, and SO_4^{2-}, the equilibrium is expressed by

$$\frac{(K^+)_i}{(K^+)_o} = \frac{(H^+)_i}{(H^+)_o} = \sqrt{\frac{(Ca^{2+})_i}{(Ca^{2+})_o}} = \sqrt[3]{\frac{(La^{3+})_i}{(La^{3+})_o}}$$
$$= \frac{(Cl^-)_o}{(Cl^-)_i} = \frac{(OH^-)_o}{(OH^-)_i} = \sqrt{\frac{(SO_4^{2-})_o}{(SO_4^{2-})_i}} \quad (18)$$

Because they are always present in aqueous systems, H^+ and OH^- are included. In equation (*18*) the corresponding $Z + Y$ may be substituted for $(M)_i$ and X for $(M)_o$.

Equation (*18*) may be regarded as equivalent to the mass-action formula, the equilibrium constant being equal to 1. Thus, for Na and K [cf. (*7*)]

$$\frac{(Na)_i \cdot (K)_o}{(Na)_o \cdot (K)_i} = 1$$

or by introducing concentration, f, signifying the activity coefficient

$$\frac{[Na]_i \cdot [K]_o}{[Na]_o \cdot [K]_i} = \frac{f_{K_i} \cdot f_{Na_o}}{f_{Na_i} \cdot f_{K_o}} = \alpha_{K,Na} \quad (19)$$

The equilibrium quotient α is a measure of the relative adsorption energy or replacing power of the two ions. Thus, for two ions, A and B, $\alpha_{A,B} > 1$ if A is more weakly adsorbed than B; $\alpha_{A,B} = 1$ if A and B are equally firmly adsorbed; and $\alpha_{A,B} < 1$ if A is more firmly adsorbed than *B*. Numerous experiments have conclusively shown that $\alpha_{A,B}$ and $k_{A,B}$ of equation (*7*) vary more or less with the mole fraction, the variation being the greater the more unequal the chemical properties of the ions and the greater the selectivity of the exchanger for any of the ions. For natural exchangers, it is only within limited concentration ranges that the quotient has proved to be fairly constant.

The preceding formulas were obtained from the chemical potential of the diffusible electrolytes. The equilibrium condition of one diffusible ion species is given by the electrochemical potential ($\bar{\mu}$), related to the chemical and the electrical potential by

$$\bar{\mu}_M = \mu_M + zF\psi \quad (20)$$

where z = valency of M, and F = Faraday.

For a clay suspension containing KCl, equation (*20*) may be written for the potassium ion

$$\bar{\mu}_K = \mu^0_K + RT \ln(K) + F\psi \qquad (21)$$

and for the chloride ion

$$\bar{\mu}_{Cl} = \mu^0_{Cl} + RT \ln(Cl) - F\psi \qquad (22)$$

Summation of equations (*21*) and (*22*) gives

$$\bar{\mu}_K + \bar{\mu}_{Cl} = \mu^0_K + \mu^0_{Cl} + RT \ln(K)(Cl) \qquad (23)$$

If we write

$$\frac{\bar{\mu}_K + \bar{\mu}_{Cl} - \mu^0_K - \mu^0_{Cl}}{RT} = a$$

we may write equation (*23*) as $(K)(Cl) = e^a$

From this expression it appears that the product of the ion activity is a constant, the magnitude of which is dependent on the temperature and on the electrochemical potentials of the two ions, and thus on the structure of the diffuse double layer, ion concentration, ion interaction, and interaction with the exchanger, both in electrostatic and nonelectrostatic ways.

Figure 4.3 gives a schematic picture of the ion distribution around a negative particle according to the Donnan equilibrium.

Obviously the activity of the counterions decreases rapidly from the particle surface outward, whereas the activity of the free salt increases in the same direction. For a salt MA, the ion activity distribution at any plane parallel with the Y axis is such as to satisfy the Donnan equation

$$(Z_M + Y_M)Y_A = X_M \cdot X_A$$

A measure of the unequal ion distribution is given either by the anion ratio of X_A to Y_A or the cation ratio of $Z_M + Y_M$ to X_M, interrelated in equation (*17*). The unequal ion distribution may be summarized as follows:

For negative colloids: $(M)_i > (M)_o$, $(A)_i < (A)_o$ and $pH_i < pH_o$.

For positive colloids: $(M)_i < (M)_o$, $(A)_i > (A)_o$ and $pH_i > pH_o$.

For all colloids, irrespective of charge: $\sum[M]_i + \sum[A]_i > \sum[M]_o + \sum[A]_o$.

For a qualitative evaluation in this respect of equation (*17*), the activity of the adsorbed ions, Z_M, may be replaced by $m(\text{CEC}/v)$, m denoting grams of exchanger in the system, CEC the exchange capacity in meq/g of exchanger, and v the volume of micellar solution. Thus, for a salt MA, equation (*17*) becomes

$$\frac{X_A}{Y_A} = \frac{\dfrac{m \cdot \text{CEC}}{v} + Y_M}{X_M} \qquad (24)$$

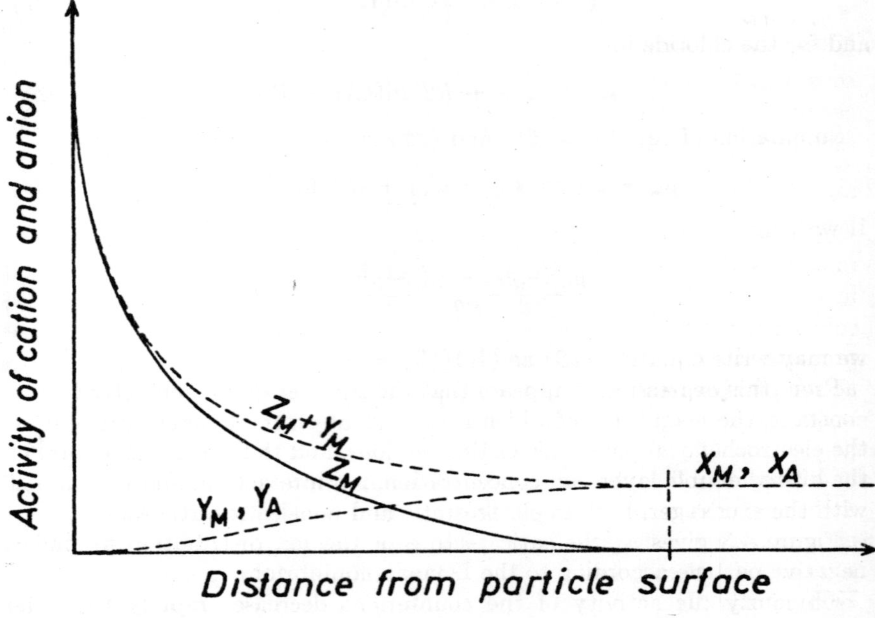

Figure 4.3. Schematic representation of the ion activity distribution according to the Donnan equilibrium around a negatively charged particle, saturated with the ion M and in equilibrium with the salt MA. Z refers to activity of counterion; Y, activity of free salt in inner solution; and X, activity of salt in outer solution.

From equation (*24*) it may be concluded that the unequal ion distribution increases with dilution of the system, that is, X_M decreases proportionally to the dilution, while $m \cdot \text{CEC}/v$ changes only slightly, and reaches its maximum if no salt MA is added. In this case, proteolysis takes place until the equilibrium conditions are fulfilled, thus for M^+

$$\text{RM} + \text{HOH} = \text{RH} + \text{M}^+ + \text{OH}^-$$

and

$$(Z_M + Y_M)Y_{OH} = X_M \cdot X_{OH}$$

This proteolysis can be minimized by alcoholic treatment of the soil, a method often applied in soil analyses. The uneven ion distribution diminishes with increasing salt concentration (that is, salts suppress the Donnan distribution), but grows with the exchange capacity. Consequently, for clay minerals it should decrease in the order: montmorillonite > illite > kaolinite, since the exchange capacity diminishes rapidly in this order.

A consequence of the unequal ion distribution is the increase in the anion concentration when salt solutions are added to clay minerals and to soils which do not adsorb the anions in question. This phenomenon is called negative adsorption and is dealt with in connection with anion exchange.

Osmotic Pressure and Membrane Potential

Another consequence of the unequal ion distribution is the higher osmotic pressure in the micellar than in the intermicellar solution—resulting in an osmotic pressure difference between a clay gel and its equilibrium solution and in the swelling of many colloids. The difference in osmotic pressure is proportional to the difference in concentration of osmotically active particles and may thus be formulated

$$P = RT(Z + 2Y - 2X) \tag{25}$$

A further indication of a heterogeneous ion arrangement between the exchanger phase and the equilibrium solution is given by the existence of a potential difference between a clay suspension and its equilibrium solution, if measured as the chain: Hg/calomel/KCl/suspension/solution/KCl/calomel/Hg. This so-called membrane or Donnan potential is of the same magnitude as the concentration potential but of opposite sign, the latter being obtained as the difference in potential between suspension and solution, each one measured separately. Consequently, if two electrodes, reversible to any of the diffusible ion species, are inserted in a clay suspension and in its equilibrium solution, separated by a membrane, no potential difference will be found. The same thing happens in a system without membrane consisting of a gel with supernatant liquid, for example, H bentonite. This fact is also to be expected, since the systems are at equilibrium and the membrane and concentration potentials thus counterbalance each other.

The two potentials have been calculated from:

$$E + \frac{RT}{zF} \ln \frac{X_A}{Y_A} = \frac{RT}{zF} \ln \frac{Z_M + Y_M}{X_M} \tag{26}$$

or if KCl is present

$$E = \frac{RT}{F} \ln \frac{(\mathrm{Cl})_o}{(\mathrm{Cl})_i} = \frac{RT}{F} \ln \frac{(\mathrm{K})_i}{(\mathrm{K})_o}$$

Relationship Between the Donnan and Gouy Treatments

The application of the Donnan membrane equilibrium to the study of ion-exchange reactions in clays and soils has long been a subject of discussion and has also raised some criticism. But use of the Donnan theory has been rapidly progressing since Mattson's pioneering work in 1929 (22). One advantage of the Donnan equilibrium over most of the other exchange formulas is that it describes the distribution of all the ions present, cations as well as anions.

It has been asserted that the Donnan theory assumes a homogeneous volume charge distribution in the clay suspension, whereas the Gouy theory assumes a heterogeneous distribution (Figure 4.2). The two theories should therefore be more or less incompatible (2). As a matter of fact this statement is not true, since the Donnan theory, as applied to colloid particles, also postulates the existence of a diffuse double layer, as does the Gouy theory, and a heterogeneous space arrangement of the exchangeable ions around the clay particles. In ion exchange, we are generally less interested in the ion distribution between a suspension and its equilibrium dialyzate than in the partition between the exchanger phase and its equilibrium solution. Analyses and calculations are therefore based on a separation of the exchanger phase and the equilibrium solution. As a matter of fact, *the two theories represent only different ways of solving the same problem, the Donnan theory, unlike the Gouy, makes no assumption about the structure of the exchanger.*

ACTIVITY OF EXCHANGEABLE IONS

As has been mentioned, all the theoretical ion-exchange formulas of the mass-action type require activity instead of concentration. But the importance of the activity concept reaches far beyond this particular sphere. Knowledge of the activity of the exchangeable ions is fundamental for a proper understanding of the general physicochemical behavior of soils and the interaction between plant roots and soil. Nutrient uptake by plants, loss of nutrients by leaching, fixation of ions in nonexchangeable form, weathering, formation of clay minerals, soil profile development, and other processes are almost certainly intimately connected with the ion activity conditions in soils.

The activity of ions present in the true soil solution phase may be experimentally determined or calculated in the same way as for pure solutions. In computing the activity coefficients, the writer has suggested consideration of the ionic diameter by applying the more complete Debye-Hückel formula

$$-\log f_{M} = \frac{a \cdot z_{M}^{2} \cdot \sqrt{I}}{1 + b \cdot d \cdot \sqrt{I}} \tag{27}$$

where a and b are constants, I refers to ionic strength, and d is the effective

TABLE 4.2. MOST PROBABLE HYDRATION NUMBERS ACCORDING TO ULICH, AND MEAN EFFECTIVE DIAMETERS (d) IN ÅNGSTRÖM AS GIVEN BY KIELLAND

Ion	H^+	Li^+	Na^+	K^+	Rb^+	Mg^{2+}	Ca^{2+}	Ba^{2+}
H_2O/ion.....	4	6	4	2.5	2	9–13	8–10	6–8
d............	9	6	4.3	3	2.5	8	6	5

mean diameter of the hydrated ion species or, in calculation of the mean activity coefficient, the closest approach of the oppositely charged ions. Certain d values are given in Table 4.2. The calculated activity coefficient will in this way include the effect of the ion hydration and thus yield higher f values with increasing hydration.

On the other hand, determination of the activity of the adsorbed ions has so far proved very difficult. Qualitatively it is possible to draw certain conclusions about the variation of f when an ion migrates from the intermicellar solution toward the particle surface. From the description of the diffuse double layer it is clear that, in systems with low salt concentration, the ionic strength increases rapidly in this direction and reaches its maximum at the very surface of the particle. According to the theory of ionic interaction, the activity coefficient of the counterions may be assumed to decrease in the same direction, attaining a minimum value close to the surface. This is particularly true for these ions, as there are indications that oppositely charged ions diminish the activity of a given ion species even more than does the total ionic strength. Therefore, f will be a function of the ionic environment and the nature of the exchanger. A small f means high bonding energy and high replacing power of adsorbed ions. The formation of an innermost firmly adsorbed counterion layer similar to the Stern layer may take place.

As a consequence of the unequal surface charge density and structure of the clay particles, it follows that the activity coefficient and the bonding energy of an exchangeable ion may vary from one spot to another, for example, for an ion held on a plane surface, on an edge or a corner, or between inner surfaces of the layer minerals. This fact helps to explain the variation of $k_{A,B}$ of the mass law with the mole fraction often exhibited in exchange experiments. It should be noted that the exchangeable ions are here regarded as completely ionized, and their activity is thus determined by the volume concentration of the micellar solution and the mean f value.

According to equation (*27*) the decrease of f with rising ionic strength is faster, the higher the valence and the less the diameter of the ion. This formula is, of course, not strictly valid for the exchangeable ions, but the inferences drawn may show the general trend. As a matter of fact, experimental measurements of ion activity in clay suspensions have shown a much greater reduction of the activity of divalent than of monovalent cations. Not considering other possible factors influencing the activity of exchangeable ions, equation (*27*) allows the assumption that the reduction of f is in the order: Li < Na < K < Mg < Ca < Ba < La, when these ions move from the intermicellar solution to the exchanger surface. For a 4 to 5 per cent Putnam clay suspension, Marshall found that the activity ratio of Ca and K was 0.15 when the molar ratio was 1.5.

By assuming $f = 1$ for an ion species when present alone and by application of the Gibbs-Duhem law, Skogseid, Sillén, and others calculated from equilibrium measurements on resins the variation of the activity coefficients of an ion pair with the molar fractions. Based on the arbitrary assumption of $f = 1$, this yields, however, only relative values, which therefore have limited significance.

Activity from the Donnan Equilibrium

Experimental determination of the activity of single ion species seems to be more promising. Several procedures have been developed. An indirect method was employed by Schuffelen and Loosjes (37) for activity measurements in suspensions. For determination of, say, potassium, the following relation was utilized:

$$pH_{susp.} - pH_{sol.} = pK_{susp.} - pK_{sol.} \qquad (28)$$

From the pH measured in suspension and solution and the pK calculated or determined in the solution, pK of the suspension was calculated.

This method is a direct application of the Donnan equilibrium, which appears by a comparison of equations (*28*) and (*18*). The crucial point of Schuffelen's procedure is the accuracy and interpretation of the suspension pH. As is well known, the pH of clay and soil suspensions is influenced by the ratio of solid to water, the so-called suspension effect of Wiegner and Pallmann. They established by potentiometric and inversometric measurements a linear relationship, within certain limits, between the hydrogen ion concentration and the content of colloids in the suspension. The significance of the calculated $(K^+)_{susp.}$ is naturally subject to the same uncertainty as the pH value. The inconsistent opinions of several investigators as to the significance of potentiometric measurements of ion activities in clay and soil suspensions and the influence of liquid junction potentials make the situation somewhat confusing. To what extent the $(K^+)_{susp.}$ is relevant to natural soils is an unsolved and still more delicate problem.

The Membrane Electrode Method

A direct method for ion activity determinations in suspension has been developed by Marshall (21). He uses specially prepared clay-membrane electrodes for measuring the activities of Na, K, NH_4, Mg, Ca, and Ba with a saturated calomel electrode as reference electrode. The clay membranes are reversible either to cations in general or to monovalent cation only. Because of lack of specificity, the measurements have been limited to homoionic systems and to systems containing monovalent and divalent cations. By such measurements, reported in numerous papers, Marshall and co-workers determine the ion activity as a function of the degree of saturation (titration curves of H clays, K—Ca and K—Mg clays), the degree of ionization* of exchangeable ions, and the mean free bonding energy ΔF. The bonding energy for a monovalent ion M is obtained from the formula

$$\Delta F_{\mathrm{M}} = RT \ln \frac{[\mathrm{M}]}{(\mathrm{M})} \tag{29}$$

(M) meaning measured activity and [M] the total concentration, that is, the sum of M in ionized and nonionized condition. Using the relation $(\mathrm{M}) = f_{\mathrm{M}} \cdot [\mathrm{M}]$, equation (*29*) is changed to

$$\Delta F_{\mathrm{M}} = RT \ln f_{\mathrm{M}}^{-1} \tag{30}$$

Marshall calls f_{M} the fraction active. Figure 4.4 illustrates these relationships.

Other investigators working along the same line as Marshall have used membranes of exchange resins and collodion for activity measurements.

Membrane Potential Measurements

Marshall's method for activity measurement is limited to systems containing one or two cation species. But in natural clays and soils, which always contain several exchangeable ion species, the possibility of determining all the cations present is of great value. Peech and Scott (31) have attempted to do this by measuring the membrane potential (E) between a montmorillonite suspension and its equilibrium solution in the manner described. The solution is analyzed in the ordinary way and from the knowledge of the ionic strength, $(\mathrm{M})_{\mathrm{sol.}}$ is calculated. Then $(\mathrm{M})_{\mathrm{susp.}}$ is obtained from the following equation:

$$E = \frac{RT}{zF} \ln \frac{(\mathrm{M})_{\mathrm{susp.}}}{(\mathrm{M})_{\mathrm{sol.}}} \tag{31}$$

* Marshall considers the exchangeable ions as only partly ionized.

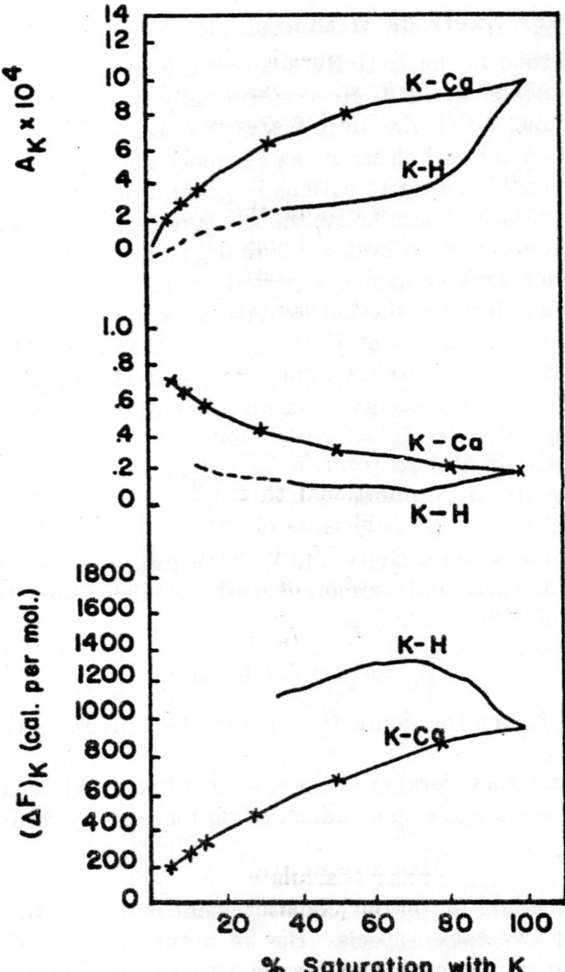

Figure 4.4. Comparison of K-H and K-Ca systems for 0.5 per cent Wyoming bentonite. (*Upper pair*) K^+ activities; (*middle pair*) fractions of K^+ active (f); (*lower pair*) $(\Delta F)_K$ (21).

By assay of the solution for other cations present, their activity in the suspension may be computed by means of the Donnan equation.

Marshall's and Peech's procedures are subject to the same sources of error as Schuffelen's method. They work with very dilute suspensions, and the measured activity is a function of the intermicellar solution and the exchanger phase rather than of the latter alone. The bearing of the results upon natural soils is therefore questionable. This is particularly true of systems containing ions of different valence where the ion distribution is markedly affected by dilution of the system.

Other Procedures

Another procedure is to determine the relative amounts of ions simultaneously released in a minute exchange. The writer has shown theoretically and experimentally that the relative release of the two homovalent ions, A and B, by a minute exchange is given by

$$\frac{A_o:A_t}{B_o:B_t} = \alpha_{A,B} \tag{32}$$

where A_o, and B_o refer to amounts replaced and A_t, and B_t to total amounts present in moles. Formula (*32*) means that the simultaneously replaced amounts are proportional to the activity of the adsorbed ions modified by the activity coefficients of the ions in the solution. This appears more clearly if $f_{B_o} \cdot f_{A_i} / f_{A_o} \cdot f_{B_i}$ is substituted for $\alpha_{A,B}$ [cf. equation (*19*)] and both sides of equation (*32*) are multiplied by $A_t:B_t$, yielding

$$\frac{A_o}{B_o} = \frac{f_{B_o}}{f_{A_o}} \cdot \frac{f_{A_i} \cdot A_t}{f_{B_i} \cdot B_t}$$

The terms f_{B_o} and f_{A_o} are easily calculated in dilute solutions. If the activity of one of the ions, for example H, can be measured, the other ion is computable without analysis of the exchanger phase. This method is applicable to systems with several monovalent ions.

The most indisputable procedure at present is probably to analyze both solution and exchanger phase, calculate f_o according to Debye-Hückel and apply the Donnan equilibria [equation (*19*)] for computation of the f_i ratios. By this procedure, only activity ratios of the adsorbed ions are obtained, but monovalent and divalent ions may be included, and disputable potentiometric measurements are not necessary.

It should be mentioned that Low (19) discusses two kinds of activity functions, the activity (as treated above) and the total activity. He considers the latter to be of more significance in soil systems than the former.

RATE OF ION EXCHANGE

Experiments by Way, Gedroiz, Hissink, Wiegner, and others have shown that the rate of cation exchange in soils is generally rapid, requiring only a few minutes for equilibrium (3). The exchange is a surface reaction and proceeds just as fast as ions from the solution are supplied to the exchanger surface either by diffusion or shaking. A condition for rapid attainment of equilibrium is, of course, that the soil be moist before the exchange, as the moistening process may take some time. But a rapid exchange is feasible only if the exchangeable ions are directly accessible to the ions in solution. If the exchange has to be preceded by diffusion of the ions to the inner surfaces of the clay particles or the interior of aggregates with narrow pores, the exchange may require considerably more time. The rate is very rapid for kaolinitic mineral, the exchange taking place mainly on the edges. In montmorillonitic minerals most of the exchange occurs between the sheets, and the rate of exchange is therefore likely to be dependent on the interlayer spacing or degree of swelling, that is, on the nature of the exchangeable ions. In swelled montmorillonite the rate is high, though probably not so high as in kaolinite. In illite the rate of exchange is slower because of the exchange that takes place along the edges between the narrow interlayer surfaces. This exchange is slow and is difficult to distinguish from fixation and defixation processes.

In this connection it should be pointed out that to reach equilibrium in the neutralization of acid clays, humus, and soils with alkali requires much longer time than does ordinary ion exchange, often many hours or even days. It probably involves two separate reactions: first, a rapidly proceeding exchange of H and Al ions and neutralization of the acidity, consuming the major part of the alkali; and second, a slow interaction of the alkali and the silicates, leading to a partial destruction of the lattice at high pH values. In organic matter, pH increase enhances oxidation processes and production of acid groups.

In ion exchange it has sometimes proved difficult to reach a true equilibrium state. For the oppositely directed exchange

$$\text{(I) R—A} + \text{B} = \text{R—B} + \text{A}, \quad \text{and} \quad \text{(II) R—B} + \text{A} = \text{R—A} + \text{B}$$

the equilibrium state is characterized by the quotients $k'_{A,B}$ and $k''_{A,B}$ which, however, may not attain the same value within a reasonable time. This phenomenon is called hysteresis and has been encountered to varying degrees with soils, clay minerals, permutite, and even resins. Wiegner and co-workers (43) studied the simultaneous release of Ca and NH_4 from Ca—NH_4 clays prepared either by adsorption of NH_4 on Ca clay or adsorption of Ca on NH_4 clay, the procedure otherwise being the same. Both NH_4 and Ca proved to be more difficult to replace if present initially upon the clay than if introduced later. The properties of the exchanger, therefore, seem to be affected by the ion with which it was originally saturated, particularly on drying. Hysteresis occurs more frequently for heterovalent than for homovalent ions. Working with bentonite, Vanselow (41) found strong hysteresis for Ca—NH_4 but none for Ba—Ca and Ba—Cu.

For ions showing hysteresis, the ion species first present on the exchanger becomes partly fixed and not altogether available for exchange. High temperature and rehydration have often proved to be a means of reducing or overcoming hysteresis. No fully satisfactory explanation has yet been given.

INFLUENCE OF TEMPERATURE

Temperature has been found to increase slightly the rate of the rapid ion exchange. For the slower exchange, temperature is likely to have a greater effect, since ion diffusion and, for anions, processes of chemical nature may be involved. In the isotopic exchange

$$RH_2P^*O_4 + H_2PO_4^- \rightleftharpoons RH_2PO_4 + H_2P^*O_4^-,$$

in two clay soils and one organic soil, a great increase was found (45) in the rate of exchange when the temperature was raised from 20 to 45°C.

It has been suggested that ordinary cation exchange be performed at elevated temperature to speed up the rate and overcome hysteresis effects. But such a procedure is not generally advisable, as an elevation of the temperature may induce fixation of the ions to varying degrees, thereby changing the ion proportions directly taking part in the equilibrium. The exchange capacity and the organic matter may also undergo certain changes, and the solubility of certain constituents increases.

The rate of anion exchange is investigated but little and is mostly limited to phosphate. The exchange of such anions as Cl^-, NO_3^-, and SO_4^{2-} is probably rapid. By means of radioactive phosphate and exchange of phosphate and arsenate in kinetic experiments, it has been demonstrated (20, 45) that the exchange of phosphate may be divided into two reactions: (*a*) a quick exchange of more loosely bound phosphate, and (*b*) a very slow exchange of firmly bound or fixed phosphate.

CATION EXCHANGE

Effect of Dilution, Exchange Capacity, and Valence

It is of theoretical as well as practical interest to study the effect on ion distribution of variation of either the solution concentration or the exchange capacity, under otherwise constant conditions. This implies in both cases a change in the relative activity of exchangeable ions and ions in solution, expressed by $(Z + Y):X$.

Many experiments have shown that dilution of a soil-water system containing monovalent and divalent cations displaces the equilibrium in such a direction that the adsorption of divalent ions increases whereas the

* Radioactive isotope.

adsorption of monovalent ions decreases, that is, the ratio of divalent to monovalent on the exchanger increases. Raising the solution concentration reverses the process, that is, the ratio of divalent to monovalent ions on the exchanger decreases. In systems with monovalent ions, dilution has proved to have a very slight effect or none at all, depending on the kind of ions. Theoretically the same inferences are easily drawn from the Donnan formula or from other equations based on the mass law.

Effect of Dilution and Valence

For a system containing Na and K salts, the equilibrium is given by $(K)_i : (K)_o = (Na)_i : (Na)_o$; and for K and Ca by $(K)_i : (K)_o = \sqrt{(Ca)_i} : \sqrt{(Ca)_o}$. If concentration is substituted for activity and the amounts of ions on the exchanger are expressed in moles, these equations read

$$\frac{K_i}{Na_i} = \alpha_{Na,K} \cdot \frac{[K]_o}{[Na]_o} \tag{33}$$

and

$$\frac{K_i}{\sqrt{Ca_i}} = \alpha_{Ca,K} \cdot \sqrt{v}\, \frac{[K]_o}{\sqrt{[Ca]_o}} \tag{34}$$

v meaning volume of micellar solution, related to K_i and Ca_i by

$$[K]_i = K_i : v, \quad \text{and} \quad [Ca]_i = Ca_i : v$$

If an exchanger is equilibrated with solutions of various concentrations but having a constant $[K]_o : [Na]_o$, it is evident from equation (*33*) that $K_i : Na_i$ changes only with variation of $\alpha_{Na,K}$, that is, with an unequal variation of f_K and f_{Na}. On dilution, f_{K_i} and f_{Na_i} undergo only slight changes, and the dilution effect, if any, is therefore likely to be determined by an unequal variation of f_{K_o} and f_{Na_o}.

Inspection of equation (*34*) shows that dilution of the system, that is, increase in $(Z + Y) : X$, increases the ratio of Ca_i to K_i, the magnitude of which is determined by the following three factors:

(a) **The Unequal Valence.** On dilution $[K]_o$ decreases much faster than $\sqrt{[Ca]_o}$, the so-called valence effect.

(b) **The Unequal Variation of f_{K_o} and f_{Ca_o}.** On dilution, the relative increase of f_{Ca_o} is much greater than for f_{K_o}, while f_{Ca_i} and f_{K_i} are only slightly changed.

(c) **The Volume of the Micellar Solution.** Though not experimentally demonstrated, a large v may involve less dilution effect. There are experiments indicating that v varies both with kind of ion and with kind of exchanger.

Effect of Exchange Capacity and Valence

The other possibility of changing the relative concentration of exchangeable ions and ions in solution is given by variation of the inner solution concentration, holding the outer solution constant. This may be achieved by:

(a) Using Exchangers of Different Exchange Capacity. There seems to be an approximate correlation between exchange capacity and micellar solution concentration, as would be expected from the theory of the diffuse double layer. Thus the concentration of the micellar solution and $(Z + Y):X$ diminish rapidly in the order: resin > montmorillonite > illite > kaolinite, leading to a great decrease of $Ca_i:K_i$ in the same direction.

(b) Change in the pH of the System when the Degree of Ionization Varies Simultaneously. The charge of the clay minerals originating from ionized-SiOH groups is likely to vary between pH 5 and 8, and the ionization of —COOH and —OH of the organic matter is also controlled by the acidity, which thereby affects the micellar solution concentration in both cases. Data from Mattson and Larsson (24) support the theory. Working with bentonite, they found that the adsorption of NH_4 and Ca from 1:1 chlorides of varying acidity was as follows:

pH	6.81	4.75	3.80	3.00
$\frac{Ca}{NH_4}$	8.4	7.9	7.5	6.7

The lowering of the pH is likely to have gradually decreased the Z value, resulting in a lower $(Z + Y):X$ and, thus, in a decrease of the $Ca_i:NH_{4_i}$. In soils, the type of clay mineral has, however, a far greater effect than the pH on the distribution of ions of dissimilar valence, since CEC may vary from 2 to 5 meq for kaolinite, to 80 to 140 meq/100 g for montmorillonite.

In Table 4.3 data are given from an investigation of the influence of CEC and solution concentration on the distribution of K and Ca in montmorillonite, illite, and kaolinite. The exchange capacities, decreasing in

TABLE 4.3. RATIO Ca TO K ON CLAYS AND Ca TO NH_4 ON RESIN AFTER EQUILIBRATION WITH CHLORIDES OF DECREASING CONCENTRATION

Normality (Ca, K, NH_4)	0.1	0.01	0.001	0.0001
Kaolinite	—	1.83	5.0	11.1
Illite	1.10	3.38	8.1	12.3
Montmorillonite	1.47	(10.0)	22.1	38.8
Resin	3.27	10.8	35.0	89.9

the same order, proved to be 81.0, 16.2, and 2.3 meq/100 g, respectively. The clays were equilibrated at pH 7 with isonormal chlorides of K and Ca of the concentrations indicated, separated by membrane filtration from the solutions, and washed free from chlorides with alcohol. For comparison, the results are given from similar experiments with a resinous exchanger having an exchange capacity of 250 meq/100 g. The table shows in the horizontal direction the influence of dilution, and in the vertical direction the influence of exchange capacity, on the distribution of K and Ca. Summarized, the data show conclusively and in accordance with the theory that decrease of the solution concentration or increase in the exchange capacity, or both, favor the adsorption of divalent ions at the expense of monovalent ions and vice versa.

Implications Regarding Plant Nutrients in Soils

These exchange properties may have important bearings on the distribution and behavior of plant nutrients in soils. On moistening, by rain or irrigation, the soil solution becomes more dilute, leading to greater adsorption of divalent and equivalent desorption of monovalent ions. On drying, the opposite process takes place. The ratios

$$\frac{[\mathrm{Ca} + \mathrm{Mg}]_i}{[\mathrm{K} + \mathrm{Na}]_i} \quad \text{and} \quad \frac{[\mathrm{K} + \mathrm{Na}]_o}{[\mathrm{Ca} + \mathrm{Mg}]_o}$$

increase on moistening but decrease on drying. To shake a soil with water for determination of free salts originally present is therefore not feasible. The degree of displacement of the ion equilibrium depends on salt content as well as amount of water added. Soils rich in salts may exhibit considerable change in equilibrium, whereas those poor in salts undergo only negligible variation.

Valence and dilution also affect the distribution of the exchangeable ions between the various constituents of the soil exchange material. Disregarding specific effects, such as the geometric fit of ions in the clay mineral structure, one may conclude that the montmorillonitic minerals and humus matter, having high exchange capacities, adsorb relatively more Ca and Mg and relatively less K and Na than the kaolinitic minerals. The illitic minerals occupy an intermediate position. The different ratios of Ca_i to K_i shown by the clay minerals in equilibrium with the same solution (Table 4.3) support these conclusions. It is of interest that Schachtschabel (33) found the soil organic matter to adsorb Ca and NH_4 in the ratio 92 to 8 from a solution equivalent in Ca and NH_4. Consequently, there are reasons for believing that, in addition to the nonuniform distribution in the interphase soil–soil solution and between the different constituents of the soil exchange complex, the cations have a more or less uneven arrangement on the surface of the soil particles, caused by an uneven charge density on the individual clay minerals. Actually, the ion distribution in soils is a three-dimensional problem and not (as generally treated) a one-dimensional problem. Also the anions have a similarly heterogeneous distribution.

Experiments have indicated that the uptake of nutrients by plant roots is to some extent influenced by the exchange capacity of the clay minerals and of the root acidoids. Elgabaly and Wiklander (9) found that excised barley roots absorbed Ca and Na from kaolin and bentonite (saturated with Ca and Na = 1:1) in the milliequivalent ratios of

Na—Ca kaolin	Ca:Na = 0.72
Na—Ca bentonite	Ca:Na = 0.45

Because of its higher exchange capacity, bentonite exercises a greater competition for Ca and less for Na than does kaolin with the plant roots.

Mattson found similar results in pot experiments with barley plants grown for 3 weeks in nutrient solution, kaolin, and bentonite.

To what extent and in what direction variations of the moisture content of the soil affect the uptake of monovalent and divalent ions by plants is an unsolved problem. Contradictory results have been reported. The kind of influence may depend on the ratio of the exchange capacities of the roots and the soil, the exchange capacity of the roots naturally referring to the active root acidoids and, thus, influencing their "micellar" solution concentration. When this ratio exceeds 1, an increase in the soil moisture content should favor the absorption of divalent ions if taken up from the intermicellar solution; with a ratio less than 1, the absorption of monovalent ions should be favored.

Influence of Mole Fraction and Complementary Ions

In plant nutrition it is of great interest to find out to what degree the availability of individual plant nutrients in soils is influenced by the relative concentrations and nature of the other ions present. Because of the general dominance of Ca among the exchangeable cations, the behavior of ions present in small quantities is of special importance.

If a soil is saturated with K, Na, H, Mg, and Ca, the sum of which is equal to CEC, and we are interested in, say, K, the other ions are frequently called *complementary ions.* The degree of K saturation, D_K, is given by the mole fraction but is often expressed as percentage saturation. Theoretical considerations and experiments have established that the replaceability of an ion depends not only on the kind of ion itself but also on the nature and amount of the complementary ions.

Theoretical Approaches

From investigations covering many different ion combinations and various natural and synthetic exchangers, the general relationship has been outlined between the exchangeability of an ion, the degree of saturation with this ion, and the nature of the complementary ions (44, 46). This was done for ions with concentrations from moderate down to extremely small values. For the theoretical treatment two methods were applied:

(1) At small replacements the amount of an ion A released is proportional to the activity fraction and the total release. Thus, for the homovalent ions A, B, and C, the expression for the replacement achieved by addition of another ion reads:

$$R_A = \frac{(A)}{(A) + (B) + (C)} \cdot R_t \qquad (35)$$

R_A meaning moles of A released and R_t the total release. In this formula, which is similar to that used by Bray for soils (5), activity may be directly replaced by activity coefficients and moles because of the homovalency. A divalent ion enters the formula with its square root.

(2) For the simple case of only two homovalent ion species, the Donnan equilibrium can be used for computation of any percentage release. When the exchanger is saturated with the ions A and B in amounts A_t and B_t moles, the replacement of A in moles is given by:

$$\frac{A_o}{A_t - A_o} = \alpha_{A,B} \frac{R - A_o}{B_t - R + A_o} \qquad (36)$$

A_o being the amount of A replaced and R the total amount replaced, that is $R = A_o + B_o$.

Under the condition that $A_t \ll B_t$, or that the exchange is very small, $R \ll A_t + B_t$, formula (36) may be simplified. The theoretical conclusions from the foregoing formulas may be summarized as follows:

(1) The percentage release of an ion(A) may decrease, be unchanged, or increase with diminishing D_A, depending on the relative bonding strength of A and the complementary ions.

(2) With diminishing mole fraction, the percentage release of A, associated with a homovalent ion B, decreases if A is more firmly adsorbed than B, that is if $\alpha_{A,B} < 1$; is unchanged if the bonding strength of A and B is the same, that is, if $\alpha_{A,B} = 1$; and increases if A is more weakly adsorbed than B, that is, if $\alpha_{A,B} > 1$.

This means that in the first case it will become more and more difficult to replace the ion A, the lower its concentration; in the second case the replaceability is independent of the concentration; and in the third case the replaceability increases with diminishing concentration.

(3) If a complementary ion is substituted with a more firmly adsorbed ion, the replaceability of A increases, but if substituted with a more weakly adsorbed ion, the replaceability of A decreases. If possible specificities are disregarded, the replaceability of an ion should therefore grow in the order

$$Li < Na < K < Mg < Ca < Ba < La$$

when these ions function separately as complementary ions. This statement is verified in experiments with a resinous exchanger (44). Further, NH_4 proved to become easier to exchange with diminishing degree of NH_4 saturation when combined with K, Mg, Ca, Ba, and La, respectively, the increase being in this order. When NH_4 was combined with Li and Na, the exchangeability of NH_4 decreased in the order Li > Na. This behavior is in agreement with the theory and with the fact that K, Mg, Ca, Ba, and La are more firmly adsorbed, but Na and Li more weakly adsorbed, than NH_4 on this exchanger.

(4) At very low degrees of saturation with A, D_A approaching zero, the percentage replacement of A attains a certain limit value, determined by the nature of the ion A, the complementary ions, and the exchanger. Using radioactive potassium and strontium in the combinations K*—Na, K*—Ba, and Sr*—Na, Sr*—Ba on a resin, experiments could be carried out with mole fractions of K and Sr ranging from 1 to very small values. The results were in good agreement with the foregoing conclusions (Figure 4.5). The trend of the curves is explained by the higher bonding strength of K than of Na, $\alpha_{K.Na} < 1$, and the lower bonding strength of K than of Ba, $\alpha_{K.Ba} > 1$.

Experiments were also performed with permutite and an illitic clay, saturated with Ca—NH_4 and Ca—Mg, the results being on the whole as depicted in Figure 4.5.

Because of fixation, defixation, and solubility processes, a study of these problems in natural soils involves difficulties. One of these is the exact determination of the amount of an ion species taking part in the exchange equilibrium. Disturbances from such processes are particularly confusing at a low concentration of the ion in question. Among the cations present in soils, Ca shows a decreasing and Na an increasing replaceability with decreasing degree of saturation, while Mg, K, and NH_4 take an inter-

* Radioactive isotope.

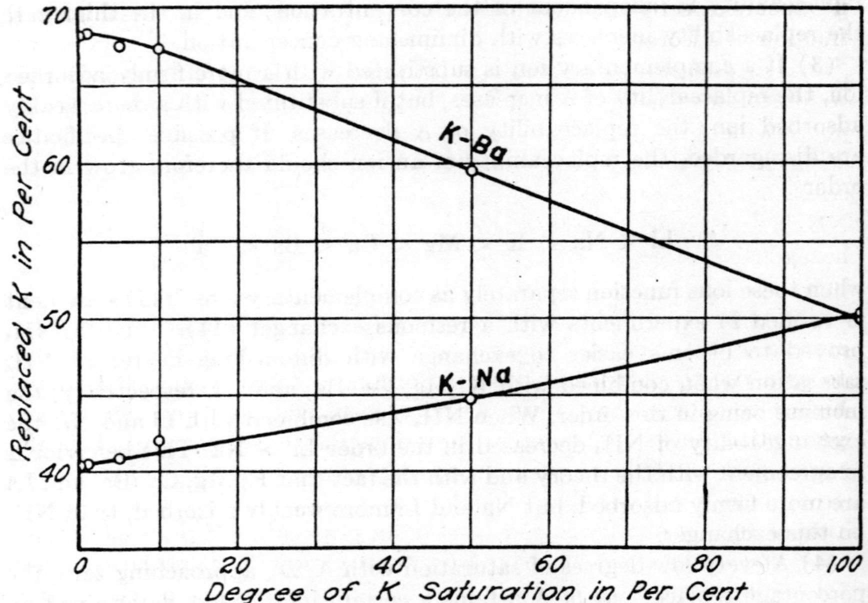

Figure 4.5. Percentage replacement of K, on addition of a constant amount of HCl, from resin saturated in various proportions with K*-Na and with K*-Ba, respectively (radioactive isotope).

mediate position. Further investigations on clays and natural soils are needed.

Influence of the Exchange Material

It is a well-known fact that ion-exchange equilibria vary a great deal with the nature of the exchange material. The following factors are important:

1. *The surface charge density or the exchange capacity of the soil colloids affecting especially the distribution of heterovalent ions.* This is the most important of the exchange material factors.

2. *The nonuniformity of the surface and of the volume charge density of the individual clay minerals*, probably still more accentuated by the existence of clay-humus complexes, resulting in a varying relative concentration over the particle surface of all the cations having unequal activity coefficients (f) in adsorbed conditions, is another factor. If for the monovalent ions A and B, the equilibrium is written

$$\frac{[A]_i}{[B]_i} = \frac{f_{B_i}}{f_{A_i}} \cdot \frac{(A)_o}{(B)_o}$$

it appears directly that the lower the f_{A_i} in relation to f_{B_i}, the higher the fraction of adsorbed A.

3. The geometric fit of ions into the mineral structure. If an ion fits very well into an exchange spot, it will become more firmly adsorbed than ions not having this size. And the more such exchange spots there are, the stronger will the dominance of this particular ion be over other ions at equal outer concentration. Thus, the diameters of the nonhydrated K^+, NH_4^+, and H_3O^+ are about 2.7 Å, and therefore, they fit well into the hexagonal cavities of the oxygen sheets of layer clay minerals, resulting in a firm bonding of these ions.

4. Polarization interaction between the adsorbed ions and exchanger surface. Ions in the innermost layer of the diffuse double layer are likely to give rise to polarization effects, forming a Stern layer and leading to higher bonding energy, lower activity coefficient, and stronger replacing power than those of the other counterions in the double layer. This effect varies with the kind of ion and probably also with the exchanger. To this group belong van der Waals-London forces supposed to act in the adsorption of certain organic ions as well as in physical adsorption of inorganic ions; the latter is assumed by Birrel and Gradwell (1) to be responsible for the apparent high CEC shown by soils containing allophane and certain amorphous oxides.

5. Formation of complex chelates and other types of bondings between certain cations and organic components of the soil (18). Copper ions are known to be bound mainly in this way. Experiments have shown that cations may be held very firmly and in large quantities by organic matter, the bonding strength and adsorption capacity varying greatly with the kind of ion and type of organic matter.

6. Clogging of exchange spots, thereby reducing the exchange capacity. If an ion is too big to enter between the basal or interior surfaces of the minerals the surface accessible to this ion will be reduced and the replacing power relative to that of a smaller ion will appear too low. Clogging may also occur more directly by adsorption of large, flat organic ions blanketing more than one exchange spot, as suggested by Hendricks (14) for montmorillonite.

Relative Replacing Power

Many experiments have evidenced that *there is no single universal order of the replacing power of cations.* Theoretically this is also to be expected. For the alkali ions, the replacing power has often been found to increase in the same order as the lyotropic series: Li < Na < K < Rb < Cs, and for the alkaline earth ions: Mg < Ca < Sr < Ba. The position of the monovalent and divalent ions in a combined series, however, has proved

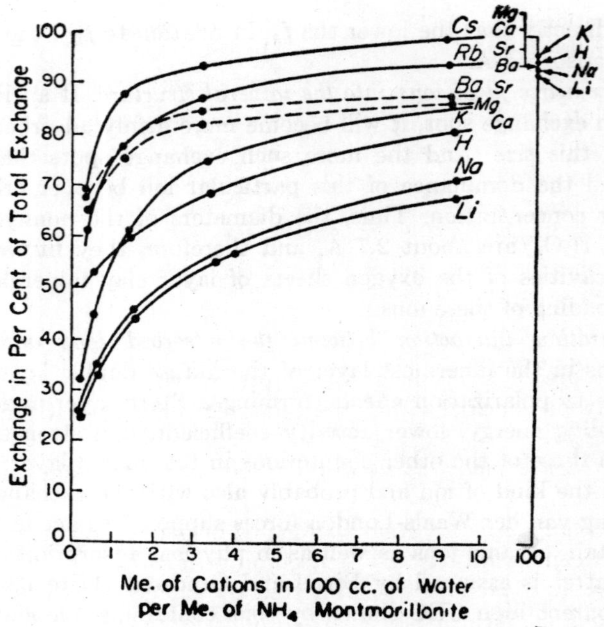

Figure 4.6. Replacement of NH_4 from NH_4-montmorillonite by various chlorides (33).

to vary considerably with the nature of the exchanger and the concentration of the solution. For H, the position in relation to the other ions varies still more. These conditions are illustrated by the extensive investigations of Schachtschabel (33) on the release of NH_4 from various minerals by different chlorides, from which Figure 4.6 and the following series of the relative replacing power are taken.

Concentration (N)	
	NH_4 Kaolin
0.005	Li < Na < H < K < Rb < Mg = Ca = Sr = Ba < Cs
0.02	Li < Na < H < K < Mg = Ca < Sr = Ba < Rb < Cs
	NH_4 Biotite
0.005	Li < Na < Mg < Ca < Sr < K < Cs < Ba < Rb < H
0.1	Li < Mg < Na < Ca < Sr < Ba < Cs < Rb < K < H

It is evident that the kind of material is important for the relative replacing power and that the adsorption of monovalent ions in relation to that of the divalent increases with concentration. In soils the relative replacing power often follows the order: $M^+ < M^{2+} < M^{3+}$.

In replacement of H from humus by neutral salts the writer (44) obtained:

Concentration (N)	H Humus
0.01 and 0.1	Li < Na < K < Mg < Ca < Sr < Ba < La
1.0	K < Na < Li < Mg < Ca = Sr = Ba < La

In the dilute series, the order of the relative replacing power agrees with that of the lyotropic series with the effect of valence: $M^+ < M^{2+} < M^{3+}$, which is to be expected in the absence of specificities.

Gieseking and Jenny (13) obtained on Putnam clay (beidellite) the following ease of entry by exchange:

$$\text{Li} \lesseqgtr \text{Na} < \text{K NH}_4 < \text{Rb} < \text{Cs} < \text{H} \quad \text{and}$$

$$\text{Mg} \lesseqgtr \text{Ca} < \text{Li} < \text{Ba} < \text{La} < \text{Th}$$

From the previous discussion of factors influencing ion exchange equilibria it may be concluded that the relative replacing power of cations is determined by:

1. The exchange material: its content of various minerals and organic matter.

2. The exchange capacity, and the degree of neutralization in relation to the solution concentration, as discussed on page 187.

3. The nature of the ions as manifested by valence, size in nonhydrated and hydrated conditions, and polarizability and polarizing power. The bonding strength increases with the valence. Ion hydration plays an important role in adsorption; the less hydrated the ion, the more tightly it is held. This agrees well with the theory that the activity coefficient is affected by the radius of the hydrated ion [cf. equation (*27*)]. Determination of the ion hydration has given rather varying results depending on the procedure used. Certain hydration numbers are given in Table 4.2.

Some doubt is thrown on the ion hydration theory by the work of several authors. Strong support is given to the importance of the hydration concept in ion adsorption, however, by the fact that increase in temperature and addition of alcohol—both having a dehydrating effect—increase the replacing power in the order: Cs < K < Na < Li. In agreement with this, the adsorption of alkali ions on ultramarine at 130° was found by Jaeger to follow that sequence.

It has been shown (4, 27) that certain polyvalent ions are adsorbed either as polyvalent or as monovalent ions, the excess charge of the ions in the latter case being neutralized by accompanying exchangeable anions such as OH^- (from salt proteolysis), Cl^-, and NO_3^-. In this way cations such as $BeOH^+$, $CuOH^+$, $CuCl^+$, $ZnOH^+$, $ZnCl^+$, $FeOH^{2+}$, $Fe(OH)_2^+$, and $Al(OH)_2^+$ are found to be adsorbed. Only in dilute acid solutions is Cu adsorbed as Cu^{2+}. Adsorption of Zn has been found (8) to decrease the cation-exchange capacity and increase the anion-exchange capacity of certain minerals.

Indeed, colloids may gradually lose their negative charge and acquire a positive charge by adsorption of certain positive ions.

Large organic cations, for example methylene blue, are very tightly held. Complete replacement is not possible with ordinary metal cations but may be accomplished with organic ions of the same or larger size.

4. The concentration of the solution. The lower the concentration of the ion, the greater is the replacing power of an ion species in relation to its concentration, which is indicated by the exchange curves (Figure 4.6). The replacing power of polyvalent ions relative to that of monovalent ions is greater at low solution concentrations than at high ones. This becomes more marked as the exchange capacity increases, and is a natural consequence, as shown by the valence effect in the mass-action law discussed on page 185.

5. The composition of the exchangeable ions and the mole fractions in the manner described on page 189.

Of particular interest is the behavior of the H ion. It may be assumed that the replacing power of H is intimately connected with the acid strength of the exchanger; H is potential-determining. The weaker the acid properties, the smaller the active fraction of the potential acidity and the stronger the replacing power of H. On exchangers containing only strong acid groups, such as sulfonic resin, the adsorption of H is largely determined by the same factors as the alkali ions, and H should therefore have a position close to Na (cf. the hydration numbers in Table 4.2). On an NH_4-saturated sulfonic acid resin the following order of the replacing power was found (44):

$$\text{Li} < \text{H} < \text{Na} < \text{K} < \text{Mg} < \text{Ca} < \text{Sr} < \text{Ba} < \text{La}$$

On permutite and clay the position of H shifted to the right, yielding for the clay at 0.04 N: K < Ca < Ba < La < H. But the replacing power of H on clay and soils may also depend on the degree of neutralization. The higher the degree of neutralization, the stronger the replacing power. Therefore, the efficiency of H in replacing plant nutrients may decrease with falling pH. The beneficial effect of liming on the cation nutrient uptake may be partly explained on this basis.

ANION EXCHANGE

Anion exchange on clay minerals and soils has been far less studied than cation exchange, and our knowledge is therefore in some respects rather limited. The effect of concentration, mole fraction, and complementary ions on the distribution of exchangeable anions seems to be similar to that for cations. In the low pH region, the mineral lattice is unstable, resulting in dissolution of Al, Fe, Mg, Mn, the ions being held partly in exchangeable form. Some anions easily form difficultly soluble or complex compounds with these cations, which greatly complicates the study of the anion exchange. Fixation of certain anions, particularly phosphate, is often encountered.

Negative Adsorption

If a dilute neutral solution of KCl is added to dry montmorillonite, the concentration of Cl in the equilibrium solution will show an increase. This happens to any anion and to any clay and soil having no adsorbing capacity of the anion at the prevailing pH. This so-called negative adsorption is easily explained by the assumption of an unequal ion distribution in the diffuse double layer according to the Donnan equilibrium and the Gouy theory [cf. Figure 4.3 and equation (*24*)]. Contributing factors are the hydration of the exchangeable ions and the mineral surface.

Mattson (22) studied negative adsorption in Sharkey clay saturated with various cations in equilibrium with the corresponding chlorides and found the negative adsorption of Cl^- to decrease according to the sequence: Na > K > Ca > Ba.

A bentonite, Na-saturated and equilibrated with various Na salts, showed negative adsorption in the order:

$$Cl^- = NO_3^- < SO_4^{2-} < Fe(CN)_6^{4-}$$

These results may be explained by application of equation (*17*) for equilibrium with a salt M_nA_m.

$$\frac{X_A}{Y_A} = \left(\frac{Z_M + Y_M}{X_M}\right)^{\frac{n}{m}}, \qquad \text{or} \qquad \frac{X_A}{Y_A} = \left(\frac{f_{M_i}}{f_{M_o}} \cdot \frac{[M]_i}{[M]_o}\right)^{\frac{n}{m}}$$

Since $(Z_M + Y_M):X_M > 1$, it follows that $X_A:Y_A$ and hence the negative adsorption increases with the valence (n) of the anion but diminishes with increasing valence (m) of the cation. The influence of homovalent cations is reflected to some extent by $f_{M_i}:f_{M_o}$, a ratio decreasing for the ions used by Mattson in the order: Na > K and Ca > Ba. Thus the theoretical conclusions seem to be in good qualitative agreement with Mattson's findings.

The negative adsorption of anions is also influenced by the nature of the exchanger. Equation (*24*) indicates that the negative adsorption increases with the amount of colloid and the exchange capacity and may decrease with increasing volume of the micellar solution, when the conditions are otherwise the same. Further, equation (*24*) indicates that the negative adsorption decreases with increasing salt concentration. Addition of salts suppresses the unequal ion distribution.

Working with a carboxyl resin, the writer found that raising the pH, that is, increasing the exchange capacity, resulted in a great increase in negative adsorption of phosphate. Schofield (36) measured the negative adsorption of Cl^- on montmorillonite and kaolinite and calculated, by application of the Gouy theory, the specific surface of the clays. He obtained for montmorillonite 470 m^2/g and for kaolinite 16 m^2/g. This procedure requires a uniform distribution of the surface charge, a doubtful presumption, especially for kaolinite. The value for montmorillonite agrees fairly well with values obtained by Eriksson from swelling data and by others from water adsorption, but is lower than those usually obtained by the ethylene glycol method.

The negative adsorption may have important bearings on the behavior of nutrient anions in the soil. It counteracts the anion adsorption by pushing the anions out of the micellar solution, thereby increasing the losses by leaching.

Adsorption and Exchange of Anions

Contrary to the cation exchange, the capacity for holding anions increases with the acidity. In the neutral region certain anions such as Cl^- and NO_3^-, unlike PO_4^{3-} and AsO_4^{3-} which are bound at higher as well as at lower pH, are adsorbed very slightly or not at all by many soils. Table 4.4 and Figure 4.7 give some data for anion sorption. Obviously, in the Nipe colloid: (*a*) the adsorption of the anions increases greatly with falling

TABLE 4.4 SORPTION OF Cl, SO_4, AND PO_4 BY NIPE (KAOLINITIC) AND SHARKEY (MONTMORILLONITIC) SOIL COLLOIDS IN RELATION TO pH*

Nipe				Sharkey		
pH	meq Sorbed per 100 g colloid			pH	meq Sorbed per 100 g colloid	
	Cl	SO_4	PO_4		Cl	PO_4
7.2	0.0	0.0	31.2	6.8	0.0	22
6.7	0.3	2.0	41.2	5.6	0.0	36.5
6.1	1.1	5.5	46.5	4.0	0.05	47.4
5.8	2.4	7.1	50.8	3.2	0.1	64.0
5.0	4.4	10.5	66.1	3.0	0.1	73.5
4.0	6.0	—	88.2	2.8	0.4	100

* After Mattson, data partially recalculated.

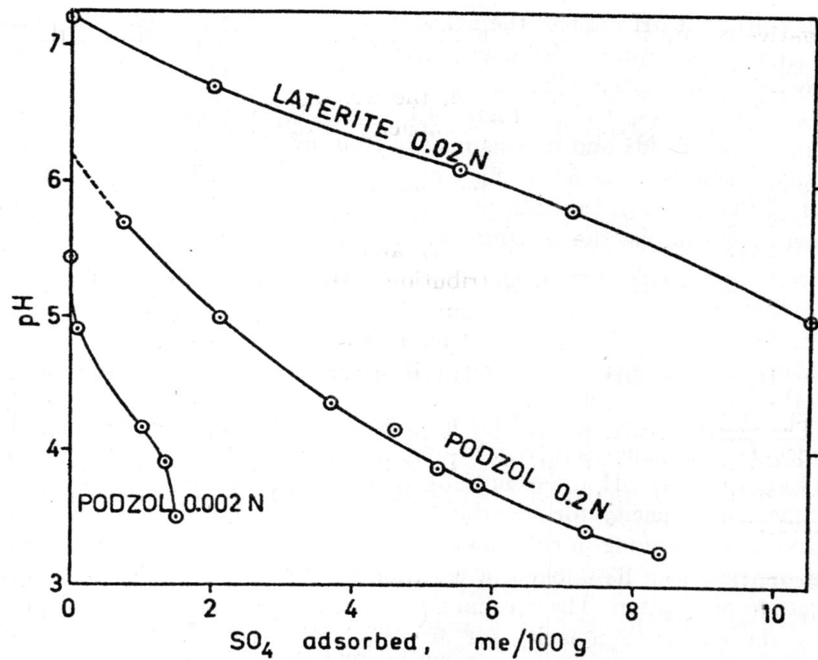

Figure 4.7. Influence of pH and salt concentration on sorption of SO_4 by laterite and podzol B_2.

pH; (*b*) Cl and SO_4 are not bound at pH above 7; and (*c*) the bonding strength is of the order: Cl $<$ $SO_4 \ll PO_4$. Adsorption of Cl and PO_4 by the montmorillonitic Sharkey clay showed the same trend, but the Cl retention was very slight and did not occur above pH 4.0. At corresponding pH values, the adsorption of PO_4 by the Sharkey clay was also less than by the Nipe. Wyoming bentonite (montmorillonite) had no sorption capacity at all for Cl. Figure 4.7 shows that SO_4 sorption increases rapidly with falling pH and concentration of sulfate in solution. It also increases with falling $SiO_2:R_2O_3$, that is, podzol $<$ laterite.

These and other results indicate that the anion adsorption is intimately connected with the nature of the soil colloids (type of clay minerals, contents of hydrous oxides, and organic matter) and with the pH of the system. For the aforementioned colloids the ratio acidoids:basoids, as manifested by the ratio cation-exchange capacity:anion-exchange capacity, increases rapidly in the order Nipe $<$ Sharkey $<$ bentonite, leading to a marked decrease of the anion-holding power.

Lowering of pH induces the activation of the basic groups by increased acceptance of protons; for example, in water, R—OH + HOH = R—OH_2OH; in presence of HCl, R—OH + H^+Cl^- = R—$OH_2^+Cl^-$; R—NH_2 + H^+Cl^- = R—$NH_3^+Cl^-$ (cf. Figure 4.1). The OH originates in clay minerals from broken bonds and in soils mainly from hydrous oxides of Fe and Al. Among the anions bound in this way may be mentioned Cl^-, NO_3^-, SO_4^{2-}, and $H_2PO_4^-$. They function as counterions and are exchangeable with other anions in just the same manner as cations are exchanged:

$$R\text{—}OH_2^+Cl^- + NO_3^- \rightleftharpoons R\text{—}OH_2^+NO_3^- + Cl^-$$

and

$$R\text{—}OH_2^+Cl^- + OH^- \rightarrow R\text{—}OH_2OH + Cl^- \rightarrow R\text{—}OH + H_2O + Cl^-$$

The $H_2PO_4^-$ adsorbed in this way is said by Mattson to be saloid-bound. It occurs in pedalfer soil (rich in hydrous oxides of Fe and Al and having no calcite) at low pH and essentially below the equi-ionic point.

Since the capacity for adsorbing Cl^-, NO_3^-, and SO_4^{2-} is slight at the pH values prevailing in cultivated soils, these ions are easily lost by leaching. As has been mentioned, a contributory factor to this process is the negative adsorption. The mechanism of anion exchange outlined explains why the exchange capacity for anions compared to that for cations is small in montmorillonite (few broken bonds in relation to isomorphous substitutions) and why the difference in the two capacities is less in kaolinite (both cation and anion exchange located to broken bonds). It further explains the higher anion-exchange capacity of soils than of clay minerals (soils contain hydrous oxides) and the great reduction of this capacity if the soils are treated with Tamm's acid oxalate solution (39), which dissolves these oxides. The mechanism also explains the exchange alkalinity, that is, the increase in pH of soil suspensions on addition of neutral salts, for example, Na_2SO_4, exhibited by many soils.

$$2R\text{—}OH_2OH + SO_4^{2-} = (R\text{—}OH_2^+)_2SO_4^{2-} + 2OH^-$$

The exchange alkalinity has been found to increase with the replacing power of the anions and the content of basoids in the soil.

There is little doubt, however, that several of the other anions, such as phosphate, become bound by mechanisms quantitatively more important in natural soils than is the foregoing. Thus, exchange of lattice ions probably occurs. Fluorine, for example has about the same size as OH and may therefore displace OH isomorphously from clay minerals and hydrous oxides in the following way: R—OH + F^- = R—F + OH^-. There is evidence that the phosphate ion, though not having the same size as OH.

is also bound as a structural nondiffusible unit by displacement of lattice OH. The similarity in bonding of hydroxyl and phosphate may explain the strong displacement of phosphate exerted by alkali. Because of differing bonding energy, the substitution of OH can occur at higher pH than the exchange of the counterion OH, and like the latter it will increase with falling pH. Phosphate adsorbed in this way occurs particularly in pedalfer soils and falls under the term "colloid-bound" introduced by Mattson. Since this type of bonding takes place also above the isoelectric point, it can contribute to the cation-exchange capacity according to the valence, that is, whether $H_2PO_4^-$, HPO_4^{2-} or PO_4^{3-}. Arsenate is similarly bound.

Hendricks has suggested another method of interaction. The PO_4 and AsO_4 ions are about the same size as the silica tetrahedron and may extend the lattice by adsorption on the edges of the silica tetrahedron layer.

Phosphate is also retained in soils by precipitation. It forms difficultly soluble compounds with Fe and Al at low pH, more soluble compounds with Ca and Mg at slightly acid and alkaline reactions, and difficultly soluble compounds with these two ions in the alkaline region. The effectiveness of ions such as fluoride, oxalate, citrate, and tartrate in releasing sorbed phosphate from soils should be attributed mainly to the formation of soluble complexes (38), especially with Fe and Al, rather than to true ion exchange. It has been pointed out by Williams (47) that fluoride forms strong complexes with Fe and Al, especially the latter, and that neutral ammonium fluoride extracts considerable amounts of Al from soils. Humic acids take part in anion exchange, as evidenced by their power of releasing adsorbed phosphate, but the exact mechanism is not known.

Because of their nature, some of the mentioned mechanisms of anion sorption should not be considered as true ion exchange according to the meaning given to cation exchange. The kinetics of isotopic phosphate exchange support this conclusion.

It has been suggested that the anion-exchange capacity be defined as the amount of phosphate bound at pH 4 (Piper) and 5.7 (Dean and Rubins, 7) and also that the clay colloids be characterized as to their mineralogical make-up and degree of weathering by the ratio of the cation- and anion-exchange capacities. Schoen (34) has given the following average values for these ratios: montmorillonite, 6.7; illite, 2.3; and kaolin, 0.5. The anion-exchange capacity as defined would be more adequately termed *sorption capacity*. Obviously it will greatly exceed the exchange capacities for such anions as Cl, NO_3, and SO_4 (cf. Table 4.4). For the study of anion exchange and the application of exchange formulas, it is important to make a distinction so far as possible between the true anion exchange and other sorption processes.

Salt Effects on Phosphate Sorption

The retention, solubility, and availability to plants of phosphate have been extensively investigated because of their agricultural importance. It is now known that pH and the content of neutral salts influence these properties in a way and to an extent which vary with the nature of the soil material. But the results obtained by different investigators have frequently proved inconsistent and difficult to explain. The often-demonstrated effect of neutral salts on phosphate adsorption and solubility has some puzzling features. Salt may increase or decrease the solubility, the effect in the latter case being contradictory to what might be expected from the relation between the solubility of crystalline compounds and the ionic strength. Addition of salts diminishes the activity coefficients, thereby increasing the solubility of ordinary slightly soluble salts. Further, the anions of the salt, at least at low pH, must be expected to have a displacing effect on the phosphate. Mattson and co-workers (25) studied the salt effect and advanced a theory that seems to explain the relationship of the salt effect, pH, and soil properties, a very brief summary of which follows.

Mattson makes the assumption that the solubility or, more exactly, the concentration of phosphate in the outer solution in pedalfer soils is determined by (*a*) a solubility product of the hydrous oxide-bound phosphate, keeping the concentration ($Y_{H_2PO_4}$) in the inner solution constant and (*b*) a Donnan distribution of the diffusible phosphate ions between the inner and outer solutions, and (*c*) the kind and contents of salts present. If a soil is above the isoelectric point, which is generally the case, it behaves like a cation exchanger and $X > Y$. For a soil containing a certain amount of KCl, the H_2PO_4—Cl equilibrium can be written $X_{H_2PO_4}:Y_{H_2PO_4} = X_{Cl}:Y_{Cl}$. On dilution, Y_{Cl} decreases faster than X_{Cl} and the ratio therefore increases, cf. $(Z + Y)Y = X^2$. As long as $Y_{H_2PO_4}$ is constant, the corresponding increase of the phosphate ratio must be caused by an increase of $X_{H_2PO_4}$. This means that progressive addition of water to the soil increases the concentration of phosphate in solution until $Y_{H_2PO_4}$ is no longer a constant. The reaction taking place is a release of adsorbed phosphate: adsorbed P $\rightarrow Y_P \rightarrow X_P$. On the addition of neutral salt the reaction goes in the opposite direction: adsorbed P $\leftarrow Y_P \leftarrow X_P$, that is, the salt induces phosphate adsorption and suppresses the solubility. The theory also explains why the magnitude of the positive salt effect on the phosphate adsorption is related more to the content of acidoids than to basoids. Z is determined by the acidoids, and a high Z means a great difference between X and Y and a large effect of the salt on $X_{H_2PO_4}: Y_{H_2PO_4}$. For the same reason the positive salt effect on the phosphate adsorption is large at high pH, decreases to zero with falling pH, and then

TABLE 4.5. SOLUBILITY OF PHOSPHATE AS AFFECTED BY ADDITION OF WATER TO SOIL SUSPENSIONS*

Phosphate*	Water: soil						
	2	8	32	64	128	256	512
P_2O_5, mg/l							
Podzol A_1	0.074	0.080	0.096	0.102	0.103	0.099	0.077
Lanna	0.066	—	0.131	0.131	0.096	0.080	0.057

* Podzol A_1, uncultivated; Lanna, cultivated soil (after Mattson *et al.*).

reverses into a negative effect. This phenomenon is explained by the fact that Z decreases in the same direction and that the saloid-bound phosphate increases with the acidity. At a certain pH, depending on the basoid and acidoid contents of the soil, the direct replacement of the saloid-bound H_2PO_4 by the anion of the salt balances the positive salt effect, and the phosphate adsorption is not altered by the addition of salt. At pH values below this point, the replacement of H_2PO_4 is greater and the salts exert a negative effect on the phosphate adsorption, that is, the solubility rises. According to this theory, phosphate adsorption and solubility in pedalfer soils are closely related not only to the pH and the salt content but also to the amphoteric properties of the soil. At a certain pH the phosphate adsorption shows a maximum.

In pedocal soils (containing calcite), neutral salts have an effect on the phosphate adsorption similar to that in pedalfer soils, the main difference being that the adsorption maxima and the points of inversion of the salt effect occur at widely different pH values corresponding to the isoelectric points of the compounds formed.

A few data on the effect of dilution on the phosphate solubility are given in Table 4.5. The figures show a rare phenomenon, namely, an increase in the phosphate concentration produced by addition of water to the suspensions, the maximum for podzol A_1 appearing after a 128-fold dilution. If the proper explanation of these results is given by the theory of Mattson and coworkers, it means that the redistribution of H_2PO_4 by the process adsorbed P $\rightarrow Y_P \rightarrow X_P$ overshadows the effect of the increasing suspension volume as long as $Y_{H_2PO_4}$ is constant. The maximum of phosphate concentration indicates the point where $Y_{H_2PO_4}$ has dropped so far that the mere effect of the dilution itself outweighs the Donnan effect.

EXCHANGE BETWEEN SOLID PHASES

The ion-exchange material of soils is very heterogeneous, being composed of various minerals, organic matter, and living plant roots. The many chemical, physical, and biological processes going on disturb the ionic equilibria and hence induce exchange reactions between the soil constituents. The exchange may take place in two ways: first, as an ordinary ion exchange via the intermicellar solution, the equilibrium being ruled by the mass-action law; and second, by direct contact and intermingling of the double layers. The latter mechanism has been suggested by Jenny (15) and given the name *contact exchange.* Every ion held at a negative spot on the clay surface is supposed to have a certain oscillation volume around this center. The oscillation volume increases with ion hydration and decreases with valence, and is therefore correlated with the bonding strength. If the oscillation volumes overlap, the ions may migrate along the exchanger surface—surface diffusion, or exchange with ions on an adjacent-particle contact exchange. Jenny argues that nutrient absorption by plants is essentially a contact exchange between plant roots and soil colloids. Much experimental evidence, though not always consistent, supports this assumption. But, it seems difficult to bring into line with the contact-exchange theory the fact that an intermingling of two diffuse double layers is opposed by a repulsion due to an increased osmotic pressure in the intermingling zone. A direct contact between exchanger surfaces may be facilitated if the surface-charge density is small and the counterions are polyvalent, thus more easily forming linkages between the two particles. The contact-exchange theory has raised a great deal of critical discussion.

Ion exchange between solids is now extensively utilized for estimation of the replaceable cations in soils and for saturation of clay and humus colloids with hydrogen and other ions. Resinous exchangers, saturated with hydrogen for the replaceable-cation estimations and for making titration curves, and with the desired cation for the colloid saturation procedure, are used.

A method for determining available phosphate, described by Møller and Mogensen (29), is also based on cation exchange between solid phases. The soil is shaken with sodium permutite (NaR), and the calcium-bound phosphate is supposed to be released by an exchange process represented as follows:

$$6\,NaR + Ca_3(PO_4)_2 = 3\,CaR_2 + 2\,Na_3PO_4$$

REFERENCES

1. Birrel, K. S., and Gradwell, M., *J. Soil Sci.* 7, 130–147 (1956).
2. Bolt, G. H., and Peech, M., *Soil Sci. Soc. Am. Proc.* 17, 210–213 (1953).
3. Borland, J. W., and Reitemeier, R. F., *Soil Sci.* 69, 251-260 (1950).
4. Bower, C. A., and Truog, E., *Soil Sci. Soc. Am. Proc.* 5, 86–90 (1940).
5. Bray, R. H.. *J. Am. Chem. Soc.* 64, 954–963 (1942).

6. Damsgaard-Sørensen, P., "Kationombytning i Jorden," Diss. København, 1941.
7. Dean, L. A., and Rubins, E. J., *Soil Sci.* **63**, 377–387 (1947).
8. Elgabaly, M. M., *Soil Sci.* **69**, 167–173 (1950).
9. Elgabaly, M. M., and Wiklander, L., *Soil Sci.* **67**, 419–424 (1949).
10. Eriksson, E., *Soil Sci.* **74**, 103–113 (1952).
11. Freundlich, H., "Kapillarchemie," Bd. 1, Leipzig, 1930.
12. Gapon, E. N., *J. Gen. Chem. U.S.S.R.* **3**, 144–152, 153–158 (1933).
13. Gieseking, J. E., and Jenny, H., *Soil Sci.* **42**, 273–280 (1936).
14. Hendricks, S. B., *J. Phys. Chem.* **45**, 65–81 (1944).
15. Jenny, H., *J. Phys. Chem.* **40**, 501–517 (1936).
16. Kerr, H. W., *J. Am. Soc. Agron.* **20**, 309–335 (1928).
17. Krishnamoorthy, C., and Overstreet, R., *Soil Sci.* **69**, 41–53 (1950).
18. Lewis, T. E., and Broadbent, F. E., *Soil Sci.* **91**, 393–399 (1961).
19. Low, P. F., *Soil Sci.* **71**, 409–418 (1951).
20. McAuliffe, C. D., Hall, N. S., Dean, L. A., and Hendricks, S. B., *Soil Sci. Soc. Am. Proc.* **12**, 119–123 (1947).
21. Marshall, C. E., *Trans. Intern. Congr. Soil Sci. 4th Congr. Amsterdam* **1**, 71–82 (1950).
22. Mattson, S., *Soil Sci.* **28**, 179–220 (1929).
23. Mattson, S., *Soil Sci.* **32**, 343–365 (1931).
24. Mattson, S., and Larsson, K. G., *Soil Sci.* **61**, 313–330 (1946).
25. Mattson, S., Williams, E. G., Eriksson, E., and Vahtras, K., *Kgl. Lantbruks-Högskol. Ann.* **17**, 64–91 (1950).
26. Mehlich, A., *Soil Sci.* **66**, 429–445 (1948).
27. Menzel, R. G., and Jackson, M. L., *Soil Sci. Soc. Am. Proc.* **15**, 122–124 (1950).
28. Møller, J., "Studier over Ionbytningsprocessen med saerligt Henblik paa Agrikulturkemien," Diss. København, 1935.
29. Møller, J., and Mogensen, Th., *Soil Sci.* **76**, 297–306 (1953).
30. Pauli, W., and Valkó, E., "Elektrochemie der Kolloide," Vienna, J. Springer, 1929.
31. Peech, M., and Scott, A. D., *Soil Sci. Soc. Am. Proc.* **15**, 115–119 (1950).
32. Russel, E. W., "Soil Conditions and Plant Growth," New York, Longmans, Green and Co., 1950.
33. Schachtschabel, P., *Kolloid-Beih* **51**, 199–276 (1940).
34. Schoen, U., *Z. Pflanzenernähr. Düng. Bodenk.* **63**, 97–119 (1953).
35. Schofield, R. K., *J. Soil Sci.* **1**, 1–8 (1949).
36. Schofield, R. K., *Trans. Brit. Ceram. Soc.* **48**, 207–213 (1949).
37. Schuffelen, A. C., *Trans. Intern. Soc. Soil Sci. Dublin*, **1**, 180–188 (1952).
38. Swenson, R. M., Cole, C. V., and Sieling, D. H., *Soil Sci.* **67**, 3–22 (1949).
39. Tamm, O., *Medd. Stat. Skogsförs.-anst.* **27**, 1–20 (1932).
40. Vageler, P., "Der Kationen- und Wasserhaushalt des Mineralbodens," Berlin, 1932.
41. Vanselow, A. P., *Soil Sci.* **33**, 95–113 (1932).
42. Verwey, E. J. W., and Overbeek, J. Th. G., "Theory of the Stability of Lyophobic Colloids," Amsterdam, Elsevier Publishing Co., Inc., 1948.
43. Wiegner, G., *Trans. Intern. Congr. Soil Sci. 3rd. Congr.* **3**, 5–28 (1935).
44. Wiklander, L., *Kgl. Lantbruks-Högskol. Ann.* **14**, 1–171 (1947).
45. Wiklander, L., *Kgl. Lantbruks-Högskol. Ann.* **17**, 407–424 (1950).
46. Wiklander, L., *Acta Agr. Scand.* **1**, 190–202 (1951).
47. Williams, E. G., *Trans. Intern. Soc. Soil Sci. Dublin* **1**, 31–47 (1952)

CHAPTER 5

Soil Organic Matter

JAMES L. MORTENSEN AND FRANK L. HIMES

Department of Agronomy
The Ohio State University, Columbus

Soil organic matter is no doubt one of the most complex materials existing in nature. In addition to the organic constituents present in undecayed plant and animal tissues, soil organic matter contains living and dead microbial cells, microbially synthesized compounds, and an endless array of derivatives of these materials produced as the result of microbial activity. Soil organic matter probably contains most, if not all, of the naturally occurring organic compounds. Some components of soil organic matter are no doubt distinctive to the soil environment, particularly those involving inorganic-organic complexes. The brown-colored, high-molecular-weight component of soil organic matter, commonly called humic acid, may have characteristics considerably dissimilar to brown-colored polymers from other natural sources.

The task of identifying and characterizing all of the components present in soil organic matter is formidable. The variety of structures which occur in natural products is infinite (110, 130), and thus there is little invitation to detailed chemical study. But it cannot be assumed that if a compound is present in very low concentration in the soil it must be of little consequence. Soil organic matter must be chemically characterized if practical questions regarding its role in soil formation and productivity are to be answered. Recent progress in the extraction, separation, identification, and characterization of components of soil organic matter is most encouraging. Application of the techniques developed by biochemists, colloid chemists, and coal chemists, particularly for the characterization of high-molecular-weight polymers, has been of great value in the characterization of soil organic matter. Ziechmann and Parvelke (206) have reviewed the application of some of these procedures to studies on soil organic matter. Much of the information is as yet fragmentary, however, and the concentrated effort of many organic matter chemists will be required before the chemistry of soil organic matter is finally elucidated.

EXTRACTION

Since the chemistry and physicochemical properties of soil organic matter are best studied in the absence of the inorganic matrix of soil, organic matter must first be dissolved by an aqueous or nonaqueous extractant (115). Presently available extraction procedures are empirical and leave much to be desired; complete extraction, however, seems to be an impossibility. Some of the difficulty probably results from the inherent insolubility of cell wall components in plant and microbial residues. Extraction of any sort of fraction representative of the whole also seems impossible, since components of soil organic matter are not equally soluble in a single extracting agent. Since most extraction procedures probably produce artifacts, organic matter extracted from the soil may be very different from that which exists in the soil. Since the organic matter content of most agricultural soils is less than 5 per cent, extraction yields are low, and large amounts of soil must be extracted to obtain sufficient organic matter for identification and characterization procedures. In spite of these difficulties, considerable information on the chemistry of soil organic matter can, however, be obtained on fractions, which can be solubilized by different extractants.

Proximate Analysis

The proximate analysis method developed by Waksman and co-workers (195) is based on classical solubility principles involving attraction between solute and solvent. Even though not too specific in extraction of well-defined components, this non-isolative procedure has been useful in some organic matter studies. General information on the quantity of different classes of compound, particularly during the process of plant residue decomposition, has been obtained by proximate analysis. Table 5.1 illustrates the kind of information which can be obtained by this procedure.

Most of the organic matter is present in the residue, which is further subdivided into two rather vague groups, the lignin-humus complex and the organic nitrogenous complexes. It cannot be assumed, however, that all the carbon in the residue is lignin or its transformation products, or that fractions dissolved by ether, alcohol, and acid hydrolysis do not contain nitrogen. The fractions labeled hemicellulose and cellulose undoubtedly

TABLE 5.1. PROXIMATE CHEMICAL COMPOSITION OF SOIL ORGANIC MATTER FROM DIFFERENT SOILS (195) ON THE BASIS OF THE TOTAL SOIL ORGANIC MATTER (C × 1.72)

Ether-soluble material (fats, waxes) %	Alcohol-soluble material (resins) %	Hydrolysis with 2% HCl (hemicelluloses) %	Hydrolysis with 80% H_2SO_4 (cellulose) %	Residue (lignin-humus complex) %	N in residue by Kjeldahl (organic nitrogenous complexes) %	Sum of the constituents accounted for %
3.56	0.58	5.44	3.55	43.37	33.78	90.28
4.71	1.53	8.60	5.22	40.81	34.74	95.61
1.94	3.10	12.59	5.36	35.18	35.77	93.94
0.80	0.82	5.53	4.12	41.87	37.35	90.49
1.02	0.88	6.96	3.50	42.05	33.25	87.66
0.46	0.84	8.54	2.83	42.83	33.36	88.86
0.52	0.63	10.66	3.38	46.50	33.13	94.82
0.62	0.61	8.61	3.64	49.29	30.38	92.75

contain other components. It is apparent that the proximate procedure has limited value in the identification and characterization of the many components present in soil organic matter.

Adsorption of Organic Matter on Clay

In addition to differential solubility problems, extraction is made difficult by the adsorption of organic matter on mineral constituents of the soil. Due to their tremendous surface area and preponderance of reactive adsorption sites, clay minerals adsorb organic matter components in such a way as to render them insoluble in common extractants. Studies with mined clays and single adsorbates have shown that polar compounds and non-ionic compounds are adsorbed on clay, particularly on the basal surfaces of expanding clays. Aliphatic molecules are oriented with the plane of their chains both parallel and perpendicular to the clay surface (71). Some organic molecules are apparently held to the clay surface through C—H···O (clay-mineral surface) bonds (19, 45, 103), although any adsorbate capable of supplying electrons to the incomplete p orbitals of adsorbed or lattice aluminum could be adsorbed as a carbonium ion (81). Exposed lattice aluminum and iron are apparently involved in the adsorption of some carbohydrates on clay (43). Compounds containing electronegative elements may be adsorbed by a hydrogen-bonding mechanism. Methylation of hydroxyl groups on organic compounds generally increases adsorption on clay. Multilayers of polar molecules, particularly those soluble in water, are no doubt rather easily desorbed from clay by aqueous extractants.

Organic cations and polycations, such as proteins, below the isoelectric point are adsorbed on clay surfaces by a cation-exchange mechanism. These coulombic forces are probably supplemented by van der Waals forces and C—H···O bonds between the organic molecule and the clay mineral surface. Many types of organic cations and polycations are adsorbed on the basal plane surfaces of expanding-type clays in an orientation that causes minimum expansion of the layers (63). Entropy considerations suggest why exchange by small ions is impossible. The use of a basic extractant removes adsorbed amphoteric compounds. An increase in basicity renders the organic matter electronegative and prevents extensive adsorption by coulombic repulsion forces.

In most agricultural soils which are near a neutral pH, the soil organic matter has a net negative charge, and cation-exchange reactions with clay are probably inoperative. Negatively charged organic anions and polyanions are apparently linked to the clay surface through polyvalent inorganic cations and ionized carboxyl groups (clay—M—OOCR). The inorganic cations can be exchangeable cations; interlayer, adsorbed aluminum or iron hydroxide polymers; or constituents of the clay lattice. Positive charges due to exposed lattice-edge aluminum and aluminate surfaces have been shown (113, 194) to be adsorption sites for negatively charged, carboxylated polymers. Extent of adsorption is governed by pH and the species and concentration of salts which effect the charge of the clay and the charge and configuration of the polymer coils. Negatively charged polymers are apparently not adsorbed on the basal surfaces of expanding clays. Desorption of model negatively charged polymers has not been extensively studied, but it appears that reagents which decompose the clay lattice, such as phosphate, fluoride, strong base, and hydrofluoric acid, may be most useful.

Few studies have been made on the adsorption on clay of organic matter extracted from soil, but the presence of polyvalent exchange or lattice metal ions and acidity increases adsorption (90, 160).

Organo-mineral Gels

Another factor which affects the solubility of organic matter is the formation of salts and gels with metal ions (116). The organo-mineral colloids present in soil are a complex mixture of floccules and gels (192) whose stability depends on the pH and oxidation-reduction status of the soil (3). Mutual coagulation and peptization reactions occur during the interaction of organic matter with aluminum and iron. The interaction is strongly pH-dependent, since iron and aluminum can exist in several forms. These organo-mineral gels are transformed into water-soluble hydroxy- complexes by partial hydrolysis in a slightly alkaline medium. In a strongly alkaline medium they may be completely hydrolyzed (88).

Pretreatment

Plant roots and undecomposed leaves and stems are usually not considered as soil organic matter and can be removed by hand before extraction. Dispersion of the soil in carbon tetrachloride or cold water can be of assistance in floating off low-density materials, but these liquids solubilize some organic matter components.

Some of the organic compounds present in soil, such as small quantities of amino acids (125), exist in the free, monomeric form, but the greater part of soil organic matter is adsorbed on soil minerals or composed of high-molecular-weight, organo-mineral gels. Hydrocarbons are present as complex monomeric mixtures. With the exception of fats, waxes, and resins which are soluble in benzene:methanol (10:1), chloroform, and ethanol, the organic matter in surface soils is quite insoluble in organic reagents (30). Extraction of the soil with these reagents is often performed as a pretreatment procedure prior to extraction with aqueous solutions. Other pretreatment procedures, such as ball-milling the soil, sulfacetolysis, or densimetric fractionation (111) to remove "unhumified" or unadsorbed material (176), hydrofluoric-acid–hydrochloric-acid mixtures, and sodium dithionite (35), and decalcification with 0.1 *N* hydrochloric acid, have in some instances increased the efficiency of extraction. Pretreatment procedures, however, solubilize some organic matter, the chemistry of which may be very different from that solubilized by the major extractant.

Aqueous Extractants

Aqueous extractants have had wide usage as extractants for soil organic matter. The extractant must wet the entire soil, but continual shaking does not seem necessary. The efficiency of extraction using ultrasonic sound has not been much greater than that obtained by ordinary shaking. A small amount of the organic matter in soil, chiefly some of the polysaccharides, can be extracted with warm water, but the yield is very low. With the exception of hydrofluoric acid-hydrochloric acid mixtures (147), soil organic matter is quite insoluble in mineral acids. Hydrofluoric acid apparently attacks mineral components releasing adsorbed organic matter. Weak organic acids such as formic acid (123) and lactic acid (30) have been used as extractants, but they are not as efficient as salts of organic acids which chelate metal ions (20, 191). Carbonate, borate, fluoride, and sulfite salts have also been used with some success (191). Extraction with sodium sulfite apparently reduces and precipitates iron with a release of sodium humate. Reagents such as 8 *M* urea and a mixture of 6 *M* lithium bromide and 2.5 *M* guanidinium chloride, which rupture hydrogen bonds, may solubilize some fractions of organic matter.

The most useful extractants for soil organic matter appear to be 0.1 *M* sodium pyrophosphate and 0.1 *N* or 0.5 *N* sodium hydroxide (20, 35). Much more organic matter is put into solution by these extractants than by milder extractants, although some auto-oxidation and production of artifacts probably occurs, particularly with sodium hydroxide. Decarboxylation reactions occur readily (44). Repeated treatments with dilute solutions of hydrofluoric acid and sodium hydroxide can apparently extract most of the organic matter from some soils (26, 35). Less than 25 per cent of the carbon in most soils is extracted by sodium pyrophosphate, but physicochemical properties of the solubilized organic matter probably undergo less alteration with this extractant than with alkali or acid. Pyrophosphate probably solubilizes organo-mineral gels by chelating entrapped metals. Pyrophosphate is also bound by some organic compounds (120).

Increasing the pH of extractants increases extraction efficiency, but concomitant hydrolysis and auto-oxidation produce artifacts. The beneficial effects of both chelation and alkali have been combined by the use of "Dowex" A-1 chelating resin in the sodium form in water as an extractant (27). Efficiency of extraction is comparable to 0.5 *N* sodium hydroxide, and an extract devoid of contaminating extractant is obtained, although the ash content of the extract is quite high.

Table 5.2 shows that some nitrogen is extracted from soil by pretreatment procedures. Some of the nitrogen extracted by the hydrochloric-acid–hydrofluoric-acid mixture is fixed ammonia. Increasing the pH of pyrophosphate extractants increases extraction efficiency, but efficiency is generally about half that of sodium hydroxide.

Hydrolysis, Oxidation, and Hydrogenation

Concentrated mineral acids, such as 6 *N* hydrochloric acid and 6 *N* nitric acid, can be used to hydrolyze organic matter for the study of monomeric components. Hydrolysis, however, is usually incomplete, and certain

TABLE 5.2. EFFECT OF PRETREATMENTS ON THE AMOUNTS OF NITROGEN EXTRACTED BY SODIUM PYROPHOSPHATE AND SODIUM HYDROXIDE (35)

Pre-treatment	% Total nitrogen extracted*				
	Pre-treatment	0.15 *M* $Na_4P_2O_7$			0.5 *N* NaOH
		pH 7	pH 9	pH 11	
None	0	19.5	27.9	31.5	60.6
0.1 *N* HCl	2.0	18.2	24.0	25.1	68.8
0.16 *M* Na_2O_4	11.7	—	20.6	—	52.1
HCl·HF†	12.7	25.7	33.8	35.8	66.6

* Mean of 5 soils.
† 1 ml HF (48%) to 100 ml 0.1 *N* HCl.

polymeric components, some of which seem to form during the hydrolysis procedure, remain in an insoluble form. Degradation procedures using alkaline nitrobenzene, alkaline permanganate, alkali fusion, hydrogen peroxide, reduction and hydrogenation have also been used to obtain monomeric components of soil organic matter.

Chelating Extractants

The B-horizon of podzol soils present a special situation for extraction procedures. Organic compounds such as acetylacetone, cupferron, 8-hydroxyquinoline, ethylenediaminetetraacetic acid (EDTA), which are capable of chelating metals (36, 53, 105, 148), have been successfully used to extract the organic matter from these soils. The organic matter present in these soils has probably been precipitated or flocculated in place by iron and aluminum. Complexing these metals renders the organic matter soluble in water.

Extraction Difficulties

In addition to differential extraction efficiency, different extractants solubilize organic matter having different physicochemical properties (53, 196). These differences have not always been recognized, since many investigators have been content to characterize their preparations by color, nitrogen content, and carbon-nitrogen ratio. Effects of shaking time, pH, temperature, soil-extractant ratio, presence of oxygen, and light on the composition of extracted organic matter have not been adequately evaluated.

By carbon-14 dating methods, Broecker and Olsen (28) have recently shown that soil organic matter extracted with alkali had an average age of 50 to 250 years. The unextracted organic matter remaining in the soil had an average age of 2000 years. Aqueous reagents apparently extract organic matter of rather recent origin. The chemistry of the insoluble portion may be very much different from that of the soluble portion. Further work similar to that of Broecker and Olsen, using different extractants, appears to be imperative.

Auto-oxidation no doubt alters many of the properties of extracted organic matter (21, 73). Production of humic-like substances may occur during the extraction process. For example, amino-pyrocatechols exhibit an usually pronounced tendency to auto-oxidize to polymeric forms in alkaline solution.

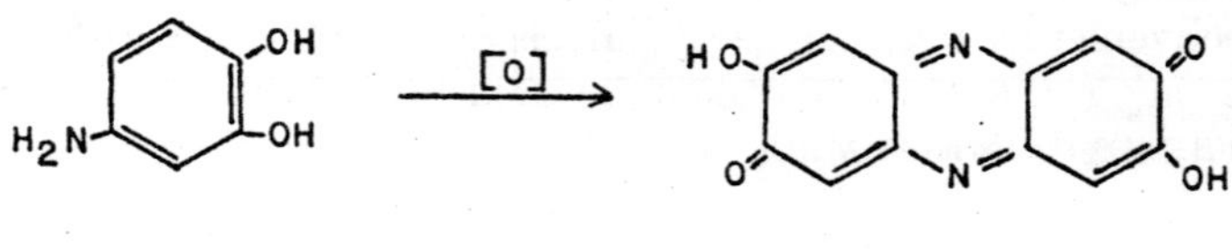

Metals are known to catalyze oxidation and polymerization reactions (101). Photochemical oxidation no doubt occurs to some extent. Such difficulties may be circumvented by performing extractions under nitrogen and special light that has no photochemical effect. Addition of reducing agents to extracting solutions may be of some value (35).

A common difficulty with all extractants is that the organic matter is contaminated with salts, clay, and metals, particularly aluminum and iron (134). It is unfortunate that so much work on the chemistry of organic matter has been performed on preparations which have been highly contaminated with inorganic compounds.

SEPARATION AND FRACTIONATION

The composition of the soil-organic-matter extract depends upon the nature of the extractant, and the extract generally consists of the extractant itself, mineral matter from the inorganic matrix of the soil, salts, complexed metals, and a heterogeneous mixture of organic compounds. The inorganic constituents must be removed if the physicochemical properties of the organic compounds are to be studied. The organic compounds are also conjugated together, and some studies may require preliminary dispersion. Desalting and dispersion of soil organic matter is difficult and not always successful, but careful studies require purification of crude extracts. Iron and aluminum are difficult to remove from organic matter extracts. Separation and fractionation should be conducted at 4°C to prevent microbial alteration. Toluene may be added to the extracts if it does not contaminate the preparation under study.

The volume of the extracts is best reduced by evaporation in a flash evaporator at 50°C or by freeze drying. The process of freezing and thawing destroys the colloidal properties of some preparations, causes coagulation, and permits rapid filtration. Evaporation by heat causes hydrolysis and decarboxylation and usually results in a preparation which is difficult to rewet and redissolve.

Removal of Clay, Salts, and Metals (168)

High-speed centrifugation in the cold (4°C) can help remove clay particles. Treatment with a mixture of hydrofluoric acid (0.3 *M*) and hydrochloric acid (0.3 *M*) for 10 minutes at 4°C (114) may assist in the removal of clay. Treatment with this acidic mixture precipitates some material and may cause some hydrolysis.

Desalting of extracts containing high-molecular-weight compounds is most conveniently performed by dialysis, using cellulose tubing. Repeated dialysis against 0.05 *N* hydrochloric acid and distilled water is usually required to reduce the ash content of an organic matter preparation below

10 per cent. Some preparations may require repeated dialysis against weak acid, weak base, and water to obtain adequate desalting. Dialysis against an EDTA solution has been of value. The EDTA is then removed from the preparation by dialysis against water. Rate of dialysis can be increased by pressure dialysis, electrodialysis, or electrical desalting procedures. Passage of the extract through an anion-exchange resin (OH^- form) and then a cation-exchange resin (H^+ form) is often helpful (72).

Desalting procedures usually cause separation of the high-molecular-weight organic components from the low-molecular-weight components. The latter will be present in the dialyzate and can be desalted by electrical desalting, liquid-liquid extraction techniques, selective ion-exchange reactions on columns of ion-exchange or ion-retardation resins (132), and gel filtration (58). Some separation and fractionation also occurs during ion-exchange reactions and gel filtration. Figure 5.1 shows that a sodium pyrophosphate extract of peat soil can be separated into four fractions by gel filtration. Conductivity measurements showed that several of the fractions had been effectively desalted.

Some organic-matter preparations can be desalted by liquid-liquid extrac-

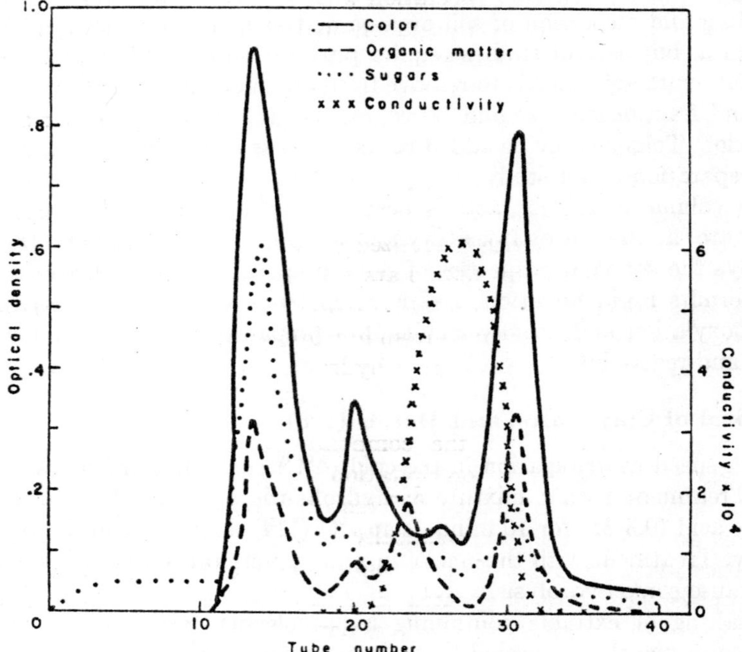

Figure 5.1. Gel filtration (sephadex G-25) of crude $0.1M$ $Na_2P_2O_7$ extract of peat (12- to 18-inch depth) (33).

tion procedures (105). The organic compound of interest is extracted with an appropriate solvent, or metals can be extracted into a nonaqueous solvent containing an appropriate chelating compound (202). Chelating compounds, however, are usually difficult to remove from organic-matter preparations and may be carried down as contaminants during subsequent fractionation procedures. Countercurrent fractionation procedures have not been applied to organic-matter preparations.

Organic Combinations

Organic matter in aqueous extracts is no doubt present as a complicated, disorganized system of many chemical constituents of various sizes and shapes. Some of the molecules are large and fibrous or globular and probably adsorb various small, micromolecular impurities or monomeric constituents. Some molecules are no doubt conjugated together by primary structural bonds, covalent bonds such as ether or ester linkages, hydrogen bonds, and various kinds of coulombic linkages. This motley fabric most often has been studied *in toto*, but much could be learned by careful degradation procedures. Careful acid or enzymatic hydrolysis, rupture of hydrogen bonds, fractional precipitation, or dissolution with various salts and solvents may be of value in degrading complex soil-organic-matter preparations. Such procedures must be mild or intramolecular bonds might be broken.

Some of the components which are adsorbed or conjugated to the high-molecular-weight brown polymeric material in soil-organic-matter extracts may be rather easily removed. Serial extraction of an electrodialyzed, lyophilized sodium hydroxide extract of peat with ethyl ether, ethyl acetate, methanol, and water have yielded a number of compounds, positive to ninhydrin and coupled with diazotized reagents, which could be separated by paper chromatography (33). Fats and waxes can be rather easily removed with appropriate solvents. Polysaccharides and nitrogen-containing compounds have been removed from humic acid by hydrolysis in 2 *N* HCl, followed by extraction with copper hydroxide in ethylenediamine (143).

Precipitation

Partial fractionation of the components in organic-matter extracts has been accomplished by precipitation with acid and metal salts or by taking advantage of solubility differences in organic solvents. Such procedures do not precipitate chemically definable substances, since co-precipitation and adsorption of other components on the precipitated material takes place (10, 166). Repeated dissolution and precipitation helps to improve the "purity" of such preparations.

The classical procedure for fractionation of organic-matter extracts involves precipitation with acid and dissolution of part of the precipitated material with alcohol (Figure 5.2). Names for each fraction have been in common use, but these fractions are a nondescript mixture of many kinds of chemical compounds. Humic acid or fulvic acid (145) so prepared cannot be considered distinct chemical entities, since the separation by precipitation is probably dependent upon the ability to form hydrogen bonds. The physical and chemical properties of these fractions cannot be explained in terms of molecular structure. The components that have been found in these fractions will be discussed later. From three-fourths to one-half the soluble organic matter occurs in the humic fraction, the proportion being lower for forest soils than for grassland soils. The fulvic acid percentage generally increases with depth in the soil profile. Fulvic acid content of the organic matter in the B_h horizon of podzol soils is usually greater than 75 per cent.

German workers (138) have fractionated alkali-soluble organic matter extracts by coagulation with 2 *N* sodium chloride at pH 7. "Gray humic acid" coagulates while "brown humic acid" stays in solution. Differences in the chemical composition of these fractions have been reported. Some Japanese workers made a preliminary separation using magnesium sulfate as a precipitant before precipitating humic acid with mineral acid. Frac-

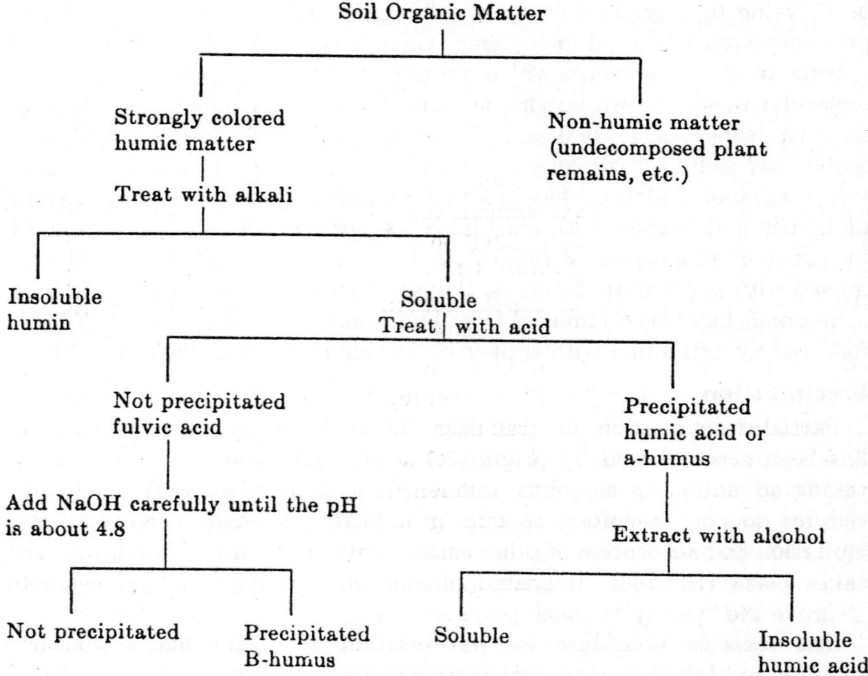

Figure 5.2. Fractionation of soil organic matter.

tionation by partial precipitation with metallic ions has been attempted (44, 46, 174), but the fractions so prepared have not been characterized. Schreiner and Shorey (151, 159, 162, 165) successfully used such precipitation procedures to isolate and purify a number of compounds from soil organic matter.

Fractionation of some organic-matter extracts has been accomplished by precipitation with ether (60, 123), cetyl and laurylpyridinium chloride (36, 40), "Cetavlon" (41, 124), acetone (60, 112), and quaternary ammonium compounds (40).

Chromatography

Chromatographic separations of crude preparations have been attempted, but the results have been somewhat disappointing. Chromatography of complex polymeric mixtures is usually not too successful. Satisfactory solvent systems have not been found which will elute tightly adsorbed crude components from cellulose, alumina, starch, and other column packing materials. Methylation of crude extracts has increased the efficiency of chromatographic separation (12).

Paper chromatography of humic acid generally results in the separation of one brown-colored fraction from the point of application (133). Most of the sample remains near the origin. Diffuse fluorescent zones (yellow-green under ultraviolet light) can be seen near both components. This fluorescent material is present in most organic-matter preparations, but very little is known concerning its identity. Paper chromatographic separation of components in fulvic acid has been reported (92, 109).

Chromatography on alumina has resulted in some separation of humic materials, but elution is quite difficult (69, 75, 91, 141). Humic acids from Japanese soils have been fractionated on alumina columns into four brown-colored components by elution with sodium hydroxide solutions (66). Barton and Schnitzer (12) separated a methylated extract of a B_h horizon into six components on alumina. Columns of alumina or silica gel can be used to remove humic materials from waxes in benzene:methanol extracts. Some fractionation on silica gel has been reported (108). Columns of starch (91), ion-exchange resins (39), cellulose (39, 100, 119, 174) and Sephadex (39, 118) have also been used. Figure 5.1 shows that Sephadex can be used to separate crude extracts into components having gross differences in molecular size. A "green humic acid" from the A_2 horizon of podzol soil (100) and polysaccharide components (119) have been successfully separated on cellulose. The "green humic acid" appears to be a derivative of polymerized porphyrin nuclei.

Components in fulvic acid and hydrolyzates of humic acid have been separated on charcoal into four fractions reportedly containing (*a*) free sugars and amino acids, (*b*) phenolic glycosides, (*c*) polysaccharides, and (*d*) materials rich in nitrogen, containing pentoses and organic phosphorus (60). Phenolic glycosides have not been identified in fraction (*b*). Colloids separated by this procedure are heterogeneous with respect to both charge and particle size (187).

Electrophoresis

The separation of substances by means of an electric current depends primarily on differences in the charge density of the material under study. Components in crude preparations of soil organic matter are negatively charged and polydisperse with respect to size and charge density, so that sharp separations are not always possible.

Paper electrophoresis of crude preparations generally results in the separation of two brown-colored components (39, 94, 136, 202; see Figure 5.3). The component of lowest mobility remains near the point of application and may be uncharged or strongly adsorbed on the paper. Ash content seems to determine the mobility of this component. Kononova and Titova (94) have shown that the component of greatest mobility contains iron in a negatively charged complex. Complexes of $Sr^{90}(Y^{90})$ and Fe^{59} with organic matter have been separated by paper electrophoresis (67, 118). Some brown material is usually distributed on the paper between the major components. A component which fluoresces (yellow-green) under ultraviolet light can be seen near the leading edge of the component of greatest mobility. When this fluorescent component is extracted from the paper and subjected to paper electrophoresis it is immobile, suggesting that it is adsorbed on the brown-colored, mobile component. Bluish, rose and yellow-orange fluorescence have also been seen on paper electrophoretograms of humic acid (78).

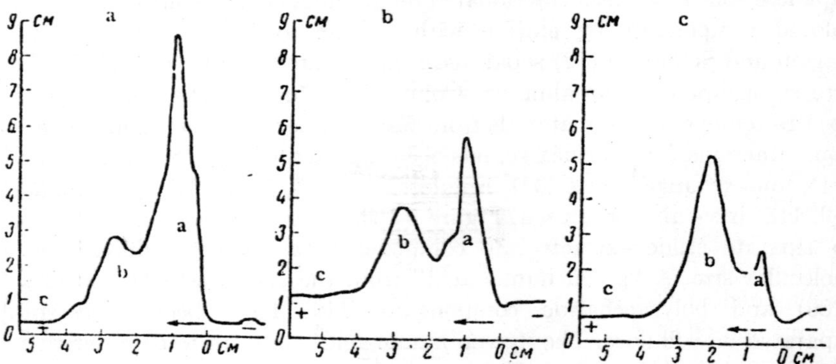

Figure 5.3. Character of humic acid distribution on phoregrams (94) (curves recorded using a densitometer) from (*a*) chernozem, (*b*) sod-podzolic soil, (*c*) a humus illuvial horizon of strongly podzolic soil; zones (*A*) at the start (*B*) brown-moving, and (*C*) fluorescent.

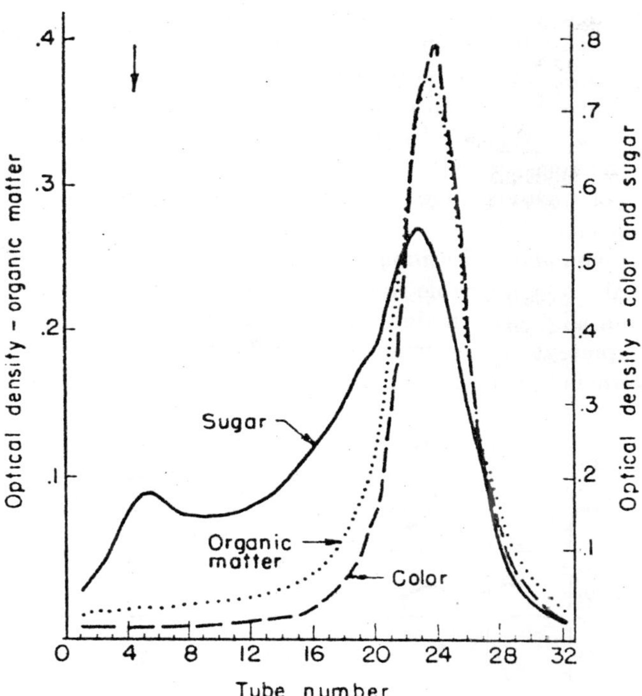

Figure 5.4. Electrophoretic separation of polysaccharides in pyrophosphate extract of soil (phosphate buffer) (114).

Continuous-flow paper electrophoresis has been used to characterize and separate components in crude preparations of organic-matter extracts*. Polysaccharides in water extracts have been separated into four uronic-acid-containing constituents by this procedure (112, 114). Uncharged polysaccharides can be separated from the major brown-colored component by continuous-flow paper electrophoresis (Figure 5.4). Repeated electrophoresis of this brown-colored component would probably result in some purification. Such a procedure should be much better than separation by acid precipitation. Even though sodium pyrophosphate extracts contain one electrophoretically separable brown-colored component, fractions having high electrophoretic mobility contained more hydrogen-bonded NH groups and COO^- groups and fewer OH and CH groups than fractions of low mobility (114). Some differences in electrophoretic mobility of the brown-colored component in different soil types has been reported (87).

* See references 30, 80, 87, 112, 114, 140, 196, and 197.

Fulvic acid (Figure 5.5) and hymatomelanic acid can be fractionated into several components by continuous-flow paper electrophoresis (196, 197).

Mackenzie and Dawson (104) have separated sodium pyrophosphate extracts of peat soils into three fractions each of polysaccharide, organic matter, and colored material by column electrophoresis on cellulose (Figure 5.6). High exchange-capacity values for muck seem to be associated with the amount of highly charged material rather than with an increase in exchange capacity.

Using free-boundary electrophoresis, Stevenson *et al.* (178, 179) have shown that the nondialyzable fraction of a sodium pyrophosphate extract consists essentially of a major mobile constituent with two additional components present in very small concentration (Figure 5.7). Considerable electrophoretic heterogeneity was noted.

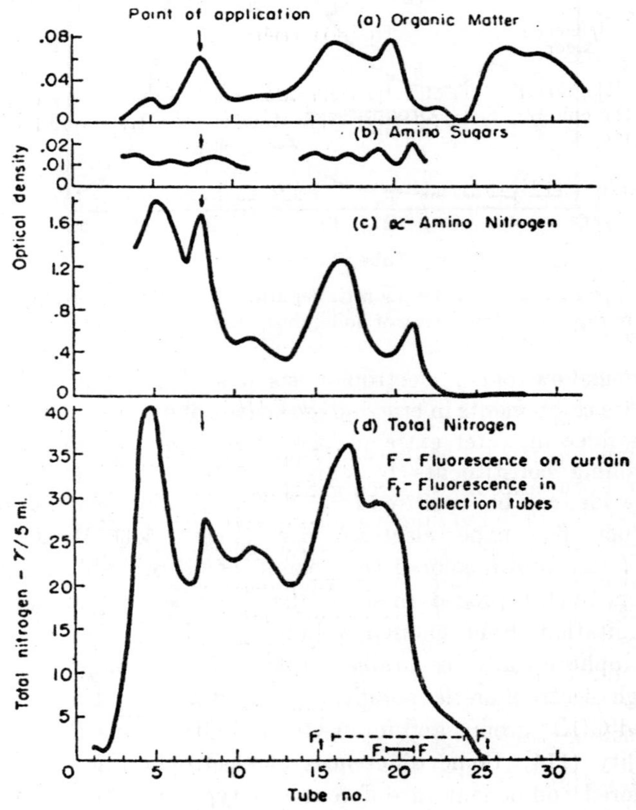

Figure 5.5. Electrophoretic fractation of NaOH extract (fulvic acid) at 500 v and 0.05 ma in borate buffer (196).

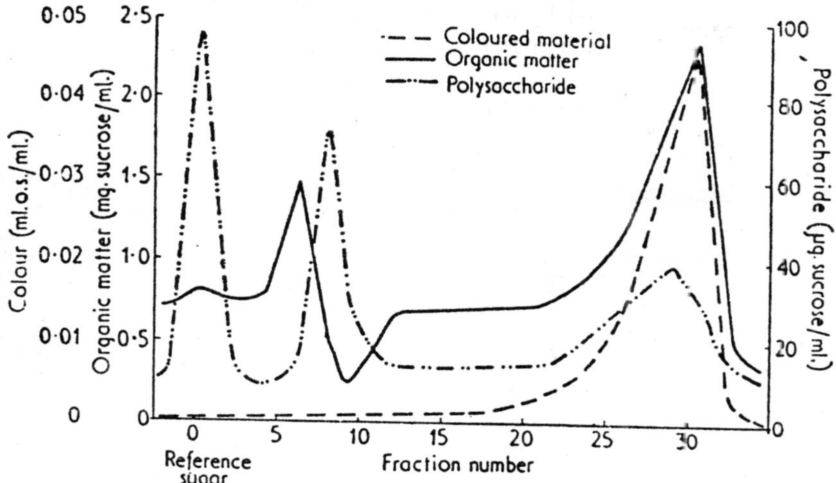

Figure 5.6. The distribution of polysaccharides, colored material, and organic matter from the column electrophoresis separation of an extract obtained from a muck soil (104).

Free-boundary electrophoresis of soil polysaccharides has demonstrated he heterogeneity of these compounds (13).

)iffusion and Sedimentation

Diffusion (39, 178) and sedimentation (141, 146, 178, 179) have been sed to characterize organic-matter polymers, but these methods are of mited value as separation procedures.

CHEMICAL COMPOSITION

Since soil organic matter is derived from the metabolic products and ssues of plants, animals, and microorganisms, its chemical composition is xtremely complex. If extraction, isolation, and identification procedures ere sufficiently sensitive, every naturally occurring organic compound ould, no doubt, be detected in soil organic matter.

The chemical composition of soil organic matter can be studied from the iewpoint of gross features, such as functional-group analysis, spectroscopy, r physicochemical properties (29, 55, 56). It must be remembered, however, that the presence of specific compounds contribute to the over-all roperties of soil organic matter.

unctional Groups

Functional-group analysis of crude preparations is quite difficult because inaccessibility of the reagents to the entire surface of the organic-matter

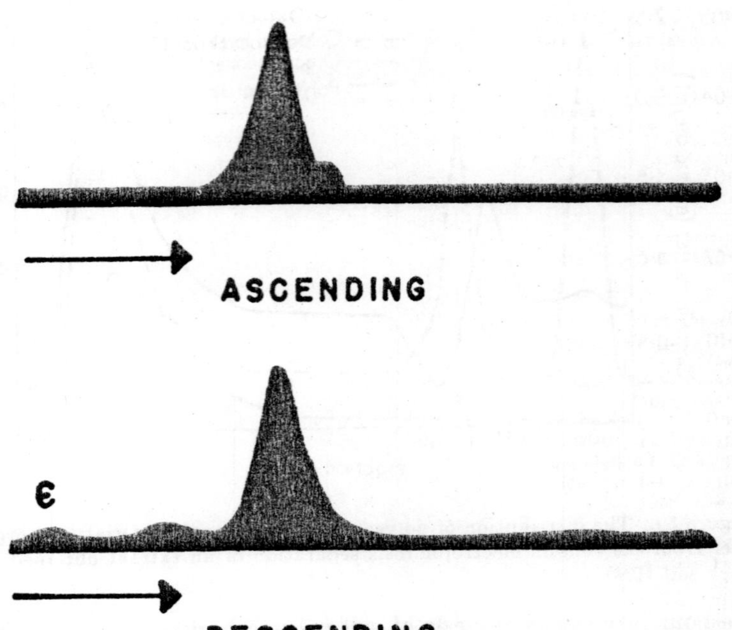

Figure 5.7. Representative electrophoresis patterns of colloidal organic matter extracted from Miami clay (179); analysis was made at pH 9.0 and ionic strength 0.1; time was 68 minutes.

mixture. Even though results often are not reproducible, valuable informa tion about the peripheral groups can be obtained by this procedure.

Methods for functional group analysis have been summarized by Va Krevelen and Schuyer (97) and used most extensively in recent years b investigators in the Canada Department of Agriculture (12, 203). Tabl 5.3 shows that dialyzable compounds in a podzol soil soluble in 0.5 *N* sodiu hydroxide contained less carbon and hydrogen and more carboxyl, alcohol hydroxyl, and carbonyl groups than nondialyzable components. Practicall all the oxygen in the dialyzable components could be accounted for i these groups. Differences in the distribution of functional groups with dept in the profile have been noted (95, 121, 203). Present evidence sugges that free amino groups are not present in soil organic matter (25, 170 Hydrolysis is required to release amino acids from polymeric forms. Son nitrogen in humic acid is stable against hydrolysis and unaffected b methylation and acetylation (59, 74), and may be present in heterocycl form.

Determination of functional groups, molecular weight, and element analysis permits calculation of molecular formulas. Even though su

TABLE 5.3. CHEMICAL PROPERTIES OF THE ORGANIC MATERIALS EXTRACTED FROM THE A_0 AND B_h HORIZONS OF A PODZOL SOIL (DRY ASH-FREE BASIS (203)

Fraction of 0.5 *N* NaOH extract	Elementary composition (%)				Functional groups (meq/g)						
	C	H	N	O + S	Total acidity	—COOH	Total OH	Phenolic OH	Alcoholic OH	C=O	OCH_3
A_0—dialyzate passed over "Amberlite" IR-120	49.93	4.67	1.27	44.13	8.9	6.1	7.4	2.8	4.6	3.1	0.3
A_0—dialyzed	55.55	5.18	2.19	37.08	5.6	2.1	6.9	3.5	3.4	1.0	0.5
A_0—dialyzed, ppt. with HCl, washed, dialyzed	58.13	5.17	2.14	34.56	5.7	2.8	5.9	2.9	3.0	3.0	0.5
B_h—dialyzate passed over "Amberlite" IR-120	48.38	3.28	0.60	47.74	11.8	9.1	7.6	2.7	4.9	1.1	0.3
B_h—dialyzed, adjusted pH 8.0 (NaOH), settled, ppt. with HCl, washed	56.69	4.94	2.47	35.90	6.6	3.7	—	2.9	—	—	0.4
B_h—fulvic acid (HCl free) of dialyzed adj. to pH 8.0 (NaOH), settled, ppt. with HCl	46.94	3.48	2.31	47.27	10.8	7.2	6.4	3.6	2.8	—	0.2

data are dependent upon the methods of extraction and analytical measurement they are of value in characterizing some organic-matter preparations. Schnitzer and Desjardins (150) reported molecular formulas of C_{75} H_{33} O_{17} N_3 $(COOH)_3$ $(OH)_{12}$ $(CO)_2$ and C_{21} H_{12} $(COOH)_6$ $(OH)_5$ $(CO)_2$ for the electrodialyzed and deionized organic matter from the A_o and B_h horizons, respectively, of a podzol soil. Hosoda and Tokata (74) reported the molecular formula of C_{265} H_{258} O_{76} (OCH_3) $(COCH_3)$ $(OH)_{12}$ $(COOH)_{12}$ for a humic acid preparation. Molecular formulas of impure humic acids are of limited value, since these preparations contain a heterogeneous mixture of many kinds of compounds.

Spectroscopy

Absorption spectra of crude and chemically undefined organic compounds present in soil are difficult to interpret, but spectrophotometry has been used with some success to corroborate functional-group analysis. Sharp absorption bands characteristic of pure compounds are obscured in crude humic preparations. Chromatographically pure compounds yield much better spectra than crude preparations. Spectrophotometry can assist in the identification of unknown compounds (51, 144).

Table 5.4 lists the assignments of common absorption bands in the infra-

TABLE 5.4. CHARACTERISTIC INFRARED GROUP FREQUENCIES (APPROXIMATE) FOUND IN ORGANIC MATTER PREPARATIONS (117)

Wavelength (μ)	Wave number (cm^{-1})	Group
2.90	3400	OH stretch*
3.03	3300	NH stretch
3.25	3100	NH_3^+ stretch
3.42	2940	CH stretch
5.7–5.8	1750	COOH and ester C═O
6.0–6.1	near 1680	COO^- antisymmetric stretch*
6.0–6.2	1650–1620	C═O stretch (Amide I)*
6.27	1600	Aromatic C═C
6.5–6.6	1570–1515	NH deformation + C═N stretch (Amide II)
6.62	1510	Aromatic C═C
6.85	1460	$OC—H_3$
6.9–7.0	near 1435	CH bend
7.1	near 1400	COO^- symmetric stretch
7.7	near 1300	C═N stretch + NH deformation (Amide III)
7.89	1267	Aromatic C—O—Me
8.1	near 1230	C—O, ester linkage
8.14	1230	Phenolic C—OH
8.5–10.5	1170–950	C—C, C—OH, C—O—C typical of glucoside linkages and polymeric substances
9.67	1035	$O—CH_3$
11.9	840	Aromatic C—H

* Also given by water in KBr pellets.

red, and Figure 5.8 shows some representative infrared spectra of organic-matter preparations. Infrared spectra of humic acid solubilized by different extractants are similar (149), but polymeric materials separated by electrophoresis (114) or altered by humification, degradative, or preparative procedures (54, 99, 137) show some differences. Aromatic structure is generally difficult to observe in infrared spectra of humic preparations, but some investigators have correlated several absorption bands with the presence of aromatic compounds (54, 86, 137). Infrared spectra indicate that humic preparations usually contain a highly unsaturated structure having hydroxyl, amide, and carboxyl functional groups.

Absorption spectra of humic preparations in the ultraviolet and visible wavelength regions (Figure 5.9), show little detail (57, 98). A small shoulder is usually seen between 250 and 279 mμ, and absorption increases toward shorter wavelengths. The visible spectra of a chromatographically separated "green" humic acid from a podzol soil suggested the presence of porphyrin nuclei (100). Extinction coefficient is pH-dependent, but preparations having higher extinction coefficients are thought to contain more condensed aromatic rings than preparations of lower extinction coefficients. Kononova (93) has shown that the extinction coefficient for humic acid preparations from various soils was in the order: ordinary Chernozem > dark-gray Forest soil > Chestnut soil > light Sierozem > Sod-podzolic soil > Kraznozem > strongly podzolic soil. Extinction coefficients for the fulvic acid fraction from these same soils were similar.

Electron paramagnetic resonance spectra show that humic acid contains stable organic-free radicals of the order of 10^{18} radicals/g (128, 177). The radical may be a mixture of a semiquinone of a catechol-resorcinol-type copolymer (I) and a quinhydrone-type radical (II).

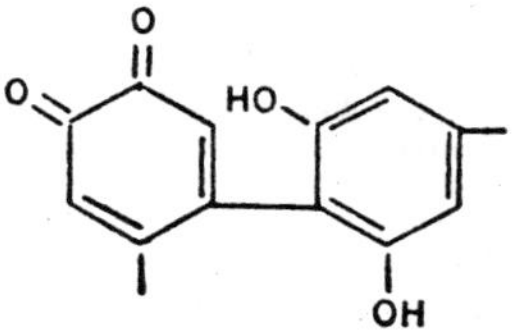

I 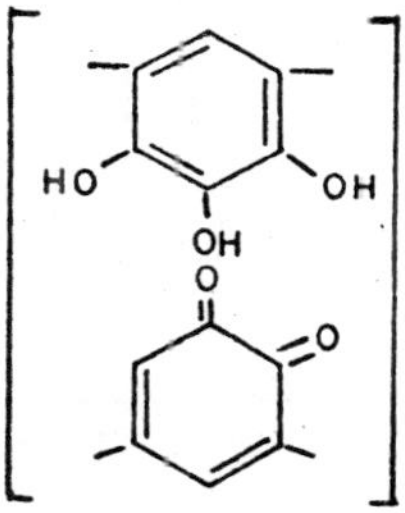 II

Nuclear magnetic resonance studies of chromatographically separated fractions of organic matter soluble in 0.5 N sodium hydroxide surprisingly revealed the absence of aromatic and olefinic hydrogen (12). Content of hydrogen species agreed with functional group analysis by chemical methods.

The Carbohydrate Fraction

Total carbohydrate content of soil organic matter has been estimated to be between 5 and 20 per cent (107). Carbohydrates occur in living microbial cells and roots, in extracellular form, and in decomposing plant residues and their degradation products. The carbohydrate fraction probably undergoes continual degradation and synthesis, since such compounds generally serve as excellent carbon sources for the microbial milieu which exists in the soil. Glucose and plant carbohydrates are quite rapidly incorporated into soil polysaccharides (89).

Small quantities of free monosaccharides have been detected in the soil, but such compounds have only transitory existence. They are rapidly metabolized or adsorbed on clay surfaces. The only reference to the presence of fructose in soil organic matter was made following chromatographic studies of a cold-water extract of a Norwegian forest soil (4).

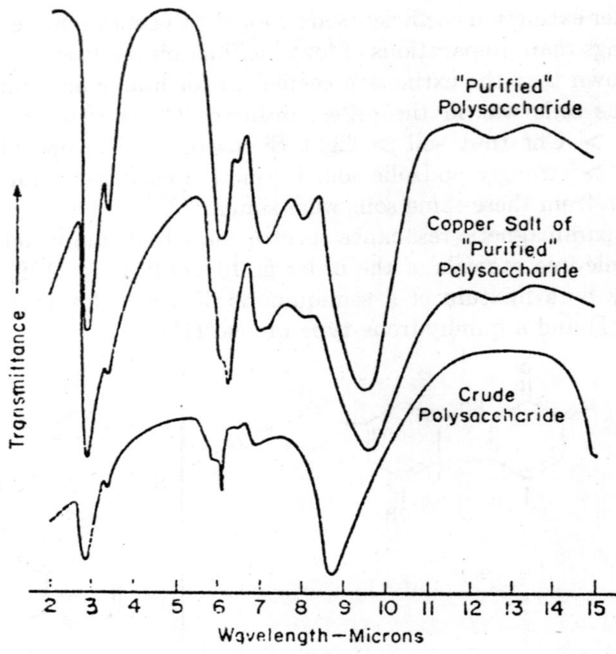

Figure 5.8. Infrared spectra of polysaccharides (112).

Total carbohydrate content of soil is most often determined following hydrolysis with acid. Hydrolysis causes some losses, and yield values must be appropriately corrected. Some of the uronic acids are decarboxylated by most hydrolysis and extraction procedures. In fact, soil organic matter contains considerable material which decarboxylates rather easily (46). Boiling soil with 12 per cent hydrochloric acid cannot be considered a reliable method for determination of uronic acids. Following hydrolysis, interfering chromogens are chromatographically removed, and constituents are determined colorimetrically: uronic acids with carbazole; hexoses with anthrone; pentoses with orcinol or aniline acetate; and amino sugars with Erhlich's reagent (76, 183). Quantitative results are difficult to obtain because of differences in extinction coefficient of each sugar in the mixture. Individual monosaccharides can be separated by column and paper chromatography (51, 61, 175). Hexose content (up to 15 per cent) is higher than pentose (up to 9 per cent) and uronic acid content (up to 4 per cent). The amino sugars, chiefly glucosamine and galactosamine, contain up to 25 per cent of the total nitrogen in soil. Muramic acids have not been reported as constituents of soil organic matter. Since bacterial cell walls contain these compounds, they should be present, but specific hydrolysis and identifica-

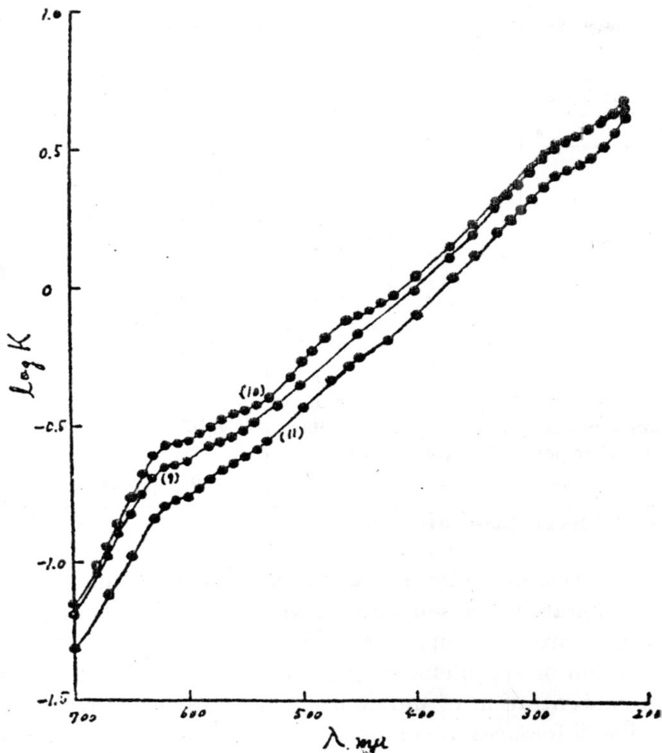

Figure 5.9. Absorption spectra of L-humic acids extracted from podzol and diluvial soils (98).

tion procedures will be required to confirm their presence in soil organic matter. Table 5.5 shows the carbohydrate composition of some surface Canadian soils. Carbohydrate content of B horizons is usually higher than that of A horizons.

Most of the carbohydrates in soil are probably present as polysaccharides (140). Various extractants have been used (128), but the yield is quite low (less than 2 per cent of the soil organic matter) and most of the polysaccharide remains in the soil (1). Polysaccharides can be detected in the humic, fulvic, and humin fractions of 0.5 *N* sodium hydroxide extracts. Most of the carbohydrates can be removed from humic acid preparations by hydrolysis with 2 per cent hydrochloric acid. Further removal of cellulose has been accomplished with Schweitzer's reagent and copper hydroxide in ethylenediamine (143). Purification procedures involving chromatography *,

* See references 13, 60, 107, 112, and 124.

TABLE 5.5. CARBOHYDRATE COMPOSITION OF FOUR CANADIAN SOILS*

Carbohydrates	Mg. sugar per g. soil			
	Grenville	Lennoxville	Scott	Lacombe
High R_f	0.78	0.86	0.49	0.47
Rhamnose	1.32	1.17	0.74	0.67
Ribose	0.21	0.22	0.13	0.14
Xylose	1.46	2.00	1.56	1.34
Arabinose	1.65	2.40	1.50	1.28
Mannose	2.11	2.10	2.15	1.58
Glucose	6.93	9.10	5.05	4.60
Galactose	2.58	2.80	2.42	2.54
Uronic acid	2.7	3.6	1.6	1.4
Hexosamine	8.16	6.22	5.01	4.73
Total carbohydrate	27.99	30.47	20.65	18.75
Carbohydrate C as per cent of total C	12.1	12.6	14.4	12.0

* Taken from Sowden and Ivarson (144).

precipitation, electrophoretic and ultracentrifugation studies have shown that such preparations of soil polysaccharides are not homogeneous (13, 112, 122, 198). Chromatography on DEAE-cellulose has resulted (119, 189) in the separation of deproteinized polysaccharides into five fractions. Differences in concentration of uronic acid and nitrogen were apparent, but no qualitative differences in components could be detected.

Table 5.6 illustrates some properties of crude polysaccharide preparations. Hydrolyzates of crude preparations contain amino acids, no doubt the result of incomplete deproteinization. Rare 4-O-methyl-D-galactose and 2-O-methyl-L-rhamnose have been isolated from a peat polysaccharide (48–51). Infrared spectra, titration curves, and viscosity measurements suggest that soil polysaccharides contain amino, guanidino, and carboxyl groups and complexed metals; are hydrogen bonded; and exhibit electrostatic and ionization reactions characteristic of carboxylated polyelectrolytes. The molecular weight of crude preparations has been estimated to be from 10,000 to 450,000 (13, 112).

Soil polysaccharides no doubt play an important role in the cation-exchange capacity, structural stability, nitrogen availability, carbon metabolism, and complexing of metals in soil.

The Nitrogen Fraction

The nitrogen content of the surface layer of many soils varies between 0.02 and 0.4 per cent (16). Practically all of this nitrogen, 92 to 96 per cent (185), is found in organic combinations. The carbon-nitrogen ratio of agri-

Table 5.6. Characterization of Polysaccharides Isolated from Various Soils

<table>
<tr><th>Factor</th><th>British* soil</th><th>Scottish† soil</th><th>Indiana‡ soil</th></tr>
<tr><td>Equivalent weight</td><td>1185</td><td>1000</td><td>1945</td></tr>
<tr><td>Nitrogen, %</td><td>0.34</td><td>1.6</td><td>0.34</td></tr>
<tr><td>Methoxyl, %</td><td>0</td><td>2.0</td><td>2.4</td></tr>
<tr><td>Reducing sugar, %</td><td>80</td><td>—</td><td>—</td></tr>
<tr><td>Uronic anhydride, %</td><td>15.8</td><td>20.1</td><td>9.1</td></tr>
<tr><td>Amino sugars, %</td><td>0</td><td>0</td><td>Trace</td></tr>
<tr><td>Component sugars as per cent of total</td><td></td><td></td><td></td></tr>
<tr><td>Glucose</td><td>20.8</td><td rowspan="2">36.0</td><td>21.2</td></tr>
<tr><td>Galactose</td><td>20.0</td><td>16.6</td></tr>
<tr><td>Mannose</td><td>21.9</td><td rowspan="2">29.8</td><td>18.5</td></tr>
<tr><td>Arabinose</td><td>11.7</td><td>10.4</td></tr>
<tr><td>Xylose</td><td>23.6</td><td>10.3</td><td>12.6</td></tr>
<tr><td>Ribose</td><td>1.5</td><td>4.5</td><td>Trace</td></tr>
<tr><td>Rhamnose</td><td>0</td><td>11.0</td><td>14.2</td></tr>
<tr><td>Fucose</td><td>0</td><td>0.7</td><td>0</td></tr>
<tr><td>Unknown</td><td>0</td><td>7.0</td><td>6.5</td></tr>
</table>

* From Forsyth (77, 153).
† From Duff (154, 155, 156).
‡ From Whistler and Kirby (150).

culturally important surface soils is approximately eleven to one, and becomes narrower with depth. Some of this change has been attributed to an increase in the amount of ammonium ions fixed within the lattice of clay minerals (185, 186). The nitrogenous organic compounds in the soil are derived from plant and animal residues, root exudates, and microbial cells (139). Most of these nitrogenous compounds are throught to be present as polymers or adsorbed on clay minerals. Soils contain very small quantities of free amino acids (126). Most of the identified nitrogen compounds from soil extracts have been found in hydrolyzates.

More than fifty ninhydrin positive compounds have been chromatographically separated from acid hydrolyzates of whole soils (197, 205). The amino acid content of several surface soils is shown in Table 5.7. All the soil hydrolyzates appear to contain the same amino acids, but some quantitative differences due to rotation (181, 205), depth (182), and climate (34) have been reported.

Acid hydrolysis solubilizes about 75 per cent of the nitrogen in most soils. The insoluble portion may contain nitrogen fixed as ammonia in clay minerals and in heterocyclic and aromatic rings, or combined in insoluble browning reaction products (70) formed during hydrolysis. Most of the ammonia fixed by lignin preparations is not solubilized by acid hydrolysis.

From one-third to one-half of the nitrogen in the acid hydrolyzates is

TABLE 5.7. AMINO NITROGEN COMPOUNDS IN ACID HYDROLYSATES OF SOIL AS REPORTED BY DIFFERENT INVESTIGATORS

Compound	Mg/100 g. Soil		
	Flanagan silt loam (181)	Lacombe (171)	Hoytville clay loam (205)
Cysteic acid	16	—	1
N-Acetyl glucosamine	*	—	P
Glucosamine	*	—	*
Galactosamine	*	—	P
Aspartic acid	67	146	182
Serine	73	48	
Threonine	81	79	80
Glutamic acid	94	117	137
Methionine sulfone	*	—	10
Hydroxyproline	*	—	33
Glycine	74	81	80
Alanine	77	46	109
β-alanine	17	—	10
α-aminobutyric acid	12	—	2
α-ε-diaminopimelic acid	10	—	P
γ-amino-*n*-butyric acid	3	—	P
Valine	62	58	81
Methionine sulfoxide	23	—	6
Proline	52	40	48
Methionine	*	12	—
Isoleucine	48	34	33
Leucine	63	39	58
Ornithine	9	—	35
Cystine	*	*	—
Lysine	70	—	69
Histidine	7	—	13
Tyrosine	23	26	17
Arginine	17	—	12
Phenylalanine	24	36	34

* Identified but quantity not estimated.
P = Provisional identification.

present in the α-amino form. Most of the remaining nitrogen can be released as ammonia. These results suggest that the nitrogen in acid hydrolyzates is derived from proteins. The ammonia comes from amides, such as asparagine and glutamine (172). Proteins have not been isolated from the soil but studies on "lignoproteins" have been reported (79). Proteins are no doubt conjugated with other organic compounds (59, 170), but the "lignin-protein" theory of humus formation lacks definite proof. Free amino groups have not been detected in soil organic matter (170). The presence of α,ϵ-diamino pimelic acid (180) indicates that some of the protein in soil organic matter is synthesized by soil organisms.

Acid hydrolysis of the humic acid fraction does not solubilize all of the nitrogen. Acid hydrolysis of either whole soil or humic acid yields the same amino acids. The brown humic acid fraction has a higher proportion of the nitrogen in the form of acid soluble and α-amino nitrogen than the gray humic acid fraction (23). Extensive purification procedures do not remove all of the nitrogen in humic acid (59, 68), but evidence for the presence of heterocyclic nitrogen as as yet circumstantial.

Fulvic acid nitrogen undergoes acid hydrolysis easier and yields more ammonia than the humic acid fraction (185). Most of the ammonia is probably derived from amides (42) and other unidentified compounds (185, 187). The fulvic acid hydrolyzate contains large quantities of β-alanine, small quantities of arginine and lysine (22), and other unidentified compounds (185, 187). Stevenson (187) accounted for the nitrogen in the fulvic acid fraction hydrolyzates as follows: amino acids (one-fourth of total N) in proteinaceous constituents of the fulvic acid fraction help to account for the narrowing of the carbon-nitrogen ratio of soil organic matter with increasing depth (185).

One group of non-proteinaceous compounds in soil organic matter which has been studied is the amino sugar group. The amino sugars are found in a large number of biological tissues. Interest in aminopolysaccharides arises from their nitrogen content and their ability to produce water-stable aggregates. Hexosamines have been extracted from the soil by acid hydrolysis (183), and may account for 5 to 11 per cent of the nitrogen in the soil (24, 173, 184). In some soils, the concentration of amino sugars has been found to be greater in the B-horizon than in the A-horizon. Sowden (173) has determined the total hexosamine content by the Elson and Morgan method and the glucosamine and galactosamine contents by the Eastoe method. Since the sum of the glucosamine and galactosamine is about equal to the total quantity of amino sugars (Table 5.8), the other amino sugars are probably present in very small quantities, if at all. The ratios of glucosamine to galactosamine did not vary much among samples within the profile, but the ratios varied greatly between soil types.

Nucleic acid derivatives were first found in soils in 1910 by Schreiner and Shorey (156, 157). Cytosine, xanthine, and hypoxathine were isolated by chemical extraction and precipitation procedures and identified by crystal structure and derivative formation. Wrenshall and co-workers (200, 201) later found xanthine, guanine, adenine, and uracil, again using crude chemical methods. Anderson (6, 7, 9) identified deoxyribonucleic acid derivatives in the hydrolyzate of humic acid fraction. The quantity of

TABLE 5.8. HEXOSAMINE CONTENT OF VARIOUS SOIL MATERIALS

Great soil group location series	Profile	Glucosamine (mg/g)	Galactosamine (mg/g)	Total Eastoe (10) (mg/g)	Total Elson and Morgan (11) (mg/g)	Glucose galactose (mg/g)	Hexose-N total N (%)	Total N (%)	pH
Black* Lacombe, Alta. (18)†	Cultivated (0–6 in.)	2.93	1.80	4.73	4.74	1.6	7.4	0.50	5.7
	1st humate fraction of group 2A colloids of above	4.25	3.10	7.35	7.00	1.4	3.2	1.76	—
Dark brown* Scott, Sask.†	Virgin (0–6 in.)	3.11	1.90	5.01	5.28	1.6	9.0	0.45	6.0
Podzol Lennoxville, Que. (18)†	A_2	4.35	1.87	6.22	6.10	2.3	10.0	0.48	4.4
	1st humate fraction of group 1 colloids of above	11.0	5.80	16.80	18.00	1.9	5.6	2.41	—
Podzol* Cape Breton, N.S. (19)	A_{00} (6–3 in.)	5.00	1.96	6.96	6.83	2.6	4.4	1.22	3.5
	A_{01} (3–1 in.)	6.96	2.72	9.68	10.20	2.6	5.5	1.40	3.6
	A_{02} (1–0 in.)	2.72	0.91	3.63	3.66	3.0	5.0	0.57	3.5
	B_{21} (8–22 in.)	1.15	0.50	1.65	1.80	2.3	24.1	0.06	5.4
Podzol* Montreal, Que. (13)*	Upper humus layer	4.80	1.18	5.98	5.64	4.1	5.3	0.86	4.4
Podzol Liverpool, N.S. (17)† Port Hebert series	A_0	6.41	3.58	9.99	10.21	1.8	8.4	0.94	4.8
	B_{21}	1.50	0.94	2.44	2.63	1.6	5.7	0.35	5.0
Grey-brown podzolic Wellington Co., Ont. (26) Guelph Series	A_2 (0–4 in.)	2.15	1.32	3.47	3.48	1.6	6.6	0.41	7.2
Brown forest* Ottawa, Ont. (13)†	Upper humus layer	5.09	3.07	8.16	7.53	1.7	7.4	0.81	7.0
Ottawa, Ont. (26)† Greenville series	A_2 (0–4 in.)	2.56	1.15	3.71	4.06	2.2	9.8	0.31	6.7

* Series not specified.
† Near this location.

cytosine exceeded that of guanine, and thymine usually exceeded adenine (Table 5.9).

The Phosphorus Fraction

Soil organic matter contains many phosphorus compounds, accounting for 15 to 80 per cent of the total phosphorus in the soil*. Many of the studies on organic phosphorus compounds in the soil have been based on indirect methods. A comprehensive review of these methods has been made by Black and Goring (15).

The extraction methods of Mehta *et al.* (106) and Kaila and Virtanen

* See references 77, 82, 85, 96, 135, 161, and 193.

TABLE 5.9. TOTAL CARBON, NITROGEN, AND ORGANIC PHOSPHATE IN SOILS; AND PURINE AND PYRIMIDINE CONTENT OF SOIL HUMIC ACIDS (9)

Soil	% C	% N	Mg. organic P/100 g soil	μmoles/100 g soil			
				Adenine	Guanine	Cytosine	Thymine
Calcareous Shelby sand	2.5	0.23	44	2.2	5.6	7.8	3.7
Fine sandy loam	3.3	0.23	57	3.0	3.7	8.6	3.3
Medium loam	2.3	0.22	53	3.1	4.8	9.4	1.9
Clay loam, poorly drained	2.3	0.20	31	2.9	4.4	7.3	3.8
Clay loam, poorly drained	8.0	0.43	45	3.7	5.1	18.5	6.9
Clay loam, poorly drained	5.0	0.36	91	9.8	16.8	23.9	10.7
Peat	46.1	1.32	29	2.9	3.6	5.3	2.9
Clay loam, well drained	4.6	0.36	100	3.4	5.4	5.5	4.1

(83) are frequently used. Both of these methods use the difference between the total amount of phosphorus and the amount of inorganic phosphorus extracted as the amount of organic phosphorus. It has been suggested that pretreatment of the soil increases the accuracy of the organic phosphorus determinations. Anderson (8) used a mild alkaline solution as pretreatment before using the Mehta (106) method. Dormaar and Webster (47) used an organic chelating agent as pretreatment for the Kaila-Virtanen method.

Ignition procedures have also been used to determine the amount of organic phosphorus in the soil (102, 135). The ignition method usually gives a higher value for the quantity of organic phosphorus than the extraction methods (Table 5.10). Wide variations in ignition temperature, concentration of acid, and time of extraction cause little change in the value for organic phosphorus (135). Kaila (84) has suggested the use of an average of ignition and extraction methods would give the most reliable estimate of organic phosphorus content.

The accuracy of a given method is dependent upon the nature of the soil. The organic-phosphorus fraction of the soil must be characterized more completely before more satisfactory methods of determination are developed. This characterization will involve the identification of the organic phosphorus compounds and their reactions with the inorganic compounds of the soil.

The sources of organic-phosphorus compounds in the soil are living organisms and their residues. The types of organic phosphorus compounds are phospholipids, nucleic acids, metabolic phosphates, phosphoproteins, and inositol phosphates.

Evidence for the presence of phospholipids in soils was first presented by Shorey (164), who reported choline, a hydrolysis product of lecithin. Goring and Bartholomew (62) have questioned the efficiency of previous

TABLE 5.10. INORGANIC, ORGANIC, AND TOTAL PHOSPHATE EXTRACTED BY FOUR METHODS FOR ESTIMATING TOTAL SOIL ORGANIC PHOSPHATE

Extraction methods*	Mg P/100 g soil		
	All soils (mean)	Acid soils (means of 22 soils)	Calcareous soils (means of 12 soils)
Inorganic phosphate			
a	51	51	50
b	52	51	53
c	63	65	57
d	60	62	57
Organic phosphate			
a	56	67	35
b	47	59	24
c	45	53	29
d	49	59	29
Total phosphate†			
a‡	106	118	84
b	99	110	77
c	107	118	86
d	109	121	86

* Results were obtained by (*a*) ignition method of Saunders and Williams; (*b*) mild extraction method of Saunders and Williams; (*c*) vigorous extraction methods of Mehta *et al.*, and (*d*) Anderson.
† Mean total soil phosphate: acid soils, 121; calcareous soils, 87.
‡ Inorganic P from ignited soil.

extraction procedures. The phospholipid fraction, however, probably accounts for only a small part of soil organic phosphorus, since no phospholipids have been isolated from soil (52).

The presence of nucleic acids was suggested by Shorey (163) after the identification of some nitrogen bases was made. The estimations of the amounts of nucleic acids present are quite varied, 0.6 to 65.0 per cent*. Nucleic acids probably account for only a few per cent of the total organic phosphorus (2). Anderson (7, 9), using modern techniques of paper chromatography and spectroscopy, has identified the purine bases, guanine, and adenine, and the pyrimidines, cytosine, thymine, and uracil, in the hydrolyzates of humic acid from Scottish soils. Methods of extraction of purine and pyrimidine bases are discussed. The proportions of the bases indicated that they were of the GC type found in some bacteria. Assuming they were derived from bacterial DNA, nucleic acid would only account for 0.6 to 2.4 per cent of the total soil organic phosphorus (9).

The metabolic phosphates, such as the phosphorylated sugars, might be expected to occur in soil. Since these are readily metabolized by any of the soil flora, the concentrations are likely to be very small. By paper

* See references 7, 9, 18, 96, 169, and 200.

chromatographic techniques on citric acid extracts of soil, Robertson (129) found spots with R_f values similar to that of glucose-1-PO_4. He suggested that such phosphorus compounds may occur in very small amounts in soils.

The phosphoproteins have received little consideration in soil organic phosphorus investigations.

Yoshida (204) isolated crystalline inositol from soil organic-phosphorus materials, thus establishing for the first time the presence of inositol phosphates in the soil. Since then, the presence of inositol phosphates in soil has been indicated by a number of workers*. Inositol phosphate is predominantly in the hexaphosphate form. Small amounts of the inositol esters with less than six phosphate groups have been found by Andelson (5).

Column chromatographic procedures for the separation and estimation of inositol hexaphosphate have been used by Smith and Clark (167), Caldwell and Black (32), and Thomas and Lynch (190). The chromatographic procedures give a much lower estimation of the inositol hexaphosphate phosphorus than values based on the precipitation methods of Bower (17) and Peterson (127). An explanation of this can be found in the work of Anderson (10). While investigating the fractionation of soil organic phosphorus by precipitation procedures, Anderson found that at pH 2.7 to 3.0, the amount of phosphorus precipitated exceeded the amount of inositol hexaphosphate present. Therefore, other phosphate esters were being precipitated as well as inositol hexaphosphate.

It has been shown by Thomas and Lynch (190) and Williams and Steinbergs (199) that the C:P ratio decreases with depth in the soil. The data of Thomas and Lynch (190), however, also indicate that the proportion of organic phosphorus present as inositol hexaphosphate decreases in lower horizons. This would indicate the presence of larger proportions of unidentified phosphorus esters.

The compounds that include the major portion of soil organic phosphorus still remain to be identified. If the maximum values for inositol phosphates and nucleic acids are taken, at least 50 per cent of the soil organic phosphorus occurs in unidentified compounds.

The organic phosphorus compounds in the soil include those compounds that are from plant tissue, unaltered and altered, and those compounds that have been synthesized by microorganisms. After the incubation of soil with labeled inorganic phosphorus, the activity was associated with the inositol hexaphosphate fraction (167). In another incubation experiment, an increase in the amounts of *meso*-inositol hexaphosphate and another isomer of inositol hexaphosphate occurred (31). Cosgrove (38) has reported evidence

* See references 5, 18, 38, 127, 167, and 201.

that *dl*-inositol hexaphosphate occurs in the soil. This compound has not been found in plant or animal tissue. Anderson's (7) work on derivatives of nucleic acid in soils shows that the relative proportion of nitrogen bases found in DNA is of a GC type, unlike that found in plants or animals but found in some bacteria. The evidence that many of the organic phosphorus compounds in the soil are of microbial origin is, therefore, accumulating.

Plants utilize the phosphorus in the organic compounds either by absorption of the organic molecule or by absorption of the phosphate ion after mineralization of the organic molecule. Phytin and lecithin have been shown to be chemically available forms of phosphorus (37, 131, 188). Organisms capable of mineralizing organic phosphorus have been isolated (188). Birch (14) found that the incorporation of plant residues in the soil did not increase the amount of extractable phosphorus during the first three months incubation. Calcium, ferric, and aluminum salts of phytin are very insoluble (201). The low solubility of these salts has been suggested as being responsible for the accumulation of phytin in soils. Adsorption of various organic phosphorus compounds on soil clays, clay minerals, and hydrates sesquioxides, as reported by Anderson (11), also play a role in their mineralization and availability.

Conclusion

The studies on a few components of soil organic matter have been reviewed. Many components have been studied in less detail because of many reasons. These constituents of soil organic matter include the aromatic compounds, sulfur compounds, free radicals, and coordination compounds. So far, no universal solvent has been found for dissolving soil organic matter. Similarly, no solvent has been found that is capable of dissolving all organic-nitrogen compounds or phosphorus compounds or other broad groups of compounds. Since the commonly used solvents dissolve a variety of organic and inorganic compounds from the soil, the separation, purification, and identification of one or more of the constituents in the extract is often tedious and complex. Yet, since the soil probably contains most types of organic compounds synthesized by biological organisms, the identification and characterization of these compounds will require the ingenuity and time of many chemists.

REFERENCES

1. Acton, C. J., Paul, E. A., and Rennie, D. A., *Can. J. Soil. Sci.* 43, 141 (1963).
2. Adams, A. P., Bartholomew, W. V., and Clark, F. E., *Soil Sci. Soc. Am. Proc.* 18, 40 (1954).
3. Alexandrova, L. N., *Trans. Intern. Congr. Soil Sci. 7th Congr.* 2, 74 (1960).
4. Alosaker, E., and Michelson, K., *Acta Chem. Scand.* 11, 1794 (1957).
5. Anderson, G., *J. Sci. Food Agr.* 7, 437 (1956).

6. Anderson, G., *Nature* **180**, 287 (1957).
7. Anderson, G., *Soil Sci.* **86**, 169 (1958).
8. Anderson, G., *J. Sci. Food Agr.* **11**, 497 (1960).
9. Anderson, G., *Soil Sci.* **91**, 156 (1961).
10. Anderson, G., *J. Soil Sci.* **12**, 276 (1961).
11. Anderson, G., and Arlidge, E. Z., *J. Soil Sci.* **13**, 216 (1962).
12. Barton, D. H. R., and Schnitzer, M., *Nature* **198**, 217 (1963).
13. Bernier, B., *Biochem. J.* **70**, 590 (1958).
14. Birch, H. F., *Plant and Soil* **15**, 347 (1961).
15. Black, C. A., and Goring, C. A. I., "Agronomy," Vol. 4, p. 123, New York Academic Press, Inc., 1953.
16. Black, C. A. "Soil-Plant Relationships," p. 179, New York, John Wiley & Sons, Inc., 1957.
17. Bower, C. A., *Soil Sci.* **59**, 277 (1945).
18. Bower, C. A., *Iowa Agr. Expt. Sta. Res. Bull.* **362** (1949).
19. Bradley, W. F., *J. Am. Chem. Soc.* **67**, 975 (1945).
20. Bremner, J. M., and Lees, H., *J. Agr. Sci.* **39**, 274 (1949).
21. Bremner, J. M., *J. Soil Sci.* **1**, 198 (1950).
22. Bremner, J. M., *J. Sci. Food Agr.* **11**, 497 (1952).
23. Bremner, J. M., *Z. Pflanzenernähr. Düng. Bodenk.* **71**, 63 (1955).
24. Bremner, J. M., *J. Agr. Sci.* **46**, 247 (1955).
25. Bremner, J. M., *J. Agr. Sci.* **48**, 352 (1957).
26. Bremner, J. M., and Harada, T., *J. Agr. Sci.* **52**, 137 (1959).
27. Bremner, J. M., and Lin Ho, C., *Am. Soc. Agron. Agronomy Abstracts*, p. 5, 1961.
28. Broecker, W. S., and Olsen, E. A., *Science* **132**, 712 (1960).
29. Burges, A., *Sci. Proc. Roy. Dublin Soc.* **1**, 53 (1960).
30. Burges, A., *Trans. Intern. Congr. Soil Sci. 7th Congr.* **2**, 128 (1960).
31. Caldwell, A. G., and Black, C. A., *Soil Sci. Soc. Am. Proc.* **22**, 293 (1958).
32. Caldwell, A. G., and Black, C. A., *Soil Sci. Soc. Am. Proc.* **22**, 296 (1958).
33. Chahal, K. S., Ph.D. dissertation, The Ohio State University, 1963.
34. Charles, J., Soubies, L., and Gadet, R., *Compt. Rend. Acad. Sci. Paris* **247**, 1229 (1958).
35. Choudri, M. B., and Stevenson, F. J., *Soil Sci. Soc. Am. Proc.* **21**, 508 (1957).
36. Coffin, D. E., and Delong, W. A., *Trans. Intern. Congr. Soil Sci., 7th Congr.* **2**, 91 (1960).
37. Conrad, J. P., *J. Agr. Res.* **59**, 507 (1939).
38. Cosgrove, D. J., *Nature* **194**, 1265 (1962).
39. Coulson, C. B., Davies, R. I., and Khan, E. J. A., *J. Soil Sci.* **10**, 271 (1959).
40. Davies, R. I., Coulson, C. B., and Suna, C., *Chem. Ind.*, 1544 (1957).
41. Davis, R. J., and Clapp, C. E., *Appl. Microbiol.* **9**, 519 (1961).
42. De, P. K., *J. Proc. Insti. Chem. (India)* **28**, 353 (1956).
43. De, S. K., and Rastogi, R. C., *Z. Pflanzenernähr. Düng. Bodenk.* **98**, 121 (1962).
44. Deuel, H., and Dubach, P., *Helv. Chim. Acta* **41**, 1310 (1958).
45. Deuel, H., *Trans. Intern. Congr. Soil Sci. 7th Congr.* **1**, 38 (1960).
46. Deuel, H., Dubach, P., and Mehta, N. C., *Sci. Proc. Roy. Dublin Soc.* **1**, 115 (1960).
47. Dormaar, J. F., and Webster, G. R., *Can. J. Soil Sci.* **43**, 35 (1963).
48. Duff, R. B., *Chem. Ind.* 1104 (1952).
49. Duff, R. B., *J. Sci. Food Agr.* **3**, 140 (1952).
50. Duff, R. B., *Chem. & Ind.* 1513 (1954).
51. Duff, R. B., *J. Sci. Food Agr.* **12**, 826 (1961).

52. Dyer, W. J., and Wrenshall, C. L., *Soil Sci.* **51**, 323 (1941).
53. Evans, L. T., *J. Soil Sci.* **10**, 110 (1959).
54. Farmer, V. C., and Morrison, R. I., *Sci. Proc. Roy. Dublin Soc.* **1**, 85 (1960).
55. Felbeck, G. T., *Agronomy Abstracts*, p. 16, American Society of Agronomy, 1961.
56. Felbeck, G. T., *Agronomy Abstracts*, p. 24, American Society of Agronomy, 1962.
57. Flaig, W., Scheffer, F., and Klamroth, B., *Z. Pflanzenernähr. Düng. Bodenk.* **71**, 33 (1955).
58. Flodin, P., Dissertation, Uppsala, 1962.
59. Forsyth, W. G. C., *J. Agr. Sci.* **37**, 132 (1946).
60. Forsyth, W. G. C., *Biochem. J.* **41**, 176 (1947).
61. Forsyth, W. G. C., *Biochem. J.* **46**, 141 (1950).
62. Goring, C. A. I., and Bartholomew, W. V., *Soil Sci. Soc. Am. Proc.* **14**, 152 (1950).
63. Grim, R. E., "Clay Mineralogy", pp. 250 New York, McGraw-Hill Book Co., Inc., 1953.
64. Gupta, U., *Soils Fertilizers* **25**, 255 (1962).
65. Hance, R. J., and Anderson, G., *J. Soil Sci.* **13**, 225 (1962).
66. Hayashi, T., and Nagai, T., *J. Faculty of Agriculture, Tottori Univ. Japan* **2**, 97 (1956).
67. Himes, F. L., Tejeira, B., and Hayes, M. H. B., *Soil Sci. Soc. Am. Proc.* **27** (in press).
68. Hobson, R., and Page, H., *J. Agr. Sci.* **22**, 497 (1932).
69. Hock, A., *Z. Pflanzenernähr. Düng. Bodenk.* **50**, 1 (1937).
70. Hodge, J. E., *Agr. Food Chem.* **1**, 928 (1953).
71. Hoffman, R. W., and Brindley, G. W., *Geochim. et Cosmochim. Acta* **20**, 15 (1960).
72. Hori, S., and Okuda, A., *Soil Plant Food* **7**, 4 (1961).
73. Horner, L., in Lunberg, W. O., ed., "Autooxidation and Antioxidants," New York, Interscience Publishers, Inc., 1961.
74. Hosada, K., and Takota, H., *J. Sci. Soil Tokyo* **28**, 64 [Abstract in *Soil and Plant Food* **3**, 197 (1957)].
75. Hoyoshi, T., and Nagai, T., *Soil Plant Food* **5**, 153 (1959).
76. Ivarson, K. C., and Sowden, F. J., *Soil Sci.* **94**, 245 (1962).
77. Jackman, R. H., *Soil Sci.* **79**, 293 (1955).
78. Jacquin, F., *Soils Fert.* **26**, 642 (1963).
79. Jenkinson, D. S., and Tinsley, J., *Sci. Proc. Roy. Dublin Soc.* **1**, 141 (1960).
80. Johnson, H. H., *Soil Sci. Soc. Am. Proc.* **23**, 293 (1959).
81. Jurinak, J. J., and Volman, D. H., *J. Phys. Chem.* **63**, 1373 (1959).
82. Kaila, A., *Valt. Maatalousk. Julk.* **129**, 118 (1948).
83. Kaila, A., and Virtanen, O., *Maataloust. Aikak.* **27**, 104 (1955).
84. Kaila, A., *Maataloust. Aikak.* **34**, 187 (1962).
85. Kaila, A., *Soil Sci.* **95**, 38 (1963).
86. Kasatochkin, V. I., Kononova, M. M., and Lil'berbrand, O. I., *Dohl. Akad. Nauk* **119**, 785 (1958) [*Soils Fertilizers* **21**, 1638 (1958)].
87. Kaurichev. I. S., Fedorov, E. A., and Shnabel', I. A., *Soviet Soil Sci.* **10**, 1050 (1960).
88. Kawaguchi, K., and Kyuma, K., *Soil and Plant Food* **5**, 54 (1959).
89. Keefer, R. F., and Mortensen, J. L., *Soil Sci. Soc. Am. Proc.* **27**, (in press).
90. Khan, D. V., *Koklady Akod. Nauk. S.S.S.R.* **81**, 461 (1951) [(C.A. **46**, 4156 (1952)].
91. Kononova, M. M., Bel'chikova, N. P., and Nikiforov, V. K., *Soviet Soil Sci.* p. 285 (1959).
92. Kononova, M. M., and Bel'chikova, N. P., *Soviet Soil Sci.* pp. 1149 (1960).
93. Kononova, M. M., "Soil Organic Matter," New York, Pergamon Press, 1961.

94. Kononova, M. M., and Titova, N. A., *Soviet Soil Sci.*, 1230 (1961).
95. Kosaka, J., and Honda, C., *Soil Plant Food* **2**, 59 (1956).
96. Kosaka, J., and Abe, Kazuo, *Soil Plant Food* **3**, 95 (1957).
97. Krevelen, D. W. van, and Schuyer, J., "Coal Science," New York, Elsevier Publishing Co., 1957.
98. Kumada, K., *Soil Plant Food* **1**, 29 (1955).
99. Kumada, K., and Aizawa, K., *Soil Plant Food* **3**, 152 (1958).
100. Kumada, K., and Sato, O., *Soil Sci. Plant Nutr.* **8**, 31 (1962).
101. Kyuma, K., and Kawaguchi, K., *Am. Soc. Agron. Abstracts* p. 25, 1962.
102. Legg, J. O., and Black, C. A., *Soil Sci. Soc. Am. Proc.* **19**, 139 (1955).
103. MacEwan, D. M. C., *Trans. Faraday Soc.* **44**, 349 (1948).
104. Mackenzie, A. F., and Dawson, J. F., *J. Soil. Sci.* **13**, 160 (1963).
105. Martin, A. E., and Reeve, R., *J. Soil Sci.* **8**, 268 (1957).
106. Mehta, N. C., Legg, J. O., Goring, C. A. I., and Black, C. A., *Soil Sci. Soc. Am. Proc.* **18**, 443 (1954).
107. Mehta, N. C., Dubach, P., and Deuel, H., *Advances in Carbohydrate Chemistry* **16**, 335 (1961).
108. Meinschein, W. G., and Kenny, G. S., *Anal. Chem.* **29**, 1153 (1957).
109. Miklaszewski, S. *Part I. Zesz. Nauk wyz. Skol. rol. Wroclaw Rol. No. 14*, **119** (1958) [(*Soils and Fert.* **22**, 991 (1959)].
110. Miller, M. W., "The Pfizer Handbook of Microbial Metabolites," New York, McGraw-Hill Book Co., Inc., 1961.
111. Monnier, G., Ture, L., Jeanson-Luusinang, C., *Ann. Agron. Paris* **13**, 55 (1962).
112. Mortensen, J. L., *Trans. Intern. Congr. Soil Sci. 7th Congr.* **2**, 98 (1960).
113. Mortensen, J. L., *Clays Clay Minerals* **9**, 530 (1962).
114. Mortensen, J. L., and Schwendinger, R. B., *Geochim. Cosmochim. Acta* **27**, 201 (1963).
115. Mortensen, J. L., *Soil Sci. Soc. Am. Proc.* (in press).
116. Mortensen, J. L., *Soil Sci. Soc. Am. Proc.* **27**, (in press).
117. Mortensen, J. L., Anderson, D., White, J. L., *Soil Sci. Soc. Am. Proc.* (in press).
118. Mortensen, J. L., and Marcusiu, E. C., *Soil Sci. Soc. Am. Proc.* **27**, (in press).
119. Muller, M., Mehta, N. C., and Deuel, H., *Z. Pflanzenernähr. Düng. Bodenk.* **90**, 139 (1960).
120. Nelson, C. A., Hummel, J. P., Swenson, C. A., and Friedman, L., *J. Biol. Chem.* **237**, 1575 (1962).
121. Norman, A. G., *Soil Sci.* **56**, 223 (1943).
122. Ogston, A. G., *Biochem. J.* **70**, 598 (1957).
123. Parsons, J. W., and Tinsley, J., *Soil Sci. Soc. Am. Proc.* **24**, 198 (1960).
124. Parsons, J. W., and Tinsley, J., *Soil Sci.* **92**, 46 (1961).
125. Paul, E. A., and Schmidt, E. L., *Soil Sci. Soc. Am. Proc.* **24**, 195 (1960).
126. Paul, E. A., and Schmidt, E. L., *Soil Sci. Soc. Am. Proc.* **25**, 359 (1961).
127. Pederson, E. J. N., *Plant Soil* **4**, 252 (1952).
128. Rex, R. W., *Nature* **188**, 1185 (1960).
129. Robertson, G. J., *J. Sci. Food Agr.* **9**, 288 (1958).
130. Roderick, W. R., *J. Chem. Educ.* **39**, 2 (1962).
131. Rogers, H. T., Pearson, R. W., and Pierre, W. H., *Soil Sci. Soc. Am. Proc.* **5**, 285 (1940).
132. Rollins, C., Jensen, L., and Schwartz, A. N., *Anal. Chem.* **34**, 711 (1962).
133. Saeki, H., and Azuma, J., *J. Sci. Soil Manure, Japan* **28**, 392 (1958) [Abstract in *Soil and Plant Food* **4**, 45 (1952)].
134. Saeki, H., and Azuma, J., *Soil Plant Food* **6**, 49 (1960).

135. Saunders, W. M. H., and Williams, E. G., *J. Soil Sci.* **6,** 254 (1955).
136. Scharpenseel, H. W., *Z. Pflanzenernähr. Düng. Bodenk.* **88,** 97 (1960).
137. Scharpenseel, H. W., and Albersmeyer, W., *Z. Pflanzenernähr. Düng. Bodenk.* **88,** 203 (1960).
138. Scheffer, F., *Trans. Intern. Congr. Soil Sci. 5th Congr.* **1,** 208 (1954).
139. Scheffer, F., *Der Organisch Gebundine Stickstoff des Bodens, seine Verwertbarkeet (Auch Harnstoff), Handbuch der Pflanzenphsiologie* **8,** 179 (1958).
140. Scheffer, F., Ziechmann, W., and Schluter, H., *Z. Pflanzenernähr. Düng. Bodenk.* **85,** 32 (1959).
141. Scheffer, F., Ziechmann, W., and Scholz, H., *Z. Pflanzenernähr. Düng. Bodenk.* **85,** 50 (1959).
142. Scheffer, F., Ziechmann, W., and Pawekka, G., *Z. Pflanzenernähr. Düng. Bodenk.* **90,** 58 (1960).
143. Scheffer, F., and Kickuth, R., *Z. Pflanzenernähr. Düng. Bodenk.* **94,** 180 (1961).
144. Scheffer, F., and Kickuth, R., *Z. Pflanzenernähr. Düng. Bodenk.* **94,** 189 (1961).
145. Schlichting, E., *Z. Pflanzenernähr. Düng. Bodenk.* **61,** 97 (1953).
146. Schluter, H., *Z. Pflanzenernähr. Düng. Bodenk.* **84,** 169 (1959).
147. Schnitzer, M., and Wright, J. R., *Can. J. Soil Sci.* **37,** 89 (1957).
148. Schnitzer, M., Wright, J. R., and Desjardins, J. G., *Can. J. Soil Sci.* **38,** 49 (1958).
149. Schnitzer, M., Shearer, D. A., and Wright, J. R., *Soil Sci.* **87,** 252 (1959).
150. Schnitzer, M., and Desjardins, J. G., *Soil Sci. Soc. Am. Proc.* **26,** 362 (1962).
151. Schreiner, O., and Shorey, E. C., *J. Am. Chem. Soc.* **30,** 1295 (1908).
152. Schreiner, O., and Shorey, E. C., *J. Am. Chem. Soc.* **30,** 1599 (1908).
153. Schreiner, O., and Shorey, E. C., *J. Am. Chem. Soc.* **31,** 116 (1909).
154. Schreiner, O., and Shorey, E. C., *J. Am. Chem. Soc.* **32,** 1674 (1910).
155. Schreiner, O., and Shorey, E. C., *J. Biol. Chem.* **8,** 381 (1910).
156. Schreiner, O., and Shorey, E. C., *J. Biol. Chem.* **8,** 385 (1910).
157. Schreiner, O., and Shorey, E. C., *U.S.D.A. Bur. Soils Bull. #74,* 1910.
158. Schreiner, O., and Shorey, E. C., *J. Am. Chem. Soc.* **33,** 78 (1911).
159. Schreiner, O., and Shorey, E. C., *J. Am. Chem. Soc.* **33,** 81 (1911).
160. Sen, B. C., *Indian Chem. Soc.* **37,** 793 (1960).
161. Sen-Gupta, M. B., and Cornfield, A. H., *J. Sci. Food Agr.* **13,** 655 (1962).
162. Shorey, E. C., and Lathrop, E. C., *J. Am. Chem. Soc.* **32,** 1680 (1910).
163. Shorey, E. C., *Biochem. Bull.* **1,** 104 (1911).
164. Shorey, E. C., *U.S.D.A. Bur. Soils Bull.* **88,** 1913.
165. Shorey, E. C., *Proc. 1st Intern. Congr. Soil Sci.* **3,** 264 (1927).
166. Singh, S., and Singh, P. K., *Proc. Nat. Acad. Sci. India* **29A,** 378 (1960) (*Chem. Abstracts* **56,** 1785).
167. Smith, D. H., and Clark, F. E., *Soil Sci.* **72,** 353 (1951).
168. Smith, I., "Chromatographic Techniques," p. 36, New York, Interscience Publishers, Inc., 1958.
169. Sokolov, D. K., *Pochvovedenie* **1948,** 502.
170. Sowden, F. J., and Parker, D. I., *Soil Sci.* **76,** 201 (1953).
171. Sowden, F. J., *Soil Sci.* **80,** 180 (1955).
172. Sowden, F. J., *Can. J. Soil Sci.* **38,** 147 (1958).
173. Sowden, F. J., *Soil Sci.* **88,** 138 (1959).
174. Sowden, F. J., and Deuel, H., *Soil Sci.* **91,** 44 (1961).
175. Sowden, F. J., and Ivarson, K. C., *Soil Sci.* **94,** 340 (1962).
176. Springer, U., *Z. Pflanzenernähr. Düng. Bodenk.* **32,** 129 (1943).
177. Steelink, C., and Tollin, G., *Biochem. Biophys. Acta* **59,** 25 (1962).

178. Stevenson, F. J., Marks, J. P., Varner, J. E., and Martin, W. P., *Soil Sci. Soc. Am. Proc.* **16**, 69 (1952).
179. Stevenson, F. J., Van Winkle, Q., and Martin, W. P., *Soil Sci. Soc. Am. Proc.* **17**, 31 (1953).
180. Stevenson, F. J., *Soil Sci. Soc. Am. Proc.* **20**, 201 (1956).
181. Stevenson, F. J., *Soil Sci. Soc. Am. Proc.* **20**, 204 (1956).
182. Stevenson, F. J., *Soil Sci. Soc. Am. Proc.* **21**, 283 (1957).
183. Stevenson, F. J., *Soil Sci.* **83**, 113 (1957).
184. Stevenson, F. J., *Soil Sci.* **84**, 99 (1957).
185. Stevenson, F. J., *Soil Sci.* **88**, 201 (1959).
186. Stevenson, F. J., and Dhariwal, A. P. S., *Soil Sci. Soc. Am. Proc.* **23**, 121 (1959).
187. Stevenson, F. J., *Soil Sci. Soc. Am. Proc.* **24**, 472 (1960).
188. Szember, A., *Plant Soil* **13**, 147 (1960).
189. Thomas, R. L., and Mortensen, J. L. (unpublished).
190. Thomas, R. L., and Lynch, D. L., *Can. J. Soil Sci.* **40**, 113 (1960).
191. Tinsley, J., and Salam, A., *J. Soil Sci.* **12**, 259 (1961).
192. Tyulin, A., Th., *Soil Sci.* **45**, 343 (1938).
193. Ulrich, B., and Benzler, J. H., *Z. Pflanzenernähr. Düng. Bodenk.* **70**, 220 (1955).
194. Vander Watt, H. V. H., and Bodman, G. B., *Clays Clay Minerals* **9**, 568 (1962).
195. Waksman, S. A., and Stevens, K. R., *Soil Sci.* **30**, 97 (1930).
196. Waldron, A. C., and Mortensen, J. L., *Soil Sci. Soc. Am. Proc.* **25**, 29 (1961).
197. Waldron, A. C., and Mortensen, J. L., *Soil Sci.* **93**, 286 (1962).
198. Whistler, R. L., and Kirby, K. W., *J. Am. Chem. Soc.* **78**, 1755 (1956).
199. Williams, C. H., and Steinbergs, A., *Australian J. Agri. Res.* **9**, 483 (1958).
200. Wrenshall, C. L., and McKibbon, R. R., *Can. J. Res.* **B 15**, 475 (1937).
201. Wrenshall, C. L., and Dyer, W. J., *Soil Sci.* **51**, 235 (1941).
202. Wright, J. R., Schnitzer, M., and Levick R., *Can. J. Soil Sci.* **38**, 14 (1958).
203. Wright, J. R., and Schnitzer, M., *Trans. Intern. Congr. Soil Sci. 7th Congr.* **2**, 120 (1960).
204. Yoshida, R. K., *Soil Sci.* **50**, 81 (1940).
205. Yong, J. L., and Mortensen, J. L., *Res. Circ. 61, Ohio Agr. Expt. Sta.*, 1958.
206. Ziechmann, W., and Parvelke, G., *Landwirtsch. Forsch.* **12**, 17 (1959).

CHAPTER 6

Biochemistry of Soil

I. L. STEVENSON

Microbiology Research Institute
Research Branch, Canada Department of Agriculture
Ottawa, Canada

With the exception of certain autotrophic microorganisms, the green or chlorophyll-bearing plants are the only living forms capable of producing organic matter from elements and mineral compounds. Using the energy of the sun and the simple nutrients derived from the atmosphere, soil, or sea, plants are able to synthesize carbohydrates, proteins, and other organic substances that make up the bulk of their structure.

Plants or their residues, in turn, form the food supply for animals. In addition to utilizing the plants as an energy source, animals are able to synthesize new organic compounds, largely protein or fat in nature. Some plant constituents, such as the celluloses, hemicelluloses, lignins, and waxes, are utilized by animals only to a limited extent and are excreted relatively unchanged.

Ultimately all plant and animal residues are returned to the soil and subjected to decomposition by microorganisms. As a result of these microbiological processes, the elements which were originally consumed by plants for organic synthesis are returned to circulation for re-utilization. Fortunately plant and animal residues do not undergo rapid and complete mineralization. A large proportion of these organic deposits are first modified by the soil flora to form resistant brown or black organic complexes. These complexes, along with the constituents of microbial cells and other resistant products of metabolism make up the organic matter or humus fraction of the soil. Because of the formation and accumulation of this humus, a part of the elements essential for organic life, especially carbon, nitrogen, phosphorus, sulfur, and potassium, is stored in an unavailable form. The humus fraction in effect acts as a reservoir; its gradual decomposition by microorganisms results in the liberation of a slow but continuous stream of elements essential for new plant synthesis. In this way the major cycle of life is completed, from the soil back to the soil. Numerous secondary cycles occur in which one or more elements are concerned; in all transformation, however, microorganisms play a major role.

MICROORGANISMS OF THE SOIL

The microscopic plant and animal world of soil is represented by the bacteria, actinomycetes, fungi, algae, and protozoa. While numerous methods have been devised to estimate total numbers of these groups in soil, all have certain inherent limitations (141). Such methods are, nevertheless, satisfactory for comparative purposes, and we find a remarkable similarity in the qualitative distribution of microorganisms in different soils, along with distinct differences in their quantitative relationships.

Bacteria are the most abundant as well as the most diversified group of microorganisms*. Numerically they are the largest group, ranging from as few as two million per gram in a poor sandy soil to hundreds of millions per gram in a soil where there is active decomposition of organic matter. The major bulk of bacteria is confined to the top 6 inches of soil, and in terms of biomass they have been estimated anywhere from 1 to 2 tons per acre. Numerous factors influence the number of bacteria which are present in a soil at any one time. Thus, the total relative numbers fluctuate widely, depending on soil type and such environmental variables as soil treatment, temperature, moisture content, aeration, pH, and the organic and inorganic nutrient supplies (2, 78, 113, 141).

Since bacterial activity in soil is not necessarily related to the taxonomy of the groups involved, it is preferable to classify them from a physiological or functional standpoint. As a major division, bacteria can be divided into *autotrophic* and *heterotrophic* forms on the basis of their sources of energy and carbon. The obligate autotrophs and the less-restricted facultative autotrophs are the most self-sufficient of the bacteria; they are able to fulfill their carbon requirements from carbon dioxide and their energy from the oxidation of inorganic substances or simple compounds of carbon. Although autotrophy is confined to a relatively few species of bacteria, the transformations carried out are of immeasurable economic importance, particularly with respect to the oxidation of nitrogen and sulfur compounds to assimilable forms. Other forms are concerned with the oxidations of iron, manganese, hydrogen, and carbon monoxide.

* See references 2, 31, 63, 71, 78, 113, 137, and 141.

Heterotrophic bacteria comprise the vast majority of soil microorganisms. They depend on the oxidation of organic compounds for their energy and are responsible for the decomposition of cellulose, hemicelluloses, starches, sugars, proteins, fats, waxes, and other plant and animal remains that find their way to soil. These organisms vary markedly in structure, physiology, function, and importance (40). The diversity of the functions of the heterotrophs, along with the overlapping of activities of apparently differing groups, has made it exceedingly difficult to propose a rational method of classification. Morphological (39) and nutritional methods (88) have been suggested, but both are subject to numerous limitations, mainly due to a lack of knowledge of the organisms themselves. The heterotrophs can be divided most conveniently on the basis of their nitrogen requirements into the nitrogen-fixing organisms and those that require nitrogen in a combined organic or inorganic form. The nitrogen-fixing organisms are relatively few in numbers, but the three representative groups, namely, the symbiotic nitrogen-fixing organisms and the aerobic and anaerobic nonsymbiotic fixers, play an important part in the soil economy.

The actinomycetes appear universally in soils, and are second in abundance only to the bacteria*. Plating estimates give values of 10^5 to 10^8 colonies per gram of arable soils, and these forms make up from one-tenth to one-half the total soil population. Despite their lower numbers the actinomycete population probably equals the bacteria in weight and thus adds appreciably to the biomass. The actinomycetes are not as subject to seasonal variations as the bacteria, but they are influenced markedly by the soil-organic-matter content, moisture level, pH, and temperature. Like the majority of bacteria, the actinomycetes are heterotrophic and thus dependent on organic substrates. In this respect they are very adaptable, and different species are able to utilize a wide range of foodstuffs. They are, however, extremely slow growers, and in view of this characteristic it is difficult to view them as active competitors with bacteria in the early stages of organic-matter decomposition. The fact that they are found in larger numbers during the latter stages of decomposition has suggested that their main role in soil is the breakdown of the more-resistant components of plant and animal tissues remaining after the initial attack. In the same vein, they have been implicated in the breakdown of the humus fraction of soil with the liberation and recirculation of nitrogen and other essential nutrients in soil.

The fungi, like the majority of the bacteria, are heterotrophic, and therefore their existence in soil is closely connected with the decomposition of organic matter†. Fungi are found in soil in three distinct groups: (*a*) the

* See references 2, 31, 71, 78, 113, 141, and 144.

† See references 2, 31, 37, 71, 78, 107, 141, and 147.

unicellular yeasts or yeast-like forms, (*b*) the higher mushroom fungi and those associated with the mycorrhiza of higher plants, and (*c*) the free-living filamentous forms. The majority of fungi isolated from soils are of the filamentous type and these may appear in many diversified forms (37). Although numerous in soils, it is difficult to assess the number of fungi accurately, since they exist in soil both in the form of vegetative mycelium and as reproductive spores. Standard plate counts place the total fungi between 10^4 and 10^6 viable units per gram of soil, but it is doubtful if these figures represent more than a small percentage of the true number. Despite their relatively low numbers, the fungi, because of their larger cell size, probably equal the bacteria and actinomycetes in terms of microbial protoplasm. Unlike the bacteria and actinomycetes, the fungi have a high acid tolerance and under conditions of extreme acidity these organisms may become the predominant flora. At less-acid reactions of pH 4.5 through to alkalinity, the fungi are less competitive and their numbers become more dependent on the source of organic matter present. As heterotrophs the fungi are exceptionally well-equipped to undertake the rapid decomposition of virtually all of the major plant constituents—cellulose, hemicelluloses, pectin, starch, and even the lignins. They also play an active role in transforming the nitrogen constituents of soil. Although noted particularly for their degradative abilities, the fungi also participate in the synthesis of a wide range of complex compounds, a number of which have been implicated in the formation of the soil humus.

Algae are universally distributed in soils but are relatively few in numbers as compared to the bacteria, actinomycetes, and fungi (2, 82, 113, 116). Numbers range from as few as one to forty thousand per gram of soil. The algae are characterized by their photosynthetic pigments and predominantly photoautotrophic nutrition. Under conditions of reduced light intensity, many of the soil-inhabiting forms behave heterotrophically and utilize organic substances. Algae are particularly dependent on high moisture levels and their low numbers in cultivated land is generally due to insufficient moisture. As photosynthetic forms, they are dependent on sunlight and carbon dioxide, although the latter rarely becomes limiting. Their need for sunlight is reflected in the fact that the algae are concentrated in the surface layers, although they have been known to survive at considerable depths. It is doubtful if the algae contribute appreciably to the fertility of normal arable soils in the face of bacterial, fungal, and actinomycete competition. Undoubtedly their major role is an outcome of their synthetic rather than their degradative abilities. In instances where soils are, or have become, barren, the algae are the primary invaders, and through the gradual accumulation of cell residues form an organic base suitable for recolonization by higher forms (57). Algae are also of considerable economic importance in the rice-growing areas in the Far East. Here their main role is the provision of a low-level nitrogen supply through nitrogen fixation and by supplying oxygen to the submerged roots of the plants (148).

Whereas the bacteria, actinomycetes, fungi, and algae constitute the microflora of the soil, the protozoa are the predominant microfauna present (2, 76, 116, 141). The protozoa are the simplest forms of animal life and, with few exceptions, are microscopic and acellular. They are universal in their distribution in soils where the population ranges from a few hundreds to half a million per gram. Protozoa obtain their nutritional requirements either from preformed organic and inorganic substrates or by direct feeding upon microbial cells. These latter forms, which predominate in soil, are quite selective as to the organisms they will ingest. Thus, in addition to algae and other protozoa, only certain bacteria are susceptible to attack, while the remainder, along with the actinomycetes and yeasts, appear unsuitable as a nutrient source. Despite the relatively high numbers of protozoa there is very little information as to the role these organisms play in soil. There is little doubt that the direct feeding of the protozoa on other microscopic forms along with the selectivity of their feeding habits does have some influence on the biological equilibrium of the soil (128). Whether this predatory effect ever reaches the detrimental stage is doubtful, since the influence of protozoa on bacterial numbers has not been related to a corresponding decrease in microbial transformations. On the other hand, there is little or no evidence that the protozoa themselves have the capability of carrying out metabolic activities which could conceivably benefit soil fertility. As such, the protozoa remain a virtually unknown quantity in soil.

THE ECOLOGY OF MICROORGANISMS IN SOIL

Soils are composed of minerals, organic matter, water, air, and a living population. The inanimate constituents, along with water, form a natural medium in which bacteria, actinomycetes, fungi, algae, and protozoa thrive. The microbiological balance in soil is referred to as a "dynamic equilibrium"; as such, it is not static but subject to continuous fluctuation due to environmental factors. Such influences as season, temperature, moisture, cropping system, and fertilizer treatment, may temporarily upset the equilibrium in soil, but it rapidly reestablishes itself in a modified form to cope with the changed conditions. In addition to physical and chemical factors, the microbiological balance in soil is dependent upon interrelationships of the micro-population itself. Soil inhabitants do not exist in an isolated state but as part of a complex community in which the individuals are subject to the associative and antagonistic effects of neighboring organisms*.

Many variations of associative effects have been recognized. Classical examples include the symbiotic relationships between leguminous plants and the root-nodule bacteria, and the living together of algae and *Azotobacter*. In the latter instance the *Azotobacter* fix nitrogen in return for an energy source produced by the algae. Unilateral associations in which one group of microorganisms benefits other groups by furnishing substrates necessary for energy are perhaps the most common. The majority of decomposition processes in soil are sequential reactions in which successive groups of organisms utilize the end products produced by the preceding flora. Thus, cellulose is initially broken down by specialized organisms to sugars and organic acids. These products in turn are metabolized by other forms which lacked the ability to initiate the attack on cellulose. Similarly, the soil nitrogen cycle involves the participation of successive groups of organisms engaged in ammonification, nitrite formation, and nitrate production. Side associations also exist which link these examples, and the ramifications of such relationships are enormous. Nutritional associations are exemplified by the detection of vast numbers of soil organisms dependent upon preformed vitamins and growth factors, along with an equal abundance of organisms capable of elaborating these substances (83, 87). Associative effects also exist where the growth of one group may favor another by improving the environment. Thus the development of anaerobic organisms can be stimulated through oxygen consumption by the concomitant growth of aerobic organisms.

Antagonistic interrelationships, wherein one group of organisms has an adverse effect on the activities of another group, are also very common. Microorganisms may be forced to compete for some limiting nutrient in the soil, a situation which generally results in the survival of the most rapidly growing species. The production of organic and inorganic acids by various groups undoubtedly influences the soil reaction, making it unsuitable for growth of other forms. Many organisms are capable of elaborating toxic byproducts as well as antibiotics and, although it is difficult to assess their relative importance, there is little doubt that the inhibitory effects of such compounds markedly influence neighboring organisms (26). Parasitism and predation are important and perhaps the most dramatic of the interactions between species. Bacteria are particularly prone to attack and are subject to the feeding habits of protozoa, myxobacteria, and myxomycetes. On the other hand, bacteria commonly feed upon fungus mycelium, while certain streptomycetes may digest both fungi and bacteria. Some fungi in turn are

* See references 2, 9, 63, 78, 84–86, 137, 138, 141, 143, and 153.

capable of parasitizing other fungi and feed at the expense of their victims, while others are actively predacious, capable of ensnaring and feeding on amoebae and nematodes (45).

In addition to the associative and antagonistic effects within the micropopulation, it is becoming increasingly apparent that the growing plant exerts the most significant effect on the microbiological equilibrium in soils (78, 85, 132). In view of the increased density of the microbial flora in the rhizosphere, it is understandable that the interactions between microorganisms are intensified at or near the plant roots. Quantitative and qualitative changes in the microbial balance in this root zone may be due to the direct or indirect effects of root excretions. The growing plant has been found to stimulate specific groups of microorganisms which, in turn, have a beneficial effect on other microorganisms through the production of growth-promoting substances.

The majority of associative and antagonistic phenomena taking place in soil are localized to the extent that only organisms lying in close proximity to one another are affected. Soil then, is not a uniform environment but rather an intricate, interwoven, biological system composed of a myriad of micro-environments. These micro-environments are created through the deposition of organic matter or the individual interactions of microorganisms, or they are induced through the proximity of plant roots. Each localized community harbors a rich microbial population where survival is the keynote and where the associative and antagonistic effects are intensified. Therefore, the microbial equilibrium which becomes established in any soil is the result of the sum total of the innumerable individual reactions and interactions which are taking place.

BIOCHEMICAL ACTIVITIES OF MICROORGANISMS

The metabolism of all living organisms, regardless of their natural habitat, consists of many integrated chemical reactions catalyzed by enzymes. In order to reproduce itself, a microorganism must find raw materials of living matter and a suitable source of energy with which to transform them into cellular protoplasm. The nutrients available to the soil flora are many and varied, and although no one organism is capable of attacking them all, very few compounds, organic or inorganic, can withstand eventual transformation in the soil. Many of the available foods in soil are extremely complex molecules, which, before they can permeate into the cell, must be broken down into their constituent units through the elaboration of extracellular enzymes. The substances that a given organism can utilize for synthesis and energy are determined by its enzymatic constitution, but in all cases energy is derived through the oxidation of organic or inorganic compounds, or, in special cases, through photosynthesis. The energy released from the oxidation reactions arises through the breaking and rearrangement of the chemical bonds of the substrate molecules. The electron transfers accompanying such changes involve concomitant energy transfers and, unless harnessed, the energy released by these reactions would be dissipated as heat. Since the biological system is extremely economical, it is usual to find an endergonic reaction or energy-requiring system coupled to an exergonic reaction where energy is released. The energy produced by the latter system is not directly transferable, but is first transformed into a special form of chemical energy held in high-energy bonds, such as the pyrophosphate bonds of adenosine triphosphate (ATP). The first major step in the energy transformations culminates in the formation of ATP at the expense of the free energy of the degradation of foodstuffs. ATP or similar high-energy compounds thus form the principal links between the energy-yielding reactions and the energy-requiring reactions involved in the biosynthesis of the structural components of the cells*.

The majority of microorganisms obtain their energy through oxidation-reduction reactions in which oxidation is accomplished through dehydrogenation with a subsequent transfer of hydrogen (or electrons) through a series of mediators to its final acceptor. Biological oxidation, in which molecular oxygen acts as the ultimate hydrogen acceptor, is termed *respiration*. Respiration yields the greatest amount of energy from a substrate and serves as the principal source of energy for the growth of most aerobic bacteria. It is not a single-step process but an orderly sequence of reactions in which the energy released from the individual steps is used in the formation of energy-rich bonds (ATP). The net result in the decomposition of glucose to carbon dioxide and water *via* one of the established pathways of metabolism (glycolysis plus the tricarboxylic acid cycle) is the formation of 38 molecules of ATP (Table 6.1)†. Most autotrophic microorganisms derive their energy from the aerobic oxidation of specific inorganic substrates such as hydrogen, hydrogen sulfide, sulfur, ammonia, and nitrite (Table 6.1). Such reactions are directly linked to oxygen and, although there is little known about their energy transformations, there is no doubt that ATP or similar high-energy compounds are generated during electron transport (79, 99, 108). The energy values given in Table 6.1 indicate the potential free energy produced during the reactions. In actual fact, the organisms are able to utilize only a portion of this energy, depending on

* See references 36, 81, 106, 129, 135, and 149.

† While glycolysis yields 2 molecules of ATP, comparative energy values for the hexosemonophosphate oxidative pathway, the 6-phosphogluconate splitting reaction, and the hexose cycle are still in some doubt.

TABLE 6.1. ENERGY-YIELDING REACTIONS OF MICROORGANISMS

Organisms*	Nutrition and conditions†	Example reactions	Energy‡ (kcal)
		Reaction type = respiration	
Many	Hetero-aerob	glucose + $6O_2 \rightarrow 6CO_2 + 6H_2O$	688.5
Many	Hetero-aerob	2 pyruvate$^-$ + $5O_2 + 2H^+ \rightarrow 3CO_2 + 2H_2O$	547.0
Nitrosomonas	Auto-aerob	$NH_4^+ + 1\frac{1}{2}O_2 \rightarrow NO_2^- + 2H^+ + H_2O$	66.5
Thiobacillus thioxidans	Auto-aerob	$Na_2S_2O_3 + 2O_2 + H_2O \rightarrow Na_2SO_4 + H_2SO_4$	211.1
		Reaction type = fermentation	
Many	Hetero-anaerob	glucose → 2 pyruvate$^-$ + $2H^+$	45.8
Yeasts	Hetero-anaerob	glucose → 2 ethanol + $2CO_2$	62.8
Lactobacilli	Hetero-anaerob	glucose → 2 lactate$^-$ + $2H^+$	50.0
Clostridia	Hetero-anaerob	glucose → butyrate$^-$ + 2 acetate$^-$ + $3H^+$	61.4
		Reaction type = anaerobic respiration	
Many	Hetero-anaerob	5 glucose + $24KNO_3 \rightarrow 30CO_2 + 12N_2 + 24KOH + 18H_2O$ glucose + $8KNO_2 \rightarrow 6CO_2 + 4N_2 + 8KOH + 2H_2O$	
Desulfovibrio	Hetero-anaerob	glucose + $3H_2SO_4 \rightarrow 6CO_2 + 6H_2O + 3H_2S$	
Thiobacillus denitrificans	Auto-anaerob	$5S + 6KNO_3 + 2H_2O \rightarrow K_2SO_4 + 4KHSO_4 + 3N_2$	

* For *Nitrosomonas* and *Thiobacillus thioxidans*; see reference 135; for the other organisms, see references 32 and 79.

† Hetero = heterotrophic; auto = autotrophic; aerob = aerobic; and anaerob = anaerobic.

‡ Free energy of the reaction. Organisms benefit to varying degrees depending on the efficiency of the biological system.

the efficiency of the system. The 38 molecules of ATP produced during the complete oxidation of a molecule of glucose represent a yield of about 380 kcal or a conversion of approximately 57 per cent of the energy available (81). Similarly, in the oxidation of ammonia by species of *Nitrosomonas*, only about 6 to 14 per cent of the total energy is made available (66).

A second class of energy-yielding reactions is *fermentation*. Here, organic substrates serve as both the hydrogen donor and hydrogen acceptor. The majority of heterotrophic anaerobes and facultative anaerobes are able to carry out the anaerobic reduction of carbohydrates and related compounds to products which cannot be further decomposed by the enzyme system of the cell except by the intervention of molecular oxygen. The fermentation of most sugars involves glycolysis or a variation of this metabolic pathway (36). Glycolysis consists of the phosphorylation of hexose sugars with an eventual split into three carbon fragments with the formation of pyruvic acid. During the formation of pyruvic acid, energy is made available by the building up and subsequent rupture of the various phosphate bonds formed. Pyruvic acid in turn serves as the branching point for a host of transformations which are characterized by a variety of end products. Thus, the formation of formic, acetic, lactic, propionic, butyric and succinic acids, ethanol, propanol, butanol, acetylmethylcarbinol, butylene glycol, acetone and molecular hydrogen are all fermentation products, many of which are characteristic of the groups of organisms carrying out this anaerobic process. Examples of fermentative reactions are given in Table 6.1, where it will be noted that from an energy standpoint the reactions are less efficient than those of respiration. Much of the original energy of the substrates is still retained in the byproducts, and it is noteworthy that fermentations which differ widely in their end products release approximately the same amount of energy.

Anaerobic respiration designates those oxidation-reduction reactions in which inorganic substrates are used as hydrogen acceptors in the place of molecular oxygen. In the oxidation of organic or, in some instances, inorganic compounds, many anaerobic and facultative anaerobic soil forms are capable of reducing various forms of sulfur, nitrogen, and carbon dioxide. Such forms are of great importance to the soil economy—particularly those concerned with sulfur transformations and denitrification (Table 6.1). Since the reduction of inorganic substrates requires an expenditure of energy, the energy made available under these conditions is considerably less than that released on the oxidation of the same substrate under aerobic conditions.

Regardless of whether an organism obtains its energy through respiration, fermentation, or anaerobic respiration, it is usual to find such reactions closely linked with energy-requiring sequences involved in the biosynthesis of cellular protoplasm. In order to grow, an organism must acquire thousands of different organic compounds that are an integral part of its living material. These include: the proteins, which are necessary in the formation of enzymes and as structural components; the nucleic acids, which control the development and heredity of the cell; and the lipids and carbohydrates, which act as both structural components and reserve foodstuffs. Heterotrophic microorganisms are able to use organic nutrients or organic compounds formed in the course of metabolism as the building blocks of biosynthesis. Autotrophic organisms, on the other hand, require more complex synthetic pathways in order to synthesize their organic requirements from simple inorganic nutrients. Despite this necessity for total synthesis of organic constituents by autotrophs, many of the basic transformations carried out are identical to those found in heterotrophic organisms.

The synthesis of polysaccharides such as starch or glycogen from glucose, involves an initial phosphorylation in which energy is derived from ATP.

$$\text{Glucose} + \text{ATP} \rightarrow \text{Glucose phosphate} + \text{ADP}$$

Subsequent reactions involve the rearrangement and condensation of the glucose phosphate molecules to form long-chain polymers with the release of inorganic phosphorus. The energy utilized in the polymerization is held in the bonds

$$n\text{ATP} + \underset{\text{Glucose}}{nC_6H_{12}O_6} \rightarrow \underset{\text{Starch or glycogen}}{(C_6H_{10}O_5)n} + n\text{ADP} + nH_3PO_4$$

and can be released and re-used by the organisms when hydrolysis of the polysaccharide occurs. Carbon dioxide fixation also provides an important mechanism by which the chain length of an organic molecule can be increased. One such reaction is the fixation of carbon dioxide by pyruvic acid to form oxaloacetic acid. These reactions are important in the synthesis of di- and tricarboxylic acids necessary in the formation of amino acids.

$$\underset{\text{Pyruvic}}{CH_3\cdot CO\cdot COOH} + CO_2 \rightarrow \underset{\text{Oxaloacetic}}{COOH\cdot CH_2\cdot CO\cdot COOH}$$

The first stage of protein synthesis consists of the formation of amino acids through addition of ammonia (amination) to certain organic acids. In this way, key intermediates for the production of most of the naturally occurring amino acids are formed. Glutamic acid is formed by the reductive amination of α-ketoglutaric acid through the action of a reduced coenzyme ($DPNH_2$) and the enzyme glutamic dehydrogenase. Glutamic acid in turn, can be transaminated in the presence of oxaloacetic acid to form aspartic acid.

$$\underset{\alpha\text{-Ketoglutaric}}{COOH\cdot CH_2CH_2\cdot CO\cdot COOH} + NH_3 + DPNH_2 \rightarrow \underset{\text{Glutamic}}{COOH\cdot CH_2CH_2\cdot CHNH_2\cdot COOH} + DPN + H_2O$$

$$\text{Glutamic acid} + \underset{\text{Oxaloacetic}}{COOH\cdot CH_2\cdot CO\cdot COOH} \rightarrow \alpha\text{-Ketoglutaric acid} + \underset{\text{Aspartic}}{COOH\cdot CH_2\cdot CHNH_2\cdot COOH}$$

Glutamic acid and aspartic acid can both give rise to families of other amino acids through a number of biosynthetic pathways. In all, five such families exist.

Protein molecules consist of extremely long chains of amino acids connected by peptide linkages in which the carboxyl and amino groups are joined. Unlike many of the polysaccharides, proteins are mixed polymers in which a variety of amino acids are used in the construction of the chain. The synthesis of proteins from simple amino acids is a complex process which is still not fully understood. It results not only in the formation of peptide bonds but also in the exact arrangement of the amino acid residues in the finished polymer. Most proteins formed by the living cell are enzyme proteins, each of which has a definite function. It is generally conceded that the function is dependent on specific sequences of amino acids in the protein chain. This arrangement of the amino acids is controlled by the ribonucleic acid (RNA) fraction of the cell which acts as the protein-forming system. The first step in protein synthesis involves the activation of the various amino acids through the mediation of ATP. These activated amino acid residues are then transferred to, and combined with, RNA, where a series of reactions occurs, resulting in the release of the formed protein molecule (5).

The synthesis of fats, nucleic acids, and other cellular components occurs in a fashion similar to that of carbohydrates and protein; again at the expense of the energy and metabolic intermediates formed by the catabolic activities of the cell (36, 106, 129, 149).

In the foregoing, the question of energy source and substrates for cellular synthesis has been considered from the standpoint of the organism itself. It is only incidental to the organisms concerned that, in providing for their welfare, they affect the fertility of the soil. Some organisms exert their greatest effects in soil as a direct result of their energy-yielding mechanisms—as in the activities of bacteria concerned with nitrification, denitrification, and sulfur transformations. On the other hand, many microorganisms exert a profound influence on the nutrient status of a soil, even before absorption of their specific nutrients. Under natural soil conditions the amount of readily diffusible carbonaceous and nitrogenous material available to microorganisms is strictly limited. The majority of nutrients present are in the form of large insoluble molecules (proteins, celluloses, hemicelluloses, nucleic acids), which must be hydrolyzed or predigested prior to absorption and eventual transformation. In this way, microorganisms prevent the accumulation of plant, animal, and microbial remains in soil, and effectively initiate the recycling of carbon, nitrogen, and other essential elements. Microorganisms generally affect the breakdown of such complex molecules through the elaboration of extracellular hydrolytic enzymes (120). These enzymes are specific in nature, capable of catalyzing a single transformation of one or a few closely related substrates. Many extracellular enzymes are adaptive in nature, being produced by the microorganisms only in the presence of a particular substrate. The decomposition of organic molecules in soil is not confined to the action of these extracellular enzymes. In certain instances, high-molecular-weight substances can be broken down by close contact between the substrate and the organism through the action of surface enzymes. Such a situation exists in the breakdown of cellulose by *Cytophaga*. Some evidence has also been produced which suggests that intracellular enzymes released on lysis of the parent cell are also responsible for certain extracellular transformations (120).

Regardless of the nature of the enzymes involved, it is apparent that the activity of the soil micropopulation is, in reality, a reflection of its enzymatic make-up. In this respect it is essential that the activities of these catalysts be considered in the light of conditions prevailing in the soil—conditions that differ markedly from those existing in laboratory culture.

ENZYME ACTIVITY IN SOIL

A large part of microbial protoplasm is composed of enzyme protein, and it is the effect of environmental changes on these biological catalysts that controls the activities of microorganisms. In a heterogenous system such as soil, enzyme activity is subject to an incalculable interplay of biological, chemical, and physical factors.

Although free enzymes are undoubtedly present in soil as the result of formation of extracellular enzymes or through the lysis of microorganisms, little is known as to the site of their activity, or their durability in such a complex medium. It is unlikely that an appreciable amount of extracellular protein exists in natural soils at any one time, since proteins would be rapidly hydrolyzed by various microbes. In order to study enzyme activity in soil apart from the living microorganisms, it is necessary to treat the soil in such a way as to prevent (*a*) the further assimilation of products of enzyme reaction, (*b*) the liberation of intracellular enzymes, (*c*) inactivation of free enzymes present, and (*d*) the further synthesis of extracellular enzymes (41).

Kiss (75) has presented an exhaustive review on soil enzymes, and only a few pertinent observations concerning the detection of free enzyme activity will be mentioned at this point. The experimental demonstration of enzyme activity in a system sterile with respect to microorganisms has proven difficult, since most heat and chemical treatments are likely to denature enzyme protein in addition to changing the chemical and nutritional properties of soils (121). The use of toluene has met with mixed success. Saccharase, amylase, urease, proteinase, β-glucosidase, and α- and β-galactosidase activity has all been demonstrated in soil-toluene systems (75). As an antiseptic, toluene is not ideal, for data have been presented which demonstrate the multiplication of organisms in its presence (38). Toluene has also been shown to accelerate the release of enzymes by the rupturing of living and dead cells (67). The efficacy of chloramphenicol and ethylene-oxide sterilization is dubious, in that the former, like toluene, may be used as a metabolite whereas the latter is known to inactivate enzymes.

The difficulties inherent in heat or chemical sterilization of soil have now been overcome, in part, through the use of ionizing radiation (95). The application of high-intensity electron beams for the sterilization of soils presents several distinct advantages: (*a*) autolysis of the microorganisms is kept at a minimum, since the soils can be irradiated in the dry state; (*b*) any change in the physical or chemical properties of soil is small; and (*c*) the technique permits differential sterilization, in that fungi, bacteria, viruses, and enzymes, can be eliminated in turn by varied exposures. With this technique it has been possible to demonstrate residual urease and phosphatase activity in sterilized soils (50, 94, 95). It still remains to be shown whether this residual activity in irradiated soils takes place within the dead cells or on soil particles. The possibility exists that the observed urease and phosphatase activity is the result of diffusion of the substrate molecules into the dead cells where it is acted upon by intracellular enzymes unaffected by the irradiation. In this respect, it is important to note that irradiated sterilized soil can still respire—suggesting that the bactericidal effect of irradiation is due to inhibition of enzyme synthesis rather than to the inactivation of enzymes already present in the cell (109).

Despite the failure to localize the site of residual enzyme activity, it is reasonable to assume that extracellular enzyme activity in soil takes place on or near colloidal particles in soil. In a strongly adsorbing environment such as soil, it is important to consider the effect of adsorption from the standpoint of both substrate and enzyme. Conflicting reports regarding the effects of adsorption on the physiological activities of microorganisms have appeared in the literature (49). Many instances have been reported in which the presence of clay minerals or other adsorbents has resulted in the protection of adsorbed proteins from enzymatic hydrolysis*. Insoluble carbohydrates, crop residues, casein, dextrins, and organic phosphorus compounds may be similarly protected from the action of soil microorganisms or added enzymes (16, 18, 89). It has been suggested that the protec-

* See references 16, 18, 48, 51, and 111.

tive action of clay minerals is due either to the inactivation of adsorbed enzymes or to the inaccessibility of substrate molecules in the adsorbed state (48). In the latter instance it is notable that minerals of the montmorillonite type exert the greatest protective effect. Studies have shown that substrates adsorbed on the surface of kaolinite or montmorillonite are readily accessible to enzyme attack, but after drying and re-wetting, the montmorillonite complexes are more resistant (92). This would suggest that the protective action of expanding lattice structures is due to factors other than adsorption (49).

The protective action afforded by mineral complexes for adsorbed substrates undoubtedly plays a major role in the conservation of organic matter in the soil (77). The presence of adsorbent surfaces also has a distinct stimulatory effect on microbial activity, due to improved aeration, removal of toxic substances, or concentration of nutrients (49). Microorganisms are rarely found free in the liquid phase of soil. Bacteria in particular are attracted to colloidal particles, a situation which gives rise to segregated colonies in favorable microecological sites. In this respect, the adsorption and concentration of nutrients out of the relatively dilute soil solution plays a major role in the development of microenvironments in soil.

In such a situation it is difficult to conceive of enzyme activity in the soil solution as opposed to activity at solid-liquid interfaces. In this respect, it should be remembered that distances that are ecologically unimportant for larger organisms become highly significant for soil microbes. Evidence has now been produced which strongly suggests that a great deal of enzyme activity in soil is the result of enzyme-substrate-adsorbent complexes (92). Such a situation would greatly benefit the organisms in the microenvironment from the standpoint of concentration and adsorption of nutrients, ready exchange of enzymes, and retention of hydrolytic products for further metabolism by the organisms concerned.

ORGANIC MATTER TRANSFORMATIONS

The organic fraction of the soil is a complex of substances whose composition is determined in part by the plant and animal residues added to the soil, and to a greater extent, by the transformations of these substrates through biological, physical, and chemical means. As a result, a vast array of compounds is present: those representing the components of organic residues undergoing decomposition and the metabolic byproducts of the microorganisms utilizing such compounds as a source of energy. Included in this extremely diverse group one finds: carbohydrates and related compounds (mono- and disaccharides, cellulose, hemicelluloses, pectins, pentosans, mannans, polyuronides, uronic acids, organic acids, alcohols, hydrocarbons, aromatics); proteins and their derivatives (amino acids, amides, amino sugars, nucleoproteins, purine and pyrimidine bases); and lignins, fats, tannins, and their various decomposition products*. In addition to the intermediary products of organic matter breakdown and secondary synthetic products of microbial action, there exists in soil an accumulation of the more-resistant products of decomposition—the soil humus. The native organic fraction of soils is made up of a heterogenous mixture of polymerized aromatic molecules, polysaccharides, bound amino acids, uronic acid polymers, and various organic phosphorus compounds (77). By definition, humus is a complex mixture of amorphic and colloidal substances arising from modified plant materials and synthesized microbial tissue. The characterization of humus is far from complete, and concepts concerned with its formative processes are, to a large extent, speculative. Early theories considered that lignin complexed with protein of microbial origin provided the main source of humic substances in soil. The similarity between the aromatic structure of lignin and that found in humic acids, coupled with the relative resistance of lignin to enzymatic breakdown, lends strong support to such a theory (142). More recent studies, however, suggest that compounds of non-lignin origin may also be a source of aromatic "structural units." In this respect the byproducts of the activities of various fungi, actinomycetes, and bacteria are particularly significant (16, 18, 54, 77). Microorganisms frequently give rise to compounds of an aromatic or similar nature through the conversion of many diverse organic compounds, including carbohydrates (111). Such aromatic structures are readily oxidizable and can be condensed with amino acids or peptides with the formation of substances similar to those found in the soil humus fraction. There is little doubt at the present time that the "nuclei" of humus arise both from altered lignin compounds and through the synthesis of aromatic compounds by microorganisms. Microorganisms are thus implicated in all phases of humus formation. They effect the decomposition of the original plant and animal residues to simpler compounds and, at the same time, are responsible for the alteration of lignin and tannin molecules which will become part of the structural units of humus. The products of synthesis and metabolism of microorganisms (amino acids, peptides, amino sugars, uronic acids, etc.) participate in the formation of the functional (side) groups of humus. Finally, the synthesis and condensation of aromatic molecules by certain microorganisms suggests their participation in the formation of the humus itself.

Figure 6.1 illustrates a schematic representation of organic matter decomposition and humus formation in soils.

* See references 16, 18, 29, 55, 77, 96, and 142.

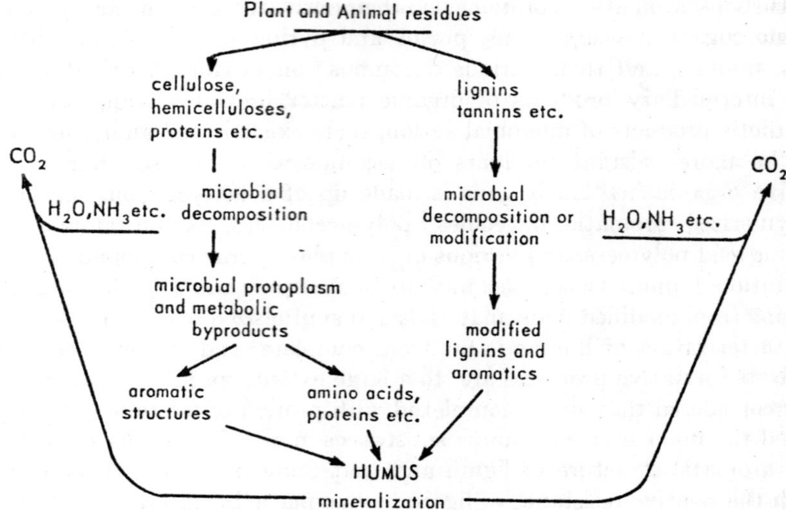

Figure 6.1. Organic matter decomposition and formation of humic substances in soil.

The presence of humus in soil exerts a pronounced influence on the physical, chemical, and biological activities taking place. Humus improves soil structure, drainage, and aeration, increases the water-holding, buffer, and exchange capacity, influences the solubility of soil minerals, and serves as a source of energy for the development of microorganisms (19). In this latter respect, humus can also be considered as a storehouse of nutrients essential to plant growth. During the slow microbial decomposition of the soil humus, there is a gradual release, with subsequent mineralization, of carbon, nitrogen, sulfur, phosphorus, and other elements immobilized during the formative processes.

Thus in soil we find two major decomposition processes occurring simultaneously: (*a*) the breakdown of fresh organic substrates, and (*b*) the decomposition of the native soil organic matter, or humus. Plant and animal residues added to soil are initially broken down to their basic structural units by extracellular enzymes. These units in turn are absorbed and oxidized by microorganisms in order to derive energy. Despite the dissimilarity of such compounds as polysaccharides, proteins, and aromatic substances it is noteworthy that after their initial degradation the metabolic sequences taking place in the microbial cell follow the same general pathways in the production of energy. As a result of these activities the constituent chemical elements in the residues (carbon, nitrogen, phosphorus, and sulfur) are liberated in forms available to the plant. At the same time, however, appreciable amounts of these elements are being immobilized during the formation of new microbial tissue or metabolic byproducts. These in turn are subject to decomposition with further mineralization. While decomposition processes in soil proceed, microbial activity gradually decreases as the more resistant structures and byproducts of microbial activity accumulate. As a result, appreciable quantities of essential nutrients remain temporarily immobilized as these resistant materials are complexed into the soil humus fraction.

Despite the fact that the soil organic fraction is being continually transformed through the activities of microorganisms, it remains remarkably stable with regard to amount and nature. This state of equilibrium is due to the balance maintained between the mineralization and immobilization processes. Thus, one finds the elements making up the plant and animal residues undergoing a cyclical alteration in soil—existing either in the free inorganic state or in a combined organic state, in living organisms or in the soil humus. Inasmuch as carbon, nitrogen, phosphorus, and sulfur, together with water, make up the greater part of plant and animal tissue, the cyclical transformations of these elements have a special significance from the standpoint of availability to the plant.

THE CARBON CYCLE

Carbon, as the major element in organic structures, is metabolized initially in the form of carbon dioxide. Since the carbon dioxide supply in the atmosphere is limited, it is essential that the carbonaceous materials produced by plants and animals eventually become completely decomposed with the formation of this gas. The carbon cycle is concerned, therefore, with carbon dioxide, its fixation, and regeneration. Carbon dioxide is converted to organic carbon mainly through the photosynthetic activities of the higher plants and, to a lesser extent, by microorganisms.

Animals in turn feed upon these preformed carbon compounds, converting them to animal tissue and respiring carbon dioxide in the process. Eventually both plant and animal remains are deposited in the soil where the microbial population plays the dominant role in the continuing cycle. In the soil these residues undergo decomposition through the activities of successive groups of microorganisms. Initially, activity is high as the readily decomposable residues are oxidized to carbon dioxide and water. As the available material decreases, evolution of carbon dioxide lessens until the products of microbial metabolism and the more-resistant fractions of the organic residues are incorporated into the native soil organic fraction. The cycle is completed and carbon is made available with the final, slow decomposition of the soil humus.

The nature of the heterotrophic flora found in soils is determined to a large extent by the amount, type, and availability of the organic matter. Many diverse groups of microorganisms participate; these include aerobic and anaerobic bacteria, actinomycetes, filamentous fungi, and higher fungi (2, 78). The decomposition of carbonaceous materials of one kind or another is rarely a unique feature of these organisms, and most are capable of transforming a wide range of substrates. Microbial activity, in turn, is governed by a number of environmental conditions, as soils differing in their physical and chemical characteristics vary considerably in their decomposing powers. Other major factors affecting organic matter transformations include temperature, aeration, moisture, pH, and the presence or absence of more readily available substrates (2, 31, 141). Inasmuch as carbohydrates make up the bulk of organic residues added to the soil, the availability of inorganic elements is a critical factor.

About 90 per cent of the dry matter of plants is composed of carbohydrates of varying complexities. Of these, cellulose, hemicelluloses, lignins, and starches form the largest part, with lesser amounts of pectins, inulin, gums, waxes, and related substances also being present. In considering the decomposition of these varied polysaccharides, it is important to realize that the critical sequence in all cases is the initial conversion of the long, complex molecules to simple easily metabolized derivatives. In most instances these conversions are catalyzed by relatively specific extracellular enzymes.

In the following sections the decomposition processes governing cellulose, hemicelluloses, lignins, starch, and pectin are discussed in greater detail. The breakdown of chitin has also been included, since an appreciable quantity of this carbohydrate also finds its way into soil.

Cellulose

Cellulose occurs abundantly in nature, primarily as the principal constituent of the cell walls of most plants. It is often found in close association with hemicelluloses, lignin, pectin, and other polysaccharides. Cellulose is a long-chain polymer of glucose joined by β-1,4-glucosidic linkages and containing between 1500 and 9500 sugar residues. In natural celluloses, the individual glucose chains are aligned in a number of ways to form complex organized fibrils with both crystalline and amorphous areas (58, 124, 150). The resistance of cellulose to decomposition appears to be related to the ability of enzymes to penetrate between adjacent chains within these structures.

Cellulose is broken down by the extracellular enzymes of microorganisms which cleave the β-1,4-glucoside linkages in a nonspecific manner (58, 120). The number of steps involved in the hydrolysis of cellulose to glucose is still a matter of controversy. It has been shown that a single cellulase is capable of hydrolyzing native cellulose to glucose (66). On the other hand, the dimer cellobiose frequently occurs as the end product of cellulase activity, and it is only through the action of a second enzyme, cellobiase (β-glucosidase), that cellobiose is split into glucose units. Evidence has also been presented which suggests that cellulase activity is the result of more than one enzyme (151). In this connection, at least two enzymes are involved in the production of low-molecular-weight compounds from native cellulose; the first enzyme (C_1), produces long, linear chains which are hydrolyzed by a second enzyme (C_4). This multienzyme system may explain the failure of many organisms to initiate the attack on native cellulose, although they can hydrolyze partially degraded material.

Cellulose is rapidly decomposed in soils by diverse groups of microorganisms. This cellulolytic population includes aerobic and anaerobic mesophilic bacteria, thermophilic bacteria, filamentous fungi, basidiomycetes, actinomycetes, protozoa, amoebae, and numerous lower animal forms (2, 58, 71, 78). It is doubtful if any of these microorganisms require cellulose specifically as a substrate, most being capable of utilizing a variety of carbohydrates.

Hemicelluloses

Apart from cellulose, the hemicelluloses form one of the major constituents of the native plant cell. Hemicelluloses refer collectively to less water-soluble cell-wall polysaccharides, excepting cellulose and the pectins. Two distinct types of hemicelluloses are recognized: (*a*) the polyuronides, composed of repeating units of sugars and uronic acids, and (*b*) the cellulosans, which are characteristically polymers of hexose or pentose sugars, although occasionally uronic acids are present. In nature, xylose polymers occur most frequently as polyuronides in association with glucuronic acid or as the cellulosan (xylan), made up of repeating units of xylose. While many hemicelluloses are of these types, others have been found to contain arabinose, galactose, mannose, and galacturonic acid (58, 150).

Few studies have been made of specific extracellular enzymes catalyzing the hydrolysis of hemicelluloses, although these substances, like cellulose, are readily decomposed by a wide variety of microorganisms (2, 71). Enzymatic hydrolysis results in the eventual liberation of the simple sugars or uronic acids of the polymers, all of which are actively metabolized by the soil flora. Frequently, the synthetic activities of these microorganisms give rise to new hemicelluloses of microbial origin in the form of excretion products, capsular material, or other cellular constituents which are recycled as part of the soil organic fraction (96).

Lignins

Lignin represents the third most abundant constituent of plant tissues where it is found in the middle lamella in close association with the cell wall carbohydrates. It is presumed that lignins do not occur in the plant in the free state but rather as chemically combined lignin-carbohydrate complexes. Lignins do not represent a uniform family of compounds, as individual lignins differ chemically, depending on their source.

The lignin molecule contains carbon, hydrogen, and oxygen, but the structure is aromatic as opposed to the carbohydrate nature of cellulose and hemicelluloses. Basically, lignin is a polymer built up from substituted derivatives of phenylpropane. Hydrolysis of many plant lignins frequently yields such structures as vanillin, syringaldehyde, coniferaldehyde, and *p*-hydroxybenzaldehyde (77). It is generally accepted that the basic units of lignin are typified by a phenylpropionic structure containing hydroxyl and methyoxyl groups. Polymerization of these basic units results in the formation of chainlike molecules, which, in turn, are condensed to form the complex branched structures found in native lignin. Detailed studies on the structure of lignins and the nature of the linkages involved have been fully reviewed by Brauns (14, 15).

Although lignins are extremely resistant, their gradual disappearance from the soil attests to the fact that they are capable of being metabolized. Studies on the decomposition of this substrate, both in soil and *in vitro*, have been seriously hampered by the lack of critical analytical techniques for determining residual lignin, and by the still incomplete knowledge of its structure and properties. Many investigations can be cited to show that soil microorganisms, particularly the higher fungi, degrade lignin both in soil and in laboratory culture (15, 16, 71). Nevertheless, virtually nothing is known of the intermediary metabolism of the degradative process.

It has been pointed out that in the formation of humus, the lignin molecule undergoes modification. It has been suggested that the lignin molecule is first split oxidatively with the formation of its primary structural groups (vanillin, syringaldehyde, etc.). These compounds then lose their methoxyl groups, are oxidized to quinones, which are then polymerized and condensed with amino acids to form humic acids (53, 54). It is unlikely that these structural units always participate in the formation of humus, since the soil harbors a large population of microorganisms capable of assimilating these aromatic compounds. Many soil fungi are capable of oxidizing vanillin, ferulic acid, syringaldehyde, and *p*-hydroxybenzaldehyde (65, 77, 113). Similarly, many soil bacteria are capable of decomposing a wide range of aromatics with an ultimate splitting of the ring structure resulting in the formation of easily metabolized, low-molecular-weight organic acids (77, 113, 130).

Starches

Starches are the most common reserve foodstuffs of plants and are found principally in the seeds, fruits, tubers, roots, and stems. They vary considerably in their physical properties when isolated from different species, although chemically they have definite similarities. Starches are not generally single substances, but rather mixtures of two structurally different polymers of glucose. The amyloses are linear, thread-like molecules in which glucose units are joined by α-1,4-glucosidic linkages, while in the amylopectins, the glucose is joined by the same linkage, but the molecules are branched with side chains attached through α-1–6-glucosidic bonds (120, 150).

Starches are rapidly decomposed by a great many soil microorganisms through the elaboration of extracellular enzymes known collectively as amylases. The enzyme β-amylase completely hydrolyzes amylose to maltose, but stops at the 6-1 branch points of amylopectin, leaving a residual unhydrolyzed dextrin. The α-amylases are characterized by their reduction of the starch molecule to large units containing approximately six glucose residues. Neither the α- nor β-amylases can hydrolyze the α-1,6 linkages—a process which is carried out by special debranching enzymes. The final product of starch hydrolysis, maltose, is converted by α-glucosidase to glucose which is rapidly metabolized.

Pectins

Pectins are found most abundantly in the intercellular layers and in the primary and secondary cell walls of all higher plants. The molecules of these substances are polymers of galacturonic acid units in which some of the carboxyl groups are esterified with methyl groups (120, 150). Pectins are commonly broken down by two distinct enzymes: pectinesterase, which hydrolyzes the methyl ester linkage to yield pectic acid and methanol; and polygalacturonidase, which destroys the linkages between the galacturonic acid units of either pectin or pectic acid to release galacturonic acid.

Many bacteria, fungi, and actinomycetes are capable of bringing about the rapid hydrolysis of pectic substances in soil (2). The metabolic activities of these organisms are important in the retting of flax and in the decay of fruits and vegetables. In addition, they have been implicated in the wilt diseases of plants.

Chitin

Chitin differs from the majority of polysaccharides found in soil in that its basic unit consists of an amino sugar. Chitin is found in the cell walls of lower plants and fungi, but the major source in soil is in the form of insect remains. The structure of chitin is similar to that of cellulose (a β-1,4-linked chain of hexose molecules), but in place of glucose, N-acetylglucosamine serves as the repeating unit. Decomposition of this material is through the elaboration of extracellular chitinase. Some evidence suggests that only one enzyme is implicated in the formation of N-acetylglucosamine. On the other hand, the frequent formation of N,N′-diacetylchitobiose necessitates the action of a second enzyme, chitobiase, for the final hydrolysis (2, 120, 150). N-acetylglucosamine is readily metabolized by the soil flora.

THE NITROGEN CYCLE

The availability of nitrogen is of prime importance to growing plants, since they are dependent on an adequate supply of nitrate and ammonia for synthesis of their nitrogenous constituents. Animals, in turn, depend on plants for their energy and synthetic processes. When plant and animal remains are added to soils, their nitrogenous constituents undergo numerous transformations, many of which are biologically opposed. The net result is that only a small proportion of the total nitrogen in soil is present in an available form at any one time. The cycle of nitrogen as it exists in nature is illustrated in Figure 6.2.

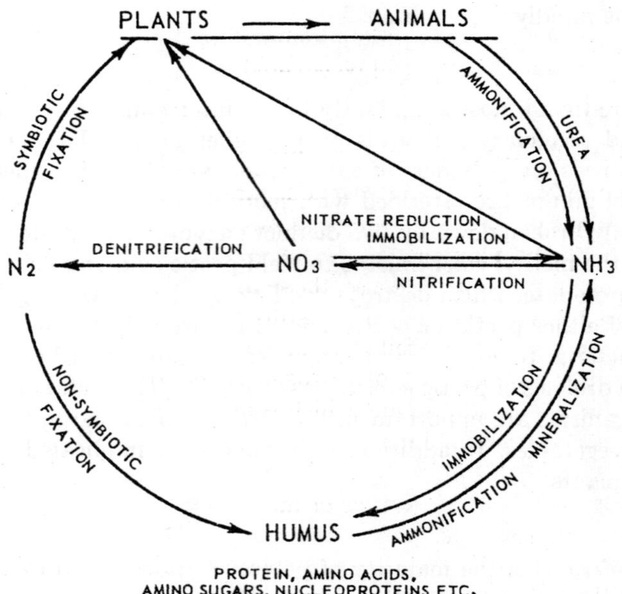

Figure 6.2. The nitrogen cycle.

Mineralization of nitrogen is concerned with the degradation of the organic nitrogenous compounds of freshly added residues, as well as those in organic complexes of the soil humus. In mineralization the initial reduction to NH_4 constitutes ammonification, while the oxidation of this compound to nitrate is termed nitrification. In opposition to mineralization, a large proportion of the ammonia and nitrate may be immobilized through the biological activities of microorganisms utilizing these compounds in the synthesis of protoplasm. Eventually much of this protoplasm and the nitrogenous by-products of metabolism enter the stable soil humus fraction. In view of the fact that active mineralization implies active growth of the soil flora, it is clear that the two processes—mineralization and immobilization—occur simultaneously.

Nitrate may be lost from the soil through biological denitrification, whereby nitrates are reduced to free nitrogen, which is unavailable to plants and most microorganisms. Certain groups of microorganisms have the ability to fix atmospheric nitrogen. The first group, or symbiotic nitrogen-fixers, are able to utilize free nitrogen when growing in symbiosis with the roots of certain plants, while the nonsymbiotic organisms are able to carry out this process independently. In the latter instance, nitrogen gas is transformed and incorporated as microbial tissue, which is eventually mineralized.

The processes of mineralization, immobilization, nitrification, denitrification, and nitrogen fixation, will be covered in greater detail in the following sections.

Mineralization and Immobilization of Nitrogen

A large proportion of the nitrogenous material found in the soil, or added to soil as plant or animal residues, is in organic combination. In order to be utilized by the plant these organic residues must be first mineralized to ammonia or nitrate. Because these assimilable forms of nitrogen are readily lost through leaching or volatilization, it is essential that this nutrient be conserved in the soil in some way. During or subsequent to mineralization, a large amount of nitrogen is incorporated and temporarily immobilized in microbial tissue, which is then complexed in the soil organic fraction. This soil organic fraction, in turn, functions as a reserve of this important element, whence it can be gradually mineralized and recycled through the continuing activities of microorganisms.

Nitrogen accounts for about 5 per cent of the soil organic fraction and approximately 98 per cent of this element occurs in the organic form. Inorganic nitrogen rarely exceeds 2 to 3 per cent of the total soil nitrogen (16, 19). While the identification of all of the nitrogenous constituents of soil is not complete, there is very good evidence that 24 to 37 per cent of the nitrogen in soils is in the α-amino form, which suggests the proteinaceous character of this fraction (16, 17, 20, 96). Approximately 3 to 10 per cent of the soil nitrogen occurs in nucleic acids, while values of 5 to 10 per cent have been established for amino sugars (17, 20, 21, 30). These substances account for roughly one-half the total nitrogen in soils. No other compounds have been identified in significant quantities, although it is possible that the unidentified fractions exist as highly resistant condensation products or strongly adsorbed complexes of lignin and ammonia or amines (21, 30).

Since the bulk of nitrogenous compounds in soil are protein in nature, their decomposition constitutes a major phase of the mineralization process. In soil, the degradation of proteins is accomplished by innumerable heterotrophic microorganisms. Very few organisms fail to utilize some complex form of nitrogen, and it is uncommon to find that nitrogen is not mineralized even under the most extreme conditions. As with most macromolecules, proteins are broken down through the elaboration of extracellular enzymes. Proteolytic enzymes attack the protein molecule by hydrolyzing the peptide linkages (56). The initial products of hydrolysis are long, amino acid chain fragments or peptones, which, on continued hydrolysis, yield peptides and eventually free amino acids. Amino acids may then be: (*a*) metabolized by microorganisms (immobilization), (*b*) transformed by microbial enzymes with the formation of ammonia, (*c*) adsorbed by clay minerals or incorporated in the humus fraction, and (*d*) utilized by plants (52). The amount of free amino acids that can be detected in soils is very small, attesting to the fact that these compounds are rapidly immobilized or transformed. Ammonia, in turn, can be: (*a*) oxidized to nitrates (nitrification), (*b*) utilized by higher plants, (*c*) immobilized by non-nitrifying organisms, and (*d*) adsorbed by clay minerals or organic matter (62, 98).

In addition to proteins, nucleic acids provide a major portion of the nitrogenous substances available to microorganisms. While small amounts of these compounds are derived from plant and animal tissues, they are to a large extent of microbial origin. Both ribonucleic acid (RNA) and deoxyribonucleic acid (DNA) are polymers of mononucleotides which consist of a purine or pyrimidine base, a pentose sugar, and phosphate. Many bacteria and fungi elaborate extracellular deoxyribonuclease and ribonuclease which cleave the large polymers of nucleic acids into their constituent mononucleotides. Dephosphorylation of the mononucleotides with a subsequent hydrolysis of the ribosides or deoxyribosides yields sugar and free purines or pyrimidine bases (56). These nitrogenous bases are readily metabolized by a variety of soil organisms to form ammonia, amino acids, organic acids, and urea. Urea may also enter the soil as fertilizer or as an excretory product of animal metabolism. Again a wide range of microorganisms is capable of elaborating the enzyme urease which hydrolyzes urea with the eventual liberation of ammonia.

Mineralization of nitrogen from fresh organic residues or from the more stable humus fraction is dependent on a number of environmental factors. Physical and chemical conditions of the soil habitat, such as moisture, pH, aeration, temperature, the total nitrogen status of the soil, and the inorganic nutrient supply, will govern the activities of the soil flora and the velocity of mineralization. In addition, the adsorption of organic nitrogen compounds, particularly proteins on clay colloids, has a distinct protective effect, as mentioned previously (16, 18, 48, 111). Both mineralization and immobilization of nitrogen in soil are markedly affected by the carbon:nitrogen ratios of added organic residues. The active decomposition of organic matter implicates an actively multiplying microflora with a simultaneous assimilation of nitrogen for its growth processes. If the amount of nitrogen mineralized is to exceed that immobilized as microbial protoplasm, the decomposing tissue must have a nitrogen content in excess of the microbial requirements. As a general rule, the critical ratio falls between 20 and 25:1—wider ratios will favor immobilization; narrower ratios, mineralization (29, 62). Thus, if protein, having a narrow C:N ratio is added to soil, a large part of the nitrogen will be recovered in mineral form, since it is in excess of the requirements of the decomposing flora. On the other hand, carbohydrates, having a wide C:N ratio, will result in the almost complete immobilization of available nitrogen as microbial protoplasm. Although the C:N ratios of added residues are prime factors in mineralization, the degree of resistance of these substrates to biological decomposition is also important. In substrates having a similar C:N ratio, the disappearance of inorganic nitrogen is associated with their degree of resistance (22). The more available the substrate, the greater the proliferation of the attacking flora and, consequently, the greater the immobilization. The nitrogen requirements for the decomposition of more-resistant materials, such as lignins, is considerably lower, since there is less proliferation and since only small amounts of microbial protoplasm are synthesized (2, 62).

The addition of fresh organic residues to soil also influences the recycling of the nitrogen reserves of the soil humus fraction. Until the advent of labeling techniques, it was impossible to distinguish between the products of added organic matter and those of the soil organic fraction (70). The labeling of the carbon and nitrogen fractions has made it possible to show that the mineralization of soil humus is greatly accelerated in the presence of added residues (16, 28). Such studies indicate that the soil organic fraction is not stable, as frequently implied, but is being continually recycled, the rate being dependent on the presence or absence of an energy source necessary to support a vigorous microbial population. Under ordinary circumstances when the C:N ratio of soil is low, the supply of readily available energy material is limited, resulting in slow mineralization of nitrogen. Such an arrangement is advantageous in that the main nitrogen reserves of soil are held in an insoluble form which cannot be leached away but which can be slowly mineralized to support plant growth.

Nitrification

From the time nitrogen is assimilated by plants until its liberation, through mineralization, as ammonia, the nitrogen atom remains in a reduced state. In most cases, however, nitrogen is used by the plant in its most oxidized form—nitrate. The capacity of soil microorganisms to convert ammonia to nitrate is of utmost significance in providing growing plants with this assimilable form of nitrogen. The oxidative sequence involved in the conversion of ammonia to nitrate is termed nitrification. In soil this is generally brought about by two highly specialized groups of aerobic autotrophic bacteria which derive their energy from the oxidation of their specific inorganic nitrogen compounds. The first group, the *Nitrosomonas*, oxidizes ammonia to nitrite; the second group, the *Nitrobacter*, oxidizes nitrite to nitrate.

$$\textit{Nitrosomonas:}\ NH_4^+ + 1\tfrac{1}{2}O_2 \rightarrow NO_2^- + 2H^+ + H_2O$$

$$\textit{Nitrobacter:}\ NO_2^- + \tfrac{1}{2}O_2 \rightarrow NO_3^-$$

Full details of the genera responsible for these transformations are given elsewhere (2, 42, 97).

Information with regard to the metabolic sequences involved in the oxidation of nitrogen from its most reduced form to its most oxidized state, is still fragmentary. In the oxidation of ammonia to nitrite by *Nitrosomonas*, hydroxylamine appears as the most likely oxidative product of ammonia, followed by the formation of hyponitrite. Evidence for the latter product is still lacking and it cannot be said at this time whether or not hyponitrite or some empirically similar compound actually exists in the pathway (35, 42, 80, 117).

$$\textit{Nitrosomonas:}\ NH_3 \xrightarrow{\frac{1}{2}O_2} \underset{\text{Hydroxylamine}}{NH_2OH} \xrightarrow{-2H} \underset{\text{Hyponitrite}}{\tfrac{1}{2}HON{:}NOH} \xrightarrow{\frac{1}{2}O_2} HNO_2$$

In the oxidation of nitrite to nitrate, there is no suggestion of intermediary nitrogen compounds. It is still not known whether the conversion is a simple dehydrogenation as it would appear (42, 80).

$$\textit{Nitrobacter:}\ HNO_2 \xrightarrow{H_2O} HON(OH)_2 \xrightarrow{-2H} HNO_3$$

The fact that nitrite is rarely found to accumulate in soils, even under conditions of rapid nitrification, suggests that the two responsible groups occupy the same environment. In this respect, the activities of both the ammonia and nitrite oxidizers in soil are affected by a number of environmental conditions, particularly those influencing pH or aeration (2, 56, 71, 97). The optimum pH for growth of nitrifying organisms tends to be related to the reaction of the soil from which they were isolated. Strains isolated from soils having a low pH are generally more acid-tolerant than those from an alkaline environment. Although nitrification will proceed in soils with a pH as low as 4.5, the rate is reduced under such conditions. The addition of lime frequently brings about a marked increase in the rate of nitrification.

As strict aerobes, nitrifying organisms are particularly sensitive to such factors as excessive moisture or soil structure which may limit the ready access of oxygen. Under anaerobic or extremely acid conditions, ammonia may accumulate, because the ammonifying population is less sensitive to these adverse environmental factors than are the nitrifiers.

Until recently, the process of nitrification has been considered a unique feature of two relatively small groups of autotrophic bacteria, with *Nitrosomonas* and *Nitrobacter* being the dominant genera. It is now becoming apparent that many heterotrophic organisms might be implicated in this important soil process. Considerable information has accumulated which demonstrates that numerous bacteria, actinomycetes, and fungi are capable of producing nitrite or nitrate from ammonia and organic nitrogen compounds in pure culture (3, 42, 52). Unlike the autotrophic nitrifiers, heterotrophic organisms rarely accumulate more than a few ppm of nitrite or nitrate. In this respect it should be remembered that *Nitrosomonas* and *Nitrobacter* are completely dependent on the oxidation of ammonia and nitrite for the energy necessary for cell synthesis. These oxidations are, consequently, extremely efficient. Heterotrophs, on the other hand, do not depend on this oxidation for growth, and the appearance of nitrite and nitrate is noted only after active growth has occurred.

Since the capacity for nitrate production is one of the more important microbiological processes in soil, it is difficult to conceive that it is solely dependent on the activities of a few related autotrophic organisms. The wide variety of environments in which nitrification occurs, many of which are opposed to the growth of *Nitrosomonas* and *Nitrobacter*, suggests that heterotrophic nitrification may be important. It is difficult to assess the comparative significance of autotrophic and heterotrophic nitrification in soils. Although the autotrophs have the more efficient nitrifying system, the greater numbers of heterotrophs able to participate in this process may well be of equal ecological importance.

Denitrification

In contrast to nitrification, one finds numerous microbiological transformations occurring in which microorganisms compete effectively with growing plants for available nitrates. Denitrification, the most important of these transformations, is of considerable economic importance and, under adverse conditions, may account for significant losses of nitrogen from the soil. The processes in which nitrates are utilized by microorganisms fall into two general classes: (*a*) nitrate assimilation or assimilatory nitrate reduction, and (*b*) nitrate respiration or dissimilatory nitrate reduction (99). Nitrate assimilation denotes the biological reduction of nitrate to the ammonia or amino level, with the products being used for the biosynthesis of nitrogen containing cellular constituents. In nitrate respiration, nitrate is used as the terminal electron acceptor in place of oxygen by a variety of microorganisms, under anaerobic conditions. Nitrate respiration is carried out by many microorganisms in the absence of oxygen, and two distinct groups have been noted (139). The first group includes those organisms which reduce nitrate to nitrite, and allow the product to accumulate; the second group, or true denitrifiers, completely reduce nitrate to nitrogen and nitric or nitrous oxides. Thus, the term denitrification has been reserved to describe microbial processes whereby nitrate is reduced to gaseous products. Both nitrate assimilation and nitrate respiration involve reductive pathways. The end products of nitrate respiration (denitrification) are however, volatilized and lost from the soil, whereas the process of nitrate assimilation results in the nitrogen remaining in the soil as microbial protoplasm, and, therefore, the nitrogen is still potentially available to the plant.

The metabolic activities of the denitrifiers have been the subject of a number of recent reviews (43, 99, 100, 139), and the biochemical sequences involved are believed to be essentially as shown in Figure 6.3.

The first step involves the conversion of nitrate to nitrite, mediated by nitrate reductase and followed by the reduction of nitrate to a hypothetical

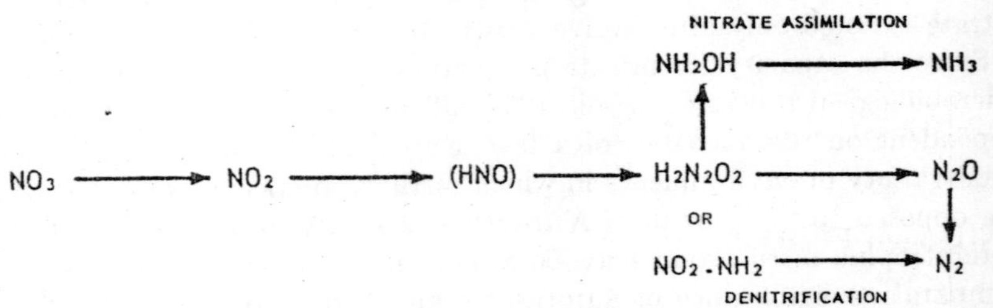

Figure 6.3. Proposed biochemical pathways for denitrification and assimilatory nitrate reduction.

nitroxyl intermediate (HNO). The latter step and the subsequent dimerization to hyponitrite or nitramide are only postulated, as are the final steps involving the formation of nitrous oxide and nitrogen gas. It is believed that assimilatory nitrate reduction follows the nitrification sequence in reverse, although different enzyme systems may be responsible.

The organisms responsible for denitrification in soil are all facultative species which utilize nitrate as an alternate electron acceptor in the absence of oxygen. In most instances these organisms will oxidize the same organic substrates under either condition. The capacity for denitrification appears limited to certain bacteria and so far no fungi or actinomycetes have been implicated. The heterotrophic denitrifiers comprise numerous genera and usually occur in large numbers in most fertile soils (2, 43, 56, 71). Under normal aerobic conditions these organisms are primarily concerned with the decomposition of organic matter. Their presence, however, indicates the denitrifying potential of a soil.

Included among the denitrifiers are a number of autotrophic organisms. *Micrococcus denitrificans*, a facultative autotroph, will develop anaerobically while oxidizing hydrogen at the expense of nitrate. *Thiobacillus denitrificans* oxidizes sulfur while reducing nitrate in the absence of oxygen (56, 139).

The rate at which denitrification occurs in soil is subject to the influence of a number of environmental factors (23). It has been well established that denitrification occurs only when the supply of oxygen required by soil organisms is restricted, generally by excessive moisture. In this respect the critical moisture level has been found to be about 60 per cent of the water-holding capacity. Below this level practically no denitrification takes place; above this level denitrification increases rapidly. Denitrification by the heterotrophic flora is also dependent on the presence of an adequate supply of a suitable organic substrate. Nitrogen losses under waterlogged conditions are directly related to the amount and availability of the organic material present. In poor soils or in the absence of an available energy source, denitrification is greatly reduced. Soil reaction and temperature influence nitrogen losses by influencing the denitrifying population as well as by affecting the volatilization of the gaseous nitrogen products. The presence or absence of growing crops may also influence the rate of denitrification—presumably through competition with the denitrifying flora for available nitrate.

Nitrogen Fixation

Plant growth, leaching, and the activities of denitrifying organisms all tend to remove the soil nitrogen so necessary for continued existence. Although the nitrogenous constituents of plants are decomposed and again made available, denitrification would eventually deplete the soil of this element if nitrogen gas were biologically inert. Fortunately, atmospheric nitrogen can be used by a number of free-living microorganisms and in symbiotic associations with higher plants. In this way large amounts of free nitrogen are initially fixed into microbial protoplasm, to be recycled eventually in the soil.

The free-living microorganisms capable of utilizing molecular nitrogen include, in addition to the classical *Azotobacter* and *Clostridium* species, a wide range of heterotrophic bacteria, autotrophic organisms, various photosynthetic bacteria, and a number of blue-green algae (2, 56). Despite the proven ability of these organisms to incorporate atmospheric nitrogen into their cells, the economic importance of this nonsymbiotic fixation in soils has still not been established. The efficiency of nitrogen fixation by the *Azotobacter* and *Clostridium* is very low. Nonsymbiotic fixation is a growth-linked process, and both groups require the presence of appreciable amounts of available organic matter in order to fix relatively small amounts of nitrogen (44, 105). Thus, the fixation of significant amounts of nitrogen in the soil by *Azotobacter* would depend on the presence of a large population of rapidly proliferating cells and a substantial amount of readily available organic matter (approximately 50 to 200 pounds for every pound of nitrogen fixed). Inasmuch as the recorded populations of *Azotobacter* and other free-living organisms in soils are generally low, it is difficult to consider these nonsymbiotic nitrogen fixers as important factors in the nitrogen economy of the soil. The importance of the *Clostridium* can be discounted for similar reasons.

It is probable that nonsymbiotic nitrogen fixation is significant in situations where algae are the predominant flora. The photoautotrophic nutrition of these forms precludes the need for large amounts of available organic matter. In this respect, nitrogen fixation by algae has been implicated in maintaining the nitrogen levels of paddy fields during the cultivation of rice (148).

The classical example of symbiotic nitrogen fixation is the association between bacteria of the genus *Rhizobium* and plants of the *Leguminosae*. Most legumes differ from other plants in that they use, in addition to the soluble nitrogen of the soil, the atmospheric nitrogen made available by *Rhizobium* species living in symbiosis with them in root nodules. Mature nodules consist of a central core containing nondividing resting cells of *Rhizobium* (bacteroids) surrounded by host tissue. It is this central area of bacterial tissue that is the site of nitrogen fixation (4, 105). Detailed information on the infection process and on nodule formation can be found elsewhere (104, 105).

Unlike nitrogen fixation by *Azotobacter* and *Clostridium* species, which is a growth-linked process dependent upon an external energy source, symbiotic fixation is mediated by resting cells of *Rhizobium* utilizing a source of energy supplied internally through the host. The nitrogen-fixing ability of the rhizobia appears to be completely dependent on this symbiotic relationship. As yet there is no evidence of *Rhizobium* species participating directly or indirectly in the fixation of atmospheric nitrogen in the absence of their host plant. In this respect, nitrogen fixation by legume bacteria is directly related to the concentration of hemoglobin present in the nodules. While the exact function of this pigment is still unknown, it undoubtedly plays a major role in the nitrogen metabolism of the *Rhizobium* species in the plant (7, 102).

Leguminous plants benefit markedly from their symbiotic relationships with *Rhizobium* and consequently the soils in which they are grown also benefit when these crops are turned under. In the field, symbiotic fixation accounts for much greater quantities of nitrogen than fixation by free-living organisms. Whether this is due to a more efficient fixation by the *Rhizobium*, or simply due to a larger mass of fixing tissue present in nodules under a crop, cannot be said—strict comparisons of the two systems are not practical. There is no doubt, however, as to the significance of symbiotic fixation in the economy of soil nitrogen. Conservative estimates of the annual fixation of all legumes average between 80 to 100 pounds per acre per year and much higher values have frequently been recorded (4).

The question of whether the nitrogen fixed in the symbiotic relationship is retained by the legume throughout the growing season is still controversial. Various studies have shown that crops grown in association with legumes give higher yields, particularly in the case of grasses and cereals (4). This beneficial effect has been attributed to the excretion of nitrogenous products into soil from the nodulated areas of the legumes. Present-day views suggest that this excretion of combined nitrogen occurs at times when the rate of nitrogen-fixation exceeds the requirements of the protein-synthesizing facilities of the plant (146).

Symbiotic fixation is not confined to plants of the legume family. Eight genera of the Angiosperms, comprising some 190 species of widely distributed trees and shrubs, also produce nodules. Investigation of a number of these species revealed that they do fix appreciable quantities of atmospheric nitrogen and that this fixation is confined to the nodules (4, 11, 12). This non-leguminous fixation is presumed to be symbiotic, although the responsible organisms have not been isolated.

Regardless of whether the ability to fix atmospheric nitrogen resides in a free-living microörganism or in root nodule bacteria, there appears to be considerable similarity in the over-all mechanisms of fixation. Fairly precise evidence has established ammonia as the key intermediate between the conversion of inorganic nitrogen and that point where nitrogen is assimilated into organic combination (154). This evidence is the result of investigations of both the nonsymbiotic nitrogen-fixers (*Clostridium*, *Azotobacter*, algae, photosynthetic bacteria) and the symbiotic systems. The main findings of these studies are: (*a*) that cells rapidly assimilate ammonia after previously being supplied with molecular nitrogen; (*b*) that highly labeled ammonia is excreted by cells after exposure to N_2^{15}; and (*c*) that glutamic acid is the first labeled amino acid formed after a short term exposure to N_2^{15} or $N^{15}H_4^+$. Hydroxylamine has been suggested as a possible alternative to ammonia. There is, however, considerable evidence which discounts this compound as the key intermediate, even though it undoubtedly enters into the over-all scheme. Hydrazine is another suggested intermediate and, as with hydroxylamine, it should not be completely ignored. Although a reductive sequence is generally considered the logical step in the formation of ammonia, some thoughts have been given to an oxidative mechanism (119). In this respect, nitrogen could be first oxidized to the nitrous oxide or hyponitrite level before entering the reductive pathway. A tentative scheme of the various pathways thought to be implicated in the fixation of molecular nitrogen is given in Figure 6.4.

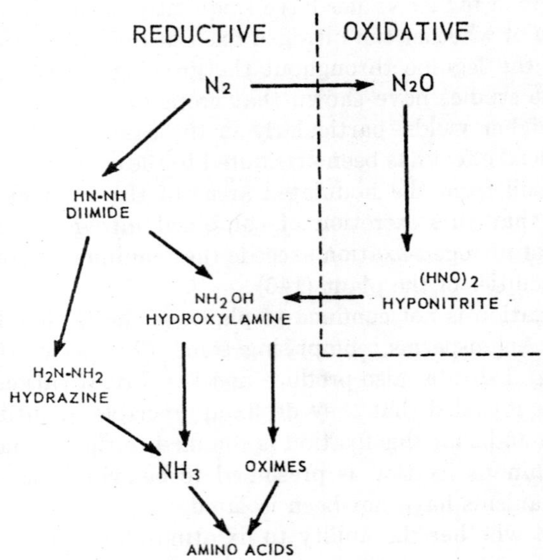

Figure 6.4. Proposed biochemical pathways for the fixation of atmospheric nitrogen.

At the present time very little is known of the enzyme systems involved in nitrogen fixation, and until these are fully explored the exact nature of the fixation process will remain obscure.

SULFUR TRANSFORMATIONS

The cycle of sulfur in nature is very similar to that of nitrogen. Sulfur compounds undergo many transformations as a result of activities of plants, animals, and microorganisms and, to a lesser extent, nonbiological action (33, 131, 152). Plants build up their sulfur-containing components primarily through the assimilation of sulfates from soil, although small amounts may be obtained through the absorption of S-amino acids or the direct assimilation of sulfur dioxide from the atmosphere (73). Animals in turn utilize plants or their residues, converting the reduced sulfur compounds to sulfates for incorporation into organic molecules. When plant and animal remains are returned to the soil, they are decomposed by microorganisms with the liberation of sulfur in inorganic forms. These inorganic sulfur compounds may be oxidized to sulfates or reduced to hydrogen sulfide by many physiologically diverse organisms.

The cycle of sulfur in soil can be divided into a number of distinct phases: (*a*) mineralization—the decomposition of large organic molecules into simple inorganic sulfur compounds; (*b*) immobilization—the assimilation of these compounds into microbial tissue; (*c*) oxidations—the conversion of inorganic forms of sulfur (sulfides, thiosulfates, polythionates, and elemental sulfur) to sulfate; and (*d*) reductions—the conversion of sulfates and other oxidized forms of sulfur to hydrogen sulfide. The general outline of the sulfur cycle as it exists in soil is illustrated in Figure 6.5.

Mineralization and Immobilization of Sulfur

Plant and animal residues introduce a wide variety of sulfur-containing compounds into the soil as substrates for microbial activity. Prominent among these are proteins containing the S-amino acids—cystine, cysteine, and methionine, the peptide glutathione, glucosides, alkaloids, thiamine, biotin, thioctic acid, thiocyanates, mercaptans, taurine, thiourea, ethereal sulfates, thiosulfates, sulfates, and sulfides. The initial decomposition of organic sulfur compounds is effected by numerous heterotrophic organisms and involves the countless reactions necessary to disrupt the variety of sulfur linkages. Little is known of the individual steps involved in the overall mineralization process, and generalizations can only be made as to the end products formed on the basis of environmental conditions. Under anaerobic conditions decomposition will result in the accumulation of mercaptans and hydrogen sulfide. In the presence of atmospheric oxygen, al-

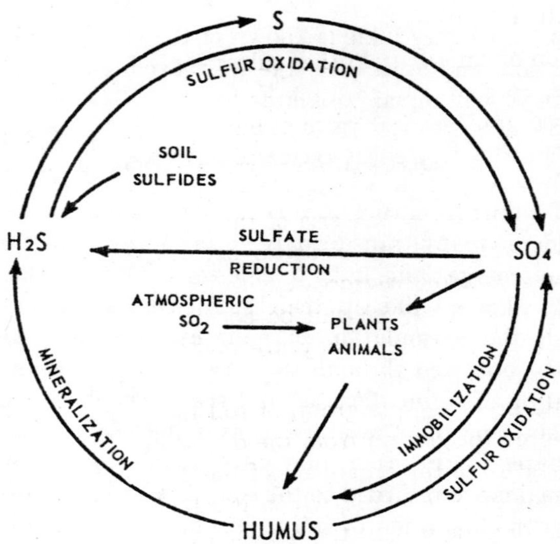

Figure 6.5. The sulfur cycle.

though sulfides may be found initially, they are rapidly oxidized to sulfate or other inorganic forms.

During the active decomposition of organic residues, a proportion of the sulfur is immobilized as microbial protoplasm. Microorganisms are capable of utilizing a wide range of inorganic sulfur compounds (sulfates, sulfites, thiosulfates, elemental sulfur) for their own growth processes. This assimilatory sulfur reduction generally involves the enzymatic reduction of these compounds to hydrogen sulfide prior to incorporation into cellular materials (108). Conservative estimates indicate that between 50 and 70 per cent of the soil sulfur is present in the soil organic fraction (77, 145). Little is known of the sulfur compounds contained in the soil humus, but presumably they are primarily of microbial origin, although modified plant and animal tissues are also present. As with other elements tied up in the soil humus, sulfur is slowly released from this fraction as it undergoes decomposition.

Under certain conditions microorganisms will compete effectively with growing plants for the available sulfur present. This is particularly true when high carbon residues are added to the soil with subsequent development of a rapidly proliferating flora. In general, carbon:sulfur ratios exceeding 50:1 will result in microbial immobilization (145).

Inorganic Sulfur Oxidation

The formation of sulfate and other sulfur compounds from more reduced forms of sulfur, both organic and inorganic, is a common characteristic of numerous microorganisms. Four groups have been implicated in these oxidations: (*a*) the chemoautotrophic bacteria of the genus *Thiobacillus;* (*b*) a poorly defined group of heterotrophic bacteria, fungi, and actinomycetes; (*c*) the colorless filamentous bacteria of the genera *Beggiatoa*, *Thiotrix*, and *Thioplaca;* and (*d*) photosynthetic sulfur-bacteria belonging to the families *Thiorhodaceae* and *Chlorobacteriaceae*. Only the first two groups are commonly found in soil and these will be considered here. The latter forms may occur in flooded soils but they are more typical in mud bottoms or strictly aquatic environments.

The thiobacilli are a unique group of soil bacteria which can obtain all the energy required for growth from the oxidation of inorganic sulfur compounds to sulfate. All of the organisms of the genus are obligate autotrophs with the exception of *T. novellus*, which is a facultative autotroph capable of developing on organic substrates. Various other physiological differences are noted between the members of this small group (122, 140). *T. denitrificans* is capable of utilizing nitrate anaerobically in place of oxygen and produces molecular nitrogen in the process. One member of the genus, *T. ferrooxidans*, is able to oxidize ferrous iron in addition to sulfur for its energy. Other differences between members of the thiobacilli include the degree of acid tolerance and the range of inorganic sulfur compounds which they are capable of oxidizing (122, 140).

Most of the thiobacilli are capable of oxidizing sulfide, elemental sulfur, thiosulfate, tetrathionate and, in some cases, sulfite to sulfate. Some typical reactions are:

T. thioxidans $$S + 1\tfrac{1}{2}O_2 + H_2O \rightarrow H_2SO_4$$

$$Na_2S_2O_3 + 2O_2 + H_2O \rightarrow 2NaHSO_4$$

T. thioparus $$Na_2S_4O_6 + Na_2CO_3 + \tfrac{1}{2}O_2 \rightarrow 2Na_2SO_4 + CO_2$$

T. denitrificans (anaerobically) $$5S + 6KNO_3 + H_2O \rightarrow 3K_2SO_4 + 2H_2SO_4 + 3N_2$$

A number of metabolic pathways have been postulated for the oxidation of sulfides to sulfates. Elemental sulfur has been considered a transitory intermediate along with polythionates (mainly tetrathionate) in the reaction sequence (140). More recently it has been projected that these compounds are not directly involved and that the main intermediate is thiosulfate (108). This proposed pathway is illustrated in Figure 6.6.

In addition to supplying sulfate for plant nutrition, the thiobacilli are

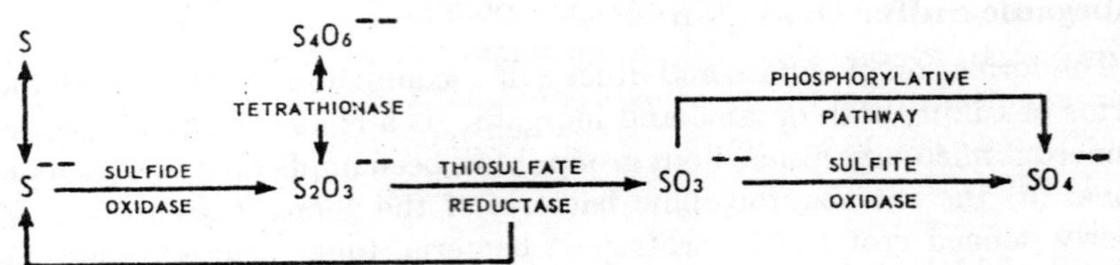

Figure 6.6. Proposed biochemical pathway for the oxidation of sulfides.

important in the treatment of plant disease, in reducing the pH of alkaline soils, and in the solubilization of soil minerals. These benefits are derived mainly from the capacity of a number of *Thiobacillus* species to produce large amounts of sulfuric acid from elemental sulfur (33, 34).

The oxidation of sulfur compounds is also carried out by various heterotrophic microorganisms. Many aerobic bacteria, actinomycetes, and filamentous fungi have been implicated in such transformations. These organisms fall into two main groups, those oxidizing elemental sulfur and thiosulfate to polythionates and those capable of converting polythionates to sulfate. Many filamentous fungi are also capable of producing sulfates from organic substrates such as cystine, methionine, and thiourea. This oxidation of sulfur compounds appears incidental to the groups concerned, and it is doubtful if much energy is liberated to the cell. The reactions of the heterotrophic organisms are generally much slower and less efficient than those of the thiobacilli. However, in view of their large numbers in soil and the correspondingly few thiobacilli present, it is possible that the heterotrophic forms are economically more important in the production of sulfate from organic matter than are the autotrophs.

Reduction of Inorganic Sulfur

The reduction of sulfate and incompletely oxidized inorganic sulfur compounds to hydrogen sulfide is relatively unimportant in well-aerated arable soils. Under these conditions many heterotrophic organisms produce sulfides during the decomposition of organic matter or through the reduction of inorganic sulfur compounds. Sulfides in turn are assimilated into microbial tissue or oxidized. Under anaerobic conditions, such as in waterlogged soils, hydrogen sulfide will accumulate at the expense of oxidized forms of sulfur. The most active agent in sulfate reduction is the bacterium *Desulfovibrio desulfuricans*. While a few related species and one species of *Clostridium* have also been implicated, *D. desulfuricans* is the most common form and is found widely distributed in nature (114, 133, 134). Because these organisms produce large amounts of sulfide they are conspicuous by the odor of hydrogen sulfide and the blackening of soils, muds, and waters through the formation of iron sulfide.

Physiologically the *Desulfovibrio* are obligate anaerobes capable of oxidizing organic substrates with the concomitant reduction of inorganic sulfur compounds. Sulfide, sulfate, thiosulfate, and tetrathionate may all be utilized as the electron acceptors, but not oxygen nor organic sulfur compounds (114, 133). Although there is lack of agreement on the number and kinds of organic compounds that can be utilized, it has been shown that simple carbohydrates, organic acids, amino acids, and other compounds are readily oxidized. Two typical reactions for *D. desulfuricans* are given which show the stoichiometric relationship between the amounts of organic compound oxidized, sulfate reduced, and sulfide produced. Acetate is a common product and is not oxidized by these organisms.

$$\underset{\text{Pyruvate}}{4CH_3\cdot CO\cdot COOH} + Na_2SO_4 \rightarrow \underset{\text{Acetate}}{4CH_3\cdot COOH} + 4CO_2 + Na_2S$$

$$\underset{\text{Lactate}}{2CH_3\cdot CHOH\cdot COOH} + Na_2SO_4 \rightarrow \underset{\text{Acetate}}{2CH_3\cdot COOH} + 2CO_2 + Na_2S + 2H_2O$$

Most strains of *D. desulfuricans* can also oxidize molecular hydrogen and are therefore capable of autotrophic growth (108, 114, 133).

$$4H_2 + CaSO_4 \rightarrow H_2S + Ca(OH)_2 + 2H_2O$$

In addition, some strains have been shown to fix atmospheric nitrogen (114).

The actual mechanism of sulfate reduction has still not been completely worked out and current thoughts on the subject have recently been reviewed (108, 114).

From an agricultural standpoint, the activities of sulfate-reducing organisms are chiefly restricted to heavy clays or waterlogged soils. Under these conditions assimilable forms of sulfur can be exhausted and toxic concentrations of hydrogen sulfide may accumulate and destroy most forms of life. In addition to the objectionable formation of hydrogen sulfide, the desulfovibrio also cause tremendous economic losses to industry each year. A number of comprehensive reviews which deal with the economic importance of these organisms have recently appeared (114, 133, 134).

PHOSPHORUS TRANSFORMATIONS

Phosphorus is an essential constituent of every living cell. It is concerned with structural compounds, nucleic acids for reproductive purposes, and the conservation and transfer of energy in the metabolic reactions taking place. Phosphorus ranks next to nitrogen in importance, and its availability to the growing crop is of prime importance in soil fertility.

The major sources of phosphorus in soil are: (*a*) the organic compounds present in animal and plant residues and products of microbial synthesis;

and (*b*) the inorganic compounds in which phosphorus is combined largely with calcium, magnesium, iron, aluminum, and clay materials. Growing plants are dependent upon soluble inorganic phosphates—chiefly as the orthophosphate ion. In comparison with other soil-derived plant-food elements, the concentration of available phosphorus in the soil solution is generally low. In this respect, the replenishment of available forms to the soil solution is closely linked with the activities of microorganisms. Transformations of phosphorus in which microorganisms are involved include: (*a*) the decomposition and subsequent mineralization of phosphorus bound in organic residues of plant and animal origin; (*b*) the altering of the solubilities of inorganic compounds of phosphorus; (*c*) the immobilization, or assimilation of phosphorus into microbial tissue with ultimate deposition in the soil humus; and (*d*) the possible participation in the oxidation and reduction of inorganic phosphorous compounds (6, 59, 110).

Mineralization and Immobilization of Organic Phosphorus

Organic phosphorus enters the soil mainly as plant or animal residues which contain phytin, phospholipids, nucleoproteins, nucleic acids, phosphorylated sugars, and related compounds. Unlike sulfates and nitrates, which are reduced when assimilated into the plant, phosphate remains unaltered and is present in organic combinations in the oxidized state.

The amounts of phosphorus present in soils in the organic form have been estimated anywhere from 25 to 80 per cent (16). In this respect, the organic phosphorus content of soils is a variable rather than a characteristic fraction of the soil organic matter, although approximate C:N:P ratios have been established (6) at 100:10:1. Studies on the nature of the phosphorous compounds in the soil humus indicate that the major proportion is present as phytin or its derivatives (40 to 80 per cent), nucleic acids (0 to 10 per cent), together with small amounts of phospholipids (0.3 per cent) (6, 8, 16, 19). While these compounds form the core of organic phosphorus in the soil, a considerable percentage still exists in unknown combination.

Nucleoproteins are actively dephosphorylated with the release of phosphate by the phosphatase enzymes of many bacteria and fungi, and it is difficult, therefore, to explain the accumulation of these compounds in soil (56). Undoubtedly a large proportion of the nucleic acids found in the soil organic fraction is an integral part of microbial cells. On the other hand, the accumulation of free nucleic acids may be due to the protective action of clay minerals. Nucleic acids and the protein moiety of nucleo-proteins are readily adsorbed on clays, and under these conditions enzymatic dephosphorylation is reduced markedly (13, 16, 64).

Phytin is not so readily metabolized in soil as are nucleic acids, although many bacteria and fungi are capable of producing phytase which dephosphorylates the inositol hexaphosphate (68). The presence of inositol, inositol monophosphate, and inositol triphosphate in soil attests to the fact that phytin is decomposed. The rate of decomposition appears to be dependent on the availability of the compound in the soil solution. In this regard, the availability of phytin or its derivatives is greatly influenced by the soil reaction. In an acid environment phytin reacts to form insoluble iron and aluminum phytates which are resistant to dephosphorylation (8, 13, 64). The addition of lime with the subsequent formation of the more soluble calcium salts results in a rapid increase in the mineralization of phytin (13, 110).

It can be seen that the biological mineralization of organic phosphorus is regulated, to a large extent, by physical and chemical factors of the soil. In most cases, the rate of mineralization of phosphorus parallels that of carbon and nitrogen, and all three elements are involved in the transformation of the soil organic fraction as a whole (8). Phosphorus mineralization is, however, somewhat more sensitive to changes in the soil pH. In this respect, an increase in the soil reaction frequently results in rates of phosphorus mineralization exceeding those of carbon and nitrogen. Soil temperature, moisture level, cultivation practices, and the nature of added organic residues, also affect phosphorus mineralization (8, 59). The extent of mineralization or immobilization depends largely on the C:P ratio of the added residues. If the C:P ratio is wide, the concentration of phosphorus may be less than that needed for the nutrition of the organisms carrying out the decomposition. Under these conditions the greater part of the phosphorus released will be immobilized. If the ratio is narrow, excess phosphorus will be liberated as inorganic phosphate. It is doubtful if microbial immobilization of phosphorus ever seriously affects plant growth. Most residues added to soil also contain a fairly high proportion of inorganic phosphorus upon which microorganisms can draw in the event of a wide C:P ratio, thus sparing the available form of soil from immobilization (8).

Solubilization of Inorganic Phosphorus

Inorganic phosphorous compounds found in soils are, to a large extent, insoluble in the soil solution. These compounds range from the relatively soluble mono- and dicalcium phosphates, which are present in low concentrations, to the relatively insoluble fluorapatite and hydroxyapatite. Iron and aluminum phosphates tend to accumulate in acid soils, whereas calcium phosphates predominate in neutral or alkaline soils.

The results of many laboratory and greenhouse studies have implicated the activities of soil microorganisms in the solubilization of inorganic soil phosphates. In general, solubilization is due to microbial production of organic or inorganic acids which effectively dissolve inorganic phosphates and render them available to the soil flora and growing plants. In the case of heterotrophic organisms the amount of phosphorus brought into solution varies with the carbon substrate used. Autotrophic organisms, primarily those concerned with the oxidation of sulfur and nitrogen, are also able to solubilize inorganic phosphates through the formation of sulfuric and nitric acids.

Many soil microorganisms produce lactic, glycolic, citric, formic, and acetic acids which can solubilize such compounds as tricalcium phosphate, synthetic apatite, and natural apatites (46, 72, 136). More recently it has been established that a number of other organic acids, including 2-keto-gluconic acid, are effective chelating agents capable of complexing with calcium, copper, nickel, manganese, iron, and aluminum salts of phosphates with resultant solubilization of phosphorus (46, 72).

The large proportion of phosphate-dissolving organisms present in soils (10 to 15 per cent) and in the rhizosphere of plants (20 to 40 per cent) implies that these organisms may play a significant role in the phosphorus economy of the soil (126, 136). However, a prerequisite of the high acid concentration necessary for solubilization is an ample supply of readily available carbohydrate—a condition that is rarely fulfilled in soil. In addition, the presence of calcium carbonate or other buffering agents would rapidly neutralize the acids formed. The presence of a larger population of phosphate-dissolving organisms in the rhizosphere, combined with available energy sources in the form of root excretions, suggests that the root zone is the most likely site of this process in soil (126, 136).

Although acid production is generally associated with solubilization of inorganic phosphates, the production of hydrogen sulfide by microorganisms has also been shown to release phosphates from insoluble iron combinations. This reaction is particularly effective under anaerobic conditions when large concentrations of hydrogen sulfide are formed (127, 136).

Little information is available on the oxidation and reduction of the phosphate ion. A number of heterotrophic organisms utilize phosphite for their own growth processes—intimating the oxidation to phosphate within the cell. A reduction process has also been shown to occur. Under anaerobic conditions, phosphate may be reduced sequentially to phosphite, hypophosphite, and phosphine. The latter observations have received little attention, but it is doubtful if the reduction of phosphate ever attains the significance of sulfur reduction or denitrification.

TRANSFORMATION OF OTHER ELEMENTS

In addition to the numerous transformations of N, S, and P achieved by microorganisms, many of the minor nutrients necessary for plant nutrition are also affected by the metabolic activities of soil organisms.* In many instances the availability of iron, manganese, potassium, selenium, tellurium, arsenic, zinc, copper, boron, calcium, magnesium, aluminum, and molybdenum is also affected, to varying extents, by microbial action. Transformations of these compounds are brought about by the same general types of reactions as for the major elements, that is, mineralization as the result of decomposition of organic plant and animal residues; immobilization due to the reduction of oxidized forms of the elements, or assimilization and immobilization in microbial protoplasm; increased availability resulting from the oxidation of reduced forms or *vice versa;* and chemical transformations occurring indirectly through the production of organic acids or chelating agents.

Iron receives preferential treatment since a number of microorganisms are specifically involved in its oxidation. The iron bacteria which deposit insoluble ferric hydroxide through the oxidation of the ferrous form are primarily aquatic organisms and range from autotrophic bacteria, which depend on this oxidation for energy, to miscellaneous heterotrophs in which the immobilization of iron is a secondary process (115). Under anaerobic conditions, sulfate-reducing organisms may also bring about a concomitant reduction of iron to ferrous sulfate. On the other hand, the production of organic acids by certain bacteria often results in the formation of soluble iron complexes. Similarly, ferric iron may be reduced to the available ferrous form by certain bacteria which utilize the ferric ion as an electron acceptor at suboptimal oxygen levels.

Microorganisms are also involved in the oxidation of manganese to an unavailable form. Although a relatively high proportion of the soil flora is capable of oxidizing the manganous ion to the unavailable manganese dioxide, it is probable that this oxidation is only significant under aerobic conditions in neutral or acid soils when the manganous ion is available. Manganese dioxide may also be reduced under anaerobic conditions in the presence of an oxidizable organic substrate. Although little manganese is actually immobilized by microorganisms, there is no doubt that the equilibrium between the manganous and manganic ion in soils is the result of a combination of both biological and chemical factors.

As with iron and manganese, selenium, tellurium, and arsenic can all be oxidized or reduced by heterotrophic soil microorganisms. Whether or not these reactions are significant in soil cannot be assessed at this time.

Potassium is a major cation in the nutrition of plants, and its mobilization in soils is of considerable importance. Microorganisms effect the availability of this element through the liberation of organically bound potassium by decomposition processes as well as by the solubilization of

* See references 2, 78, 113, 141, and 143.

insoluble forms present in soil minerals. In the latter instance, potassium is made available as a result of acid production by soil microorganisms. Immobilization of potassium during active proliferation of the soil flora may also shift the equilibrium between available and bound potassium resulting in a release of the found form (1).

The availability of zinc, copper, calcium, magnesium, aluminum, and boron has also been shown to be markedly influenced by microbiological activity in soils (2, 141, 143). The conditions under which these conversions occur have not been defined. There is no doubt, however, that the biological transformations of these elements in soils are important.

PESTICIDE TRANSFORMATIONS

The increasing number and widespread use of herbicides, fungicides, and insecticides has aroused concern from the standpoint of their accumulation in soils and their possible injurious effects on the soil microflora.* Despite the chemical diversity of these compounds, very few can withstand biological decomposition and they rarely persist in soils for any great length of time. When applied at recommended rates, the concentration of herbicides, fungicides, and insecticides present in soil does not significantly alter the microbial activities that are important to soil fertility. Improper distribution or excessive applications may upset the microbial equilibrium and, in some instances, nitrification or the nodulation of legumes. It is doubtful if these deleterious effects are significant when compared with the over-all beneficial effects of the treatments involved.

Many different kinds of microorganisms, including bacteria, actinomycetes, and fungi, attack stable pesticides in order to obtain energy and carbon for growth. In this respect, the persistence of these compounds in soil will depend to a large extent on conditions which influence the rate of microbial activity. These conditions include temperature, moisture level, soil reaction, and soil type. The presence of readily available organic matter in soil often increases the persistence of pesticides by providing an alternate substrate. Adsorption on clay minerals may also decrease the rates of detoxification of many of these compounds. In addition to biological breakdown, pesticides in soil are subject to chemical and photochemical reactions, volatilization, and loss by leaching (155).

THE OCCURRENCE OF VITAMINS AND ANTIBIOTICS IN SOIL

Vitamins and Growth Factors

It is now well known that most, if not all, of the B-vitamins are present in fertile soils, together with other microbial growth factors. Their occur-

* See references 10, 47, 90, 101, and 155.

rence depends chiefly on the synthetic abilities of microorganisms, although some may be present as a result of decomposition of plant and animal residues or liberation from the roots of growing plants. Numerous vitamins and vitamin precursors have been detected in soils or soil extracts. These include: thiamin, thiazole, pyrimidines, biotin, riboflavin, inositol, pyridoxine, nicotinic acid, *p*-amino benzoic acid, vitamin B_{12}, and the still uncharacterized terregens factor (84). Recent studies have shown that over 50 per cent of the bacterial isolates of soil and up to 80 per cent of rhizosphere isolates were capable of synthesizing one or more of these growth factors (83). In addition, many actinomycetes and fungi have also been implicated in the synthesis of these compounds. A large proportion of the soil population is also dependent on the presence of preformed growth factors. In one instance, 27 per cent of a large number of bacterial isolates from soil were shown to require one or more specific vitamins. In these studies, thiamin, biotin, and vitamin B_{12} were the most commonly required factors (87). The level of vitamins present in soil depends, to a large extent, on the prevailing conditions. Generally, however, the concentration of growth factors at any one time depends on the balance established between organisms synthesizing these compounds and organisms requiring them for growth.

It is still not possible to evaluate fully the significance of growth factors in soil. The large numbers of soil organisms that produce or require these compounds attest to their probable role in establishing the microbial equilibrium. The presence of even larger numbers of these organisms in the rhizosphere, particularly of those synthesizing vitamins, suggests their possible significance in interactions between the soil flora and root pathogens. Some evidence has shown that vitamins stimulate the germination of dormant fungal spores—a stimulation which may or may not lead to an increased disease incidence, depending on concomitant antagonistic interactions (85).

Antibiotics

Since the majority of antibiotics, as we know them, are produced by soil isolates (bacteria, actinomycetes, fungi), it is important to consider the ecological significance of these compounds in the soil habitat (26). Numerous studies have shown that antibiotics can be produced when organisms are inoculated into supplemented and unsupplemented sterile soil. Although such studies cannot be related to natural conditions, they do emphasize the fact that soil, as a medium, does possess all the requirements necessary for the synthesis of these complex growth inhibitors. Small amounts of antibiotics have also been detected in inoculated, non-sterile soils, but generally only when such soils have been extensively supplemented with organic residues.

A number of factors affect the production and ready detection of antibiotics in natural soils. These include: (*a*) the presence of a readily available carbon source, (*b*) the competition between the antibiotic-producers and other members of the soil flora, and (*c*) the adsorption and possible inactivation of antibiotics by soil colloids. The frequent failure to detect antibiotics in the absence of a large supply of available organic carbon in soil does not preclude antibiotic production in the immediate vicinity of localized deposits of a suitable carbon source. Since many microorganisms inhabit such sites, the antibiotic formed may well have a considerable effect in this restricted environment. The normal soil flora may also restrict antibiotic production or activity by competing successfully with antibiotic-producers for available nutrients, or it may metabolize the antibiotic itself. In this connection it is well known that many of the known antibiotics are highly labile and subject to biological decomposition. Antibiotics may also be chemically inactivated by adsorption on clay colloids. Most basic antibiotics are particularly susceptible to adsorption, but it should be pointed out that not all are inactivated. Some still retain their inhibitory activity even in the adsorbed state (112).

Despite the failure to detect appreciable quantities of antibiotics in soils by extraction techniques, the inhibitory action of these compounds in soil has been established by more sensitive bioassay methods. Although no direct connection between antibiotic production and the many antagonistic phenomena observed in soil has yet been shown, there is little doubt that antibiotics are involved in many instances.

Whereas antibiotics undoubtedly play a major role in the ecologica relationships within a soil, their presence in the root zone of plants may be of special significance. It appears likely that root exudates, which are largely responsible for the rhizosphere effect, might also stimulate antibiotic production. In this respect, antibiotic-producing cultures have been used with variable success to control root-invading fungi (25, 132). Antibiotics may also be absorbed by the plant roots with either inhibitory or beneficia effects. The effect of antibiotics on plants has been the subject of a recen review (27).

Toxic materials with a more general distribution than antibiotics have also been found in soils (69). These substances which inhibit bacterial and fungal growth vary qualitatively and quantitatively according to source but their occurrence in soils is widespread. As yet uncharacterized, there is little doubt that these factors are of biological origin.

GENERAL

Our knowledge of many of the transformations occurring in soil is general rather than absolute. Much of the information available on the metabolic activities of soil microorganisms is empirical, in that it is based on the results of *in vitro* studies. In this respect, evidence for specific transformations will remain fragmentary until such time as methods are devised which will permit their characterization in soil amidst concurrent chemical, physical, and biological interactions.

For ease of presentation, many of the major biological transformations have been treated as separate entities. It should be reiterated, however, that in soil, all transformations are occurring simultaneously and the dominance of one over the other will depend, to a large extent, on the soil, the flora, and the environmental conditions. By necessity, many important transformations and soil phenomena have been given only superficial coverage. It is hoped that, in these instances, the reader will avail himself of the literature cited for fuller details.

REFERENCES

1. Agarwall, R. R., *Soils Fertilizers* **23**, 375 (1960).
2. Alexander, M., "Introduction to Soil Microbiology," New York, John Wiley & Sons, Inc., 1961.
3. Alexander, M., Marshall, K. C., and Hirsch, P., *Trans. Intern. Congr. Soil Sci. 7th Congr.* **2**, 586 (1960).
4. Allen, E. K., and Allen, O. N., in W. Ruhland, "Handbuch der Pflanzenphysiologie," Vol. 8, p. 48, Berlin, Springer-Verlag, 1958.
5. Allfrey, V. G., and Mirsky, A. E., *Sci. Am.* **205**, 74 (1961).
6. Barrow, N. J., *Soils Fertilizers* **24**, 169 (1961).
7. Bergerson, F. J., *Bacteriol. Rev.* **24**, 246 (1960).
8. Black, C. A., and Goring, C. A. I., in W. H. Pierre and A. G. Norman, "Soil and Fertilizer Phosphorus," p. 123, New York, Academic Press Inc., 1953.
9. Blair, I. D., *Lincoln, New Zealand Canterbury Agr. Coll. Tech. Publ. No. 5*, 1951.
10. Bollen, W. B., *Ann. Rev. Microbiol.* **15**, 69 (1961).
11. Bond, G., in E. G. Hallsworth, "Nutrition of Legumes," p. 216, London, Butterworth's Scientific Publ., 1958.
12. Bond, G., *Soc. Exp. Biol. Symposia* **13**, 59 (1959).
13. Bower, C. A., *Iowa Agr. Expt. Sta. Research Bull. 362*, 1949.
14. Brauns, F. E., "The Chemistry of Lignins," New York, Academic Press Inc., 1952.
15. Brauns, F. E., and Brauns, D. A., "The Chemistry of Lignins," Supplement volume, New York, Academic Press Inc., 1960.
16. Bremner, J. M., *J. Soil Sci.* **2**, 67 (1951).
17. Bremner, J. M., *J. Sci. Food Agr.* **3**, 497 (1952).
18. Bremner, J. M., *J. Soil Sci.* **5**, 214 (1954).
19. Bremner, J. M., *Soils Fertilizers* **19**, 115 (1956).

20. Bremner, J. M., *J. Sci. Food Agr.* **9**, 528 (1958).
21. Bremner, J. M., and Shaw, K., *J. Agr. Sci.* **44**, 152 (1954).
22. Bremner, J. M., and Shaw, K., *J. Sci. Food Agr.* **6**, 341 (1957).
23. Bremner, J. M., and Shaw, K., *J. Agr. Sci.* **51**, 22 (1958).
24. Bremner, J. M., and Shaw, K., *J. Agr. Sci.* **51**, 40 (1958).
25. Brian, P. W., *J. Appl. Bacteriol.* **17**, 142 (1954).
26. Brian, P. W., in R. E. O. Williams, and C. C. Spicer, "Microbial Ecology," p. 168, Cambridge, Cambridge Univ. Press, 1957.
27. Brian, P. W., *Ann. Rev. Plant Physiol.* **8**, 413 (1957).
28. Broadbent, F. E., *Soil Sci. Soc. Am. Proc.* **12**, 246 (1947).
29. Broadbent, F. E., *Advan. Agron.* **5**, 153 (1953).
30. Broadbent, F. E., *Soil Sci.* **79**, 107 (1955).
31. Burges, A., "Micro-organisms in the Soil," London, Hutchison Univ. Library, 1958.
32. Burton, K., and Krebs, H. A., *Biochem. J.* **54**, 94 (1953).
33. Butlin, K. R., *Research (London)* **6**, 184 (1953).
34. Butlin, K. R., and Postgate, J. R., in B. A. Fry and J. L. Peel, "Autotrophic Micro-organisms," p. 271, Cambridge, Cambridge Univ. Press, 1954.
35. Chao, T-T., and Kroontje, W., *Soil Sci. Soc. Am. Proc.* **27**, 44 (1963).
36. Cheldelin, V. H., "Metabolic Pathways in Microorganisms," New York, John Wiley & Sons Inc., 1961.
37. Chesters, C. G. C., *Trans. Brit. Mycol. Soc.* **32**, 197 (1949).
38. Claus, D., and Mechsner, K., *Plant Soil* **12**, 195 (1960).
39. Conn, H. J., *Soil Sci.* **25**, 263 (1928).
40. Conn, H. J., *Bacteriol. Rev.* **12**, 257 (1948).
41. Drobnik, J., *Plant Soil* **14**, 94 (1961).
42. Delwiche, C. C., in W. D. McElroy, and B. H. Glass, "Inorganic Nitrogen Metabolism," p. 218, Baltimore, Johns Hopkins Press, 1956.
43. Delwiche, C. C., in W. D. McElroy, and B. H. Glass, "Inorganic Nitrogen Metabolism," p. 233, Baltimore, Johns Hopkins Press, 1956.
44. Delwiche, C. C., and Wijler, J., *Plant Soil* **7**, 113 (1956).
45. Duddington, C. L., in R. E. O. Williams, and C. C. Spicer, "Microbial Ecology," p. 218, Cambridge, Cambridge University Press, 1957.
46. Duff, R. B., Webley, D. M., and Scott, R. O., *Soil Sci.* **95**, 105 (1963).
47. Eno, C. F., *J. Agr. Food Chem.* **6**, 348 (1958).
48. Ensminger, L. E., and Gieseking, J. E., *Soil Sci.* **53**, 205 (1942).
49. Estermann, E. F., and McLaren, A. D., *J. Soil Sci.* **10**, 64 (1959).
50. Estermann, E. F., and McLaren, A. D., *Plant Soil* **15**, 243 (1961).
51. Estermann, E. F., Peterson, G. H., and McLaren, A. D., *Soil Sci. Soc. Am. Proc.* **23**, 31 (1959).
52. Eylar, O. R., and Schmidt, E. L., *J. Gen. Microbiol.* **20**, 473 (1959).
53. Flaig, W., *Verhandl, II u. IV Komm. int. Bodenk. Ges.* **2** (1958).
54. Flaig, W., *Trans. Intern. Congr. Soil Sci. 7th Congr.* **2**, 648 (1960).
55. Forsyth, W. G. C., *Chem. Ind. (London)* 515, (1948).
56. Fry, B. A., "The Nitrogen Metabolism of Micro-organisms," London, Methuen and Co., Ltd., 1955.
57. Fuller, W. H., Cameron, R. E., and Raica, N., *Trans. Intern. Congr. Soil Sci. 7th Congr.* **2**, 617 (1960).
58. Gascoigne, J. A., and Gascoigne, M. M., "Biological Degradation of Cellulose," London, Butterworth and Co., Ltd., 1960.

59. Goring, C. A. I., *Plant Soil* **6**, 17 (1955).
60. Gottlieb, S., and Pelczar, M. J., *Bacteriol. Rev.* **15**, 55 (1951).
61. Gupta, V. C., *Soils Fertilizers* **25**, 83 (1962).
62. Harmsen, G. W., and Van Schreven, D. A., *Advan. Agron.* **7**, 299 (1955).
63. Hawker, L. E., Linton, A. H., Folkes, B. F., and Carlile, M. J., "An Introduction to the Biology of Micro-organisms," London, Edward Arnold, Publ. Ltd., 1960.
64. Hemwall, J. B., *Advan. Agron.* **9**, 95 (1957).
65. Henderson, M. E. K., *J. Gen. Microbiol.* **14**, 684 (1956).
66. Hofman, T., and Lees, H., *Biochem. J.* **52**, 140 (1952).
67. Hofmann, E., and Niggemann, J., *Biochem. Z.* **324**, 308 (1953).
68. Jackman, R. H., and Black, C. A., *Soil Sci.* **73**, 167 (1952).
69. Jackson, R. M., *J. Gen. Microbiol.* **18**, 248 (1958).
70. Jansson, S. K., *Trans. Intern. Congr. Soil Sci. 7th Congr.* **2**, 635 (1960).
71. Jensen, H. L., in W. Ruhland, "Handbuch der Pflanzenphysiologie," Vol. 11, p. 707, Berlin, Springer-Verlag., 1959.
72. Johnston, H. W., *New Zealand J. Sci. Tech.* **2**, 215 (1959).
73. Jordan, H. V., and Ensminger, L. E., *Advan. Agron.* **10**, 407 (1958).
74. Kevan, D. K. M., "Soil Zoology," London, Butterworth's Scientific Publ., 1955.
75. Kiss, I., in I. M. Csapo, "Soil Science," p. 495, Bucharest, Roumania, State Agro-Silvic Publ. (in Hungarian), 1958.
76. Kitching, J. A., in R. E. O. Williams, and C. C. Spicer, "Microbial Ecology," p. 259, Cambridge, Cambridge University Press, 1957.
77. Kononova, M. M., "Soil Organic Matter," New York, Pergamon Press, Inc., 1961.
78. Krasil'nikov, N. A., "Soil Microorganisms and Higher Plants," Moscow, Acad. Sci. U.S.S.R. Publ., 1958. (Translation by Y. Halperin, 1961, available from Office of Technical Services, U.S. Department of Commerce, Washington 25, D. C.)
79. Krebs, H. A., and Kornberg, H. L., in "Ergebnisse der Physiologie, Biologischen Chemie und Experimentellen Pharmakologie," Vol. 49, p. 212, Berlin, Springer-Verlag, 1957.
80. Lees, H., in B. A. Fry, and J. L. Peel, "Autotrophic Micro-organisms," p. 84, Cambridge, Cambridge University Press, 1954.
81. Lehninger, A. L., *Sci. Am.* **205**, 63 (1961).
82. Lewin, R. A., "Physiology and Biochemistry of Algae," New York, Academic Press Inc., 1962.
83. Lochhead, A. G., *Soil Sci.* **84**, 395 (1957).
84. Lochhead, A. G., *Bacteriol. Rev.* **22**, 145 (1958).
85. Lochhead, A. G., *Trans. Roy. Soc. Can.* **52**, 17 (1958).
86. Lochhead, A. G., in "Recent Advances in Botany," Vol. 1, p. 619, Toronto, University of Toronto Press, 1961.
87. Lochhead, A. G., and Burton, M. O., *Can J. Microbiol.* **3**, 35 (1957).
88. Lochhead, A. G., *et al.*, "Qualitative Studies of Soil Microorganisms. I - XV (1938–1957)," Microbiol. Research Inst., Canada Department of Agriculture, 1959.
89. Lynch, D. L., and Cotnoir, L. J., *Soil Sci. Soc. Am. Proc.* **20**, 367 (1956).
90. Martin, J. P., and Pratt, P. F., *J. Agr. Food Chem.* **6**, 345 (1958).
91. McLaren, A. D., *Enzymologia* **21**, 356 (1960).
92. McLaren, A. D., and Estermann, E. F., *Archiv. Biochem. Biophys.* **61**, 158 (1956).

93. McLaren, A. D., and Estermann, E. F., *Archiv. Biochem. Biophys.* 68, 157 (1957).
94. McLaren, A. D., Luse, R. A., and Skujins, J. J., *Soil Sci. Soc. Am. Proc.* 26, 371 (1962).
95. McLaren, A. D., Reschetko, L., and Huber, W., *Soil Sci.* 83, 497 (1957).
96. Mehta, N. C., Dubach, P., and Deuel, H., *Advances in Carbohydrate Chemistry* 16, 335 (1961).
97. Meiklejohn, J., in B. A. Fry and J. L. Peel, "Autotrophic Micro-organisms," p. 68, Cambridge, Cambridge University Press, 1954.
98. Mortland, M. M., *Advan. Agron.* 10, 325 (1955).
99. Nason, A., *Bacteriol. Rev.* 26, 16 (1962).
100. Nason, A., and Takahashi, H., *Ann. Rev. Microbiol.* 12, 203 (1958).
101. Newman, A. S., and Downing, C. R., *J. Agr. Food Chem.* 6, 352 (1958).
102. Nicholas, D. J. D., in E. G. Hallsworth, "Nutrition of Legumes," p. 239, London, Butterworth's Scientific Publ., 1958.
103. Norman, A. G., "Biochemistry of Cellulose, Polyuronides, Lignin etc.," Oxford, Oxford University Press, 1937.
104. Nutman, P. S., in E. G. Hallsworth, "Nutrition of Legumes," p. 87, London, Butterworth's Scientific Publ., 1958.
105. Nutman, P. S., *Soc. Exp. Biol. Symposia* 13, 42 (1959).
106. Oginsky, E. L., and Umbreit, W. W., "An Introduction to Bacterial Physiology," San Francisco, W. H., Freeman and Co., 1959.
107. Parkinson, D., and Waid, J. S., "The Ecology of Soil Fungi," Liverpool, Liverpool University Press, 1960.
108. Peck, H. D., Jr., *Bacteriol. Rev.* 26, 67 (1962).
109. Peterson, G. H., *Soil Sci.* 94, 71 (1962).
110. Pierre, W. H., *J. Am. Soc. Agron.* 40, 1 (1948).
111. Pinck, L. A., Dyal, R. S., and Allison, F. E., *Soil Sci.* 78, 109 (1954).
112. Pinck, L. A., Holton, W. F., and Allison, F. E., *Soil Sci.* 91, 22 (1961).
113. Pochon, J., and De Barjac, H., "Traité de Microbiologie des Sols," Paris, Dunod, 1957.
114. Postgate, J., *Ann. Rev. Microbiol.* 13, 505 (1959).
115. Pringsheim, E. G., *Biol. Rev.* 24, 200 (1949).
116. Provasoli, L., *Ann. Rev. Microbiol.* 12, 279 (1958).
117. Quastel, J. H., and Scholefield, P., *Bacteriol. Rev.* 15, 1 (1951).
118. Reese, E. T., *Appl. Microbiol.* 4, 39 (1956).
119. Roberts, E. R., *Soc. Exp. Biol. Symposia* 13, 24 (1959).
120. Rogers, H. J., in I. C. Gunsalus, and R. Y., Stanier, "The Bacteria," Vol. II, p. 257, New York, Academic Press Inc., 1961.
121. Russell, E. J., and Russell, E. W., "Soil Conditions and Plant Growth," 8th ed., New York, Longmans, Green and Co., 1950.
122. Schwartz, W., in W. Ruhland, "Handbuch der Pflanzenphysiologie," Vol. 9, p. 89, Berlin, Springer-Verlag., 1958.
123. Singh, B. N., *J. Gen. Microbiol.* 1, 1 (1947).
124. Siu, R. G. H., "Microbial Decomposition of Cellulose," New York, Reinhold Publishing Corporation, 1951.
125. Siu, R. G. H., and Reese, E. T., *Botan. Rev.* 19, 377 (1953).
126. Sperber, J. I., *Australian J. Agri. Res.* 9, 778 (1958).
127. Sperber, J. I., *Nature* 181, 934 (1958).
128. Sprinson, D. B., *Advances in Carbohydrate Chemistry* 15, 235 (1960).

129. Stanier, R. Y., Doudoroff, M., and Adelberg, E. A., "The Microbial World," Englewood Cliffs, N. J., Prentice-Hall, Inc., 1957.
130. Stanier, R. Y., Sleeper, B., Tschuchida, M., and MacDonald, D., *J. Bacteriol.* **59**, 137 (1950).
131. Starkey, R. L., *Soil Sci.* **70**, 55 (1950).
132. Starkey, R. L., *Bacteriol. Rev.* **22**, 154 (1958).
133. Starkey, R. L., *Producers Monthly* **22**, 12 (1958).
134. Starkey, R. L., *Tappi* **44**, 493 (1961).
135. Stephenson, M., "Bacterial Metabolism," 3rd ed., London, Longmans, Green and Co., 1949.
136. Swaby, R. J., and Sperber, J. I., in E. G. Hallsworth, "Nutrition of Legumes," p. 289, London, Butterworth's Scientific Publ., 1958.
137. Thornton, H. G., *Proc. Roy. Soc. (London) Ser. B.* **145**, 364 (1956).
138. Thornton, H. G., *J. Sci. Food Agr.* **2**, 93 (1956).
139. Verhoeven, W., in W. D. McElroy, and B. H. Glass, "Inorganic Nitrogen Metabolism," p. 61, Baltimore, Johns Hopkins Press, 1956.
140. Vishniac, W., and Santer, M., *Bacteriol. Rev.* **21**, 195 (1957).
141. Waksman, S. A., "Principles of Soil Microbiology," 2nd ed., Baltimore, The Williams and Wilkins Co., 1932.
142. Waksman, S. A., "Humus," 2nd ed., Baltimore, The Williams and Wilkins Co., 1938.
143. Waksman, S. A., "Soil Microbiology," New York, John Wiley & Sons Inc., 1952.
144. Waksman, S. A., "The Actinomycetes," Vol. 1, Baltimore, The Williams and Wilkins Co., 1959.
145. Walker, T. W., *J. Brit. Grassl. Soc.* **12**, 10 (1957).
146. Walker, T. W., Orchiston, H. D., and Adams, A. F. R., *J. Brit. Grassl. Soc.* **9**, 249 (1954).
147. Warcup, J. H., *Trans. Brit. Mycol. Soc.* **34**, 376 (1951).
148. Watanabe, A., *J. Gen. Appl. Microbiol. (Tokyo)* **8**, 85 (1962).
149. Werkman, C. W., and Wilson, P. W., "Bacterial Physiology," New York, Academic Press Inc., 1951.
150. Whistler, R. L., and Smart, C. L., "Polysaccharide Chemistry," The New York, Academic Press Inc., 1953.
151. Whitaker, D. R., *Can. J. Biochem. Physiol.* **34**, 488 (1956).
152. Wiame, J. M., in W. Ruhland, "Handbuch der Pflanzenphysiologie," Vol. 9, p. 101, Berlin, Springer-Verlag., 1958.
153. Williams, R. E. O., and Spicer, C. C., "Microbial Ecology," Cambridge, Cambridge University Press, 1957.
154. Wilson, P. W., in W. Ruhland, "Handbuch der Pflanzenphysiologie," Vol. 8, p. 9, Berlin, Springer-Verlag., 1958.
155. Woodford, E. K., and Sagar, G. R., "Herbicides and the Soil," Oxford, Blackwell Scientific Publ., 1960.

CHAPTER 7

Acid, Alkaline, Saline, and Sodic Soils

LLOYD F. SEATZ AND H. B. PETERSON

Department of Agronomy, University of Tennessee, Knoxville, and Department of Agronomy, Utah State University, Logan

Soil reaction and the salt content of a soil are two factors which are of extreme importance in evaluating its production potential for most crops. Not only can practical judgments concerning land use be made from such information, but it also provides a basis for many inferences concerning soil development and past use.

The theoretical implications associated with these topics are among the most fascinating ones in soil science. Although the concept of pH was originally formulated with respect to true solutions, the basic relationships were readily transferable to colloidal suspensions and soil systems. In so doing, however, additional factors were introduced which altered the basic relationships to some degree. In addition, some interesting theoretical questions became apparent relative to the validity of accepting pH measurements as true values of H^+ ion activity. Several of these concepts are discussed in this chapter.

Our concept of the true nature of the cause of soil acidity has changed considerably within the past few years. Although aluminum was known to be a major soil constituent and was, indeed, an exchangeable ion under acid soil conditions, little effort was made to make a strong case for explaining a substantial part of soil acidity in most mineral soils as being related to the hydrolysis of the aluminum ion. The titration curves of acid clay or soil systems were used as evidence that acid soil systems were analogous to ordinary weak acids. The similarity of the titration curves of many systems with those of acetic acid led to the development of apparent pK values for soil systems. The normal buffering properties of these systems could also be explained by this analogy. The more or less continuous nature of the titration curves was explained on the basis of H^+ ions being held with different strengths of adsorption by the different acidic components of the normal soil system.

It has been shown more recently that a true hydrogen-saturated clay produces a titration curve very similar to that of a strong acid, and that only after aluminum is added to the system is the usual, well-buffered, titration curve obtained (18, 29).

The pH of soil is usually measured potentiometrically by means of a glass electrode assembly consisting of a glass electrode half-cell and a calomel half-cell with a saturated KCl bridge (10, 13). The soil is usually prepared for the measurement by adding distilled water or a neutral, unbuffered salt solution of known concentration to the soil sample in some definite weight ratio or to some degree of consistency. The usefulness of this method is based on the assumption that the potential measured is due wholly to the potential difference across the glass membrane brought about by the difference in H^+ ion activity between the solution on the inside of the glass electrode and the solution or suspension on the outside, which has an unknown H^+ ion activity. Potassium chloride is used for the salt bridge because of the equal transference numbers of the K^+ and Cl^- ions in true solution.

Recently glass electrodes which are sensitive to changes in sodium ion concentration have been perfected and these can be used to measure sodium ion concentrations in soils in a manner analogous to that used for hydrogen ion activity measurements.

ACID SOILS

The acidity in soils arises from several different sources. Each source has its specific characteristics as to intensity and quantity of acidity produced.

Humus

Humus materials in soils occur as the result of microbiologic decomposition of organic matter and exist as an intermediate step in the decomposition process. Humus possesses several types of functional groups which are capable of attracting and dissociating hydrogen ions. Among these are carboxyl, phenolic, amino, and similar-type groupings. The strength of the acid produced will depend upon the nature of the predominant groups present. Humus is a very heterogeneous material and, therefore, varies in composition from one situation to another.

Humus may also react with iron and aluminum ions to form complexes which may subsequently undergo hydrolysis to yield hydrogen ions.

During the breakdown and mineralization of organic matter and humus, the various constituents of the material are released. These would include water, CO_2, and various salts. Organic acids may also be produced during this process. The increased concentration of CO_2, the hydrolysis of acid salts, and the organic acids produced add to the total acidity of the system.

Aluminosilicates

Aluminosilicate clay minerals are considered to be layer minerals containing alumina (Al_2O_3) and silica (SiO_2) components. The charge on these minerals arises from either the isomorphous substitution of a cation of lower valence for one of higher valence in the lattice framework of the mineral, or from the dissociation of hydrogen ions from hydroxyl ions or bound water of constitution, which are structural components of the lattice (30).

The total charge exhibited by a clay mineral may be separated into the permanent charge and the pH-dependent charge. Permanent charge results from isomorphous substitution within the lattice framework. The pH-dependent charge can arise from several sources. It is possible that structural OH^- ions at corners and edges may dissociate H^+ ions within the slightly acid to alkaline pH range, since these OH^- ions are held in the lattice by valence forces to Al and Si or their substituted cations. Amorphous aluminum and iron hydroxy complexes may coat the aluminosilicate minerals, and in so doing, they may block normal exchange sites. These complexes are not readily exchangeable to unbuffered neutral salts at low pH values. However, as the pH of the system increases these complexes undergo hydrolysis with the concomitant unblocking of exchange sites on the mineral.

In acid soils an equilibrium exists between hydrogen and aluminum ions. As hydrogen ions are produced they will, in turn, dissolve some aluminum from the clay mineral lattice. In base-unsaturated clay systems both hydrogen and aluminum are present as exchangeable ions. As such a system is titrated with a base the H^+ ions will be neutralized by the first increments. Upon the addition of more base, the aluminum begins to undergo hydrolysis with the production of H^+ ions in amounts equivalent to the amounts of aluminum present. At low pH values most of the aluminum is present as the hexahydrated Al^{+++} ion. At pH values above about 5.0, hydrated hydroxyaluminum ions probably exist in exchangeable form. These reactions can be simply shown as follows:

$$Al^{+++} + H_2O \rightarrow Al(OH)^{++} + H^+$$

$$Al(OH)^{++} + H_2O \rightarrow Al(OH)_2^+ + H^+$$

$$Al(OH)_2^+ + H_2O \rightarrow Al(OH)_3 + H^+$$

The hydroxy aluminum ions shown above also tend to polymerize to produce much more complex systems than those indicated (22). No doubt all the charged aluminum and aluminum hydroxy ion species can occupy exchange sites on the clay minerals (23).

When acid aluminosilicate mineral soils are leached with a neutral unbuffered salt, such as KCl, the leachate contains both H^+ and Al^{+++} ions. It may also contain other ions, such as Mg^{++} and Fe^{+++}, which were dissolved from the mineral structure under the acid conditions. The Fe^{+++} ion may also undergo stepwise hydrolysis to produce acidity in the system.

Hydrous Oxides

Hydrous oxides are principally oxides of iron and aluminum. They may occur as amorphous particles of colloidal dimensions, as crystalline colloidal material such as gibbsite, as coatings on other mineral particles, or as interlayers between crystal lattice structures (21). These amorphous coatings may block some exchange sites and thus would tend to lower the cation-exchange capacity of the material. Under acid conditions the iron and aluminum from these oxides may be brought into solution and undergo stepwise hydrolysis with the release of hydrogen ions.

Soluble Salts

Soluble salts may be either acid, neutral, or basic salts. They may arise from several sources, such as fertilizer salts, mineralization of organic matter, and weathering of minerals. These salts undergo hydrolysis to form acids. A net acid reaction is produced if the basic portion of the salt is absorbed by plants more readily than the acidic portion. When the basic portion of the salt is ammonium ion it undergoes biological transformation in the soil to form the acid-forming nitrate ion. If reduced forms of sulfur are present they may undergo oxidation to produce acid-forming sulfate ions. Leaching, weathering, plant uptake, and biologic transformations, therefore, all play a role in determining the influence of soluble salts on the acidity of the soil.

NATURE OF SOIL ACIDITY

Acid soils can be prepared by leaching them with a weak mineral acid or by electrodialysis. In either process the basic cations are exchanged and a hydrogen-aluminum system is formed. If clay suspensions are washed rapidly with cold, dilute mineral acid or are allowed to pass through a column of H^+-saturated ion-exchange resin, a true H^+-ion-saturated clay will be formed (18). Upon aging, however, the H^+ ion in the system will react quite rapidly with the clay, releasing Al^{+++} which, in turn, becomes an exchangeable ion. Such systems are, therefore, H-Al systems, and are said to be unsaturated with respect to bases. The pH of such a system has been designated the "ultimate pH." When this system is titrated with a base, the pH rises and the basic ions exchange for exchange acidity at the colloidal surfaces.

The percentage of the total cation-exchange capacity (CEC) satisfied with basic ions is termed per cent base saturation. It is defined as:

$$\% BS = S/T \times 100$$

where BS = base saturation; S = meq of basic cations/100g soil; and T = total cation-exchange capacity. A further refinement is to consider the per cent base saturation of the permanent CEC separate from the per cent base saturation of the pH-dependent CEC.

The acidity associated with the clay mineral and organic fractions is called exchange acidity, since it is associated with the cation-exchange sites on the substance. As the acidity is neutralized the exchange sites become satisfied by base-forming ions—principally the metallic cations. Base saturation of the colloidal material increases at the expense of the exchange acidity.

Per cent base saturation is closely related to the lime requirement of soils and will be considered further in that discussion.

Titration curves of acid soils will differ with different kinds of colloidal material and different bases. At a given per cent base saturation the pH of a kaolinitic clay is higher than that of a montmorillonitic or organic system. Likewise, at a given degree of base saturation, a sodium system will have a higher pH than a calcium system (30). This is because of the greater degree of dissociation and hydrolysis of sodium than calcium from the colloidal material as would be expected from lyotropic series considerations.

FACTORS AFFECTING pH MEASUREMENT

Several factors are known to affect the pH measurement obtained for a particular soil sample:

Soil : Water Ratio

As a soil suspension becomes more dilute the pH of an acid soil rises. This phenomenon has come to be known as the "suspension effect." Several explanations have been given for this observed effect. It can be assumed that in acid soils the concentration of H^+ or Al^{+++} is greatest near the negatively charged soil colloidal particles and that a gradient exists from the particle into the soil solution. Therefore, the thicker the paste into which the glass electrode is placed, the more intimate the contact of the electrode with the compressed exchange acidity surrounding the colloidal particles, and the lower the pH value. As more water is added the ion layer would become more diffuse, until finally the electrode would be measuring the pH of the liquid phase of the suspension.

Another explanation for the "suspension effect" is that a "junction potential" is present in clay-water systems (12). As has been stated previously, the potentiometric measurement of pH requires that the transference of both the K^+ and Cl^- ions of the salt bridge takes place at the same rate. In clay systems the K^+ ion would be attracted to the negatively charged particles and, therefore, would not diffuse at the same rate as the Cl^- ion. This phenomenon would give rise to a potential difference at the junction of the salt bridge with the clay system. The thicker the suspension or paste, the greater this potential is likely to be. Since the potentiometer measures total potential difference, and since this total difference is usually interpreted to be solely due to H^+ ion activity in the system, any "junction potential" would lead to errors in the pH measurement.

To alleviate this problem some workers measure soil pH at a fairly wide soil:water ratio by placing the calomel half-cell in the supernatant liquid and the glass electrode in the sediment.

Soluble Salts

Usually as the salt concentration increases the measured pH decreases. It can be assumed that the cation of the salt will undergo cation exchange with the exchange acidity in the soil to release acid into the soil solution. This increased acidity is then measured by the glass electrode.

Two procedures have been used to eliminate this salt problem. One has been to leach out the soluble salts prior to making the pH measurement. This procedure requires a time-consuming step and does not lend itself to rapid, mass pH measurements.

The second and more commonly used procedure utilizes a salt solution rather than distilled water for making the paste or suspension. It is assumed that the salt concentration in the soil sample is negligible with respect to the amount of salt added in this solution. One tenth normal KCl and 0.01 *M* $CaCl_2$ are the solutions most commonly used. Measurements made in $CaCl_2$ have been used to calculate a "lime potential" for soils according to the formula: $\text{pH} - \frac{1}{2}\, p(\text{Ca} + \text{Mg})$ (37). This value for a particular soil is constant over a fairly wide range in conditions.

The presence of electrolytes also reduces the pH of soils in the neutral and alkaline range. If a sodium clay, for example, is dispersed in pure water a small proportion of sodium ions will hydrolyze from the diffuse layer. The hydrolysis of the sodium clay increases the hydroxyl ion concentration of the solution. Addition of an electrolyte decreases the thickness of the diffuse layer and suppresses hydrolysis. Some exchangeable sodium may be replaced by other cations, and the proportion of adsorbed sodium ions that dissociates into the outer solution is decreased. The presence of an electrolyte and the negative adsorption of anions from the clay surfaces both decrease the hydrolysis of dissociating sodium ions and reduce the resulting pH.

The effectiveness of electrolytes in lowering soil pH values increases with the valence of the cation. Thus, in dilute solution at equal concentrations, $CaCl_2$ lowers the pH of a soil more than does NaCl. This greater effect of calcium results from its higher energy of adsorption on clay particles. The salt content of the soil solution fluctuates with temperature, oxidation of organic matter, and leaching action of rain or irrigation water. Fluctuations in electrolyte concentration account for some of the changes in soil pH during the growing season.

Certain ions or salts, added or naturally occurring in soils, may greatly influence pH. Sulfides occur in large quantities in soils of limited areas. Sulfuric acid is formed through sulfide oxidation and may render the soil so acid as to make it practically barren. Sulfur added to soil is similarly oxidized to sulfuric acid. Ammonia and ammonium salts are acted on by bacteria, converting the ammonia nitrogen to nitric acid. Iron and aluminum sulfates are often added to soils as acidulants. Their action results from a hydrolysis to form sulfuric acid and the oxides of the metals.

Carbon Dioxide

A number of investigations have shown that the pH of soils containing calcium carbonate or appreciable amounts of bicarbonate or carbonate ions in the soil solution are closely associated with the carbon dioxide pressure, concentration of bicarbonate ion, and the ionic strength of the soil solution (7). These factors are interrelated in accordance with the equation:

$$\text{pH} = \text{pK}_1 - 0.5\sqrt{\mu^+} \log \frac{[\text{HCO}_3]}{[\text{CO}_2]}$$

In this relationship, pK_1 is the negative logarithm of the first dissociation constant of carbonic acid (6.26 at 24°C) and μ the ionic strength of the soil solution. The $[CO_2]$ (moles per liter) $= (P \cdot a \cdot p\text{CO}_2 \cdot 1000)/(760 \cdot 22{,}400)$ where P equals atmospheric pressure in millimeters of mercury, a the solubility of CO_2 in milliliters per milliliter of water, and pCO_2 is the CO_2 pressure in atmospheres.

In accordance with the aforementioned relationships, the pH of calcareous soils is reduced in proportion to the logarithm of the CO_2 pressure of the soil air. In absence of appreciable exchangeable sodium, the pH of a calcareous soil, which is 8.5 at the CO_2 pressure of air, falls to about 8.0 at a pressure of 0.002 atmosphere, and to about 7.5 at 0.02 atmosphere. Carbon dioxide pressures in excess of these are not uncommon in soils.

The nature of the several factors controlling soil pH indicates that the attempt to present soil characteristics, such as the availability of plant nutrients, as continuous functions of pH is an oversimplification of a complex situation (41). Fortunately, in most soils of humid regions the factors controlling pH are sufficiently uniform to permit many useful generalizations. In alkaline soils, quite different chemical situations exist when the pH is principally influenced by exchangeable calcium alone, by exchangeable calcium and an excess of calcium carbonate, or by exchangeable sodium. Evaluation of factors associated with soil pH must be based, therefore, on full consideration of the soil constituents controlling the pH situation rather than on pH per se.

LIME REQUIREMENT

Acidity in soil systems can be conveniently classified as active or potential acidity. Active acidity includes those H^+ ions that are present in the solution phase and that can be measured by normal procedures. Potential acidity may be considered to be the exchange acidity, and it makes up the bulk of the total acidity of the system. This potential acidity becomes active as active acidity is neutralized and cation-exchange processes occur which bring the potential acidity into solution. An equilibrium exists between the active and potential acidity. The relationship is governed by such factors as the nature of the colloidal material and the degree of neutralization of the system.

The lime requirement of a soil can be defined as the amount of a liming material that must be added to raise the pH to some prescribed value. This value is usually in the range of pH 6.0 to 7.0, since this is an easily attainable value within the optimum growth range of most crop plants. Several methods have been proposed for estimating lime requirement (1, 38). Of these, pH measurement of the soil is that most widely used. The basis for this method is that in acid soils a relationship exists between the pH and the percentage base saturation of the soil. Once this relationship between the percentage base saturation and the soil pH is known, it is useful for determining the lime requirement. Care should be exercised to restrict the use of a given titration curve relating pH and per cent base saturation to soils containing similar percentages of clay and organic matter of similar types.

More precise methods of determining lime requirement attempt to measure directly the potential or exchange acidity in the system. This is usually accomplished by using a strongly buffered system. One of the first of these to be used was *para*nitrophenol which is buffered at pH 7.0. When this solution is added to a sample of an acid soil it reacts and reaches an equilibrium pH value somewhat less than 7.0. The amount of pH depression is directly related to the amount of exchange acidity in the soil and can then be converted into the lime requirement by appropriate procedures. One serious difficulty with this method is that in weakly buffered soils the pH depression may be slight and the error of measurement may be a substantial part of the total measurement. By using weaker buffer solutions with such soils, greater pH depressions may be obtained from a limited amount of exchange acidity.

Triethanolamine (TEA) buffered at pH 8.2 has also been used in measuring exchange acidity. In this method the TEA and soil are allowed to react and then the suspension is filtered. The filtrate is titrated to some specified pH value and compared with the titer of an equivalent amount of TEA solution. The difference is considered to be due to exchange acidity.

The theoretical basis for measuring exchange acidity at pH 8.2 is related to the equilibrium pH value attained in systems containing free calcium carbonate. The pH of such a system depends on the partial pressure of CO_2. At the partial pressure of CO_2 found in the atmosphere (0.0003 atm) the system of $CaCO_3$-$Ca(HCO_3)_2$-H_2CO_3 has a pH of 8.2 (7). The practical implications of this relationship have several serious limitations. The CO_2 pressure in most soils is higher than that in the atmosphere and, therefore, the measured pH of most soils containing free $CaCO_3$ is less than the theoretical value. Perhaps the most serious limitation is that soils are seldom, if ever, limed to such a high pH, and it is difficult to evaluate the exchange acidity measured in this way with a reasonable lime requirement.

A method for determining exchange acidity has been proposed in which a soil sample is allowed to react with a solution of calcium acetate which is buffered at pH 7.0. The filtrate is then titrated to pH 8.3, which is a point at which the calcium acetate is not highly buffered and gives a sharp end point. This titer is compared with one for the solution alone, and the difference is calculated as the exchange acidity. This method overcomes the disadvantages of the *para*nitrophenol method, which relies on a slight pH depression, and of the TEA method, since a pH level of 7.0 is a more practical value for assessing lime requirement.

Each procedure for estimating lime requirement gives the theoretical amount of calcium that would be needed to raise the pH or percentage base saturation to a desired level. When the estimated amount of calcium required is added as limestone to soils in the field, the pH of the soil is not raised as much as predicted by the testing procedure. Therefore, a "liming factor" has been introduced to bring about the desired results. This factor varies from 1 to 3, depending on rate of limestone solution, plant uptake, and leaching during the reaction period. A "liming factor" of 1.5 to 2.0 is commonly used when the theoretical amount of lime necessary to bring the soil to a given pH is converted to the actual amount of limestone to be added in the field (36).

LIMESTONE REACTION

Lime reaction in soils is conditioned mainly by the nature and the fineness of the liming material (3). Lime is usually added to soils in the form of ground limestone. Limestones can be classified as calcitic ($CaCO_3$), dolomitic [$CaMg(CO_3)_2$], or a mixture of the two. Both of these limestones are sparingly soluble in pure water but do become soluble in water containing CO_2. The greater the partial pressure of CO_2 in the system, the more soluble the limestone becomes. Dolomite is somewhat less soluble than calcite. The reaction of limestone can be written as

$$CaCO_3 + H_2O + CO_2 \rightarrow Ca(HCO_3)_2$$

$$Ca(HCO_3)_2 \rightarrow Ca^{++} + 2HCO_3^-$$

$$H^+ + HCO_3^- \rightarrow H_2CO_3 \rightleftarrows H_2O + CO_2$$

In this way the H^+ ions in the soil solution react to form weakly dissociated water, and the Ca^{++} ion is left to undergo cation-exchange reactions with perhaps additional exchange acidity. The acidity of the system is, therefore, neutralized and the per cent base saturation of the colloidal material is increased.

The rate of solution of a slightly soluble material, such as limestone, is determined to a considerable degree by the amount of surface exposed to the reacting solution. A more finely ground limestone exposes more surface per unit weight than a coarsely ground material. For example a 40-mesh separate has four times the surface and 64 times the number of particles as an equivalent weight of a 10-mesh separate. The additional number of particles associated with fine grinding also results in more rapid solution and neutralization in soils. The limestone particle reacts with the acid soil particles and the soil solution in its immediate vicinity. After the initial reaction the solution immediately surrounding the particle will be saturated with respect to the reactants. In order for the reaction to continue, the reactants must diffuse away from the site of the reaction. When a few large particles are involved, the reaction will soon be governed by diffusion rate, while with many small particles, diffusion rate will not be such a controlling factor.

Several environmental factors also affect the rate of limestone reaction (11). The greater the amount of moisture, the more rapid is the rate of reaction. Obviously moisture must be present before the solubility reaction can occur. As moisture increases, the degree of aeration is reduced, which would result in an increase in the concentration of CO_2 in the soil air and an increase in the rate of limestone solution. The increased moisture would also allow for a greater volume of solution and, therefore, a lower concentration of reaction end products. Since the reaction is an equilibrium reaction, the accumulation of end products would reduce reaction rate over time.

Soil temperature also affects reaction rate. Limestone reacts more rapidly at high than at low temperatures. This effect is probably related to diffusion rates of end products away from the reaction sites.

The amount of exchange acidity present in the soil affects reaction rate. If a soil has a high lime requirement and if a sufficient quantity of limestone is added to neutralize the acidity present, the initial reaction will be quite rapid. However, as the acidity becomes neutralized, the rate of reaction decreases and finally, as neutrality is approached, becomes almost negligible.

LIMESTONE REACTIVITY EVALUATION

Several chemical methods have been proposed to measure the reactivity of limestone materials (3). Solvents such as acetic acid and carbonated water have been tried but do not give close agreement with the reactivity of limestone in soils. Oxalic acid and other oxalates, and EDTA solutions have also been used to measure the effective surface of limestone particles. These methods have not proved satisfactory, since surface measurement gives an indication of only the initial reaction of the material and does not measure the capability of the material for sustained reaction. For instance, a small amount of a finely divided material might have the same amount of surface for initial reaction as a larger amount of a coarser material. Over a finite period of time, however, the coarser material would have more neutralizing effect because of its greater capacity for sustained reaction. These methods that measure surface reactivity do not adequately differentiate between calcitic and dolomitic materials. Although the initial surface reactivity of these materials may be nearly the same, the rate of solution of dolomite is considerably slower than that of calcite.

An ammonium chloride digestion procedure has been proposed as a way to evaluate the neutralizing value of limestone materials. In this procedure hot ammonium chloride reacts with the limestone with the release of ammonia. This ammonia is collected in standard acid and determined by back titration with standard base. This procedure has the advantage of evaluating the sustained reaction potential of limestone and of differentiating between calcitic and dolomitic materials.

REASONS FOR LIMING ACID SOILS

Acid soils are usually made more suitable for agricultural use by liming (2, 3, 36, 41). Among the several reasons for liming soils are the following:

(*a*) Liming the soil affects the solubility and plant availability of most of the nutrient ions in the soil. In many instances the solubility products of sparingly soluble compounds are affected by the pH of the system. The nature of compounds or complexes formed in soils may be influenced by pH or lime status. In other instances, nutrient availability may be a secondary effect, as in the case of biological transformations. These relationships may be summarized as follows:

Calcium is added directly in the liming material. Magnesium is added also when the liming material contains both Ca and Mg. Soluble phosphorus-containing materials react in soils principally with Ca or Al and Fe. The most stable phosphorus-containing compound that will form in soils depends greatly on the pH of the soil system. Above pH 6.0 calcium phosphates will dominate, while below this pH aluminum and iron phosphates are the dominant compounds regulating phosphorus availability in soils. The availability of nitrogen, sulfur, and phosphorus in soils is affected to some extent by biological transformations. Many of the microorganisms which perform these functions operate more effectively near neutrality than at other pH values. The solubilities of most micronutrient elements are related to the pH of the system. With the exception of molybdenum, these elements are more soluble under acid conditions and reach a minimum solubility under slightly alkaline conditions.

The addition of liming materials to soils may influence nutrient uptake in plants by producing certain "antagonistic" effects.

Liming may bring about changes in the ionic concentration in the soil solution which would result in an alteration in plant uptake and leaching.

(*b*) Liming affects the solubility of toxic substances in the soil. This is especially true of aluminum and manganese which are more soluble under acid conditions.

(*c*) Liming may result in an improvement of soil structure, especially under conditions where organic materials are significantly related to cementation and structure stability. This effect may also have a secondary relationship to the beneficial influence of liming on microorganism activity.

(*d*) Liming influences the distribution of roots in soil. Root distribution studies show that plant roots develop more fully in limed than in unlimed zones. Liming an acid soil, therefore, will encourage an increased root system with the added advantages of nutrient and water availability that are associated with greater root development.

CALCAREOUS SOILS

Zonal soils of arid regions usually contain accumulations of lime at some point in the profile. When the zone of lime accumulation coincides with the depth of high root concentration, it may be an important factor in plant nutrition. Many soils are formed from limestone or marl parent materials. Often the calcium carbonate has not been leached from these soils to any appreciable degree. Such soils are typically calcareous throughout the profile. The content of calcium and magnesium carbonates may be as high as 60 to 70 per cent of the total soil mass. Soils referred to as "calcareous" may vary in lime content from those having a small concentration some place in the profile to those containing an appreciable amount of lime throughout the profile.

Calcium carbonate occurs in the sand, silt, and clay size fractions of calcareous soils. The proportion in each fraction varies greatly. Occasionally there may be considerable cementation where large amounts of lime have accumulated in the subsoil. Such lime hardpans are especially typical of lime accumulations formed over high water tables.

Calcareous soils encompass special problems that seem associated with the alkaline pH (8), free calcium and magnesium carbonates, reactions with plant nutrients, and interactions with plant roots.

In general, the pH of calcareous soils rises as the proportion of water to soil is increased. This may result in part from a reduction in CO_2 pressure (16, 45). The dilution of salts in the soil solution also permits greater hydrolysis of the calcium clay. The hydrolysis of calcium carbonate with its production of hydroxyl ions has been shown to be important in determining the productivity of calcareous soils:

$$CaCO_3 + H_2O \rightleftharpoons Ca^{2+} + HCO_3^- + OH^-$$

Because of this reaction and other relationships between moisture and pH, the control of excess water seems to be more important in calcareous and other alkaline soils than in soils that are acid in reaction. The build-up of hydroxyl ions from the hydrolysis of calcium carbonate is prevented by adequate carbon dioxide in the soil. With low carbon dioxide pressure, moist calcareous soils may become excessively alkaline.

The carbonates of calcium and magnesium participate directly in some reactions, controlling the solubility of phosphate and iron (33). The phosphate problems are discussed in an earlier section of this chapter.

Reactions between carbonate ions and iron may be a factor in reducing iron availability to plants. The following reactions are known to be exothermic:

$$Fe^{2+} + CaCO_3 \rightleftharpoons FeCO_3 + Ca^{2+}$$

$$4FeCO_3 + O_2 + Ca(HCO_3)_2 \rightleftharpoons 2Fe_2(CO_3)_3 + Ca(OH)_2$$

$$Fe_2(CO_3)_3 + 3H_2O \rightleftharpoons Fe_2O_3 + 3H_2CO_3$$

Since iron is probably largely assimilated by plants as the ferrous ion, moist calcareous soils would not favor iron availability for reasons in addition to the alkaline pH.

Soils high in lime are frequently productive for many ordinary field crops, including most forage crops, corn, cotton, sugar beets, potatoes, and tomatoes. Some other crops, including berries, deciduous and citrus fruits, many flowers and ornamental shrubs, sorghums, and, to a lesser extent, peas and beans, suffer from acute iron deficiency on certain high-lime soils. The typical yellowing of leaves under these conditions is termed lime-induced chlorosis and is characterized by iron deficiencies in the plant leaves.

No adequate hypothesis has been found to explain why chlorosis occurs on some high-lime soils and not on others. It is known that chlorosis is increased in severity by excess moisture and low or high temperatures in the soil. Some evidence indicates that calcium and bicarbonate ions may be more plentiful in the solution of soils that are associated with chlorosis problems than in other calcareous soils.

Leaves yellowed from lime-induced chlorosis have been found to have a high content of potassium, organic acids, and soluble nitrogen. In at least some instances calcium or magnesium may be present in greater concentrations in the expressed sap from these chlorotic leaves than in that from green leaves. Evidence is not conclusive that total iron content of chlorotic leaves is always lower than in comparable green leaves, but acid-soluble iron is always low in chlorotic leaves.

In solution culture, a chlorosis similar in appearance to lime-induced chlorosis has been induced by the bicarbonate ion. Also, the bicarbonate ion has been shown to interfere with the iron nutrition of many plants. These relationships support the hypothesis that the bicarbonate ion is an important factor in lime-induced chlorosis.

A fairly widespread philosophy holds that plant growth and nutrition are helped by the acidification of alkaline-calcareous soils. Various acidification trials indicate that lime is the first point of attack for acids added to calcareous soils, but little change in pH or general nutrient solubility is brought about until most of the lime is removed. A soil containing 5 per cent or more lime has such a high buffering capacity, however, that direct treatments for lime removal are uneconomical. The evidence is not conclusive that acid treatments insufficient to react with most of the excess lime significantly increase plant nutrient uptake of such elements as phosphorus, iron, manganese, or zinc. Nor have plant yields been significantly increased in several experiments in which acid treatments were applied to calcareous soils.

SALINE AND SODIC SOILS

In areas of low rainfall, salts formed during the weathering of soil minerals are not fully leached. During periods of higher than average precipitation, however, the more soluble salts are frequently leached from the more permeable and high-lying soils. These salt-charged waters find their way to lower-lying soils, and there, if drainage is impeded, the ground waters gradually accumulate high concentrations of salts. Upon evaporation of these waters, salt deposits are left in the soil. Repetition of this leaching-deposition process, with the numerous variations that occur in nature, results in a complex pattern of soils high in salt intermixed with normal nonsaline soils. Soils thus affected by salts generally occupy the lower positions in arid regions, positions comparable to those of the humic glei and planosol soils of humid regions. Saline soils occasionally occur along some seacoasts, and even in humid areas, because of tidal action or the presence of saline groundwater.

The soluble salts that accumulate in soils consist principally of the cations of calcium, magnesium, and sodium, and the chloride and sulfate anions. Potassium, bicarbonate, carbonate, and nitrate ions occur in smaller quantities. Borates occasionally occur in small amounts but receive considerable attention because of their exceptionally high toxicity to plants. The proportion of ions occurring in soils varies greatly. The nature of the salts present obviously depends on the composition of the rock weathered, the nature of the weathering process, and the subsequent reactions that occur as the salts are moved from site of weathering to place of deposition.

Salt may influence soils in many ways (19). The direct presence of salts is one consideration, changes in exchangeable cations on the soil colloids is another, and the indirect effects of salts on soil microbes, plant root activities, and the physical properties of soil colloids constitute a third.

The terminology applied to salt-affected (salted or alkali) soils varies from country to country. Many colloquial terms are also employed in limited areas. The designations proposed by the United States Salinity Laboratory are, however, most generally used. The terms saline and sodic are used separately and, as combined terms, saline-sodic and nonsaline-sodic. Carter (9) has prepared a useful bibliography covering the field of saline and sodic soils.

Saline soils include soils containing soluble salts in quantities sufficient to interfere with the growth of most crop plants but not containing enough exchangeable sodium to alter soil characteristics appreciably. Technically, a saline soil is defined as a soil having a conductivity of the saturation extract greater than 4 millimhos/cm and an exchangeable sodium percentage less than 15 (these limits are set as somewhat arbitrary guides). The pH is usually less than 8.5. Sodium constitutes not more than about two-thirds the total soluble cations. The principal soluble anions are chloride, sulfate, bicarbonate, and occasionally some nitrate. Measurable quantities of soluble carbonates are usually absent.

Sodic (nonsaline-alkali) soils include soils containing exchangeable sodium in a quantity sufficient to interfere with the growth of most crop plants and not containing appreciable quantities of soluble salts. The exchangeable sodium percentage is greater than 15, and the conductivity of the saturation extract is less than 4 millimhos/cm. The colloids in nonsaline-sodic soils are usually deflocculated, and drainage and aeration are poor. The pH usually ranges between 8.5 and 10.

Because of the high pH and the dominance of sodium ions in sodic soils, part of the soil organic matter is commonly brought into solution. Extracts of such soils have a characteristic dark brown color. The dissolved organic matter in the soil solution becomes deposited as a thin film on the soil surface or on the surface of soil aggregates. The prevalence of such black stains in nonsaline-sodic soils has led to the common designation of black alkali soils (24, 26).

When the soluble salts are slowly leached from a saline-sodic soil, the colloidal particles tend to disperse. The dispersed particles are gradually moved downward, and dense subsoil horizons are formed. Upon continued alternate wetting and drying, the dense subsoil material develops characteristic columnar structures. These columns are typically $1\frac{1}{2}$ to 2 inches in diameter and 4 to 6 inches in length. The tops of the columns are rounded. Soils having such subsoil structures are termed *solonetz*. Solonetz-like structures have also been observed in soils not evidencing any influence of high exchangeable sodium. At present, solonetz generally refers to a soil structural situation which, though common among sodic soils, is not restricted to any particular chemical situation.

The classification of salt-affected soils into saline, sodic, and saline-sodic groups is based on chemical properties rather than on distinct differences in soil morphology. The classification is designed to facilitate discussions of soil management and is not a pedologic system. In fact, normal nonsaline soils may be transformed to any one or all three of these types in a short time. Also reclamation may restore the original soil condition, leaving few marks to indicate past deterioration through the influence of salt.

Transition Processes in Salt-affected Soils

Salt accumulation is the first stage in the sequence of processes common to the family of salt-affected soils. It usually occurs in areas of limited rainfall, where surface or groundwater does not drain away satisfactorily. Instead, water is removed largely through evaporation or transpiration, and salt concentration is increased. Sodium salts often predominate in early stages of salinization. Calcium and magnesium salts accumulate more slowly. As the salt concentration increases, magnesium may precipitate as a carbonate while calcium may precipitate as carbonate or sulfate. This concentration and precipitation process causes calcium carbonate to accumulate in the soils and results in a gradual increase in sodium in proportion to other cations in solution.

As salts accumulate in soils, the equilibrium among cations in the soil solution and those adsorbed on the colloids continually shifts to accomodate changes in the composition and concentrations of dissolved cations. When sodium increases in the soil solution, the proportion of sodium adsorbed on soil colloids also increases. Since sodium has a lower affinity for clay than do the other cations in soil, the proportion of sodium to other soluble cations on an equivalent basis (designated as soluble sodium percentage) must be well above 50 per cent before any appreciable amount of sodium is adsorbed.

Several equations have been developed to illustrate the relationship between soluble cations in soil solutions and those adsorbed on clay (25). Although these equations are based on different physicochemical concepts, the resulting formulas are similar to those based on the law of mass action applied to the general alkalization reaction:

$$\mathrm{CaX} + 2\mathrm{Na}^+ \rightleftharpoons 2\mathrm{NaX} + \mathrm{Ca}^{2+}$$

The most simple type of equation, then, becomes:

$$\frac{(\mathrm{NaX})^2(\mathrm{Ca}^{2+})}{(\mathrm{CaX})(\mathrm{Na}^+)^2} = k$$

An equation developed from statistical thermodynamics (27, 43) gives a more constant value for k with changes in cations and adsorbents:

$$k = \frac{(\mathrm{NaX})^2}{(\mathrm{Na}^+)^2} \cdot \frac{(\mathrm{Ca}^{2+})}{(\mathrm{CaX})[1\frac{1}{2}(\mathrm{CaX}) + (\mathrm{NaX})]}$$

In these equations X stands for the clay exchange complex and the quantities for the ions in parentheses denote activities.

Such equations are difficult to use for saline and sodic soils because of the common presence of calcium and magnesium carbonates, which makes the determination of exchangeable calcium and magnesium difficult and inexact. Also cations other than just calcium and sodium are involved. Such equations have further limitations, because k is not constant for different soils. The United States Salinity Laboratory has found that the ratio

$$\frac{Na^+}{\sqrt{\frac{Ca^{2+} + Mg^{2+}}{2}}} = SAR$$

is highly correlated with the proportion of exchangeable sodium on the exchange complex. In this ratio, which has been designated the *sodium adsorption ratio*, the soluble cations are expressed on an equivalent basis. The *SAR* of a saturation extract or of irrigation water is useful in evaluating exchangeable sodium percentage or estimating the sodium hazard when a specific irrigation water is used.

As a result of changes in climate, soil, or soil management practices, salts may be leached from a saline or saline-sodic soil. In a saline-sodic soil, leaching the salts leaves a nonsaline-sodic soil. Under such soil conditions, leaching is accompanied by swelling of the colloids, reduction in soil permeability (17), and development of excessive alkalinity.

If extensive leaching of a saline-sodic soil occurs in the absence of any source of calcium or magnesium, part of the exchangeable sodium is gradually replaced by hydrogen. The resulting soil may be slightly acid, yet contain enough sodium to give it an unstable structure. Such a soil, which has previously been designated as a *degraded-alkali* soil, should now be referred to as a *degraded sodic* soil.

Leaching of saline-sodic soils containing gypsum or other materials that provide soluble calcium or leaching with water containing dissolved calcium, leads to the replacement of sodium by calcium. A normal soil results.

The sodic muds that occur on parts of the ocean bottom and under some bays contain considerable quantities of sulfides. These sulfides have accumulated over a period of time, probably as a result of sulfate reduction under the anaerobic conditions. When such areas are drained they can be reclaimed as is done in Holland (20, 44) by leaching without the addition of gypsum. The sulfides are oxidized and the resulting acid acts on the lime-like material from shells etc., to provide the calcium ions to replace the undesirable sodium adsorbed on the colloids. If such reclaimed soils are inundated a second time with sea water, they again become more or less saturated with sodium. If drained for the second time, they normally cannot be reclaimed again by leaching only; unless considerable time has lapsed, the sulfides are not likely to be present in sufficient quantities and/or the natural supply of limelike materials will be missing. When sodic muds are exposed to aerobic conditions by drainage, and the supply of lime, whether natural or added, is insufficient, the soil materials will become acid due to the oxidizing sulfides.

Evaluation and Reclamation of Saline Soils

The principal quality of saline soils injurious to plants is the high osmotic pressure of the soil solution, which reduces the availability of the water. Some plants, however, show special sensitivity to certain ions and are injured more by one salt than by another (31).

Until about twenty years ago most evaluations of saline soils were on the basis of percentage or parts per million of salt in them. Interpretation of such data required special provisions for broad differences in salt composition. For example, chloride salts were considered much more toxic to plants than were sulfate salts. Allowances were also made for differences in soil texture. Equal amounts of salt were considered more toxic in sandy soils than in loam or clay soils. These variations are largely eliminated by evaluating saline soils on the basis of the osmotic pressure of the soil solutions or on the basis of some character directly proportional to the osmotic pressure.

Direct determination of the osmotic pressure of soil solutions is tedious. The solution must be removed from the soil. With soil moisture below field moisture capacity, removal of usable amounts of soil solution requires large samples of soil and considerable energy in the form of pressure or suction. Osmotic-pressure determinations with the standard ice bath and thermometer arrangements are time-consuming. Even with modern refrigerator baths and thermistors, the process requires too much time for extensive field evaluation studies.

Obviously, measurements of osmotic pressure of soil solutions should be based on some uniform moisture condition. Field moisture capacity would be a logical point, but it is difficult to handle large numbers of samples and extract sufficient moisture for analysis unless they contain more water than this. Extraction of water from soil at the moisture saturation point has proved to be a practical and useful procedure. The saturation point represents approximately twice the water present at field moisture capacity. Water is readily removed from saturated soils, but the degree of dilution is not great enough to bring appreciably more gypsum into solution than that already dissolved under field moisture conditions.

Ascertaining the electrical conductivity of extracts from saturated soil pastes has become a standard procedure for evaluating the salt status of soils. Since the standard conductivity unit, mhos per centimeter, is too large for the convenient expression of most analytical values for soils, millimhos is commonly used. The osmotic pressure of a solution can be estimated from conductivity values by the relationship, $OP = 0.36\ EC$, when electrical conductivity is expressed in millimhos per centimeter.

Electrical conductivity standards developed by the United States Salinity Laboratory for evaluating saline soils are as follows:

Conductivity of Saturated Extracts (mmho/cm)	*Plant Growth Conditions*
Less than 2	Salinity effects mostly negligible
2 to 4	Yields of very sensitive crops may be restricted
4 to 8	Yields of many crops restricted
8 to 16	Only tolerant crops yield satisfactorily
Higher than 16	Only very tolerant crops yield satisfactorily

Crop plants differ in their tolerance to salinity. Such crops as tall wheatgrass, beets, and spinach not only are highly tolerant but usually evidence growth stimulation by salt levels that would be near lethal to sensitive plants such as peas. The basis for such differences in tolerance is not understood.

The evaluation of a salt problem requires special determinations of chloride and sodium concentrations when fruit trees are involved. Chloride in excess of about 10 meq/l in solution cultures has been found to foster excess accumulations in the leaves, with burning of the leaf tissue.

Boron is highly toxic to nearly all crop plants. Plants may be injured by concentrations as low as 1 ppm in the saturation extract. The greatest difficulty in this regard has been encountered on soils irrigated with water high in boron.

Reclamation of saline soils is effected by leaching. Since salt accumulations have usually developed because drainage is poor, the leaching operation is often difficult. Where satisfactory drainage can be economically established, leaching readily removes salt. Where drainage is impractical, saline soils are considered nonreclaimable. Because reclamation presupposes satisfactory drainage, the water transmission characteristics are most important in evaluating saline soils for agricultural purposes.

Sodic Soils

The distinguishing characteristics of a sodic soil are high content of exchangeable sodium and low permeability (6).

Exchangeable sodium determinations are too time-consuming to be used for routine evaluations of large areas of soil. The pH test is often used to indicate degree of sodium saturation of soils (15), but the relationship between pH and exchangeable sodium is complex when applied to a wide range of soils. Salt is particularly important in giving lower pH values than would be expected from exchangeable sodium percentages. The most common procedure used to circumvent this salt effect is to determine soil pH first on a paste and then on a 1:5 or 1:10 soil-water suspension. The increase in pH between the paste and the suspension is used as a criterion of exchangeable sodium. In some areas this procedure has given useful results. Extensive investigations have indicated that none on the proposed pH testing procedures gives a satisfactory indication of exchangeable sodium for all soils (15). For limited areas of soil, however, pH may be a fully satisfactory procedure. It is recommended that a preliminary investigation be made of representative soils in each geographic area. If a satisfactory relationship is obtained, the more rapid pH test is used. If the relationship is not satisfactory, the more tedious determination of cation-exchange capacity and exchangeable sodium may be necessary to evaluate reclamation problems (5).

Several quick tests have been proposed as a basis for evaluating alkali-sodic soils. The "gypsum requirement" is one of the most helpful chemical tests of this group. A small sample of soil is added to a relatively large volume of standard, nearly saturated gypsum solution. The loss in calcium from the solution, not compensated for by gain in magnesium, is used as a direct measure of gypsum required to replace the sodium.

Evaluation tests should be used to determine potential sources of calcium for replacing exchangeable sodium (39). Calcium carbonate and gypsum in soil are the most important sources. The leaching waters may also contain significant amounts of calcium.

Both field and laboratory procedures are available to evaluate the permeability of sodic soils. It is important in such tests to use the same water that will be employed for leaching. No permeability limits have been established, but if water to a depth of a few feet can be put down through a sodic soil in a few weeks, reclamation is generally feasible.

Exchangeable sodium exerts several effects which lower the productivity of soils. Sodium ions when coupled with high pH have a toxic action on roots. Under extreme conditions, roots as well as organic matter may be dissolved in part. Aeration is often a problem in sodic soils. High alkalinity, the dominance of sodium, and poor aeration all combine to give a disturbed plant nutrition situation.

High levels of exchangeable sodium promote deflocculation and swelling of clay particles. Soils of similar exchangeable sodium percentages may differ considerably in physical properties. Coarse-textured soils are generally less affected by exchangeable sodium than are fine-textured soils.

Also, soils containing clay of the expanding lattice types, such as montmorillonite, swell and disperse more with sodium adsorption than do soils containing clays with nonexpanding lattices. The degree of swelling is related to the specific surface of soils.

The clay of saline-sodic soils is usually flocculated, and the degree of swelling is low as long as the concentration of salts remains high in the soil solution. If such soils are leached with water of low salt content, dispersion and swelling of the clay upon salt removal may essentially seal off further movement of water. Such a transition is termed "freezing." Prevention of "freezing" is an important part of the reclamation procedure.

The reclamation of sodic soils is often more difficult than that of saline soils, because, in addition to the problem of leaching for salt removal, there are the problems of replacement of exchangeable sodium by calcium and the restoration of a granular physical state (28). Estimation of the amount of an amendment that should be added to supply calcium depends first on estimating the quantity of exchangeable sodium that should be replaced from the soil. Exchangeable sodium data can be used in the following relationship to estimate the quantity of amendment that must be added: amendment (tons per acre) $= 10\,NA$, in which N is the number of milliequivalents of exchangeable sodium per 100 g of the soil, and A equals the grams of the amendment required to furnish 1 g/meq of soluble calcium. The estimate is based on a soil depth of $6\frac{2}{3}$ inches.

Instead of basing the calculations on complete removal of sodium, it is often desirable to reduce the sodium to only 10 per cent of the exchange capacity. This permits satisfactory crop growth, and natural processes associated with root activity, irrigation, and leaching can complete the removal.

Most field situations do not result in a quantitative exchange of calcium from amendments for exchangeable sodium. Some tests indicate that an excess of about 25 per cent should be allowed. On the other hand, considerable calcium may be supplied in leaching water, or it may be brought into solution by root activity and organic matter decomposition after the reclamation process is carried far enough to permit these activities.

The choice of amendments for aiding in sodic soil reclamation depends upon soil characteristics as well as upon availability and costs of acceptable materials (32). The principal purpose of the amendments is to furnish soluble calcium to replace exchangeable sodium. For degraded sodic soils with slightly acid pH values, ground limestone is satisfactory. For saline-sodic and nonsaline-sodic soils, limestone is too insoluble to promote rapid reclamation. If the sodic soil contains lime, then acidulating materials, such as sulfuric acid, that react with the lime and release calcium are satisfactory.

For noncalcareous sodic soils, a relatively soluble source of calcium such as gypsum usually gives best results.

Some of the common amendments and the reactions that make them helpful are as follows:

(*1*) *Limestone on degraded sodic soil*

$$2CaCO_3 + 2HX \rightleftharpoons Ca(HCO_3)_2 + CaX$$

$$Ca(HCO_3)_2 + 2NaX \rightleftharpoons CaX + 2NaHCO_3$$

(*2*) *Gypsum*

$$CaSO_4 + 2NaX \rightleftharpoons CaX + Na_2SO_4$$

(*3*) *Sulfuric acid on calcareous sodic soil*

$$(A)\ H_2SO_4 + CaCO_3 \rightleftharpoons CaSO_4 + H_2O + CO_2$$

or

$$(B)\ H_2SO_4 + 2CaCO_3 \rightleftharpoons CaSO_4 + Ca(HCO_3)_2$$

Reaction (*3A*) would furnish one equivalent of soluble calcium for each equivalent of acid. Reaction (*3B*) would furnish two. It has been suggested that in practice about $1\frac{1}{2}$ equivalents of soluble calcium could be expected. Laboratory and greenhouse tests have indicated that only the one equivalent indicated in reaction (*3A*) can be depended on.

(*4*) *Sulfur: the first step is a biological oxidation to sulfuric acid*

$$2S + 2H_2O + 3O_2 \rightarrow 2H_2SO_4$$

In some soils, particularly those in areas with cold winters, sulfur oxidation is too slow to give satisfactory results.

In soil reclamation experiments in California, treatment with sulfuric acid brought a sodic soil into good crop production more quickly than did treatment with gypsum or sulfur. In fact, sulfur was not appreciably superior to no-amendment treatment in hastening reclamation or increasing crop yields.

Quality of Irrigation Waters

All irrigation waters contain some salt, but quantities and kinds vary greatly. Since the salt and sodic status of a soil can be greatly affected by salts in irrigation water, successful irrigation farming requires that the salt composition of all questionable waters be known and that some system for evaluating their analysis be available.

The quality of irrigation waters depends principally upon the total amount of salt present and the proportion of sodium to other cations (46, 47). As with soils, the most satisfactory method for rating the salt content of irrigation waters involves measuring electrical conductivity. Various conductivity limits have been proposed to indicate the degree of salinity problems that might be anticipated with the use of irrigation water on general crop farms.

The actual damage done by salt to plants or soils depends on the concentration in the soil solution rather than on the quantity in irrigation water. Thus, use of the same water source might lead to a severe salt problem in one soil where drainage is restricted and great concentration of salts occur, whereas under another situation of good drainage and the application of excess water to leach salt residues from the soil, crops could be grown without evidence of salt damage.

Several schemes have been proposed as ways to evaluate irrigation waters with regard to sodium. Irrigation waters having a ratio of Na to (Ca + Mg) higher than one have been suggested as leading to alkali formation. A widely used scheme is the rating of waters according to the soluble-sodium percentage $(Na \cdot 100)/(Ca + Mg + Na)$. A diagram relates soluble-sodium percentage and salt concentration for waters to certain equilibrium exchangeable-sodium percentage values that exist when soils are continuously leached with waters of specified composition. In one scheme, all waters that would cause equilibrium exchangeable-sodium percentages of 15 or greater are assumed to be unsatisfactory for general use.

A scheme proposed by the United States Salinity Laboratory has been generally approved in the western states. Sodium-adsorption-ratio values are used to separate water classes according to hazards from sodium accumulation. A logarithmic scale is used for the electrical conductivity base to establish water classes based on hazards from total salt concentration. A modification of the Salinity Laboratory scheme (40) is shown in Figure 7.1. In this modification, electrical conductivity has been extended to 6000 micromhos before waters are considered too high in salt for irrigation use under any conditions. The number of classes and the values for class limits above the 2250-micromho limit have also been changed.

According to this scheme, a water having a conductivity of 1500 micromhos and containing 5 epm (equivalents per million) of calcium, 3 of magnesium, and 7 of sodium would fall in class C3-S1. The chart indicates that the water can be used successfully with occasional leaching on soils of good permeability and crops of moderate-to-good salt tolerance. The sodium should not create a serious problem.

Waters containing carbonate and bicarbonate ions in excess of calcium plus magnesium often lead to much greater alkali formation than is indi-

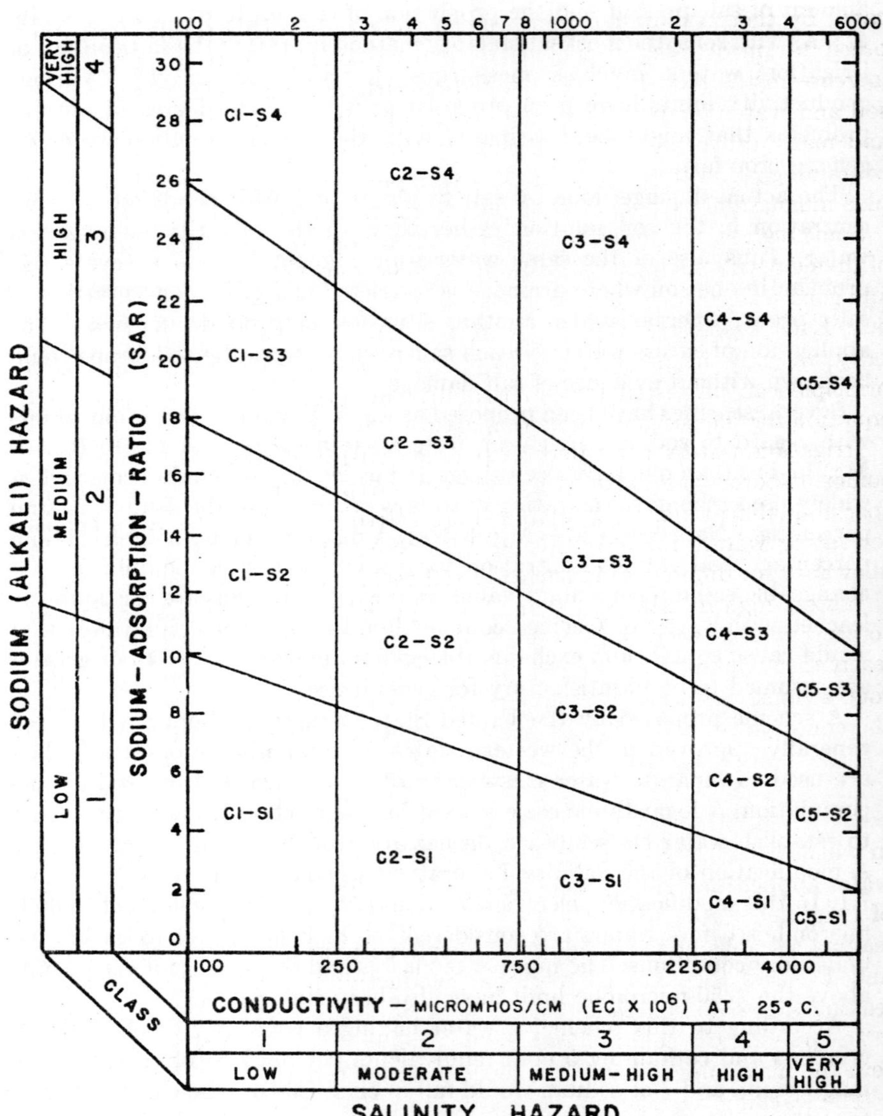

Figure 7.1. Diagram for determining the quality rating of an irrigation water from its sodium adsorption ratio and electrical conductivity. (*Modified from U. S. Salinity Laboratory Staff*)

cated by their sodium-adsorption-ratio values (14). This results from the precipitation of calcium and magnesium carbonates, with a concomitant increase in the sodium-adsorption ratio. Such carbonate effects vary with soil and water temperatures, the degree of salt concentration in the soil, and many other soil and management factors. Usually, however, waters containing residual carbonates (carbonate plus bicarbonate exceeds calcium plus magnesium) are marked as possessing special sodium hazards.

Some high-salt waters can be used to advantage as flocculants and sources of divalent cations in the reclaiming of sodic soils (34). A mixture of salty and good water can be used to leach soils if a series of applications is carried out. Each successive application has to have a lessening proportion of salty water if the procedure is to give the most rapid reclamation. This principle can be tested for practicability in specific cases by use of the equation for the sodium absorption ratio.

Irrigation waters are rated as poor or unsatisfactory if they contain sufficient boron to damage crops. Waters for special purposes may require special qualifications. Some plants are sensitive to certain ions. All schemes for rating waters for irrigation must be used liberally with considerable allowance for differences in soils and soil-management practices.

Waters high in salt content must be applied to soil in excess quantities to prevent toxic accumulations. The percentage of water applied that must leach through soil to maintain the drainage water at the bottom of the root zone at certain predetermined levels can be calculated according to the equation proposed by the United States Salinity Laboratory (35, 43):

$$LR = \frac{D_{dw} \cdot 100}{D_{iw}} = \frac{EC_{iw} \cdot 100}{EC_{dw}}$$

in which LR is the "leaching requirement" expressed in percentage of the water applied that must leach through the soil; D_{iw} is the depth in inches of irrigation water applied; D_{dw} is the depth in inches of drainage water leaving the soil; EC_{iw} is the electrical conductivity of the irrigation water, and EC_{dw} is the average electrical conductivity of the drainage water leaving the bottom of the root zone.

In applying the equation, a value is usually assumed for EC_{dw} to represent the maximum soil salinity that can be tolerated.

The leaching requirement in relation to crops having different sensitivities to salt is illustrated in Figure 7.2.

A process which could economically remove salt from water for irrigation, or in other ways make low-quality waters more satisfactory for irrigation, would greatly help to increase water supplies in arid regions. Waters low in total salt but high in sodium are often improved through addition of finely

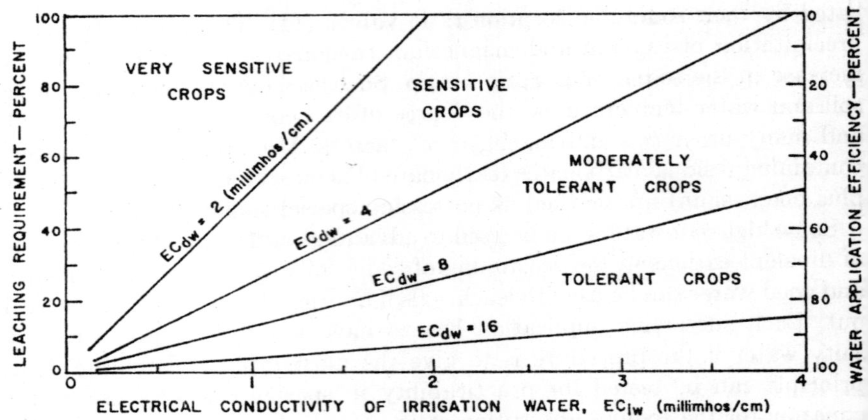

Figure 7.2. Leaching requirement as related to crop salt tolerance. (*From Reeve 1957*)

powdered gypsum. The added calcium reduces the sodium adsorption ratio. This practice is not effective with waters of medium or high salt content.

Research currently sponsored by the United States Department of the Interior (42) on the demineralization of saline waters is of great interest to irrigation agriculture. Demineralization processes being investigated include vaporization procedures, methods for crystallizing salts, adsorption and diffusion phenomena, osmosis, ion exchange, precipitation, electro-ion migration, streaming potentials, electrostatic and electromagnetic effects and use of ultrasonics and ultrahigh frequency currents. Proposals involve the use of such low-cost energy sources as sunshine, tides, and wave action. At present no process appears economically feasible for treating water for irrigation.

REFERENCES

1. Adams, F., and Evans, C. E., *Soil Sci. Soc. Am. Proc.* **26**, 355 (1962).
2. Bear, F. E., *et al.*, *Soil Sci.* **65**, 1 (1948).
3. Bear, F. E., *et al.*, *Soil Sci.*, **73**, 1 (1952).
4. Bernstein, L., "Salt-affected Soils and Plants," p. 139, Paris, UNESCO Symposium, 1960.
5. Bower, C. A., Reitemeier, R. F., and Fireman, M., *Soil Sci.* **73**, 251 (1952).
6. Bower, C. A., and Hatcher, J. T., *Soil Sci.* **93**, 275 (1962).
7. Bradfield, R., *Soil Sci. Soc. Am. Proc.* **6**, 8 (1942).
8. Buehrer, T. F., and Williams, J. A., *Arizona Agr. Expt. Sta. Bull.* **64** (1936).
9. Carter, D. L., *U.S.D.A. Agricultural Research Service* **41** (1962).
10. Clark, W. M., "The Determination of Hydrogen Ions," 3rd ed., Baltimore, Williams & Wilkins Co., 1923.

11. Coleman, N. T., Kamprath, E. J., and Weed, S. B., *Advan. Agron.* **10**, 475 (1958).
12. Coleman, N. T., Williams, D. E., Nielsen, T. R., and Jenny, H., *Soil Sci. Soc. Am. Proc.* **15**, 106 (1950).
13. Dole, M., "The Glass Electrode," New York, John Wiley & Sons, Inc., 1941.
14. Eaton, F. M., *Soil Sci.* **69**, 123 (1950).
15. Fireman, M., and Wadleigh, C. H., *Soil Sci.* **71**, 273 (1951).
16. Gardner, R., and Whitney, R. S., *Soil Sci.* **56**, 63 (1943).
17. Harris, A. E., *Soil Sci.* **32**, 435 (1931).
18. Harward, M. E., and Coleman, N. T., *Soil Sci.* **78**, 181 (1954).
19. Hayward, H. E., and Wadleigh, C. H., *Advan. Agron.* **1**, 1 (1949).
20. Hissink, D. J., *Soil Sci.* **45**, 83 (1938).
21. Hsu, P. H., and Rich, C. I., *Soil Sci. Soc. Am. Proc.* **24**, 21 (1960).
22. Jackson, M. L., *Soil Sci. Soc. Am. Proc.* **27**, 1 (1963).
23. Jenny, H., *Soil Sci. Soc. Am. Proc.* **25**, 428 (1961).
24. Kelley, W. P., "Calif. Agr. Expt. Sta. Bull **617**," 1937.
25. Kelley, W. P., "Cation Exchange in Soils," New York, Reinhold Publishing Corporation, 1948.
26. Kelley, W. P., "Alkali Soils," New York, Reinhold Publishing Corporation, 1951.
27. Krishnamoorthy, C., and Overstreet, R., *Soil Sci.* **69**, 41 (1950).
28. Larson, G. A., *Wash. Agr. Expt. Sta. Bull* **376**, 1939.
29. Low, P. F., *Soil Sci. Soc. Am. Proc.* **19**, 135 (1955).
30. Marshall, C. E., "The Colloid Chemistry of the Silicate Minerals," New York, Academic Press, Inc., 1949.
31. Hieman, R. H., *Botan. Gaz.* **121**, 279 (1952).
32. Overstreet, R., Martin, J. C., and King, H. M., *Hilgardia* **21**, 113 (1951).
33. Pratt, P. F., and Thorne, D. W., *Soil Sci. Soc. Am. Proc.* (1948), 13, 213 (1949).
34. Reeve, R. C., and Bower, C. A., *Soil Sci.* **90**, 139 (1960).
35. Reeve, R. C., Third Congress of International Commission of Irrigation and Drainage, San Francisco, Calif., Question 10R.10, 10. 175–10.187 (1957).
36. Russell, E. W., "Soil Conditions and Plant Growth," 9th ed., New York, John Wiley & Sons, Inc., 1961.
37. Schofield, R. K., and Taylor, A. W., *Soil Sci. Soc. Am. Proc.* **19**, 164 (1955).
38. Shoemaker, H. E., McLean, E. O., and Pratt, P. F., *Soil Sci. Soc. Am. Proc.* **25**, 274 (1961).
39. Thorne, D. W., and Peterson, H. B., "Irrigated Soils, Their Fertility and Management," 2nd ed., New York, Blakiston Co., 1954.
40. Thorne, J. P., and Thorne, D. W., "Utah Agr. Expt. Sta. Bull. 346," 1951.
41. Truog, E., "Mineral Nutrition of Plants," p. 23, Madison, University of Wisconsin Press, 1951.
42. United States Department of the Interior, "Demineralization of Saline Waters," Washington, D. C., 1952.
43. United States Salinity Laboratory Staff, *U. S. D. A. Handbook* **60**, (1954).
44. Van Beekon, C. W. C., *et al.*, *Netherlands J. Agr. Sci.* **1**, 153 (1953).
45. Whitney, R. S., and Gardner, R., *Soil Sci.* **55**, 127 (1943).
46. Wilcox, L. V., *U. S. D. A. Tech. Bull.* **962** (1948).
47. Wilcox, L. V., *Agr. Information Bull.* **197**, 6 (1958).

CHAPTER 8

Trace Elements in Soils

ROBERT L. MITCHELL

Department of Spectrochemistry
The Macaulay Institute for Soil Research, Aberdeen, Scotland

In the crust of the earth, the ten elements, O, Si, Al, Fe, Ca, Na, K, Mg, Ti, and P, constitute over 99 per cent of the total element content. It is with the behavior in soils of the remaining elements, whose individual contributions do not exceed 1000 ppm, that this chapter is concerned. At present, the trace elements recognized as essential for the healthy development of animals, plants, or microorganisms are Co, Cu, Mn, Mo, Zn, S, Se, B, F, and I, with the possibility that V and Cl should be added (6, 7, 23, 48). Of these, the essentiality of Co for plants and Se for animals has only recently been established; others will no doubt be added in due course. Many trace elements, whether or not they are essential, are noxious above quite low levels. Plants or animals derive these trace elements directly or indirectly from the soil, and an understanding of their behavior is, therefore, of considerable importance.

Until sensitive modern methods of analysis became available, most investigations concerned with trace element problems dealt with individual elements, and even now the information available is to a large extent restricted to those elements which are readily determined by spectrochemical (5) or other physicochemical methods. As a result of the widespread adoption of such methods, comparable results are now becoming available for soils of different types from many parts of the world. There is still a need for a comprehensive survey of many elements of possible biological significance by methods giving a hundred or a thousand times the sensitivity conveniently available at present. This requirement is illustrated by the present shortage of information on Se in deficient soils, or even on such elements as W, Ge, Nb, Au, Ag, or Hg.

It is not only because of their biological functions that the study of trace elements in soils is important. A knowledge of their behavior during soil profile development can contribute to the solution of complex pedological problems and help to indicate the nature of the processes involved. Trace element analyses of soils and plants have been applied in the search for new sources of economically important metals such as Cu, Zn, Pb, Mo, and Ni by the techniques of biogeochemical surveying. A knowledge of the behavior of the radioactive isotopes of such elements as Sr, Cs, and I is of medical importance; there have been numerous investigations of the soil content and plant uptake of these isotopes, but limitations of space prevent their discussion here.

In considering the information currently available, it must be appreciated that some of the earlier work may be of dubious value, because the methods employed may have been unreliable and because errors arising from contamination and interference effects were not fully appreciated. For instance, most determinations of water-soluble B in soils are almost certainly high if the soil samples were transported in paper containers. Many analyses have probably been carried out on samples that were not representative and probably there have been assumptions made regarding the uniformity of the soil parent material throughout the profile that are not justified.

In addition to the references cited, use has been made in this chapter of many other sources and of unpublished results obtained mainly by spectrochemical methods at the Macaulay Institute for Soil Research*.

SOURCE OF TRACE ELEMENTS

The trace element content of a soil is dependent almost entirely on that of the rocks from which the soil parent material was derived and on the processes of weathering, both geochemical and pedochemical, to which the soil-forming materials have been subjected. The more mature and older the soil, the less may be the influence of the parent rock. The effects of human interference are generally of secondary importance.

The rules governing the geochemical distribution of trace elements in igneous rocks are reasonably well understood. The processes of physical and chemical weathering that lead to the formation of sedimentary rocks are sufficiently clear to enable generalizations to be made regarding the

* A well-documented review (20A) of the behavior of biologically important trace elements has appeared while this book was in press.

probable distribution of trace elements in such rocks. A short review of these processes is necessary to explain the distribution of trace elements in different soil parent materials (16, 30).

Igneous rocks make up some 95 per cent of the crust of the earth. Of the remaining 5 per cent of sedimentary origin, about 80 per cent are shales, 15 per cent sandstones, and 5 per cent limestones. The surface distribution shows a somewhat greater proportion of sediments, as these tend to form a relatively thin skin overlying the igneous rocks from which they were derived.

The following are the contents in the crust of the earth of the eight elements which are the major constituents of the main groups of rock-forming minerals, and as such play an important part in guiding the distribution of many of the trace elements in rocks:

Element	Weight %	Volume %	Element	Weight %	Volume %
O	46.6	92.0	Ca	3.6	1.48
Si	27.7	0.80	Na	2.8	1.60
Al	8.1	0.77	K	2.6	2.14
Fe	5.0	0.68	Mg	2.1	0.56

Oxygen thus contributes almost half of the weight and occupies over nine-tenths of the volume. The most probable values for the average abundance of the less-common constituents of the lithosphere (Table 8.1) demonstrate how elements of biological importance, such as B, Co, Cu, I, Mn, Mo, Ni, Se, V, and Zn, are widely scattered throughout the order of abundance and how some are no more plentiful geochemically than many whose occurrence in soils has scarcely been reported.

TABLE 8.1. AVERAGE ABUNDANCE OF TRACE ELEMENTS IN THE CRUST OF THE EARTH*

	ppm		ppm		ppm		ppm		ppm
Ti	4400	Ni	100	Th	12	Yb	2.7	I	0.3
P	1200	Zn	80	B	10	Er	2.5	Bi	0.2
Mn	1000	Cu	70	Ge	7	Br	2.5	Tm	0.2
F	800	Li	65	Sm	6.5	Mo	2.3	Cd	0.2
S	520	Ce	42	Gd	6.4	Ta	2.1	In	0.1
Cl	480	Sn	40	Be	6	Ho	1.2	Se	0.09
Ba	430	Co	40	Pr	5.5	Eu	1.1	Ag	0.02
C	320	Y	28	Sc	5	W	1	Pd	0.01
Rb	280	Nd	24	As	5	Sb	(1)	Pt	0.005
Zr	220	Nb	20	Hf	4.5	Tb	0.9	Au	0.001
Cr	200	La	18	Dy	4.5	Lu	0.8	Ir	0.001
Sr	150	Pb	16	U	4	Hg	0.5	Rh	0.001
V	150	Ga	15	Cs	3.2	Tl	0.3	Re	0.001

* After V. M. Goldschmidt. Courtesy A. Muir, editor, and Clarendon Press, Oxford, publishers of "Geochemistry."

TABLE 8.2. DISTRIBUTION OF TRACE ELEMENTS IN IGNEOUS ROCKS, ILLUSTRATED FROM THE SKAERGAARD INTRUSION OF E. GREENLAND

Element	Original magma	Gabbro-picrite	Olivine gabbro	Hypersthene olivine gabbro	Olivine-free gabbro	Ferrohortonolite ferro-gabbro	Basic hedenbergite granophyre	Hedenbergite granophyre	Acid granophyre
	%	%	%	%	%	%	%	%	%
SiO_2	47.9	41.3	46.2	46.4	48.2	44.6	52.1	58.8	75.0
MgO	7.8	27.1	10.0	9.6	5.3	1.7	1.1	0.72	0.15
K_2O	0.19	0.13	0.12	0.20	0.14	0.35	1.38	2.39	3.85
	ppm	ppm	ppm	ppm	ppm	ppm	ppm	ppm	ppm
Ga	17	10	15	23	20	20	30	50	20
Cr	170	2000	300	175	<1	<1	<1	<1	5
V	140	150	200	225	250	<5	<5	<5	20
Mo	<1	3	<1	<1	<1	<1	<1	3	3
Li	2	1	3	2	3	2	10	30	5
Ni	170	1000	400	135	50	<2	5	8	5
Co	53	80	70	55	30	20	8	4	3
Cu	130	200	50	80	150	600	300	300	20
Sc	12	<10	10	20	30	15	10	<10	<10
Zr	50	30	50	35	50	20	200	2000	1000
Mn	800	1240	1160	700	920	1620	1500	1500	80
Y	<30	<30	<30	<30	<30	<30	200	200	200
La	<30	<30	<30	<30	<30	<30	40	50	200
Sr	350	100	300	700	600	500	300	500	150
Ba	43	20	10	25	60	50	500	600	1500
Rb	<20	<20	<20	<20	<20	<20	30	40	300

These elements are not uniformly distributed throughout igneous and sedimentary rocks, and the factors governing their distribution are primarily responsible for the differences found in soils from different parent materials. The magnitude and the course of the differentiation that can occur in igneous rocks is illustrated in Table 8.2, which presents some results from a series of igneous rocks formed by fractional crystallization of the basic (basaltic) magma which produced the Skaergaard Intrusion in East Greenland (51). These rocks vary in composition from early crystallizing, ultrabasic (Mg-rich, SiO_2-poor) through basic and intermediate stages to late crystallizing, acidic (SiO_2-rich) rocks similar in composition to granites. Although this intrusion shows some unusual features, for the present purpose it can be assumed that rather similar factors have operated to give a corresponding trace element differentiation throughout most igneous rocks. Many trace element analyses of rocks substantiate this assumption.

Significant findings in Table 8.2 include: the preference of Co, Ni, and particularly Cr for the more basic rocks; the greater abundance of V in the intermediate rocks in which the ratio of Ni to Co may fall below unity; the almost uniform distribution of Cu except in the ferrogabbro, in which the presence of sulfide has been established; the occurrence of Mo in the ultrabasic and in the more acidic rocks; and the higher contents of Ba, Rb, Y, La, and Zr in the acid rocks. Manganese is fairly uniformly distributed. Figures for B, Zn, and Se are not available for these samples, but analyses of other rock series indicate that contents of about 30 ppm B are common in basic rocks, and that in acid rocks less than 10 ppm B is a normal content. Zinc occurs in many basic rocks at 100 to 200 ppm; in acidic rocks the content may be below 50 ppm. The Se content of igneous rocks is reported to be between 0.5 ppm in basic and 0.1 ppm in acidic rocks.

The geochemical distribution of trace elements in igneous rocks can be explained by a study of the contents of the individual constituent minerals of the rocks. Table 8.3, in which one of the rocks of Table 8.2 is considered in more detail, shows that certain trace elements display very pronounced preferences for specific minerals. When the errors arising from spectrochemical determinations and from the assessment of the proportions of the different minerals making up the rock are taken into account, many of the trace elements in the rock are accounted for wholly in the three minerals examined. Exceptions are Cr, V, Sc, Zr, and Cu, which probably also occur in accessory minerals such as spinels and iron oxides. Barium in the plagioclase is probably overestimated.

TABLE 8.3. DISTRIBUTION OF TRACE ELEMENTS IN THE CONSTITUENT MINERALS OF AN IGNEOUS ROCK, ILLUSTRATED BY A HYPERSTHENE OLIVINE GABBRO FROM E. GREENLAND

Element	Constituent minerals (ppm)			Total from minerals (ppm)	Found in rock (ppm)
	Plagioclase	Pyroxene	Olivine		
Ga	55	3	3	28	23
Cr	<1	350	<1	100	175
V	10	250	<5	75	225
Mo	<1	2	10	2	<1
Li	4	2	3	3	2
Ni	<2	150	350	120	135
Co	<2	50	125	40	55
Cu	40	35	20	35	80
Sc	<10	30	<10	8	20
Zr	<10	50	<10	14	35
Mn	<10	2100	1700	940	700
Sr	2000	<10	10	950	700
Ba	110	5	5	60	25

TABLE 8.4. IONIC RADII (r) AND IONIC POTENTIALS (z/r) OF SOME CONSTITUENTS OF MINERALS

Mineral constituent	r*	z/r*	Mineral constituent	r*	z/r*
Cs^{+}	1.65	0.61	Fe^{2+}	0.83	2.4
Rb^{+}	1.49	0.67	Cu^{2+}	0.83	2.4
Tl^{+}	1.47	0.68	Co^{2+}	0.82	2.4
Ba^{2+}	1.43	1.4	Ni^{2+}	0.78	2.6
K^{+}	1.33	0.75	Mg^{2+}	0.78	2.6
Sr^{2+}	1.27	1.6	Li^{+}	0.78	1.3
La^{2+}	1.22	1.6	Mo^{4+}	0.68	5.9
Hg^{2+}	1.12	1.8	Fe^{3+}	0.67	4.5
Y^{2+}	1.06	1.9	V^{3+}	0.65	4.6
Ca^{2+}	1.06	1.9	Cr^{3+}	0.64	4.7
Na^{+}	0.98	1.0	Ti^{4+}	0.64	6.3
Cu^{+}	0.96	1.0	Ga^{3+}	0.62	4.8
Mn^{2+}	0.91	2.2	Al^{3+}	0.57	5.3
Te^{4+}	0.89	4.5	Si^{4+}	0.39	10
Zr^{4+}	0.87	4.6	P^{5+}	0.35	15
Hf^{4+}	0.86	4.7	B^{3+}	0.20	15

* r is expressed in Ångstrom units, and z equals the ionic charge. Throughout the discussion, these values of r and z/r based on Goldschmidt's initial determinations (16) are employed: more recent studies, e.g., by Ahrens (1), have suggested that more precise values are now available, but these do not alter the general trend of the argument.

The explanation of this distribution lies in the crystal structure of the rock-forming minerals. Ferromagnesian minerals, such as olivines and pyroxenes, include Mg^{2+} (ionic radius 0.78 Å) or Fe^{2+} (0.83 Å) in a silicate lattice structure that can accommodate cations of approximately 0.8 Å radius. Any other cation of suitable radius present in the crystallizing magma can equally well occupy this place in the lattice. Ga^{3+} (0.62 Å), on the other hand, takes the place of Al^{3+} (0.57 Å) in aluminosilicates, but does not occur in silicates. Barium and Sr can occur in K- and certain Ca-minerals, whereas Cr and V find a place in the more complex silicates, such as pyroxene, but not in the simpler olivines. The ionic radii, which are the main factors in this distribution, are detailed in Table 8.4.

In the formation of a crystal of a mineral species, for instance of olivine, it is not primarily a lattice containing Mg^{2+} or Fe^{2+} that is being formed, but one that can accommodate virtually any cation of appropriate radius. Goldschmidt has suggested that the size tolerance is 14 per cent. Preference is for ions of smaller radius or greater charge, the excess charge being balanced by some complementary substitution elsewhere. Another factor, sometimes called, for want of more precise definition, chemical affinity, prohibits such substitutions as Zr^{4+} for Te^{4+}. Thus divalent Ni, Co, and

Zn may take their place in ferromagnesian minerals and are found principally in ultrabasic or basic rocks, whereas monovalent Li occurs in the ferromagnesian minerals of the later crystallizing, more acid rocks; Cr in ultrabasic rocks occurs as chrome spinels, in later crystallizing rocks, it enters iron oxides; V occurs in iron oxides and pyroxenes; Mo occurs, probably replacing Si, in olivine and later as accessory molybdenite; Cu occurs in both felspars and ferromagnesian minerals, possibly replacing both Na^{+} and Fe^{2+}, until the composition of the residual magma brings about the separation of a sulfide phase, in which other chalcophile elements, such as Zn, Co, Ni, and V, may also be enriched.

These considerations serve to explain the location of the major portion of the trace elements in the unweathered minerals of a soil derived from an igneous rock. The initial stages of weathering are common to the formation of either a soil or a sedimentary rock. Physical weathering, the first and simplest stage in the breakdown, will lead generally to the formation of sediments containing some or all of the original minerals, with their trace elements. Such rocks as conglomerates and sandstones fall into this category.

Chemical weathering (21) inevitably involves a solution process, with subsequent precipitation influenced by such physicochemical factors as ionic potential (Table 8.4), pH value, oxidation-reduction potential, and colloidal conditions. These govern the differentiation in most sedimentary rocks. Some formations, including coals, oil shales, and certain limestones, also involve biological factors.

The rate of weathering of the individual constituent minerals of igneous rocks governs the rate at which the trace elements become available to the sediments being formed. The rate of breakdown is affected to some extent by the conditions prevailing, but for particle sizes above colloidal dimensions a generalized scheme can be given. This is illustrated in Table 8.5 which, in addition to presenting the approximate order of stability of the commoner rock-forming minerals, details the trace elements most likely to be present and suggests the order of release of these elements in a soil containing such minerals. It is significant that the bulk of such biologically important trace elements as Zn, Mn, Co, and Cu occurs in the more easily weathered constituents of igneous rocks.

The fate of the trace elements released during the chemical weathering of igneous rocks is of particular interest because of the great importance of sedimentary rocks, especially the shales or hydrolyzates, as soil parent materials over great areas of the earth's surface. Table 8.6 traces the process of sedimentation, showing the principal products and the trace elements commonly associated with them. This table merely indicates where relative concentration occurs, taking into account the terrestrial abundance of the

TABLE 8.5. RELATIVE STABILITY OF THE COMMONER MINERALS OF IGNEOUS ROCKS AND THE TRACE ELEMENTS WITH WHICH THEY ARE ASSOCIATED

Stability	Mineral	Major constituents	Trace constituents
Easily Weathered ↑	Olivine	Mg, Fe, Si	Ni, Co, Mn, Li, Zn, Cu, Mo
	Hornblende	Mg, Fe, Ca, Al, Si	Ni, Co, Mn, Sc, Li, V, Zn, Cu, Ga
	Augite	Ca, Mg, Al, Si	Ni, Co, Mn, Sc, Li, V, Zn, Pb, Cu, Ga
	Biotite	K, Mg, Fe, Al, Si	Rb, Ba, Ni, Co, Sc, Li, Mn, V, Zn, Cu, Ga
	Apatite	Ca, P, F	Rare earths, Pb, Sr
	Anorthite	Ca, Al, Si	Sr, Cu, Ga, Mn
	Andesine	Ca, Na, Al, Si	Sr, Cu, Ga, Mn
	Oligoclase	Na, Ca, Al, Si	Cu, Ga
Moderately	Albite	Na, Al, Si	Cu, Ga
Stable	Garnet	Ca, Mg, Fe, Al, Si	Mn, Cr, Ga
	Orthoclase	K, Al, Si	Rb, Ba, Sr, Cu, Ga
	Muscovite	K, Al, Si	F, Rb, Ba, Sr, Ga, V
	Titanite	Ca, Ti, Si	Rare earths, V, Sn
	Ilmenite	Fe, Ti	Co, Ni, Cr, V
	Magnetite	Fe	Zn, Co, Ni, Cr, V
↓	Tourmaline	Ca, Mg, Fe, B, Al, Si	Li, F, Ga
Very	Zircon	Zr, Si	Hf
Stable	Quartz	Si	

elements in question, and mention of an uncommon constituent does not necessarily imply that it is present in an amount greater than an unmentioned more common element. The general course of distribution is governed by the ionic potentials (Table 8.4). Elements with values below 2 tend to remain in ionic solution during the processes of weathering and transportation; those between 2 and 12 are precipitated in the hydrolyzates; above 12 they form soluble oxyacid anions.

The important sediments from the soil point of view are the shales, which, as already mentioned, constitute some 80 per cent of the total. Sandstones are composed almost entirely of minerals that weather with difficulty and would, *a priori*, be expected to produce soils deficient in available trace elements, quite apart from such accentuating factors as low clay content and high drainage loss. Many reports of plant and animal deficiencies are in fact from sandstone areas. In sandstones, the trace elements most likely to be observed are enumerated in Table 8.6. Limestones and dolomites are of more interest as soil additives than as soil parent materials. The importance of ore deposits is very local: in soils in the immediate neighborhood very high contents of trace elements may be found.

The hydrolyzate sediments may be divided into shales and bauxite

TABLE 8.6 DEPOSITION OF SEDIMENTARY ROCKS WITH THEIR ASSOCIATED TRACE ELEMENTS

Process of sedimentation	Major constituents	Type of product	Main rock types	Associated trace constituents
↓	Si	→ Resistates	Sandstones	Zr, Ti, Sn, Rare earths, Th, Au, Pt, etc.
	Al, Si, K	→ Hydrolyzates	Shales and bituminous shales	V, U, As, Sb, Mo, Cu, Ni, Co, Cd, Ag, Au, Pt, B, Se
			Bauxites	Be, Ga, Nb, Ti
	Fe, Mn	→ Oxidates	Iron ores	V, P, As, Sb, Mo, Se
			Manganese ores	Li, K, Ba, B, Ti, W, Co, Ni, Cu, Zn, Pb
	Ca, Mg, Fe	→ Carbonates	Limestones, dolomites	Ba, Sr, Pb, Mn
	K, Na, Ca, Mg	→ Evaporates	Salt deposits	B, I
	Na, Mg	→ Sea		B, Al, I, Br, F, Rb, Li

deposits. The shales, in turn, may be subdivided into those of essentially inorganic origin, comprising mainly, in order of abundance, the illite, montmorillonite, chlorite, and kaolinite groups of clay minerals, which also make up the colloidal fraction of most soils, and the bituminous shales of biological origin. Formation of clay minerals in a medium in which the trace elements of the original igneous rocks are present gives conditions in which the trace elements can be incorporated in the crystal lattice or adsorbed as exchangeable cations. It has been reported that in the montmorillonite structure Al can be replaced by Fe^{3+}, Mg, Zn, and small amounts of Li, Cr, Mn, Cu, or Ni. Similar substitutions can occur in the hydrous micas. These clay minerals have high cation-exchange capacities, and the power with which such trace elements as Cu, Ni, Co, and Mn are adsorbed is such that they might be expected to occur as rather difficultly extractable exchangeable cations. There would be considerable interest in establishing their exact location. Exchangeable cations would be much more readily displaced than those built into the lattices of the clay minerals, which are rather slowly weathered, being, in fact, end products of geochemical and pedochemical weathering.

The manner of formation makes it more difficult to state the normal contents of the more important trace elements for shales than it is for specific igneous rocks, but the orders of the values to be anticipated are (in ppm): B, 20 to 200; Co, 10 to 50; Ni, 30 to 150; Cr, 100 to 500; V, 100 to 250; Zn, 20 to 100; Mn, 500 to 5000; Cu, 20 to 200; Se, <0.1 to 5; and Mo, <1 to 5, with occasional values for Se and Mo as high as 100 ppm. Enrichment of certain elements, including Se, Mo, and U, is often, but not always, encountered in bituminous shales, possibly as a result of sulfide precipitation. In bauxitic and lateritic deposits, Be and Nb are among the trace elements that are relatively enriched.

The effects of metamorphism on igneous and sedimentary rocks may be briefly considered at this point. The application of pressure and heat to the argillaceous sediments results primarily in a recrystallization, during which the clay minerals are transformed into minerals more closely related to those of igneous rocks, in which adsorbed ions will become incorporated in the crystal lattice, and their availability will depend once more on the rate of weathering of the minerals concerned. Boron, for instance, is incorporated in the very resistant tourmaline crystal and becomes much less available. In slates, schists, and gneisses the same trace elements can be expected as in their parent rocks, except where thermal effects may have introduced or removed certain of the more volatile constituents. A Scottish limestone with a B content approaching 1 per cent has been observed near a basic igneous intrusion, and high contents of ore minerals of Sn, Pb, or Zn around granitic intrusions may be of similar origin. On the other hand, the high contents of such elements as Li, Rb, Cs, Be, Sc, Zr, Mo, In, W, Nb, Ta, U, Th, and the rare earths in some pegmatitic veins are of residual origin, being the last dregs from the process of fractional crystallization, disposing of those elements for which provision had not previously been made, and which had accumulated in the slowly cooling siliceous, increasingly water-rich, magma.

TOTAL TRACE ELEMENT CONTENT OF SOILS

The foregoing discussion of the distribution of trace elements in the rocks that form soil parent materials enables one to anticipate with some degree of accuracy the approximate total content of many of the trace constituents to be expected in any particular soil and to recognize those parent materials in which deficiencies or excesses may occur. This has been well illustrated by the results for over a hundred top soils of varied igneous origin from the southwest Pacific (53).

This correlation is most obvious in areas where the soils are young and have undergone little weathering. Where the soils have developed on parent materials that have undergone several cycles of geological weathering over a period measured in millions of years, or where sedimentary rather than igneous rocks are widespread, the correlation may be less obvious, because of the difficulty of determining the precise nature of the parent rock. Any estimation of the trace element content of a soil from knowledge of its parent material must refer to the soil as a whole; in some soil types the amounts in individual horizons can be modified by pedological factors.

The total trace element contents of the surface layers of nine typical Scottish arable soils from different geological parent materials (Table 8.7) reveal the operation of the principles that have been discussed (44). In the first four soils, derived directly from igneous rocks, Co, Ni, Cr, and Mn

TABLE 8.7. TOTAL CONTENTS OF TRACE ELEMENTS IN SOME SCOTTISH SURFACE SOILS ON PARENT MATERIALS DERIVED FROM DIFFERENT ROCK TYPES

Element	Total trace element content (ppm oven-dry matter)								
	Serpentine	Olivine gabbro	Andesite	Granite	Granitic gneiss	Quartz mica schist	Shale	Sandstone	Quartzite
Ag	<1	<1	<1	<1	<1	<1	<1	<1	<1
Ba	1000	2000	600	2000	3000	1500	500	400	2000
Be	<5	<5	<5	<5	5	5	<5	<5	5
Co	80	40	8	<2	10	25	20	<3	20
Cr	3000	300	60	5	200	150	200	30	250
Cu	20	40	10	<10	25	100	10	<10	40
Ga	30	50	30	25	70	40	20	15	60
La	<30	<30	25	30	200	100	50	40	30
Li	30	30	50	7	70	200	60	20	15
Mn	3000	5000	800	700	1000	3000	1000	200	1000
Mo	1	2	<1	<1	<1	5	<1	<1	1
Ni	800	50	10	10	40	80	40	15	50
Pb	20	20	40	20	70	80	20	<20	60
Rb	60	70	30	600	500	200	250	<30	300
Sc	<10	10	<10	<10	<10	10	15	<3	<10
Sn	<5	<5	<5	<5	<5	<5	<5	<3	<5
Sr	300	700	60	300	700	150	70	100	300
V	100	200	100	20	250	200	200	60	250
Y	50	30	30	30	80	100	70	25	70
Zr	1000	300	200	>1000	800	>1000	300	600	>1000

all occur in greater amount in the two soils of basic-igneous origin; V, Cu, and Sr are highest in the olivine gabbro; and Zr, La, and Rb highest in the granite. In the andesite the Co:Ni ratio of approximately one is typical of soils derived from intermediate rocks. The difference between the granite and the fifth soil from granitic gneiss, is considerable, particularly for Cr, Co, Ni, and V, and suggests that trace element contents may help to distinguish granitic rocks that may have been derived by metamorphism from sedimentary rock such as shale. The findings explain why some granitic areas carry Co-deficient stock while others are free from this disorder.

The remaining four soils are derived from rocks of sedimentary origin. All except the sandstone have been subjected to some metamorphism. The quartz mica schist appears to have been formed in a medium somewhat higher in such elements as Cu, Mn, Mo, and Be than the others. The sandstone is typically low in most constituents. Some sandstones have higher contents of Pb and other elements, derived from percolating waters following deposition: such accumulation may arise in a zone overlying a less-pervious clay layer. The quartzite soil is almost certainly somewhat contaminated with argillaceous material, and as a result the Co and Ni contents are higher than often found.

The figures in Table 8.7 give some idea of the normal total contents to be expected in soils from the commoner soil parent materials, although again it appears that Ba may have been overestimated in several of the soils. Values reported from other areas, for instance from Florida (2, 15), are of a like order. The extreme ranges reported in the literature are much wider (42) and are presented in graphical form in Figure 8.1, which indicates on a logarithmic scale usual and extreme ranges for the more important trace constituents of mineral soils of all types.

The total content can be a reasonable indication of the trace element status of a soil. For certain elements this content may vary from soil to soil, according to its origin, by more than a thousandfold, and ranges of a hundredfold are common. In consequence the total trace element content can be more useful as an index of potential availability than is the corresponding figure for a major constituent, which seldom varies from soil to soil by more than fivefold. The total content is particularly useful as an indicator of the possibility of excess of such elements as Ni or Zn, although even in such cases the availability must be considered. A study (38) of sixteen trace elements in ten New Jersey soil types and of their uptake by corn leaves has indicated that, although the total supply of trace elements in soils has a definite bearing on the amount delivered to the plant, factors which influence the available state, such as pH, mineral form, and solubility, are more important in regulating absorption.

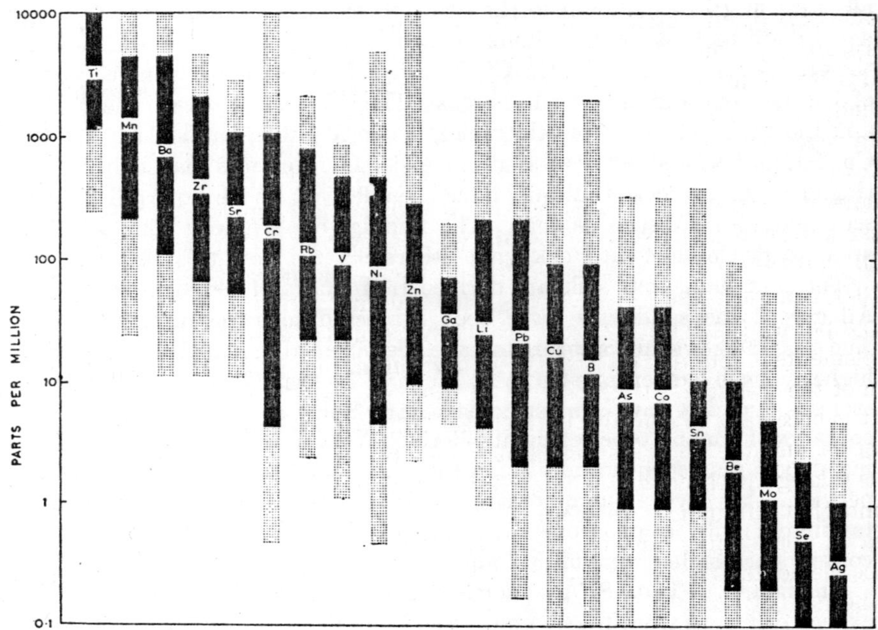

Figure 8.1. Range of contents of some trace elements commonly found in mineral soils. Stippled sections indicate more unusual values; certain extremely high values from localities influenced by ore deposits have been ignored.

LOCATION OF TRACE ELEMENTS IN SOILS

Investigation of the availability to plants of the trace elements present in soils demands some knowledge of the forms in which these elements can be expected to occur and of the proportions of the different elements present in these forms. A considerable proportion of elements that are constituents of the crystal lattices of the rock-forming minerals will occur in this form in soils unless weathering has resulted in the disintegration of these minerals. The trace elements so bound up must be considered unavailable until released by weathering. The amounts present in this form depend on the nature of the parent rock and the weathering cycles through which the material making up the soil parent material has passed. The proportion will normally be greatest in young soils derived from igneous rocks.

Distribution in Size Fractions

The contribution of the different particle-size fractions tc the total and acetic-acid-extractable contents of trace elements in four Scottish soil profiles has been examined in a series of investigations at the Macaulay Institute (9). The various fractions were separated by repeated shaking and sedimentation in water in order to minimize loss of the less firmly bound trace elements. The amounts of the clay, silt, and sand fractions obtained did not differ significantly from the results of mechanical analysis following more vigorous dispersion. The soils were freely and poorly drained profiles developed on boulder clays derived from basic igneous and granitic rocks and represent relatively young soils. They contain 75 to 90 per cent sand (2.0 to 0.02 mm), 10 to 15 per cent silt, and 5 to 10 per cent clay (<0.002 mm) on the former parent material and 5 to 10 per cent silt and 3 to 8 per cent clay on the latter. It is impossible to present the results for 20 elements in four size fractions of all horizons of four profiles: some indication of the trends must suffice.

It is necessary to consider both the actual trace element content of each fraction and the contribution it makes to the content of the whole soil, thus taking into account the relative amounts in each fraction. It has been found that throughout these profiles the sand fractions, because of their greater abundance, almost invariably contribute the major part of the trace elements. They contain 95 per cent of the Sr, and 60 to 90 per cent of the Ba, Ga, Mn, Ti, Y, and La. Only in the poorly drained profiles does the contribution of the sands occasionally fall below 60 per cent for Co, Ni, Cr, V, Li, and Rb, with contributions as low as 20 to 30 per cent for Li and Rb. The sands generally contain less than 50 per cent of the Cu, with the clay in the poorly drained profiles contributing up to 70 per cent. In the poorly drained soils the lower values from the sands can be ascribed to the breakdown of ferromagnesian and other readily weathered minerals. Most of the trace elements so released appear to be transferred to smaller-size fractions.

In the clays from the poorly drained soils, the actual Co and Ni contents may be up to 20 times, and the Cr and V contents up to 5 times, those in the sands. The differences between sands and clays in the freely drained profiles are generally less. The factor for Rb and Li is 5 to 10 times, except in the freely drained soil derived from basic igneous rock where no concentration in the clay occurs. The Cu contents of the clays are considerably higher than those of the sands even in the freely drained soils. The contents of Ba and particularly Sr are lower in the clays than in the sands. Some typical results for Sr and La are given in Table 8.8. In the surface horizons, the differences are sometimes smaller than those quoted, possibly because the presence of organic matter complicates the redistribution of the mobilized cations and also because of the effects of cropping and cultivation. This is illustrated in the La results in Table 8.8. In contrast to La, the contents of Y and Zr in the clay fractions are not greater than those in the

Table 8.8. Examples of Total Trace Element Contents in the Various Size Fractions of Two Soils

Depth (in.)	Total trace element content (ppm oven-dry matter)					
	Coarse sand	Fine sand	Silt	Clay	Whole soil	
					Calculated	Determined
Sr in a freely drained profile on basic igneous parent material						
1–4	600	800	80	20	510	400
6–10	400	800	60	6	410	600
12–15	600	1000	100	6	720	800
26–30	600	1500	400	15	1060	1000
34–38	600	1500	300	15	800	1000
La in a freely drained profile on granitic parent material						
2–6	10	30	30	15	21	30
9–13	10	40	40	14	20	30
22–26	10	30	60	150	25	30
30–34	10	30	60	150	24	30
37–41	15	30	60	200	29	30

sands by more than a factor of two, with their highest contents generall in the fine sand or silt fractions, where their unweathered primary mineral are presumably most abundant. The Ti contents are highest in the sand of the basic igneous and in the silts or clays of the granitic soil.

These findings serve to indicate the distribution of the total trace elemen contents in the size fractions of slightly podzolized soils derived directl from typical igneous rocks. Unfortunately no equally comprehensive in formation is yet available for soils of sedimentary origin or for other type of pedological weathering.

The percentage contribution of the different-size fractions to the acetic acid-extractable trace elements in the whole soil presents a somewhat dif ferent picture. Generally 40 to 50 per cent of the extractable Co, Ni, an Cr (and Fe) is contributed by the clay fraction, although this fractio contains only some 10 per cent of the total content of these elements in th freely drained soils and up to 30 per cent in the poorly drained profile Only 20 per cent of the extractable V and Ti comes from the clay, th largest contribution being rather unexpectedly from the sands, although th clays have the highest extractable V and Ti contents. Acetic-acid-extract able contents of Co and V in Table 8.9 illustrate typical results for the actu contents in different-size fractions as distinct from their contributions to th

TABLE 8.9. EXAMPLES OF ACETIC-ACID-EXTRACTABLE TRACE ELEMENT CONTENTS IN THE VARIOUS SIZE FRACTIONS OF TWO SOILS

Depth (in.)	Acetic-acid-extractable trace element contents (ppm oven-dry matter)					
	Coarse sand	Fine sand	Silt	Clay	Whole soil	
					Calculated	Determined
Co in a poorly drained profile on granitic parent material						
4–8	1.10	0.99	2.47	3.40	1.54	1.32
12–16	0.12	0.23	1.33	3.12	0.32	0.31
20–24	0.08	0.17	0.71	1.61	0.25	0.28
28–34	0.09	0.33	1.76	5.32	0.75	0.70
38–44	0.08	0.13	0.88	3.80	0.41	0.40
V in a freely drained profile on granitic parent material						
2–6	0.17	0.33	0.94	2.33	0.55	0.43
9–13	0.14	0.23	0.82	1.13	0.26	0.28
22–26	0.13	0.22	0.67	1.68	0.24	0.27
30–34	0.12	0.22	0.93	1.58	0.23	0.28
37–41	0.14	0.21	0.89	1.22	0.24	0.24

amount extracted from the whole soil. For V the contributions of all fractions generally vary between 20 and 40 per cent. Wherever results are available for extractable Zn, the highest contents are in the clays, at 10 to 20 times those of the sands.

While there are very considerable differences between the total contents of most trace elements in soils of basic igneous and granitic origins (see Table 8.7) and in the contents and contributions of the different-size fractions, there are marked similarities between them when ratios of total to acetic-acid-extractable contents are considered. On the other hand, significant differences exist between freely drained and poorly drained profiles whatever their parent material. Thus, in the whole soils the total to extractable ratios for Co and Ni are generally between 20 and 200 in the freely drained profiles and 2 and 20 in the poorly drained. The same differences apply in the sands, although all the clays show ratios between 2 and 20. The ratios for V in the whole soils (and sands) are 200 to 600 and 25 to 100, respectively; in the clays the ratios are 50 to 200 and 8 to 50. For Cr there is little difference due to the effects of drainage apart from a high ratio of 400 in the sands and whole soils from the freely drained basic igneous material. Otherwise the ratios for Cr are 50 to 200 in the whole soils, 30 to 250 in the sands, 25 to 200 in the silts, and 10 to 50 in the clays.

In order to interpret such findings, it is necessary to consider the forms in which the trace elements released from the constituent minerals of the soil by weathering may be retained in the soil. If they are admitted into the crystal lattices of clay minerals in the course of formation, they return to a relatively unavailable form. On the other hand, if they are adsorbed at exchange active sites on the surface of these clay minerals they may remain relatively readily extractable. There is now considerable evidence that trace elements may also be coprecipitated with or adsorbed by sesquioxides, chiefly amorphous ferric oxide or goethite, which form a surface coating on soil minerals or occur as discrete particles. There appears to be a considerable proportion of amorphous material in many soil clays. It is improbable that, in normal circumstances, any of the biologically important trace elements are precipitated to any significant amount as sulfides or other insoluble salts. This is supported by the finding that in sediments the precipitation of Zn as sulfide is of subsidiary importance, although it is the least readily adsorbed of the cations other than the alkalies and alkaline earths. It is possible that Pb may be immobilized as sulfate or carbonate. If release occurs in a horizon containing organic material, a portion of the trace elements may be chelated or adsorbed in some form by organic or organic-mineral complexes.

Occurrence in Clay Minerals

The clay minerals that constitute the main inorganic exchange material include illite, montmorillonite, kaolin, and occasionally vermiculite or hydrobiotite. The cation-exchange capacity ranges from 1 meq/g for montmorillonite to less than 0.1 meq/g for some kaolins. It is sufficient for the present purpose to assume that the cation-exchange capacity arises partly from excess negative charges due to substitution in the Si-tetrahedron and Al-octahedron layer lattices, these charges being neutralized by the exchangeable cations in the interlayer spaces, and partly from negative charges of terminal O ions in the lattice edges. Different binding forces at these locations could partially account for variations in the relative extraction rates of different trace element cations by different reagents.

Comprehensive studies of the behavior of trace elements in such exchange materials at the levels present in soils are lacking. It has been shown that the apparent cation-exchange capacity of a clay is greater when determined for elements such as Zn, Cu, Fe^{3+}, or Th, which form weak bases, suggesting that these elements may in fact form basic salts with the clay in which only part of the valency goes to the exchange complex. Lanthanum, a relatively strong base, gives a similar cation-exchange capacity to that obtained with alkalies and alkaline earths. This effect, if it occurred at the very low concentrations of the trace elements, might be of importance in availability. There is little doubt that most of the biologically important cations are very readily adsorbed by the exchange-active clay minerals and are correspondingly difficult to displace; the order of difficulty of displacement is approximately Cu > Pb > Ni > Co > Zn > Ba > Rb > Sr > Ca > Mg > Na > Li but the relative positions in such a series vary with concentration in the solution and the nature of the exchange active material.

In ionic solution the same general order can be expected for ease of adsorption by exchange-active material. It has however been shown that the order of cation exchange on to kaolin is Th > La > Ca > K in the absence of any complexing agent, but becomes K > Ca > La > Th in the presence of citrate or fluoride ions. It is probable that complexing agents in soils can similarly disturb the anticipated order of ease of adsorption of trace elements by clay minerals. Organic acids, such as fulvic acid, may act as such complexing agents.

On release by weathering, other trace constituents probably appear as complex anions. This group includes Mo, B, Se, and V. If not lost in drainage, these may be precipitated by, for instance, excess Ca, be bound in organic complex, or, as has been suggested for Mo, be adsorbed by anion-exchange materials. Molybdate appears to be less strongly bound in anion exchange than is phosphate or sulfate, but more firmly than are halides or borate. Investigations into anion exchange in soil minerals, particularly kaolin, have dealt almost entirely with phosphate; presumably exchange with OH groups occurs.

Occurrence in Sesquioxides

Knowledge of the trace element fraction associated with precipitated sesquioxides is based largely on a series of investigations carried out at Rothamsted Experimental Station (24, 31, 32). Extraction of surface and illuvial horizons of two grey-brown podzolic soils derived from rocks of sedimentary origin by acid ammonium oxalate resulted in the dissolution of amorphous Fe_2O_3 (but little Al_2O_3) and its associated trace elements to the extent of about 5 per cent of the soil by weight. Some 60 per cent of the total Fe and Co in the soil, but 50 per cent of the Cu and Pb, and 80 per cent of the Mn was extracted, as well as 40 per cent of the V and 20 per cent of the Cr and Ni. In addition to the trace elements associated with the iron oxide, this reagent extracts the adsorbed and some of the organically bound fractions. At least 600 times as much Fe, Cr, V, and Ti was found to be extracted by oxalate as by acetic acid. For Co, Ni, and Cu and for Pb in the surface horizons the ratio was around 100 and for Mn in the surface horizon around 10. The oxide layer coats particles of all size fractions, and most elements, with the exception of Sr and of Ga in the clay, were found to be more abundant in the extracted material than in the original soil or any of the cleaned soil fractions. The sand and silt contained most of the Ba, Sr, and Ti. Only for Cu and Ga was the content of the cleaned clay more than twice the total soil content. It would appear, therefore, at least in soils of this type derived from sedimentary rocks, that reported accumulations of trace elements in the clay fraction may be related to the oxide coating on the clay minerals. Some Russian investigations suggest that the nature of the mineral on which the coating is deposited affects its stability. The film deposited on bentonite was found to be thicker than that on kaolin, but was more readily dissolved by oxalate. This must affect the release of any trace elements sorbed by the ferruginous coating. Equally important must be any factors which hasten or retard crystallization of goethite. A linear correlation between Mo and Fe extracted from German soils at different oxalate concentrations suggests that Mo is bound largely in the iron-rich material attacked by oxalate.

It is interesting to compare the ratios of total to acetic-acid-extractable trace elements from Scottish soils derived from igneous rocks with the oxalate-to-acetic acid ratios from English soils of sedimentary origin. The soils from the poorly drained profiles contain a much greater proportion of readily extractable trace elements than do the freely drained soils from Scotland or the sedimentary soils from England. The implications of this on availability will emerge later, but it is relevant at this point to mention some further Rothamsted investigations, dealing with the effect of fermenting plant materials on mobilization or dissolution of precipitated iron and other metallic oxides, of such iron oxides containing trace elements and of trace elements in some selected soils. Such conditions cannot be considered equivalent to those arising during gleying in lower horizons, where little organic matter is present, but somewhat similar factors may operate in the surface horizons of poorly drained soils.

Preliminary studies on individual trace element oxides showed that under anaerobic conditions, in the presence of fermenting plant materials, there was little or no mobilization of Ga, Mo, Ti, or Cr. Appreciable mobilization, with more in a fermenting than in a sterile medium containing plant material, occurred with Zn, Mn, Co, Ni, and Pb, although for Pb there was evidence of subsequent precipitation as carbonate in the fermenting medium. Appreciable mobilization also occurred with Cu, but was greater in the sterile medium. With mixed precipitates, at a ratio of 1:1, Cu, Zn, and Mn were more readily, and Co and Ni less readily, mobilized than Fe. With a 50-fold excess of Fe the relative trace element mobilization was reduced.

The extent of sorption of some trace elements, particularly Co and Ni, from solution in such plant material extracts by precipitated iron oxide was found to be high compared with their coprecipitation with ferric oxide from such extracts. In precipitation tests it was found that Mo, V, Pb, and Sn, could be almost completely coprecipitated with ferric oxide, but Co, Ni, and Mn were only slightly (5 per cent) precipitated, and Cu and Zn occupied an intermediate position. In the absence of precipitated iron oxide, over 90 per cent of Cu, Zn, Co, Ni, and Mn remained in solution when the fermented extract was oxidized by allowing access of air, but 80 per cent of Fe was precipitated under such conditions from a ferrous solution. When the adsorptive power of precipitated Fe_2O_3 was tested by adding it to trace elements mobilized in the fermenting extracts, it was found that up to 50 per cent of the Cu, Zn, Co, and Ni present was removed from solution at pH 7; at lower pH values the adsorption decreased. The same effect was shown by a sand coated with ferric oxide.

Tests by dialysis indicated that almost the whole of the trace elements mobilized by fermenting plant material was in true solution, while the stability of the solutions at increasing alkalinity showed that most of the trace elements were complexed rather than in ionic solution. All the elements tested, except Co, were taken up more readily by cation-exchange resin from inorganic solution than from the fermented extract, again indicating that the metals were complexed. Molybdenum and vanadium apparently persist in their anionic forms during the fermentation and extraction. Plant fibers irreversibly sorb Cu and Ni from such fermentation solutions.

In a parallel Rothamsted study on selected soils, it was found that Fe, Co, Ni, Zn, Pb, V, and Mo were mobilized on anaerobic incubation with plant material. Chromite was unreactive but Cr was mobilized from a normal soil. Evidence again indicated that Mo remained in an anionic form. In a freely drained soil Cu was mobilized on incubation alone, but was removed by the plant fibers if plant material was added. In an organic, poorly drained soil, water-soluble Cu was released but soon disappeared; sorption on plant fibers was not considered responsible in this instance. Repeated flooding and drying increased the extractability of iron and trace elements. On subsequent oxidation the total amounts of extractable Mn, Co, and Zn were hardly affected, but there was evidence of formation of $PbCO_3$.

These investigations indicate that during soil development, an appreciable proportion of the trace elements may become associated with precipitated iron and other oxides from which they can be mobilized by fermenting plant materials. This may be more important in soils derived from sedimentary rocks which have passed through several cycles of weathering than in younger soils derived mainly by mechanical comminution from igneous rocks. The presence of such material protecting the surface of clay minerals must restrict the adsorption of cations or anions by these minerals. More information is needed before the relative importance of trace elements adsorbed by clay minerals or bound by sesquioxides can be assessed.

Occurrence in Organic Complexes

In the upper layers of mineral soils, organic materials may make an appreciable contribution to the capacity to fix trace elements by adsorption or complex formation. Such effects must predominate in peat and organic soils. The nature of the organic compounds taking part in such reactions has been intensively investigated, and it appears that the humic acid and fulvic acid fractions contribute most to the ability of soil organic matter to remove metallic ions from solution (10, 25, 29). Generally the metals involved are Fe^{3+} and Al^{3+}, as these are the most abundant of the readily complexed cations, but many of the trace elements can presumably take their place. Humic acid forms complex compounds with Al^{3+} and Fe^{3+} but not with Ca^{2+}, tending to precipitate when chelation occurs. Fulvic acid has a higher complexing capacity, and its chelates are normally water-soluble. It displays a greater capacity for the elution of cations from an exchange resin, suggesting a greater role in soil-forming processes. The proportion of fulvic acid is generally greater in the lower horizons. In fulvic acid the bulk of the O occurs in functional groups, compared with only some 60 per cent in this form in humic acid. The greater acidity of the former is due to the presence of more COOH groups, with which the major part of the cation-exchange capacity of soil organic matter is probably associated. The extent of adsorption of these cation chelating agents by soil clays and sesquioxides to form organic-mineral complexes appears to vary with pH and the nature of the exchangeable alkali and alkaline earth cations present. It is reported that divalent Zn- or Mn-EDTA chelates are not fixed by clays but that Zn- or Mn-APCA (Chelate 138) complexes are.

It has been suggested by Russian workers that in gley soils up to 90 per cent of the Fe^{2+} present forms stable Fe-organic compounds. This complexed Fe can be extracted by water-soluble organic compounds formed by the anaerobic decomposition of plant residues. Presumably certain trace elements will behave similarly. The complexes are stated to be with polyphenols, amino acids, acid polysaccharides, and fulvic acids.

Humus from the illuvial layers of podzols is flocculated in acid solution by Al^{3+} or Fe^{3+}, but Ni^{2+} is not active below pH 8. This may explain the very slight concentration of many trace elements in the B_1 horizon of podzols. In a spectrophotometric investigation of the fulvic acid fraction from peat, at least one Cu chelate, and three or four Co and Cr chelates were reported. The suggestion that to complex Cu, both SH and COOH groups are required, has not been confirmed, but there is evidence that SH groups produce very stable Cu complexes. The NH_2 group may also be involved in chelation.

For Cu the optimum acidity for complex formation is pH 2.5 to 3.5 with humic acid, and pH 6 with fulvic acid. Other investigations have shown that Cu is adsorbed as $CuOH^+$ by carbonyl groups, and as Cu^{2+} by phenyl groups, the extent being governed largely by pH, with a limiting pH below which adsorption does not occur. Complexes of Cu with humic acid from peat have been shown to contain Cu which is available to oat plants, in addition to some unavailable Cu.

The occurrence of Mo in soil organic matter has been investigated by several workers, but the form in which it is complexed has not been elucidated. The presence of plant-available organic-complexed Mo in some soils with moderate organic matter contents is suggested by high plant uptake in conditions of quite high acidity.

In an investigation of the biogeochemistry of Ge it has been found that some 32 per cent of the amount taken up from solution as a stable compound by peat was in the humic acids, 46 per cent in the fulvic acids, and the remainder in the peat residue. Maximum uptake by humic acids was obtained, as with Cu, at high acidity (pH 1 to 2.4). The existence of these Ge complexes is of interest because the only detection of Ge in Scottish soils has been in the surface organic horizons of some podzols.

There is some evidence that cations that form weak bases are, in excess of cations that form strong bases, more readily sorbed by peat than by montmorillonite. In considering the organic-matter–trace-element relationships it is interesting to note that while none of the organic material from the A_0 horizon of a podzol is dialyzable through a cellulose membrane, much of that from the B horizon is, and that 90 per cent of the latter is extracted by EDTA at pH 7.

The availability to plants of some forms of chelated trace elements has been suggested by experiments in solution culture carried out at the Macaulay Institute. It was found that Ni complexed with EDTA is not taken up, but that such trivalent cations as Ga, Cr, and Al, when so chelated, are taken up into the leaf, although this does not occur when they are present in ionic form. It appears that the nature of the chelating agent is important, the uptake of trivalent cations being reduced when chelated with DTPA.

The behavior of Co, which can probably occur in both divalent and trivalent form, is intermediate. This uptake of chelated trivalent trace elements may explain the relatively high plant uptake of such elements as Ti and V from certain soils, in which suitable chelating agents may be present (52).

EXTRACTION OF TRACE ELEMENTS FROM SOILS

The uptake of trace elements by plants is the obvious criterion of their availability. But no two species of plants growing in the same soil remove identical quantities of the various trace elements, and the relative amounts taken up by different species vary from soil to soil because of such factors as composition, pH, moisture status, fertility level, and the content of the element being considered. For instance, at low soil-Cu levels, grasses contain more Cu than clovers, while at high levels the position is reversed. Other complicating factors include seasonal changes in plant composition and variations within the plant. In exceptional instances, thousandfold variations may occur from species to species and a fivefold variation is not unusual. The availability of a trace element in a soil, as shown by the capacity of a plant to take it up, is in fact as much a function of the plant as of the soil. Even a biological extractant such as *Aspergillus niger* cannot give a consistently reliable result for all types of soil or for the requirements of different crops.

It is therefore desirable to be able to make some arbitrary assessment of availability by an extraction technique. Further, by appropriate choice of extractant some indication can be obtained regarding the form in which trace elements occur and the nature of the pedological changes taking place within the profile. The important fractions include the water-soluble, exchangeable, and organically complexed portions. The amounts coprecipitated with sesquioxides or bound in the crystal lattices of silicate minerals are probably less significant in the assessment of availability.

A number of trace elements, including Fe, Cu, Mn, Mo, Zn, B, and Co, are taken up by microorganisms (41). It is possible that the availability of trace elements to plants may be affected by assimilation by microorganisms in the soil, and also that the plant-microorganism balance may be affected by a deficiency or an excess of trace elements. Information on the full implication of these effects is lacking, but it has been suggested that certain deficiencies, notably of Cu and B, in plants may be aggravated by the level of the population of soil microorganisms.

The elements that may occur as complex anions, including Mo, B, and Se, are frequently assessed on the basis of their water solubility. Hot-water extraction generally gives a good indication of the B status of a soil, and it has been reported that much of the Mo in Mo-rich soils from California is water-soluble. In organic soils, water extraction may not be as effective as in mineral soils. There are considerable differences between the forms in which these anionic elements occur in soils. In a study of Latvian soils (35), the mobility of B was assessed by water solubility, while for that of Mo, extraction by oxalate at pH 3.3 was preferred, since it was considered that both water-soluble and adsorbed Mo is available to plants and that the hydrated oxides are active in adsorbing Mo. It was suggested that organically bound Mo becomes available only after mineralization, and that humic acids, free from sesquioxides, do not bind Mo, a finding not in accord with other evidence. The low availability of Mo in acid podzolic soils was considered to be due to its fixation by oxides of Fe and Al. Water-soluble B can range from less than 1 per cent of the total content in soils in humid areas to the bulk of the content in arid areas.

For the determination of adsorbed cations, the choice of extractant must be guided by the degree of displacement which is required. This can be illustrated by comparing the effectiveness of ammonium acetate of pH 7 and acetic acid at pH 2.5. The relative amounts of adsorbed cations displaced by these reagents depend on their positions in a series similar to that given on p. 337 and on the nature of the exchange complex. The amounts of trace elements extracted by ammonium acetate are generally quite small. Considerably larger amounts of such elements as Co, Ni, Pb, Zn, Cu, Mn, Ti, V, and Cr, as well as of Al and Fe, are extracted by acetic acid: the increase may be of the order of 10-fold. The extraction of a small quantity of Al and Fe by acetic acid suggests some dissolution of precipitated sesquioxides. There is, in fact, in many Scottish soils, little difference in the amounts extracted by acetic acid and by hydrochloric acid at pH 2.5. For the most readily displaced elements, including K, Na, Ca, and in most instances Mg, the amounts removed by all three extractants are similar. The strength with which these cations are bound is such that they can be completely displaced by ammonium ions. For the cations of the heavy metals, factors including ionic potential, hydration, and size of the ions produce much stronger binding. Soils from which more Mg is extracted by acid extractants are those derived from ultrabasic rocks containing readily weathered ferromagnesian minerals, mainly olivine, which are susceptible to attack by acids. If this effect were the explanation of increased extraction of Co and Ni by acids, in all such cases an increase in Mg would be anticipated.

The behavior of Mn might have been expected to resemble that of the alkaline earths, but as this element occurs in soils largely in the form of insoluble oxides, its increased extraction in acids can be explained. The plant availability of Mn in neutral or alkaline conditions is generally small. Ammonium-acetate- or acetic-acid-soluble Mn does not appear to be related to availability, but extraction with a neutral salt such as $Ca(NO_3)_2$ after treatment with a weak reducing agent like alcoholic quinol, or with acid phosphate, appears to give a value closely related to plant availability. In neutral or alkaline soils in Wisconsin more reliable correlation with deficiency in oats has been obtained with H_3PO_4 extraction than with a reducing extractant or with EDTA, better results being obtained with dried than with moist soils. The reducible or phosphate-extractable surface layers of manganese oxides appear to provide plant-available Mn.

In Scotland, for arable soils of pH 5 to 6 a reasonable degree of correlation between acetic-acid-extractable Co and plant uptake has been found, but treatment with small amounts of water-soluble Co (as chloride or sulfate in a phosphatic fertilizer), insufficient to cause a detectable change in the amount extracted by acetic acid, can increase the plant uptake severalfold, so that absolute correlation cannot be expected by this method on treated soils. Account must also be taken of soil acidity and organic matter content in interpreting the analytical results.

The possibility of organically complexed trace elements being taken up by plants has already been mentioned. In order to get some measure of the amount present in the soil in this form, a sufficiently powerful chelating agent is employed as extractant. Satisfactory correlation has been obtained between the uptake of Cu by red clover from Scottish soils and the amount extracted by 0.05 *M* EDTA, but not by water or acetic acid. Soils carrying Cu-deficient oats generally contain less than 0.6 ppm Cu extractable by EDTA. Correlation has been reported between the plant uptake of Zn and the amount extracted by dithizone.

It is reported that the uptake of a number of trace elements, including Cu, Zn, and Se, can be reduced by phosphate fertilization, although Fe and Mn are apparently less affected. Such effects may depend to a considerable extent on other soil conditions. Liming increases the uptake of Mo but reduces that of Co, Ni, and Mn. The availability of a number of trace elements in American organic soils at different soil pH values has been examined (26). The uptake of Mn by plants begins to fall off above pH 5.5 to a minimum at pH 7.5, above which availability increases to a second maximum above pH 8.5. The behavior of B is similar, although the fall in uptake does not commence until pH 6.0 is reached. The availability of Cu is maximal at pH 5.0 to 6.0 and falls off slowly above, and more quickly below, this range. For Zn the availability falls off slowly above pH 6, and for Mo a fall occurs from pH 5.5 downwards.

Occasionally mixed reagents have been employed; for instance an extract containing 2.5 per cent acetic acid and dithizone has shown a significant correlation between the Co content of the leaves of swamp black gum and the amount extracted from the soil.

The amounts of trace elements normally extracted from surface layers of arable Scottish soils by 2.5 per cent acetic acid at pH 2.5 are within the following ranges (in ppm): Co, 0.05 to 2.0; Cr, <0.01 to 1.0; Cu, <0.05 to 1.0; Mn, 5 to 100; Mo, <0.02; Ni, 0.1 to 5.0; Pb, <0.2 to 4.0; Sn, <0.5; Sr, 0.2 to 10; Ti, <0.1 to 1.0; V, <0.05 to 1.0; and Zn, <2 to 30. Of these, only Sr is extracted in like amount and Mo in greater amount by neutral normal ammonium acetate, Sr falling into line with Ca despite its presence at less than one-hundredth of the content. For Mn and Zn the extraction is at least three times as great in the acid reagent, in which most other elements are extracted at about ten times the amount. Results from other temperate countries suggest that, in general, similar ranges hold for this type of extractant. Limited information on tropical soils indicates that in acetic acid extracts the amounts of some trace elements may be less than the lower limits detailed above.

In Scottish soils the amounts extracted by 0.05 *M* EDTA are generally within the following ranges (in ppm): Co, <0.05 to 4.0; Cr, 0.1 to 4.0; Cu, 0.3 to 10.0; Mn, 5 to 100; Mo, <0.03 to 1.0; Ni, 0.2 to 5.0; Pb, 1.0 to 10.0; Sn, <0.5; Ti, 0.5 to 10.0; V, 0.2 to 5.0; and Zn, <3 to 20. While in most instances the levels do not differ greatly from the amounts taken out by dilute acetic acid, there are a few points which warrant comment. A much larger amount of Mo is extracted. The amount of Co extracted from freely drained soils is less than that taken out by acetic acid, while in the more organic-rich surface horizons of poorly drained arable soils, the Co extracted by EDTA may be several times that removed by acetic acid. Similar remarks apply to a greater or less extent to Ni and Zn. The amounts of Cu and Pb extracted from surface horizons by EDTA are always considerably greater, possibly because these elements occur largely in an organically complexed form, while the EDTA-extractable Mn is less, because it occurs largely as oxides.

Such limited information as is available for extraction of Scottish soils with 0.1 *N* HCl indicates that, while for trace elements the increase compared with 2.5 per cent acetic acid may be up to tenfold, the amount of Fe extracted may increase by a hundredfold.

When the ability of an extractant to diagnose deficiencies or excesses is being assessed, it is necessary to consider its performance with different soil types on different parent materials. Some extractants may be efficient for soils of one type on one parent material but fail completely with another. Similarly, in some circumstances correlation may be obtained between total soil content and plant uptake on a single soil type. Numerous instances of claims based on inadequate experience occur in the literature, and many of the contradictions can be reconciled by the appreciation of the various forms in which trace elements can occur.

TRACE ELEMENT BALANCE IN SOILS

Human interference with the trace element status of soils brings about decreases in content because of crop and stock removal or increases as a result of intentional or accidental additions. Irrigation and drainage, tillage, deforestation, and vegetation burning all effect changes which could affect trace element balance.

A summary (43) is now available of the information scattered throughout the literature on the quantities of trace elements in soil additives. It appears that the amounts present are seldom sufficient to affect the total content of a soil significantly but may on occasion modify the soluble or available fraction.

The ameliorant applied in greatest amount to soils in the podzolic zone is some form of lime or limestone, at the rate of about 2 parts in 1000 of the cultivated layer every 5 to 10 years. In general, limestones are exceptionally free from trace constituents other than alkalies and alkaline earths. Calcic limestones commonly contain about 1000 ppm Sr; dolomites generally contain less, sometimes much less, than 100 ppm; Mn occurs at levels between 100 and 2000 ppm; B seldom exceeds 10 ppm; and Co and Ni do not exceed 5 ppm except in impure limestones such as cementstones containing appreciable amounts of argillaceous material. If quartz is present it is accompanied by some Zr. A content of 10 ppm in a limestone amounts to a 0.02 ppm addition to the soil, and is therefore negligible in comparison with total soil contents.

The trace element contents of commercial fertilizers are generally quite low. The contents of the common types would be expected from the results available to fall within the ranges quoted in Table 8.10. Trace elements present in ammonium and alkali salts will generally be in a soluble form, but those in phosphates and slags may well be unavailable to plants. The values most worthy of comment are the B content of sodium nitrate and the high Mn, Cr, V, and Ti contents of slags, which might cause difficulty if present in an available form. The rate of application of fertilizers seldom exceeds one-tenth of that of limestones, although the frequency is greater. The amounts of trace elements in fertilizers must depend to a considerable extent on the manufacturing processes involved and on the source of the raw materials.

Much higher contents are often found in waste materials such as sewage

TABLE 8.10. RANGE OF TRACE ELEMENT CONTENTS EXPECTED IN THE MORE COMMONLY USED FERTILIZERS

Element	Range of element contents in fertilizers (ppm)					
	Sodium nitrate	Ammonium sulphate	Rock phosphate	Superphosphate	Basic slag	Potassium salts
B	100–500	0–10	10–50	0–20	10–100	0–500
Cu	1–20	1–10	1–50	10–100	10–100	0–10
Co	0–5	<1	1–10	1–10	1–10	<1
Ni	<1	<1	1–10	1–10	1–10	1–10
Pb	<1	<1	1–10	5–100	1–20	<1
Mo	<1	<1	1–20	1–20	1–20	<1
Zn	1–10	1–50	50–1000	50–1000	10–100	0–10
Mn	1–50	1–50	10–1000	10–1000	1%–5%	0–10
V	<1	<1	10–1000	50–2000	1000–5000	<1
Cr	<1	<1	100–500	10–500	1000–5000	1–10
Ti	<1	<1	10–1000	10–1000	500–5000	<1

sludges, in which amounts may be as great as 1 per cent Cu and Zn, around 0.1 per cent Ni, Sn, Pb, Cr, and Ba, and up to 100 ppm Co. Their composition, and that of dusts from town refuse incinerators, depends on local industrial activities, and the possibility of noxious levels must be kept in mind. A further factor to be considered is atmospheric pollution. Acetic acid extraction of surface soil (sampled to a depth of 8 inches) from the downwind side of a group of metal refineries in England removed up to 1.5 ppm Co, 8 ppm Ni, 8 ppm Pb, 1 ppm V, and 1000 ppm Zn, compared with values of 0.3 ppm Co, 2 ppm Ni, 2 ppm Pb, 0.2 ppm V, and 30 ppm Zn for the soil protected from major pollution by the direction of the prevailing wind. Ammonium-acetate-extractable Mo rose from less than 1 ppm to more than 40 ppm. Effects due to Se from factory smokes and dusts are also possible, as is danger of F from certain industrial processes, although this may be more probably a direct surface contamination of vegetation. There is some evidence of increased Pb content in the ashes of plants adjoining major highways, presumably because of the presence of tetraethyl lead in gasoline. Contents above ten times normal are reported.

In order to counteract deficiencies in soils it is common practice to add trace elements, incorporated into fertilizers in amounts sufficient to supply 0.4 pounds per acre Co, 5 pounds per acre Cu, 0.2 pounds per acre Mo, or 40 pounds per acre Mn. Two pounds per acre correspond to approximately 1 ppm. The production of commercial fertilizers containing several trace elements is not to be recommended, as all are seldom deficient simultaneously, and it is undesirable to add trace elements in available form to soils which do not require them. This is particularly so for elements like Mo and Se, which are readily taken up by plants in amounts toxic to stock.

Trace element excess may arise from the repeated application of insecticides or fungicides. Reports of high Cu contents in surface soils come from vineyards (305 to 845 ppm), apple orchards (1500 ppm in surface mat and 120 ppm in top 4 inches), and hop plantations. Arsenic in Canadian orchard soils has been shown to rise to 10 to 125 ppm, with 0.4 to 9 ppm water-soluble, compared with 3.7 to 7.9 ppm in untreated soils in the neighborhood. In the top 2 inches of an apple orchard which had received 21 sprays of phenyl mercury acetate in five years, 0.5 to 1.1 ppm Hg has been found, although no root absorption is reported.

The depletion of trace elements as a result of crop and stock removal is difficult to assess accurately. A hay crop of 3 tons an acre containing in the dry matter 0.1 ppm Co and V; 1 ppm Ni, Mo, Sn, and Pb; 10 ppm Cu; 20 ppm B; 40 ppm Zn; and 100 ppm Fe, and Mn, would deplete the surface soil by approximately one three-hundredth of these amounts. On the basis of these trace element contents and 1000 tons of soil an acre, a growing crop weighing a total of 10 tons an acre would take up trace elements from the surface soil equivalent to only 0.001 ppm Co and V; 0.01 ppm Ni, Mo, Sn, and Pb; 0.1 ppm Cu; 0.2 ppm B; 0.4 ppm Zn; and 1 ppm Fe and Mn. It is of interest to compare these values with the amounts extracted by the commonly used diagnostic extractants. Herbage contents of 0.1 ppm Co are generally obtained from soils from which more than 0.25 ppm is soluble in acetic acid, more than 250 times the 0.001 ppm actually taken from the soil by the plant. Plants showing signs of Ni poisoning contain upwards of 50 ppm, corresponding to a removal of 0.5 ppm from the soil; acetic acid may extract more than 50 ppm Ni from such soils. The amount of Mo extracted from normal soils by ammonium acetate of pH 8 is about 0.03 ppm, of the same order as the plant extraction. Acetic acid removes <0.05 to 1.0 ppm Cu from Scottish soils, again of the same order as the plant requirement, but no relationship between amount extracted and plant content has so far been established for this reagent. On the other hand, EDTA, which shows some correlation, removes 0.3 to 10 ppm Cu from the soil, much more than the plant demands. Water-soluble B in healthy soils is generally greater than 0.4 to 0.8 ppm, in reasonable agreement with plant removal. Acetic-acid-soluble Zn is approximately ten times the plant uptake. The amount of Fe extracted by 2.5 per cent acetic acid is roughly 10 to 100 ppm, much more than the plant removes. The amount of Mn extracted from nondeficient Australian soils by the quinol neutral salt technique generally exceeds 20 ppm, again much more than the plant demands for healthy growth. On the basis of 2 per cent Ca,

3 per cent K, and 0.1 per cent P in the plant, the comparative figures for these elements are 200 ppm Ca extracted by the plant from the soil, compared with 1000 ppm extracted by acetic acid or ammonium acetate; 200 ppm K compared with 40 ppm exchangeable; and 10 ppm P compared with 40 ppm extracted by dilute acetic acid.

These comparisons would appear to suggest that for several of the trace elements the extractants employed are too vigorous for diagnostic purposes. Technical difficulties in the determination of smaller quantities of certain of the elements restrict investigations along these lines, and it seems unwise at present to speculate further.

DISTRIBUTION OF TRACE ELEMENTS IN SOIL PROFILES

Many factors can influence the profile distribution of total and extractable trace elements. It is assumed that the soil parent material is uniform within the limits of accuracy of the analytical technique employed. This assumption can seldom be substantiated, and before any effect is ascribed to pedological action the possibility of nonuniformity of the parent material must be considered. When a profile is developed *in situ* on an igneous rock or on an unsorted sediment such as sandstone or loess, a considerable degree of uniformity can be expected. Reasonable uniformity can also be expected on many glacial drifts, provided the local solid geology is uniform, but mixed drifts derived from a number of rock types can be quite variable, as can soils developed on argillaceous or other water-sorted sediments or on old land surfaces. Many records of trace element contents in soil profiles appear to deal with soils developed on nonuniform parent materials.

The chief pedological factors involved in profile differentiation include the following:

(*a*) Surface enrichment of the trace elements taken up by plants, an effect especially obvious in profiles with surface horizons rich in organic matter.

(*b*) Leaching of mobilized constituents, such as Mn, Li, B, or Se, either completely out of the profile or to zones of accumulation; this is particularly important in tropical soils from humid regions.

(*c*) Translocation, in the course of soil-forming processes such as podzolization, of trace elements together with Fe and Al. Organic and organic-mineral complexes are probably involved.

(*d*) Mobilization of trace elements through breakdown of soil minerals as result of gleying.

(*e*) Mechanical translocation of clay, thereby increasing in the layers of accumulation the contents of those trace elements that are more abundant in clay than in other soil fractions.

(*f*) Surface accumulation of soluble salts, such as borates, in arid regions.
(*g*) Mobilization or fixation arising from microbiological activity.

Surface Enrichment

Effects arising from biological enrichment in the surface horizon are well established. The surface layers of most Scottish soils have a significantly higher Pb content than the underlying layers (44), and similar findings have been reported elsewhere. The absolute increase is generally greatest in soils with high Pb contents and on uncultivated soils, suggesting long-term plant accumulation rather than fertilizer contamination or atmospheric pollution, neither of which could produce increases of up to 70 ppm Pb in areas remote from industrial activity or high traffic density. It is relevant to point out that the contents of most trace elements in leaves, both of agricultural plants and of deciduous trees, fall as the leaves become senescent, but that the content of Pb generally increases. This may be a partial explanation of the greater enrichment of Pb compared with other elements. When the results from uncultivated organic-rich surface horizons are presented on the basis of mineral matter instead of oven-dry material the enrichment applies, in Scottish conditions, to Ag, Cu, Mo, Pb, Sn, and Zn, but not to Co, Ni, Cr, V, Mn, Ti, Ga, Li, Rb, Ba, or Sr. The high content of Ag and Cu in many fungi suggests that these organisms may be active in the mobilization of trace elements in surface horizons.

Podzolic Soils

Typical trace element contents for a Scottish peaty podzol with iron pan are presented in Table 8.11. This soil, developed on a reasonably uniform granitic till, shows in the total contents little evidence of trace element movement. The total Fe contents in the upper and lower A_2, B_1 (iron pan), and B_2 horizons are respectively 0.24, 1.01, 5.37, and 2.0 per cent. Variations of this order in the trace elements would readily be observed by the spectrographic technique employed. In fact, only for Ti and possibly V is there any suggestion of slight translocation. In some other granitic profiles there is a similar increase in Cr. For Co and Ni there is no evidence of accumulation in the B_1 horizon, although in other soils increases of up to double may be observed, compared with twentyfold for Fe. There is, on the other hand, taking into account the organic matter content of the H layer, definite indication of leaching of Mn, Li, and Rb from the upper layers and of some accumulation there of Pb and possibly Cu. The other variations throughout the profile are less than possible analytical and sampling variations. It appears that in this well-developed

TABLE 8.11. TRACE ELEMENT CONTENTS OF A PEATY GLEYED PODZOL WITH IRON PAN, POORLY DRAINED ABOVE AND FREELY DRAINED BELOW PAN, ON GRANITIC TILL

Horizon	H	$A_2(g)$	$A_2(g)$	B_1	B_2	B_3C	C
Depth, in.	7–9	10–14	17–21	21	21–25	31–36	40–46
Loss-on-ignition, %	65.7	5.8	9.5	16.0	6.2	3.4	3.1
Clay, %	—	6.3	11.2	—	12.3	6.3	7.6
pH	4.2	4.4	4.6	—	4.9	4.8	4.9
Element	**Total contents (ppm)**						
Ba	250	800	800	800	800	500	500
Be	8	15	15	20	20	25	30
Co	<1	<3	<3	3	3	3	3
Cr	7	25	30	50	30	20	20
Cu	3	<5	<5	<5	<5	<5	<5
Ga	10	20	40	30	30	40	30
La	50	70	70	40	70	150	150
Li	20	80	80	150	200	200	200
Mn	50	200	300	300	300	500	500
Mo	<1	<1	<1	<1	<1	<1	<1
Ni	4	8	10	10	15	8	10
Pb	50	15	20	20	30	30	20
Rb	400	1000	700	1000	700	1500	1000
Sc	3	<10	<10	<10	<10	<10	<10
Sr	100	100	100	150	100	150	150
Ti	1000	2000	3000	3000	1000	1000	2000
V	15	25	50	80	50	30	50
Y	20	50	30	40	50	30	100
Zr	200	800	800	300	300	500	800
Element	**ppm extracted by 2.5% acetic acid**						
Al	875	695	4060	5850	5550	3130	2920
Co	0.14	<0.03	<0.08	<0.12	<0.11	0.12	0.09
Cr	0.06	<0.03	0.21	0.17	<0.11	0.15	<0.06
Cu	<0.10	0.11	0.13	0.27	0.22	0.13	0.24
Fe	25	30	40	595	100	10	<5
Mn	1.4	<0.8	<0.8	0.84	3.5	4.2	<0.8
Mo	<0.03	<0.03	<0.08	<0.12	<0.11	<0.06	<0.06
Ni	0.33	0.21	0.13	<0.12	<0.11	0.14	0.09
Pb	<0.7	<0.6	<1.6	<2.4	<2.2	0.9	1.8
Sn	<0.2	<0.2	<0.4	<0.6	<0.6	0.3	<0.3
Ti	0.31	0.27	3.5	0.79	1.2	0.44	0.12
V	0.06	<0.03	0.49	<0.12	<0.11	0.08	0.06
Zn	26	<5	<10	<10	<10	<10	<10

podzol which has been in the course of formation for some ten thousand years from a glacially comminuted granitic rock, little major change in trace element content has taken place due to the processes of podzolization. The results for the amounts of trace elements extracted by 2.5 per cent acetic acid are also given in Table 8.11. The presence of large amounts of extractable Fe and Al in the B_1 horizon is expected, but the only other elements to show increases in the B_1 or B_2 horizons are Cu, Mn, Cr, and Ti, to a very much smaller extent. It is also interesting to note the higher contents of extractable Ti, V, and Cr, as well as of Al, in the lower A_2 horizon. Acetic acid extraction may be providing evidence of mobilization and impending translocation; after accumulation, possibly fixed by sesquioxides or organic-mineral complexes, the trace elements may no longer be readily extracted by dilute acid. Results of extraction by acid oxalate or chelating agents are not yet available to confirm this. The amounts involved must however be small relative to the total trace element contents.

In less strongly podzolized arable soils in Scotland similar processes have been operative. There is often a loss of Cr, Ni, and Co from the upper layers of soils derived from ultrabasic rocks. Cultivation destroys the differentiation in the upper horizons, and there is generally a slow fall in acetic acid extraction of Co, Ni, Cu, and Zn with depth. There is often an increased content of extractable Cr, V, Ti, Al, and Fe below plough depth. Manganese does not show the leaching from the surface generally observed in podzols, possibly because of the higher pH.

A series of virgin podzolic soils from the eastern United States on sandstone, carbonate rock, and shale drifts have been examined for Cu, Mn, Zn, Ni, Co, Sn, Pb, Cr, V, and Ga in the A, B, and C horizons (5). True podzols, found on grey acid sandstones low in trace elements, display leaching of all elements except Ti from the A horizon, in which the contents are only one-third to one-tenth those in the C horizon. The clays in the A and B horizons are, on the other hand, two to ten times as rich in trace elements as the whole soil of the C horizon, the clay of which is at least as rich as the upper clays, again excepting Ti. In two podzolic soils on drifts which were calcareous before leaching, the total trace element contents are much higher, and there is considerably less evidence of depletion of trace elements in the A horizon or of enrichment in the clay. Two further grey-brown podzolic soils on a red shale show little or no translocation, although there is up to fivefold enrichment in the clays of the A, B, and C horizons compared with the whole soil in the C horizon.

In two Canadian podzols on sandstone, evidence was found (56) for Pb accumulation at the surface and for leaching from the A_2 horizon of Co, Cu, Zn, Mn, Pb, and Mo, with Zn, Pb, and Cu accumulation in the B horizon, particularly in the better-drained profile. In grey-brown podzolic soils on calcareous parent materials there was less indication of accumulation of trace elements relative to sesquioxides in the B horizon. In brown forest soils on calcareous material, the total contents of the various horizons showed little or no evidence of movement of trace elements.

Similar findings are reported from Czechoslovakia, where some accumulation of trace elements is reported in the B horizon of podzols, while in brown and grey forest soils, and in chernozems and rendzinas, distribution was more uniform throughout the profile. The many reports in the Russian literature present a similar picture of slight trace-element accumulation in the B horizon of some podzols but little or no evidence of movement to the extent found for Fe or Al.

Some podzolized soils on æolian sands overlying buried soils developed on basalt or basaltic tuff, reported from South Australia (46), illustrate the interpretative difficulties with nonuniform parent materials. Some accumulation of Ni and other elements in the hardpan is claimed, but it appears possible from the analytical results that the only effect is a leaching from the surface without subsequent deposition of the leached fraction. This is illustrated by the results for total Ni contents in one ground-water podzol. These rise steadily from <2 ppm at all levels above 40 inches, through 6.0 ppm at 50 inches and 7.7 ppm at 70 inches in the pan, to 18 ppm at 90 inches, 56 ppm at 105 inches, 110 ppm at 120 inches to over 200 ppm at 170 inches. Only Mo shows what may be a true accumulation in the upper part of the hardpan. In a podzol developed entirely on olivine basalt there is a slight increase in the total contents of Co, Ni, Cu, Mo, and other elements in the horizon rich in Fe and Al. As the Zr content of this horizon shows a similar increase some other factor may be operative, but it may be an instance of translocation of Zr.

Consideration of the many results available for podzolized soils suggests that mobilization and translocation of a significant proportion of the total content of certain trace elements may be observed when the parent material is sandstone or other material in which the bulk of the content is in a potentially mobile form. In other parent materials, such as argillaceous rocks and many igneous rocks, the proportion fixed in slowly weatherable minerals may be such that translocation is not apparent from the total contents of the various horizons. Other factors, such as the nature of the vegetation and the age of the soil, may also be relevant.

The Effect of Impeded Drainage

The most marked pedological effect observed in Scottish profiles (44) has been on the content of extractable cations in soils with impeded drain-

TABLE 8.12. TRACE ELEMENTS EXTRACTED BY 2.5 PER CENT ACETIC ACID AND NEUTRAL NORMAL AMMONIUM ACETATE FROM PROFILES DEVELOPED ON BASIC IGNEOUS (OLIVINE GABBRO) PARENT MATERIAL

Depth (in.)	Trace elements extracted (ppm oven-dry matter)*										
	Extractant	Co (40 ppm)	Ni (70 ppm)	Fe (5.10^4 ppm)	Al (1.10^5 ppm)	Cr (200 ppm)	V (200 ppm)	P (2000 ppm)	Mo (1-2 ppm)	Cu (40 ppm)	Mn (6000 ppm)
					Freely drained profile						
0-8	HAc	1.3	1.3	20	2500	0.24	<0.15	3	<0.05	0.13	
	NH_4Ac	0.04	0.11	<5		0.03	0.01		0.02		0.84
12-15	HAc	0.52	0.36	20	4600	0.43	<0.09	10	<0.09	0.21	
	NH_4Ac	0.02	0.02	<5		<0.01	<0.01		<0.01		0.10
16-20	HAc	0.17	0.31	15	3600	0.25	<0.07	41	<0.07	0.28	
	NH_4Ac	0.01	0.01	<5		<0.01	<0.01		<0.01		0.10
20-30	HAc	0.14	0.15	15	3300	0.25	<0.06	162	<0.06	0.17	
	NH_4Ac	0.01	<0.01	<5		<0.01	<0.01		<0.01		0.21
36-40	HAc	0.15	0.12	15	1500	0.05	<0.03	218	<0.03	0.16	
	NH_4Ac	0.02	<0.01	5		<0.01	<0.01		<0.01		0.16
					Very poorly drained profile						
0-6	HAc	1.5	4.1	25	550	0.18	1.2	3	<0.02	0.63	
	NH_4Ac	0.08	0.22	<5		<0.01	0.02		0.02		3.3
9-12	HAc	2.5	4.2	500	840	0.14	3.8	51	<0.05	1.4	
	NH_4Ac	0.37	0.44	5		<0.01	0.03		0.05		1.6
14-18	HAc	2.2	6.7	20	475	0.25	1.1	4	<0.02	0.33	
	NH_4Ac	0.37	0.59	<5		<0.01	0.04		0.02		>10
24-29	HAc	2.1	3.9	75	410	0.14	1.7	203	<0.02	2.7	
	NH_4Ac	0.76	1.1	<5		<0.01	0.04		0.01		>10
36-40	HAc	18	22	2500	765	0.32	1.7	58		6.2	
	NH_4Ac	13	15	<5		<0.01	0.02		<0.01		>10

* Approximate total content throughout profile is shown (in parentheses) for each element.

age. Table 8.12 illustrates typical results obtained from two adjoining arable soil profiles developed on glacial till composed mainly of basic igneous (olivine gabbro) rocks. The mobilization of Co, Ni, Fe, V, Cu, and Mn in the gleyed sub-surface layers of the poorly drained profile is remarkable, particularly the fact that much of the Co and Ni (20 to 40 per cent of the total content) is soluble in ammonium acetate at pH 7. The reduction of Al solubility in gleyed horizons is another feature consistently found. Water extracts of the gleyed layers remove appreciable amounts of Co, Ni, and Fe^{2+}.

Other gleyed profiles have shown increased solubility of Mo (in ammonium acetate) and Ag and Pb (in acetic acid) in horizons with impeded drainage. Similar effects have been observed on soils derived from granites (cf. Co in Table 8.9) and other igneous rocks and from slates, in which at one location at comparable depths, acetic acid extraction of V rose from 0.08 to 4.3 ppm and of Pb from <0.1 to 1.9 ppm as a result of gleying.

This increased mobilization in subsurface gleyed horizons does not appear to be related to the influence of organic material. Examination of profiles on granitic and basic igneous parent materials has shown that in the lower gleyed horizons of poorly drained soils acetic acid and EDTA extractions remove similar amounts of all the trace elements studied, although in the surface horizons of poorly drained profiles EDTA extracts at least twice as much Co, Cr, Cu, Fe, Ni, Ti, V, and Zn. These elements are presumably present to a considerable extent in organically bound form in the somewhat organic-rich surface layers of poorly drained soils. On the other hand, much less Mn is extracted by EDTA. Acetic acid generally extracts more than EDTA from the surface layers of freely drained soils, in which the organic matter content is lower and possibly different in character.

The increased mobilization of trace elements under conditions of even very slightly impeded drainage is often so marked that changes in extractable trace element contents due to this cause may obscure other effects. Experience with Scottish profiles suggests that any change in extractable trace elements can generally be ascribed to drainage factors.

In Table 8.13 are presented the trace element contents at three depths in a Scottish scirpus-sphagnum deep peat. On the basis of mineral matter content, a twofold increase might be anticipated in the lowest layer, provided there was no mineral contamination, leaching, or species variation in the peat. Only Ti and possibly Si and B are in reasonable agreement with this expectation. The intermediate horizon shows a very marked increase in Mo, Al, Cr, and V, and to a lesser degree Cu, whereas the lower layers are equally high in Co and Ni. The lowest layer is particularly high in Fe and Mn. The low Sr content in the intermediate layer is also noteworthy, as a similar effect is shown by Ca. Apart from Al, the elements which have accumulated in the peat are those mobilized in mineral soils with impeded drainage and must presumably have been derived by lateral infiltration from soils surrounding the peat deposit.

Mechanical Translocation

Increased trace element contents in specific horizons of mineral soils might arise from size sorting, which leads to the formation of clay-rich

TABLE 8.13. TOTAL CONTENTS OF TRACE ELEMENTS IN THREE HORIZONS OF A DEEP PEAT

Element	Trace element contents in 3 horizons (ppm dry matter)		
	6–18 in. Ash = 2.3%	42–54 in. Ash = 2.9%	102–114 in. Ash = 4.9%
Co	0.3	2.1	1.9
Ni	2.3	5.8	8.2
Mo	0.6	13.5	1.4
Al	1290	10100	5680
Fe	580	1400	5210
Pb	4	9	<10
Sn	<1	<1	<1
Zn	10	<50	<50
Cr	0.5	8.1	1.9
V	1.6	21.1	5.3
Ti	95	160	204
Si	1270	1090	2660
B	4.5	5.7	14.1
Cu	6.3	11.2	7.8
Mn	2.5	10	115
Ba	9	12	11
Sr	23	5	46

layers. Since a clay fraction has been shown generally to be richer in many trace elements than are its associated sands, the total contents of these trace elements will show accumulation relative to the other horizons in clay-rich horizons formed in this way. If large changes in clay content in a profile are not accompanied by corresponding changes in the total contents of these trace elements in the clay-rich horizons, then it appears justified to assume that the clays have been formed *in situ* and have not accumulated by translocation.

Comparison of Soil Types

Over a vast area of the Russian plain, the trace element contents of the zonal soils vary within narrow limits because of the uniformity of composition of the soil parent materials (50). Greater differences occur in soils near the great mountain ranges, for instance in those on alkalic rocks on the Kola peninsula and in the red soils on basic rocks in the Caucasus. Soils on the western slopes of the Urals derived from Cu-rich Permian sediments stretch through various latitudinal soil zones and show typical profile distributions. Thus in podzols and forest soils there is evidence of Cu accumulation in the upper humic horizons and perhaps also in the illuvial B horizon. In chernozems there is marked Cu accumulation in the humus-rich upper horizons. In chestnut soils and serozems the total Cu content follows the organic matter content. Profile distribution of Li, Rb, Cs, As, F, Br, and I, in addition to the trace elements usually determined, has been studied in several Russian soil types; there is, in general, evidence of slight surface depletion from podzolic soils and accumulation in organic-rich soils in the less humid zones.

As chernozems have a high content of clay and organic matter and are saturated with Ca, the trace elements, apart from those taken up by plants, tend to remain *in situ* and not to be leached out. The chernozems and related soils, therefore, often have a higher trace element content than other soils on the same parent material, and display surface enrichment. Standard contents for a number of trace elements in Russian chernozems have been given (22) as Mn, 830 ppm; Ba, 400 ppm; Cr, 400 ppm; Sr, 400 ppm; V, 290 ppm; Ni, 49 ppm; Zn, 72 ppm; Cu, 30 ppm; Co, 6.1 ppm; and B, 12 ppm.

In eight different soil types from Israel (39), results are available for both total and extractable contents of B, I, Br, As, Zn, Cu, and Co at three arbitrary depths. Generally there is a very slight fall in total content with depth, most apparent with Cu and B, pointing to surface enrichment through plant residues, and a greater fall in the amounts of water-extractable B, EDTA-extractable Zn and Cu, and HNO_3-extractable Co. These and many other similar results from similar climatic areas confirm Russian findings.

In Australia, the total contents of trace elements in profiles from various areas, including Queensland, central Australia, South Australia and Tasmania, on very varied soil types, including the podzolic soils already mentioned, red-brown earths, krasnozems, black earths, solonetz, terra rossa, and rendzinas have been reported. In a series of red-brown earths from South Australia (28, 33) in which there is a marked discontinuity involving an increase of up to tenfold in the clay content between the A and B horizons, and a corresponding increase in trace element content, it has been concluded from the analytical results for the trace elements that an eluvial-illuvial process has not been involved, that all but an insignificant part of the clay content of the B horizons was formed in its present position, and that there had been a loss of clay from the surface layers. Other red-brown earths that have only a slight increase in clay content in the B horizons show a gradual slight fall in total trace element content down the profile.

In a study (34) of soils from several of the zonal soil types in Queensland, an attempt has been made to establish a statistical assessment of trace element behavior, using a measure called the trend (T), defined as $T =$

$(W - S)/W$, where S is the concentration of a trace element in the surface horizon and W its weighted mean in all horizons above the C horizon. The variableness (V) of a trace element in a profile is defined as $V = (H - L)/W$ where H and L are the highest and lowest contents in all horizons above the C horizon. The applicability of such a scheme must depend primarily on the uniformity of the parent material. In the profiles studied, small trends were found for heavy grey and brown soils, krasnozems, and black earths. Weighted means for the A and B horizons were greater than the C horizon contents in the black earths and less in the solodized solonetz. Other soil types gave rather irregular results for trend and variableness.

Zirconium has frequently been used as an index for evaluating loss or gain of other soil constituents on the assumption that it is all present in zircon and that it is immobile. Losses of 20 to 50 per cent of the Zr in the parent rock appear possible from some Australian results (54) for granitic soils, implying loss from zircon. In basic igneous rocks, Zr may occur in some readily weathered ferromagnesian minerals.

In a number of different Tasmanian soils (47) that have developed on dolerite, it is interesting to note little or no concentration of Co, Ni, or Mn in the clay fractions, although this occurs with Mo and Cu in some soils. Grey-brown podzolic soils show a marked increase in Cu but not in Co content in the B horizons. In krasnozems, brown earths, and black earths, there is little profile differentiation of trace elements, except in black segregations in the krasnozems, in which Co, Mn, and Zn are concentrated ten to a hundred times.

In the United States, an early investigation into the contents of nine trace elements in profiles from erosion experiment stations (40) indicated that, apart from some loss of Se, V, and Cu from surface horizons in some instances, soil-forming processes have not extensively modified the amounts present in soils from widely scattered areas in the midwestern and southern states covering podzolic, rendzina, chernozem, prairie, and lateritic types. Relatively constant profile distribution was obtained for As, Co, Ni, Zn, Ba, and Cr. Arsenic contents of 1 to 20 ppm and Se contents of 0.02 to 2.5 ppm, with abnormally high values up to 80 ppm As and 20 ppm Se on certain shales, were found; other elements fell into the normal ranges. Further investigation of Cu and Zn in these and other profiles has confirmed these findings, variations within the profiles being related to changes in clay and organic matter contents.

Tropical Soils

The trace element contents of African tropical soils are illustrated by some results from Dahomey, Angola, and Kenya. The information available does not yet enable a generalized statement of trace element behavior in tropical soils to be made.

In lateritic soils from Dahomey (37), surface accumulation of Mn and B and surface depletion of Cu, Ni, Co, V, Cr, Zn, and Ga are apparent. In a tropical ferruginous soil, depletion of all trace elements from the surface is observed, while in a poorly drained black soil, maximum contents of most trace elements occur in the 20- to 60-inch zone, but there is a marked accumulation of Sn in the relatively organic-matter-rich surface horizon. The Mn content of this soil is high, rising to 1 per cent; Cu and Mo are generally low in all the soils studied; other elements are within what would be considered the normal ranges of contents.

Trace element contents are available for 18 profiles from Angola (14), covering peats and humic gleys, black clay soils, and laterites, on various parent materials. In the organic soils most of the Co, and almost half the Zn, particularly in the surface horizons, is acetic acid extractable. The peats, at 5 ppm, have relatively high Mo contents. In black clays derived from calcareous and basic igneous rocks, higher contents of Ba, Cr, La, Li, Mo, Pb, Rb, V, Y, and Zr are found in the former, in line with their higher clay content, while much higher contents of Co and Mn occur in the latter. Little systematic variation of total content with depth is apparent. Most of the extractable contents fall slightly in lower horizons, although Cr, Ti, and V contents occasionally increase down to 30 inches, the greatest sampling depth. The lateritic soils include red soils from basalts and dolerites and yellow soils from granites or other acid igneous rocks. Higher contents of Co, Cr, Cu, Mn, Ni, Ti, and V are found in the former while the total contents in the latter are similar to those found in other soil types developed on granite, and appreciable contents of extractable Pb are recorded.

Results on acetic-acid-extractable materials only are available for some soils from Kenya (11). Those on volcanic ash from the Rift Valley floor are exceedingly low in Co, although soils on the alluvium derived therefrom contain over ten times as much Co. An interesting finding is the occurrence of high extractable Be, levels over 5 ppm being recorded in soils with low Co contents. Very high contents of extractable Co ($>$8 ppm) occur on wet hilltops, although ultrabasic rocks do not appear to be involved. Figures for Co, Be, Pb, Ni, and Zn show reduced extraction with depth, except that in some soils Pb reaches its maximum at 50 to 70 inches. The extractable Zn contents are generally high (20 to 45 ppm) in the surface horizons.

Profile investigations in the Indian subcontinent have dealt largely with Mn and show that exchangeable and water-soluble Mn contents decrease regularly with depth in most soils. Reducible Mn may be less uniformly distributed, being generally 20 to 50 per cent of the total contents, which usually lie between 200 and 1500 ppm. Smaller amounts of reducible Mn often occur in horizons with $CaCO_3$ concretions. Black soils (montmorillonitic) are well supplied with active Mn, whereas lateritic and acid soils (kaolinitic) generally have small amounts. In the Punjab, typical Mn ranges are: water-soluble, <2.5 ppm; exchangeable, <41 ppm; reducible, 61 to 258 ppm; and active, 67 to 280 ppm; these are higher in acid and noncalcareous than in alkaline and calcareous soils, with all extractable forms decreasing with depth.

BEHAVIOR OF BIOLOGICALLY IMPORTANT TRACE ELEMENTS

Certain points in the behavior of trace elements which are of significance in plant and animal nutrition have not been considered in the more general discussions. Some of the more important aspects are dealt with below.

Boron

The high content of B in argillaceous sediments of marine origin (20 to 200 ppm) is explained by the relatively large content of B (5 ppm) in sea water. Boron in the salts, clay minerals, or sesquioxide complexes of shales is more available than that in the tourmaline of igneous or metamorphic rocks. Finely ground crystals of tourmaline are not effective in preventing B-deficiency disorders in plants.

The deficiency level as shown by hot water extraction of B is generally reported to be near 0.5 ppm. Light acid soils in humid regions are likely to be deficient in available B. Higher amounts of available B are generally found in soils with high organic matter contents, suggesting that in acid conditions, organic matter can protect B from loss by leaching without rendering it unavailable. Plant uptake of B is reduced by liming, probably, apart from physiological effects, because of inorganic and organic fixation and not because of the formation of calcium borate, which is readily soluble. The joint use of water-soluble and phosphoric-acid-soluble B for diagnosis has been suggested in a Swedish investigation which emphasizes the importance of drought in reducing B availability (36).

Boron toxicity can arise in arid areas where sodium and calcium borates occur in surface soils, but may also be caused by the use of excessive amounts of boronated fertilizers, particularly with monocotyledons, which have a lower B requirement than dicotyledons and are more susceptible to B poisoning. Irrigation waters containing 2 ppm B are reported to be undesirable, and in this connection it must be remembered that sea water contains more than twice this amount. The safe limits of available B content between deficiency and excess are narrow (8).

Cobalt

Cobalt has been reported to be concerned in N fixation but naturally occurring Co deficiency in plants has not been established. In ruminant animals, production of vitamin B_{12} by rumen bacteria necessitates an adequate supply of Co (about 0.08 ppm) in the herbage dry matter.

Instances of Co deficiency on soils of very diverse types have been widely reported. The most extensive areas probably occur on soils derived from sandstones, and the symptoms are particularly pronounced if the soils are calcareous. Limestone soils but not residual soils derived from limestones are also likely to be deficient. Loess soils may be, but alluvial deposits or soils derived from shales seldom are. Among igneous rocks the most widespread areas of deficiency occur on some granitic and related rocks, such as the rhyolitic pumice of New Zealand. Many organic soils are Co-deficient. The effect of impeded drainage in mobilizing Co has already been discussed (20).

In deficient acid soils there is generally an inherently low content of Co, total contents of <0.5 to 3 ppm being commonly reported, whereas contents of up to 50 ppm are usual in soils carrying healthy stock. The amount of Co extracted from Scottish arable soils by 2.5 per cent acetic acid varies from <0.05 ppm in very deficient sands to >1 ppm in many soils derived from basic igneous rocks. The level below which pasture herbage takes up insufficient Co to carry healthy stock lies at about 0.25 ppm in mineral soils of average acidity (pH 5 to 6), but other factors, such as organic matter content, must be considered in interpreting analytical results. Indicator plants which take up Co preferentially at many times the normal level include *Nyssa sylvatica* (black gum) and *Clethra barbinervis* (an Asiatic ericaceous shrub).

Copper

Copper deficiency in cereals occurs on many types of soil, particularly those high in organic matter and those derived from arenaceous or acid igneous rocks inherently deficient in Cu. Other species of plants are also susceptible. Copper deficiency in stock may occur on low Cu herbages, but other factors, including the level of uptake of Mo and SO_4, may be involved. Simultaneous deficiencies of Co and Cu in stock or of Mn and Cu in crops are not uncommon.

Numerous soil extractants have been proposed for Cu, but the most satisfactory appears to be 0.05 *M* EDTA, which extracts from 0.1 to 10 ppm from normal soils, the deficiency level for cereals being about 0.6 ppm. The extraction of Cu with versenate has been described as essentially an exchange reaction between the Cu-soil complex and the EDTA complex. Generally 10 to 50 per cent of the total Cu is in available form. The failure of acid extractants suggests that organically bound Cu is concerned in plant uptake. When Cu is added as a top dressing to a peat soil, it may remain in the top 2 inches for at least 6 years and appears to be organically bound, in a form available to plants. In sandy soils there is deeper penetration but also reduced availability (27).

Comparison of lateritic and podzolic mountain soils from Java shows that 70 per cent of the total Cu is extractable by 2 per cent H_2SO_4 from the former but only 25 per cent from the latter, although this rises to 70 per cent on ashing, again indicating the presence of organically bound Cu (49). The extraction of Cu by neutral salts is increased when plant extracts are added to the extractant, a finding which also demonstrates the role of organic matter in controlling the behavior of soil Cu.

Manganese

Patterns of profile distribution of total Mn include surface accumulation, increase with depth, decrease with depth, and accumulation in illuvial horizons overlying calcareous layers. Manganese is essential in plant nutrition and deficiencies occur chiefly on neutral or alkaline soils. Excess Mn may restrict growth of plants, in which connection the ratio of Mn to Fe is important. There is some evidence that certain disorders in animals are related to low available Mn in the diet.

In soils the total content is of no diagnostic value. Many attempts have been made to identify the more available forms (18, 19). Acetic acid, ammonium acetate, and Ca and Mg nitrates have all been employed to assess exchangeable or available Mn, but the problem is complicated by the presence of insoluble higher oxides. Association of high pH and suitable organic material produces organic complexes of divalent Mn and leads to insufficient available Mn for susceptible plants such as cereals or peas. Plants appear to be able to extract Mn from certain oxides, depending on surface area and disorder and on soil pH. Aging of the oxides reduces availability, and addition of soluble manganese salts is not a permanent cure for Mn deficiency. Experimental findings suggest the presence of hydrated Mn_2O_3, higher oxides approaching MnO_2, and more complex forms, in soils. Healthy soils are reported to contain at least 20 to 50 ppm extractable Mn when assessed by extraction with acid phosphate or with a reducing agent followed by a neutral salt.

Manganese toxicity in plants, suggested by the presence of more than 1000 ppm Mn in the dry matter, generally occurs on very acid soils and can often be remedied by liming.

Molybdenum

Disorders due to Mo include deficiencies in plants and excesses in animals (4). Contents of <0.1 to 5 ppm are normal in soils, in good agreement with the abundance of 2.3 ppm in the lithosphere. Deficiency has been reported in numerous species of plants. Animal disorders arise from excess when plant contents exceed 5 ppm and are related to Cu and SO_4 metabolism.

In soils, Mo released by weathering from olivine, clay minerals, sulfides, or other ore minerals occurs as the molybdate anion, and there is evidence that under acid conditions sesquioxides and clay minerals can bind it in unavailable form. On reduction of soil acidity from pH 5 to pH 7 by liming, Mo uptake by plants may increase tenfold.

Water, ammonium acetate, acetic acid, hydrochloric acid, phosphoric acid, sodium hydroxide, and acid oxalate extractions have all been employed to diagnose excesses. Generally the amount extracted increases with increase in pH of the extractant. This does not hold for acid ammonium oxalate at pH 3.3, which complexes Mo. Despite its action on sesquioxide-bound Mo, this extractant has been found to give some correlation with field response for deficient soils from Australia and New Zealand (17), where such soils contain less than 0.15 ppm oxalate-extractable Mo and have a pH of 6.3 or less. It may be necessary to take account of pH in assessing deficiency. In some deficient areas heavy liming may increase availability to a satisfactory level, while in areas with relatively high Mo contents it may raise the plant uptake to a level dangerous to stock.

Uptake of Mo in amounts excessive for animals normally occurs on neutral or alkaline soils derived from shales or on soils with high organic matter contents, whereas plant deficiency is commonest on acid sandy soils. In Californian soils where Mo contents of herbage are high, water is as effective an extractant as ammonium acetate at pH 8.5, suggesting that most of the available Mo is in a water-soluble form. High uptake of Mo can occur from acid soils of pH <5 if the organic matter content is high. Presumably organic chelation protects the Mo from entry into the insoluble compounds that reduce availability in acid conditions.

Selenium

Excessive uptake of Se by plants, leading to toxicity in animals, arises in areas where shales and other sedimentary rocks containing abnormally high amounts of Se occur. Although the terrestrial abundance of Se is less than 0.1 ppm, such rocks may contain 1 to 10 ppm, with occasional values up to 100 ppm. In some oxidates, such as vanadium-uranium ores, contents of 2500 ppm have been reported. Among hydrolyzates, shales containing organic matter of vegetable origin are generally highest in Se.

Selenium excess occurs chiefly on two types of soil. It was first observed in the United States on arid soils derived from seleniferous rocks where the rainfall is insufficient to leach out water-soluble Se, present mainly as selenate (3). In one study of 500 such soils carrying vegetation containing over 50 ppm Se, all but 43 contained more than 0.5 ppm Se and almost 50 per cent contained 1 to 4 ppm Se; values over 120 ppm were found. Water-soluble Se values up to 38 ppm have been recorded, but generally they do not exceed 0.1 ppm. Normally only a small percentage of soil Se is water-soluble, but the proportion may rise to over 90 per cent, some of the insoluble Se being present as ferric selenate. Available Se appears to be that present in water-soluble form as selenate, and in organic compounds derived from the decay of plants high in Se. Seleniferous soils are characterized by certain species of plants which may require, and certainly tolerate, Se and can absorb more than 1 per cent. Such indicators include species of *Astragalus* (vetch), *Xylorrhiza* (woody aster), *Oonopsis* (golden weed), and *Neptunia* (a legume from Queensland). Effects of excessive Se on stock may give rise to either acute or chronic poisoning, resulting in alkali disease or blind staggers.

Similar symptoms have been observed in stock grazing poorly drained organic soils in various areas throughout the world, including Ireland and parts of the United States. In Ireland (13) these soils, which are alkaline and are influenced by wash from shales, contain up to 1200 ppm Se. Herbage contents as high as 500 ppm are reported and are often associated with high Mo contents. In the seleniferous areas of Utah, high U contents of up to 100 ppm have been observed in shallow peats containing up to 10 ppm Se and 30 ppm Mo as well as high V contents. These peats are alkaline with high Ca contents and the accumulation of these trace elements which normally occur in anionic form is no doubt related to the alkalinity of the waters percolating through the country rocks, which are basic lavas, limestones, and shales.

It has been found that certain disorders of animals, including muscular dystrophy, white muscle disease, liver necrosis, and infertility, can be related to Se deficiency. In most but not all instances a vitamin E deficiency may be involved, as dosing with tocopherol can be curative. Such disorders are reported from New Zealand, from western Finland, and from parts of Scotland. They occur on acid igneous and arenaceous soils which would, on the basis of their S contents, be considered likely to be low in Se. The amounts of Se involved are quite small, as plant contents of the order of 0.01 ppm Se and equally low contents of extractable Se in soils are involved. No reliable soil information is available, but it has been suggested that remedial soil treatments are of the order of 5 g Se per acre, and that 25 to 50 g Se per acre could raise the plant content above the toxicity level in plant dry matter.

In Scotland a large-scale trial involving Se dosing of lambs has given small but significant growth responses on farms on soils from arenaceous Old Red Sandstone rocks and also on acid igneous and basic igneous rocks. In none of the areas tested was there any symptom suggesting acute Se deficiency. The responses obtained in New Zealand on pumice soils were several times as great, at above 10 lb/lamb. Intermediate responses have been obtained in Oregon. In both New Zealand and Oregon, symptoms of acute deficiency are observed, and in the latter case an Se/SO_4 antagonism may be involved.

Zinc

Most minerals in which Zn occurs are readily weathered; thereafter, to a large extent, Zn is adsorbed as a divalent cation into the exchange complex and into organic combination. Zinc deficiencies of fruit trees, cereals, and herbaceous plants (45) appear to be related to soil reaction rather than to total Zn content. Correlation has been reported in Kentucky between uptake of Zn by corn plants and the amount extracted by dithizone, particularly when allowance is made for soil pH. It has also been reported that, by plotting acid-extractable Zn against titratable alkalinity to pH 5, deficient and nondeficient soils can be distinguished. Healthy plants can be grown in neutral soils containing fresh organic matter, suggesting that this provides Zn available to the plant. Some forms of decomposing soil organic matter on the other hand appear to give a relatively unavailable Zn complex. Results from various countries suggest a correlation between total and extractable Zn independent of soil pH, total contents in the range of 10 to 200 ppm giving 0.5 to 15 ppm dilute-acid-extractable, 0.4 to 40 ppm EDTA-extractable, and 0.2 to 15 ppm *Aspergillus niger*-extractable Zn.

The plant uptake of Zn is relatively high compared with soil contents, and the possibility of Zn deficiency arising from crop removal can be assessed by considering the Zn uptake of a 10 ton/acre crop containing 100 ppm Zn, a content somewhat above normal. This crop would remove 1 ppm Zn from the surface soil, of the same order as the amount extracted by the less-vigorous extractants and perhaps one-tenth to one-hundredth of the total soil content. This also gives a measure of the Zn returned to the soil in organically bound form, should the crop not be completely removed: it could vary from 0.1 to 1 ppm, depending on the weight and nature of the crop actually removed.

Restriction of plant growth due to an excess of Zn can arise in acid soils in certain areas, but is more likely to be caused by industrial contamination or by addition of Zn-rich materials such as sewage sludges. It may be possible to control Zn excess by liming.

Other Elements of Biological Interest

Arsenic occurs naturally in most soils in amounts between 1 and 70 ppm. The water-soluble content is not related to the total, and may be very low in soils with high total amounts (55). Fortunately, plant growth appears to be restricted before amounts dangerous to animals are taken up. There is often a relationship between contents of As and Se in soils, as both are concentrated in hydrolyzates and oxidate sediments.

Fluorine and iodine are of considerable importance. In the lithosphere, F is more than 2000 times as abundant as I. The bulk of the fluorine is fixed in the crystal lattices of silicate and phosphate minerals, and an abundance of 800 ppm F in the earth's crust compares with 1.4 ppm in the sea. For I, the comparative figures are 0.3 and 0.05 ppm. The content of F in soils ranges from about 10 to 1000 ppm. High contents of weatherable minerals can yield in drainage enough F to render drinking waters dangerous. The desirable water content is 0.3 to 0.5 ppm, and many waters are artificially fluorinated to this level to prevent decay of teeth. The danger level lies above 1 to 2 ppm. Little is known about soluble F in soils or uptake by plants. Whereas the chief danger with F appears to be excess, deficiency of I is known in man and domestic animals, and leads to goiter (12). Though the average I content of rocks is 0.3 ppm, the normal range for soils is 0.6 to 8.0 ppm, with exceptional values up to 70 ppm. The bulk of this is easily soluble, presumably resulting from deposition of air-borne material of marine origin, as contents decrease with distance from the sea. Chilean nitrate, containing an average of 150 ppm, is a further source of soil I.

Vanadium, whose essentiality to a green alga has been reported, occurs in greatest amount in shales and intermediate igneous rocks; increased amounts soluble in acetic acid occur in soils with impeded drainage. Relationship of plant uptake to soil content has not been studied sufficiently closely to indicate the most suitable diagnostic extractant, nor can speculation on the forms of occurrence of V in soils serve a useful purpose, although it is possible that chelated V is the form most readily taken up by plants.

Instances of acute Pb poisoning of stock are frequently reported, but can generally be ascribed to random contamination from lead paints or other artificial sources. The Pb accumulated in organic surface horizons of many soils does not appear to be readily available to plants. In the neighborhood of ore deposits and on some types of soils, such as certain sandstones and acid igneous rocks, the Pb content of vegetation can appreciably exceed the normal content of about 1 ppm in the dry matter, and it has been suggested that such contents may give rise to chronic ailments in animals or humans. Further medical investigation of this possibility is necessary.

Among the remaining elements of natural occurrence in toxic amounts, the most important is probably Ni, which is abundant in ultrabasic rocks in a form readily released on weathering. In Scotland it has accumulated in amounts toxic to cereals and root crops in poorly drained areas on or near such rocks. Contents of up to 8000 ppm total Ni and 100 ppm Ni soluble in acetic acid are found in some small infertile patches, which are generally acid in reaction. Crops, containing even then up to 50 ppm Ni, can be obtained by liming to neutrality. Other infertile areas on ultrabasic rocks throughout the world can probably be related to an excess of Ni rather than of Cr, which is not normally present in soils in a form toxic to plants.

REFERENCES

1. Ahrens, L. H., *Geochim. et Cosmochim. Acta* **2**, 155–169 (1952).
2. Allison, R. V., and Gaddum, L. W., *Soil Sci. Soc. Florida Proc.* **2**, 68–91 (1940).
3. Anderson, M. S., Lakin, H. W., Beeson, K. C., Smith, F. F., and Thacker, E., "U.S.D.A. Agricultural Handbook No. 200," 1961.
4. Bear, F. E., ed., *Soil Sci.* **81**, 159–258 (1956).
5. Bear, F. E., ed., *Soil Sci.* **83**, 1–83 (1957).
6. Bear, F. E., ed., *Soil Sci.* **84**, 59–116 (1957).
7. Bear, F. E., and Kitchen, H. B., ed., *Soil Sci.* **60**, 1–189 (1945).
8. Berger, K. C., *Advances in Agron.* **1**, 321–351 (1949).
9. Berrow, M. L., and Mitchell, R. L. (in preparation for *J. Soil Sci.*).
10. Burges, A., *Sci. Proc. Roy. Dublin Soc., Ser. A1, No. 4*, 53–58 (1960).
11. Chamberlain, G. T., *East African Agr. J.* **25**, 121–125 (1959).
12. Chilean Iodine Educational Bureau, "Geochemistry of Iodine: Iodine in Rocks, Minerals and Soils," London, 1956.
13. Fleming, G. A., *Soil Sci.* **94**, 28–35 (1962).
14. Fragoso, M. A. C., Lisbon, "Memorias da Junta de Investigacoes do Ultramar," No. 11, pp. 238, 1959.
15. Gammon, N., Henderson, J. R., *et al.*, *Florida Agr., Expt. Sta. Tech. Bull.* **524**, 1953.
16. Goldschmidt, V. M., Muir, A., ed., "Geochemistry," Oxford, Clarendon Press, 1954.
17. Grigg, J. L., *New Zealand J. Sci. Technol.* **A34**, 405–414 (1953).
18. Hammes, J. K., and Berger, K. C., *Soil Sci.* **90**, 239–244 (1960).
19. Heintze, S. G., *J. Soil Sci.* **8**, 287–300 (1957).
20. Hill, A. C., Toth, S. J., and Bear, F. E., *Soil Sci.* **76**, 273–284 (1953).

20A. Hodgson, J. F., *Advances in Agron.* **15**, 119–159 (1963).

21. Jackson, M. L., and Sherman, G. D., *Advan. Agron.* **5**, 219–318 (1953).
22. Kovda, V. A., and Vasil'yevskaya, V. D., *Soviet Soil Sci.* 1369–1377 (1958).

23. Lamb C. A., Bentley, O. G., and Beattie, J. M., ed., "Trace Elements," New York, Academic Press, Inc., 1958.
24. Le Riche, H. H., and Weir, A. H., *J. Soil Sci.* **14**, 225–235 (1963).
25. Lewis, T. E., and Broadbent, F. E., *Soil Sci.* **91**, 393–399 (1961).
26. Lucas, R. E., and Davis, J. F., *Soil Sci.* **92**, 177–182 (1961).
27. Lundblad, K., Svanberg, O., and Ekman, P., *Plant Soil* **1**, 277–302 (1949).
28. McKenzie, R. M., "C.S.I.R.O. (Australia), Div. Soils, Div. Report 9/56," 1956.
29. Manskaya, S. M., Drosdova, T. V., and Emel'yanova, M. P., *Soviet Soil Sci.* pp. 623–629 (1958).
30. Mason, B., "Principles of Geochemistry," New York, John Wiley & Sons, Inc., 1952.
31. Ng, S. K., and Bloomfield, C., *Geochim. Cosmochim. Acta* **24**, 206–225 (1961).
32. Ng, S. K., and Bloomfield, C., *Plant Soil* **16**, 108–135 (1962).
33. Oertel, A. C., *J. Soil Sci.* **12**, 242–258 (1961).
34. Oertel, A. C., and Giles, J. B., "C.S.I.R.O. (Australia), Div. Soils, Div. Report 8/58," (1959).
35. Peyve, Y. V., *Soviet Soil Sci.* pp. 941–947 (1960).
36. Philipson, T., *Acta Agr. Scand.* **3**, 121–242 (1953).
37. Pinta, M., and Ollat, C., *Geochim. Cosmochim. Acta* **25**, 14–23 (1961).
38. Prince, A. L., *Soil Sci.* **84**, 413–418 (1957).
39. Ravikovitch, S., Margolin, M., and Navrot, J., *Soil Sci.* **92**, 85–89 (1961).
40. Slater, C. S., Holmes, R. S., and Byers, H. G., *U.S.D.A. Tech. Bull.* **552**, 1937.
41. Starkey, R. L., *Soil Sci.* **79**, 1–14 (1955).
42. Swaine, D. J., *Commonwealth Bur. Soil Sci. Tech. Commun.* **48** (1955).
43. Swaine, D. J., *Commonwealth Bur. Soil Sci. Tech. Commun.* **52** (1962).
44. Swaine, D. J., and Mitchell, R. L., *J. Soil Sci.* **11**, 347–368 (1960).
45. Thorne, D. W., *Advan. Agron.* **9**, 31–65 (1957).
46. Tiller, K. G., "C.S.I.R.O. (Australia), Div. Soils, Div. Report 3/57," 1957.
47. Tiller, K. G., "C.S.I.R.O. (Australia), Div. Soils, Div. Report 6/58," 1959.
48. Underwood, E. J., "Trace Elements in Human and Animal Nutrition," 2nd ed., New York, Academic Press, Inc., 1962.
49. Vermatt, J. G., and van der Bie, G. J., *Plant Soil* **2**, 257–282 (1950).
50. Vinogradov, A. P., "The Geochemistry of Rare and Dispersed Chemical Elements in Soils," 2nd ed., New York, Consultants Bureau, 1959.
51. Wager, L. R., and Mitchell, R. L., *Geochim. Cosmochim. Acta* **1**, 129–208 (1951).
52. Wallace, A., ed., "A Decade of Synthetic Chelating Agents in Inorganic Plant Nutrition," Los Angeles, Arthur Wallace, 1962.
53. Wells, N., *J. Soil Sci.* **11**, 409–424 (1960).
54. Wild, A., *Australian J. Agr. Research* **12**, 300–305 (1961).
55. Williams, K. T., and Whetstone, R. B., *U.S.D.A. Tech. Bull.* **732**, 1940.
56. Wright, J. R., Levick, R., and Atkinson, H. J., *Soil Sci. Soc. Am. Proc.* **19**, 340–344 (1955).

CHAPTER 9

Soil Fixation of Plant Nutrients

LOUIS T. KARDOS

Department of Agronomy
The Pennsylvania State University, University Park

Fixation of plant nutrients in soils may be defined as the process whereby readily soluble plant nutrients are changed to less-soluble forms by reaction with inorganic or organic components of the soil, with the result that the nutrients become restricted in their mobility in the soil and suffer a decrease in their availability to the plant.

Of the elements presently regarded as essential for plant growth (C, H, O, N, P, K, Ca, Mg, S, B, Mn, Fe, Cu, Zn, Mo, Cl), the C, H, and O are obtained primarily from the air and water; the chlorides and anionic forms of S and N have not been regarded as "fixed" by soils when soluble salts of these elements are added.

Since the micronutrient elements B, Mn, Fe, Cu, Zn, and Mo are discussed (Chapter 8) and since there is little evidence that Mg and Ca become limiting to plant growth because of soil fixation reactions, primary consideration is given here to the fixation reactions of P, K, and NH_4-N.

PHOSPHORUS FIXATION

The types of reactions by which phosphorus becomes fixed can be placed in three general groups: adsorption, isomorphous replacement, and double decomposition involving solubility-product relations.

In all the reactions the phosphorus is involved as one or more of the ionic forms into which orthophosphoric acid may dissociate. From a consideration of the ionization constants of H_3PO_4, Buehrer (10) has calculated the relation of pH to the relative concentration of the undissociated H_3PO_4 and the three ionic species $H_2PO_4^-$, HPO_4^{2-}, and PO_4^{3-} (Table 9.1).

TABLE 9.1. CONCENTRATION OF PHOSPHATE IONS IN SOLUTION AT VARIOUS HYDROGEN ION CONCENTRATIONS*

pH	(H^+)	(H_3PO_4) (moles/l)	$(H_2PO_4^-)$ (moles/l)	(HPO_4^{2-}) (moles/l)	(PO_4^{3-}) (moles/l)
3	10^{-3}	8.6×10^{-7}	9.5×10^{-6}	1.9×10^{-8}	6.8×10^{-19}
4	10^{-4}	9.3×10^{-8}	1.02×10^{-5}	2.05×10^{-8}	7.4×10^{-17}
5	10^{-5}	9.1×10^{-9}	1.01×10^{-5}	2.02×10^{-7}	7.3×10^{-15}
6	10^{-6}	7.8×10^{-10}	8.2×10^{-6}	1.71×10^{-6}	6.2×10^{-13}
7	10^{-7}	3.1×10^{-11}	3.4×10^{-6}	6.9×10^{-6}	2.47×10^{-11}
8	10^{-8}	4.5×10^{-13}	4.9×10^{-7}	9.8×10^{-6}	3.53×10^{-10}
9	10^{-9}	4.6×10^{-15}	5.1×10^{-8}	1.02×10^{-5}	3.7×10^{-9}
10	10^{-10}	4.7×10^{-17}	5.2×10^{-10}	1.03×10^{-5}	3.7×10^{-7}

* Courtesy, T. F. Buehrer and Arizona Agricultural Experiment Station.

For these calculations he used the equilibrium constants shown in Table 9.2 and a total concentration of H_3PO_4 equivalent to one part per million.

Obviously over the pH range, 4.0 to 8.0, in which virtually all agricultural soils are found, the principal ionic species present in the soil solution are $H_2PO_4^-$ and HPO_4^{2-}. Below pH 6.71 the monovalent ion would be preponderant, whereas above this pH the divalent ion would dominate. Above a pH of 9.0 the trivalent ion becomes more important than the monovalent ion, but even at a pH of 12 the HPO_4^{2-} concentration is still relatively greater than that of PO_4^{3-}. From these relationships it is obvious that all phosphate reaction systems will be fundamentally influenced by the hydrogen ion activity in the systems.

Adsorption Reactions

Adsorption reactions may be classified, generally, into two types: chemical adsorption and physical adsorption. Both types may be characterized by the Freundlich absorption isotherm or by the Langmuir adsorption equation.

Low and Black (37) plotted the amount of phosphorus fixed by kaolinite against the equilibrium concentration of phosphorus in dilute solutions and found the data to fit a typical adsorption curve (Figure 9.1), which could be represented by a Freundlich adsorption equation having the form $x/m = kc^{1/n}$.

Low and Black further found that the degree of adsorption was increased by increasing temperature and, hence, concluded that adsorption was chemical. Since Kolthoff (31) has shown that the Freundlich equation can represent exchange reactions when applied to ions, it was suggested that the adsorption reaction is an exchange reaction between phosphate and hydroxyl ions. Low and Black concluded that, although the short-time

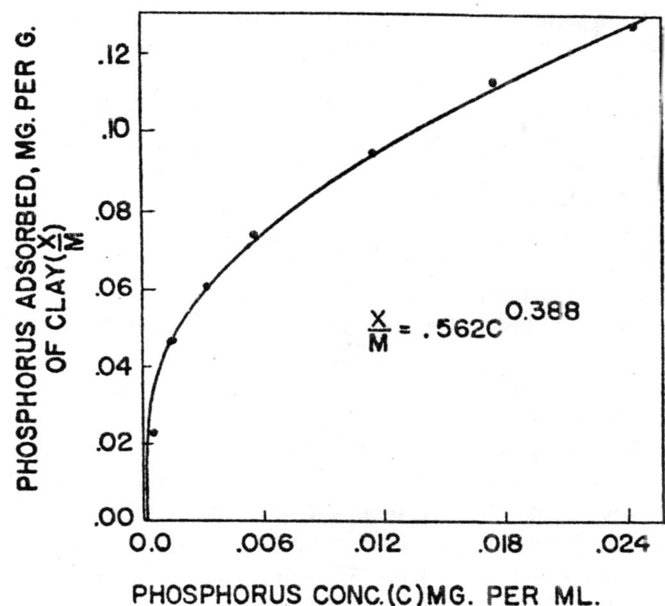

Figure 9.1. Adsorption of phosphate from dilute solutions by kaolinite (37).

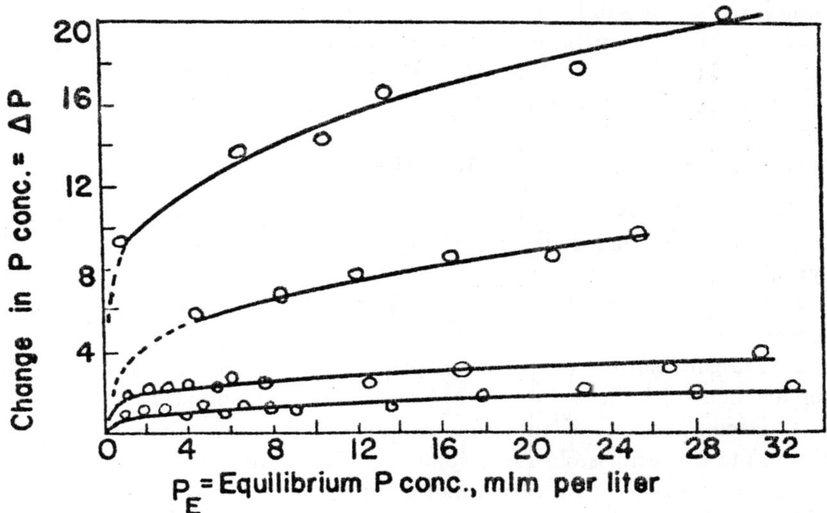

Figure 9.2. Equilibria between soil and phosphate solutions (16).

reaction ($1\frac{1}{2}$ to 3 hours) at low concentrations (up to $0.0016M$) may be an adsorption reaction, other reactions dominate at higher concentrations ($1.0M$) and after more prolonged contact (8 to 11 days).

Davis (16), in studying the sorption of phosphates by noncalcareous Hawaiian soils, found that the data for fixation could be expressed empirically by the function $\Delta P = kP_E^{1/3}$, where ΔP equals the change in concentration and P_E the equilibrium concentration of phosphate. The pH values of the systems were approximately 7.0, and the reaction was allowed to proceed for 22 days. Figure 9.2 shows the nature of the adsorption curves, which obviously resemble the Freundlich adsorption.

Since the soil was low in Ca and the systems were approximately neutral in reaction, side reactions with soluble iron or aluminum, or with exchangeable Ca, were minimized. Davis found, however, that unlike typical adsorption reactions the equilibrium was slow in attainment and not readily reversible upon dilution. He suggested that, during the long-time reaction (22 days), an absorption reaction, involving diffusion of phosphate into the solid phase and reaction of the phosphate with a component of the solid phase, may be responsible for the fixation. From a consideration of the activities of the phosphate in the solution and solid phases according to Henry's law, he was able to show that an absorption reaction, involving a diffusion process, may give results which can be expressed in terms of a Langmuir adsorption equation, but the reversibility would be influenced by the rate of the diffusion process or rate of solution of the new phosphate complexes in the solid phase.

In general, it can be concluded that the adsorption reaction is fundamentally involved in the fixation of phosphate, but that fixation is, undoubtedly, "adsorption plus."

Isomorphous Replacement Reactions

Reactions involving isomorphous replacement of components of a crystal lattice may be regarded as "adsorption-plus" reactions of three general types:

1. Continuation of the adsorption reaction through intercrystalline absorption.
2. Transformation of the adsorption reaction to one of isomorphous replacement of hydroxyl or silicate anions from the crystal lattice.
3. Decomposition of the isomorphously transformed crystal lattice as the limits of permissible isomorphous replacement are exceeded, followed by recrystallization as a new mineral compound.

MacAuliffe *et al.* (38) in studying the kinetics of the reaction:

Surface phosphate (P^{31}) + solution phosphate (P^{32}) $\leftrightarrows$

surface phosphate (P^{32}) + solution phosphate (P^{31})

deduced that two distinct reactions were involved. In the first of these the ratio P^{32} solid:P^{32} solution changed logarithmically with time; in the second, the ratio changed linearly with time. Evidence concerning the nature of this second reaction was obtained from experiments using deuterium exchange for hydrogen as a means of determining the extent of hydroxylic surfaces on several minerals. These experiments indicated that there was considerable diffusion of deuterium from the surface into crystals of gibbsite kaolinite, and halloysite. In kaolinite, 35.8 per cent of the hydroxyl groups in the crystals were exchanged after 35.9 days at 170°C; in halloysite, at the same temperature, 52.9 per cent of the hydroxyl groups were exchanged in 4.1 days. Such values may be regarded as indicative of the lability of the hydroxyl groups that might be involved in the continuing reaction of phosphate with the soil minerals.

From x-ray diffraction data for phosphated kaolinite and halloysite, Stout (51) suggested that phosphate tetrahedrons in replacing hydroxyl ions between the crystal units at the cleavage planes modified the lattice parameters sufficiently to result in disordered packages in kaolinite and to introduce a sharp diffraction line corresponding to a plane of repetition at 16.1 Å in halloysite. The latter was regarded as evidence of the holohedral character of the halloysite crystal. Although Stout did not attempt to identify the phosphated halloysite as any particular mineral species from its x-ray pattern, Cole and Jackson (13) reported its identification "as potassium aluminum taranakite," which is characterized by the formula $2K_2O \cdot 3Al_2O_3 \cdot 5P_2O_5 \cdot 26H_2O$. In his phosphation Stout used an ammonium salt rather than a potassium salt; hence his phosphation of halloysite could hardly have produced a potassium aluminum taranakite. Furthermore, the work of Haseman *et al.* (22) has indicated that ammonium systems gave NH_4Al-phosphate crystals that did not have the x-ray diffraction pattern of K-Al-taranakite. Wada (56), however, recently (1959) reported the formation of an ammonium taranakite by the reaction of ammonium phosphate with soil clays.

Stout further indicated that the reaction was primarily a displacement of hydroxyl by phosphate from a determination of the water loss in a system: clay mineral + KH_2PO_4 + dehydration at 150°C. The loss of water by this system, in excess of the sum of the H_2O lost by the clay mineral alone and the KH_2PO_4 alone, was approximately equivalent to the milliequivalents of phosphate fixed.

Additional support of the hydroxyl replacement mechanism was given by Kelly and Midgley (29), who found a direct relation between the amount of phosphate fixed and the increase in pH obtained when isohydric suspensions of various solid phases and phosphate solutions were mixed. The materials used included $Fe(OH)_3$, finely ground kaolin, and Hermon fine sand B horizon enriched with R_2O_3.

Although the theory of hydroxyl replacement mechanism is strongly supported, there is also evidence that silicate replacement by phosphate may be important in phosphate fixation.

Thus, Mattson (43), in preparing phosphosilicates of Al and Fe, found the combining capacity of the phosphate to be much greater than that of the silicate. The proximate compositions of the isoelectric precipitates he obtained were: $Al_2O_3 \cdot (SiO_2)_{0.132}(P_2O_5)_{0.762}$ at pH 5.6, and $Fe_2O_3 \cdot (SiO_2)_{0.087}(P_2O_5)_{0.840}$ at pH 4.2. Since the phosphate and silicate were added to the precipitation systems in a ratio of 3 Si:2P, it may be deduced that the bonding energy of phosphate with ferric and aluminum ions is considerably greater than that of the silicate.

Mattson suggested that desilication of products of rock weathering takes place when phosphatic solutions, such as leachings from guano beds, react with the products.

Ross and Hendricks (48), in discussing the minor constituents in clay minerals, suggested that some proxying of Si by P may occur, as evidenced by the work of McConnell on the apatite group, in which he shows that 17.31 per cent of SiO_2 proxies P_2O_5 in ellestadite. The importance of this proxying in the montmorillonite group of clays cannot be evaluated, since phosphorus was not determined in many of the clay analyses which they reported.

Definite evidence of the displacement of silicate from soil colloids by phosphate was reported by Toth (52), who found that as the phosphate fixation on each of four soil colloids increased, the amount of silica in the supernatant liquid increased.

When Low and Black (37) phosphated kaolinite at 60°C, considerable silica was released within 48 hours, but when the reaction was carried out at 45°C very little silica was released until the eighth day. After this time silica continued to be released up to 35 days (Figure 9.3). It was suggested that the silica release was the result of isomorphous replacement of silicon tetrahedrons up to about the fourteenth day, or when about 24 millimoles of SiO_2 were released by each 100 g of clay. Beyond this point the replacement was described as resulting in an unstable phosphate compound, because of an infraction of the electrostatic valence rule when two phosphorus tetrahedrons share a common oxygen ion. It should be pointed out, however, that silicon tetrahedrons on the edges and corners of the crystal,

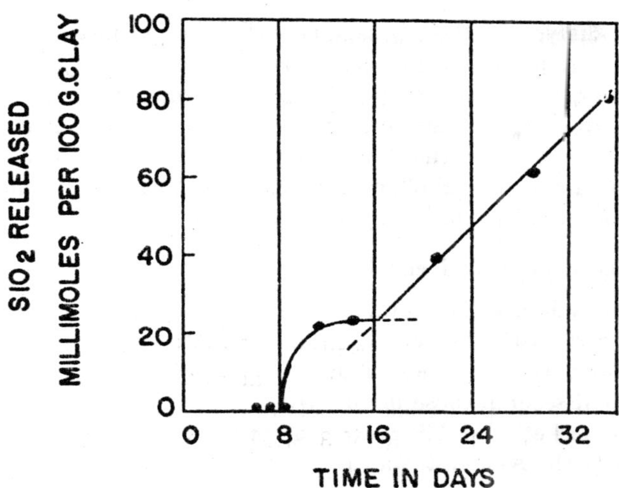

Figure 9.3. Rate of silica release from kaolinite in presence of phosphate (37).

which have two corners unshared along the edge, may be proxied by phosphorus tetrahedrons without infraction of the electrostatic valence rule. Of the two shared tetrahedral corners, one would be shared with an adjacent silicon tetrahedron and the other with an aluminum octahedron.

One may infer from these considerations that, when more edges and corners are present for each unit mass by virtue of finer natural particle size or grinding of the coarser particles, a larger amount of stable isomorphous replacement of silicon by phosphorus would occur.

Further study of the lattice configurations which the phosphate tetrahedrons may assume in replacing silica tetrahedrons or hydroxyl groups will be necessary to evaluate the importance of these mechanisms in the fixation of phosphorus.

Some of the crystalline products produced by phosphation of illite, kaolinite, montmorillonite, goethite, and limonite have been identified by Haseman *et al.* (22) as members of an isomorphous series with the general composition $(H, K, Na, NH_4)_{3-x}(Fe, Al)_{x/3}PO_4 \cdot nH_2O$, where x is greater than zero but less than 3. Some of these products showed x-ray diffraction patterns similar to that of palmerite.

Other phosphatolysis products were identified as members of the variscite-strengite isomorphous series, $Al_xFe_yPO_4 \cdot 2H_2O (x + y = 1)$.

Some of the formulas assigned to the various products might be questioned if silicon and phosphorus can proxy each other to a limited extent, since, upon analysis of the crystals, silica was found in every case where the crystals were produced from phosphation of silicate clay minerals. This silica, however, was assumed to be wholly an impurity and not a lattice constituent. In certain crystalline products, if the silica had been assumed to be a true constituent of the crystals, less of the aluminum would have had to be discarded and several of the formulas would have been similar to the phosphosilicate of aluminum prepared by Mattson.

Double Decomposition Reactions

From a consideration of solubility product relationships, several reactions may be regarded as significant in the fixation (precipitation) of soluble phosphates. Broadly, they fall into two categories: first those involving Fe and Al; and second, those involving Ca.

Although the Fe^{3+} and Al^{3+} reactions are not identical, they are similar; therefore, only the Al^{3+} systems will be considered here in detail.

The aluminosilicates and free sesquioxides may be regarded as the primary compounds supplying the Al^{3+}. Using kaolinite and gibbsite as typical examples of these two classes of compounds, we see that the hydroxyl ion concentration has a dominating influence in the solubility product relation in both cases:

Kaolinite $Al_2(OH)_4Si_2O_5$ or $Al(OH)_2(HSiO_3) \leftrightharpoons Al^{3+} + 2OH^- + HSiO_3^-$

$$K_{sp} = [Al^{3+}][OH^-]^2[HSiO_3^-]$$

Gibbsite $Al(OH)_3 \leftrightharpoons Al^{3+} + 3OH^-$

$$K_{sp} = [Al^{3+}][OH^-]^3$$

From the solubility product of the gibbsite, 1.9×10^{-33}, it can be calculated that at pH 5.0 the Al^{3+} concentration will not exceed 1.9×10^{-6} *M*. At pH 6.0 it drops to 1.9×10^{-9}, whereas at pH 4.0 it increases a thousandfold to 1.9×10^{-3}.

If phosphate ions are added to a soil system in the form of a soluble fertilizer, the quantities added at any one time do not usually exceed 200 pounds of P_2O_5 per acre. In a 2-million-pound acre furrow slice containing 20 per cent water, if the phosphate were dissolved uniformly throughout the soil water the resulting phosphate concentration would be 0.007 *M*.

Now, assuming that the phosphate will form a variscite-like compound, $Al(OH)_2H_2PO_4$, having a solubility product of 2.8×10^{-29} according to the equation $Al(OH)_2H_2PO_4 \leftrightharpoons Al^{3+} + 2OH^- + H_2PO_4^-$, we may calculate the extent by which the aforementioned phosphate concentration (0.007 *M*) exceeds the equilibrium concentration at pH 4.0 when Al^{3+} is maintained at 1.9×10^{-3} *M* by an excess of solid phase $Al(OH)_3$.

$$[H_2PO_4^-] = \frac{2.8 \times 10^{-29}}{[1.9 \times 10^{-3}][10^{-10}]^2} = 1.5 \times 10^{-6}$$

Thus the phosphate concentration would be reduced from 7×10^{-3} M to 1.5×10^{-6} M, or a reduction in concentration of 99.98 per cent.

In absence of solid phase $Al(OH)_3$, the Al^{3+} concentration may be maintained by solid-phase aluminosilicates. Apparently no studies of the solubility product of an aluminosilicate like kaolinite have been reported in the literature. But Mattson (42) prepared isoelectric precipitates of aluminosilicates for which solid-phase composition and equilibrium ion concentration data have been given. Two of these systems are as follows:

System 27

Composition of floc = $Al_2O_3(SiO_2)_{2.90} \rightleftharpoons Al(OH)_{1.55}(HSiO_3)_{1.45}$

According to the reaction $Al(OH)_{1.55}(HSiO_3)_{1.45} \rightleftharpoons Al^{3+} + 1.55OH^- + 1.45HSiO_3^-$ and the equilibrium concentration $[Al^{3+}] = 6 \times 10^{-6}$ M; $[HSiO_3^-] = 1.07 \times 10^{-2}$ M; $[OH^-] = 10^{-9.8}$ M (pH 4.2) the $K_{sp} = 1.5 \times 10^{-22}$.

System 29

Composition of floc = $Al_2O_3(SiO_2)_{2.80} \rightleftharpoons Al(OH)_{1.6}(HSiO_3)_{1.4}$.

According to the reaction: $Al(OH)_{1.6}(HSiO_3)_{1.4} \rightleftharpoons Al^{3+} + 1.6\ OH^- + 1.4\ HSiO_3^-$ and the equilibrium concentrations $[Al^{3+}] = 4 \times 10^{-5}$ M; $[HSiO_3^-] = 5.65 \times 10^{-3}$ M; $[OH^-] = 10^{-9.3}$ M (pH 4.7), the $K_{sp} = 3.59 \times 10^{-23}$.

Although these precipitates are not identical with kaolinite, it may be assumed that they resemble the amorphous aluminosilicates that have been described as present in "unpurified" clay fractions from soils. They probably more closely approximate halloysite or allophane. Low and Black (36), for example, found the equilibrium SiO_2 concentration to be 1.8×10^{-4} M when kaolinite was suspended in dilute NH_4Cl solution at pH 4.7, whereas the SiO_2 concentration in Mattson's isoelectric precipitate at pH 4.7 was 5.65×10^{-3} M. The soluble Al^{3+} concentration from well-crystallized and aged aluminosilicates, such as kaolinite, could also be expected to be less than that found by Mattson with his precipitates. The equilibrium Al^{3+} concentration in his *System 29* at pH 4.7 is slightly greater (4×10^{-5} M) than would be expected (1.52×10^{-5} M) from $Al(OH)_3$ at this pH. It is slightly greater than that found by Coleman (15) upon equilibration of a kaolinitic soil clay at pH 4.8. He found 0.6 mg of R_2O_3 in solution per gram of clay in a 50-ml volume. If all of the R_2O_3 is assumed to be Al_2O_3, this gives an Al^{3+} concentration equivalent to 1.18×10^{-5} M.

Magistad (39) found that the curve for the solubility of Al in the soil solution at various pH values practically coincided with the curve for the solubility of aluminum sulfate at the same pH value. From the nature of the reactions already described, it would be expected that relatively small amounts of solid-phase gibbsite in a soil could control the Al concentration even in the presence of large amounts of aluminosilicates. Results similar to those of Magistad, therefore, would be expected.

Lindsay *et al.* (35) calculated the Al^{3+} activities in $CaCl_2$ extracts of several acid soils by using the equation of the first stage hydrolysis of the Al-hexahydronium ion, $Al(H_2O)_6^{3+} + H_2O \leftrightarrows AlOH(H_2O)_5^{2+} + H_3O^+$, and the pH, the total Al, the ionic strength of the soil extracts, and the Boltzmann equation for the distribution of cations in proximity to a charged surface. The total Al concentration in the $CaCl_2$ extracts of one acid soil ranged from 9.2×10^{-5} *M* in 0.001 $M \cdot CaCl_2$ to 237×10^{-5} *M* in 0.1 $M \cdot CaCl_2$. However the calculated Al^{3+} activities were, respectively, 5.4×10^{-5} *M* and 26×10^{-5} *M* at these two range limits. They concluded from their data that the constancy of the pH-$\frac{1}{3}p$Al values could be predicted from the Donnan and Gouy theories and did not necessarily indicate some crystalline form of hydrous aluminum oxide having a definite solubility product.

Lindsay and Moreno (34) developed a solubility diagram for variscite [$AlPO_4 \cdot 2H_2O^-$], strengite [$FePO_4 \cdot 2H_2O$], fluorapatite [$Ca_{10}(PO_4)_6F_2$], hydroxyapatite [$Ca_{10}(PO_4)_6(OH)_2$], octocalcium phosphate [$Ca_4H(PO_4)_3 \cdot 3H_2O$], and dicalcium phosphate dihydrate [$CaHPO_4 \cdot 2H_2O$]. In constructing the diagram they assumed that Al^{3+} activity was limited by the solubility of gibbsite, Fe^{3+} activity by that of goethite, F^- activity by that of fluorite, that Ca^{2+} activity was arbitrarily that of a 0.005 *M* solution, and that the activity of H_2O was unity. The resulting solubility diagram in Figure 9.4 indicates the important effect of pH on all the systems. Although the solubility diagram is supported in general by phosphorus solubility data in acid and alkaline soils, Lindsay and Moreno (34) suggest that kinetic considerations during equilibration in soils could more often than not preclude the obtaining of solubility data which would correspond to any known solubility product.

Frink and Peech (20) later showed that crystalline gibbsite was extremely slow in dissolution in 0.01 *M* $CaCl_2$ soil extracts, and that such extracts to which gibbsite had been added remained undersaturated with respect to gibbsite, even after one month of equilibration.

It would appear from these data that the supply of aluminum in soils for reaction with phosphate may be controlled as much by the aluminosilicates and the aluminum on exchange sites as by gibbsitic aluminum. The complexity of the aluminum hexahydronium ion with its monomeric and polymeric hydroxy forms, as described by Jackson (26), increases the difficulty of interpreting solubility-product phenomena, particularly in soils containing clays with large layer charge densities like vermiculite.

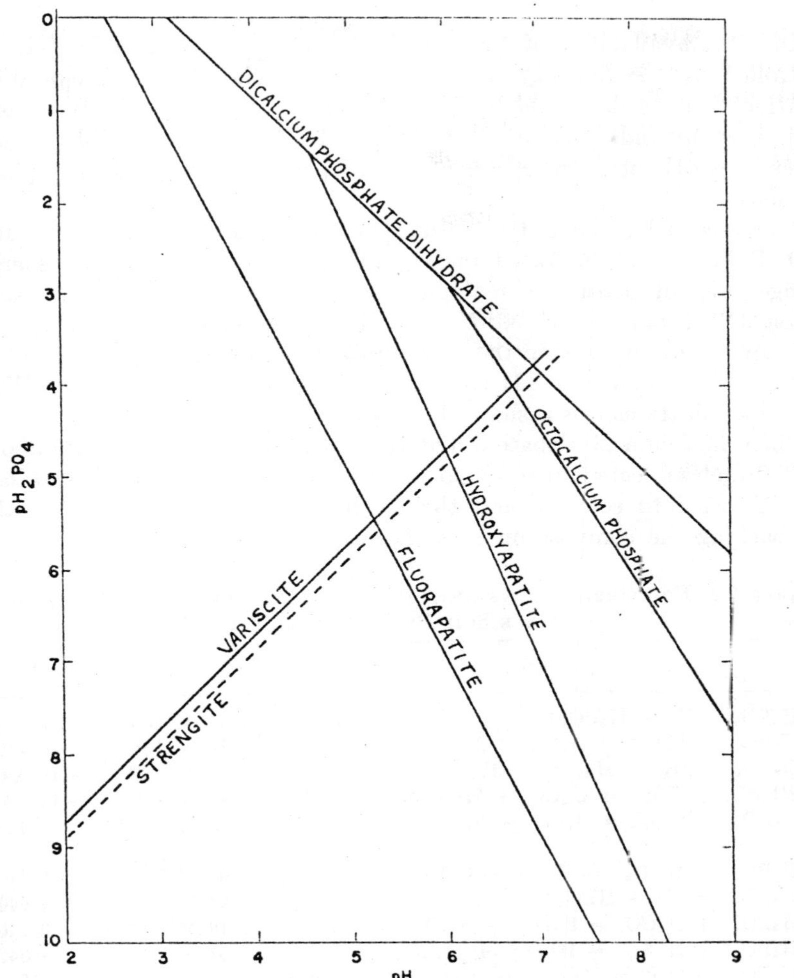

Figure 9.4. Solubility diagram for phosphate compounds at 25°C and with the pAl of gibbsite, the pFe of goethite, or the pF of fluorite and a pCa equal to 2.5 (34).

Further, as Huffman *et al.* (25) pointed out, all the complex phosphates (K and NH_4 taranakites, acid NH_4 and K ferric phosphates, and Ca ferric phosphate) dissolve incongruently in water, leaving solid residues which behaved like tertiary phosphates of Fe and Al in which the P:Al or P:Fe ratio approached 1.0.

Since the availability of Al^{3+} and Fe^{3+} for reaction with phosphate is controlled by the hydroxyl ion concentration, phosphorus precipitation (fixation) can be decreased by increasing the pH of the soil. But when calcium compounds, such as $Ca(OH)_2$ and $CaCO_3$ are responsible for the increase in pH, other reactions may take place to precipitate the phosphorus.

From a consideration of the ionization constants of H_3PO_4, H_2CO_3, and H_2O, Buehrer (10) indicated the equilibrium constants and free energy changes of soil phosphate reactions that may be encountered in soils (Table 9.2). From these equations and solubility product data, he proposed a phosphate cycle showing the relation of the calcium phosphates to one another when subject to pH changes, hydrolysis, and carbonation. With slight modifications this is shown in Figure 9.5.

When dicalcium phosphate dihydrate is held in aqueous medium above pH 5.0 and subjected to repeated extraction, more phosphorus than calcium comes into solution and the residue becomes more basic, finally approaching the composition of hydroxyapatite.

TABLE 9.2. EQUILIBRIUM CONSTANTS AND STANDARD FREE ENERGY CHANGES OF POSSIBLE SOIL PHOSPHATE REACTIONS*

Reaction	K_{eq}	ΔF
1. $H_3PO_4 = H^+ + H_2PO_4^-$	1.1×10^{-2}	2,670
2. $H_3PO_4 + HCO_3^- = H_2CO_3 + H_2PO_4^-$	3.14×10^{4}	−6,140
3. $H_3PO_4 + OH^- = H_2PO_4^- + H_2O$	1.12×10^{12}	−16,450
4. $2H_3PO_4 + CO_3^{2-} = H_2CO_3 + 2H_2PO_4^-$	6.4×10^{13}	−17,500
5. $H_2PO_4^- + HOH = H_3PO_4 + OH^-$	9.15×10^{-13}	16,430
6. $H_2PO_4^- + HCO_3^- = H_3PO_4 + CO_3^{--}$	4.91×10^{-9}	11,340
7. $H_2PO_4^- = H^+ + HPO_4^{2-}$	1.95×10^{-7}	8,940
8. $H_2PO_4^- + H_2CO_3 = H_3PO_4 + HCO_3^-$	3.18×10^{-5}	6,140
9. $H_2PO_4^- + HCO_3^- = HPO_4^{2-} + H_2CO_3$	5.57×10^{-1}	−347
10. $H_2PO_4^- + OH^- = HPO_4^{2-} + H_2O$	1.94×10^{7}	−10,000
11. $HPO_4^{2-} = H^+ + PO_4^{3-}$	3.6×10^{-13}	16,600
12. $HPO_4^{2-} + HOH = H_2PO_4^- + OH^-$	5.16×10^{-8}	10,000
13. $HPO_4^{2-} + HCO_3^- = PO_4^{3-} + H_2CO_3^-$	1.03×10^{-6}	8,170
14. $HPO_4^{2-} + HCO_3^- = H_2PO_4^- + CO_3^{2-}$	2.77×10^{-4}	4,860
15. $HPO_4^{2-} + CO_3^{2-} = PO_4^{3-} + HCO_3^-$	6.67×10^{-3}	2,970
16. $PO_4^{3-} + HOH = HPO_4^{2-} + OH^-$	2.79×10^{-2}	2,120
17. $PO_4^{3-} + HCO_3^- = HPO_4^{2-} + CO_3^{2-}$	6.67×10^{3}	−2,970
18. $PO_4^{3-} + H_2CO_3 = HPO_4^{2-} + HCO_3^-$	9.73×10^{5}	−8,170

* Courtesy, T. F. Buehrer and Arizona Agricultural Experiment Station.
Free energies expressed in calories per mol of reactant under standard conditions.

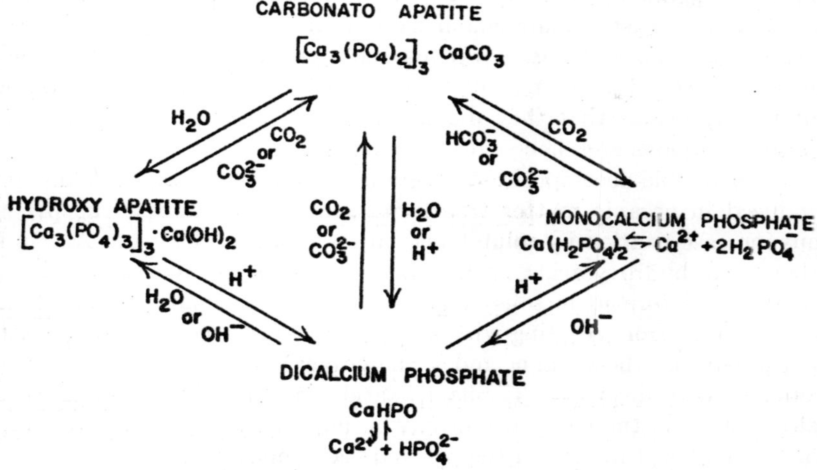

Figure 9.5. Phosphate cycle (10).

In like manner dicalcium phosphate is thermodynamically unstable in the presence of $CaCO_3$ and changes slowly to carbonatoapatite. Both the apatites are much less soluble than the dicalcium phosphate. A solubility product constant of 1.53×10^{-112} has been reported (47) for hydroxyapatite with a formulation $Ca_{10}(PO_4)_6(OH)_2$.

From the reactions involved, it is obvious that Ca^{2+} concentration and pH are the controlling factors in the precipitation of the calcium phosphates. The Ca^{2+} acts primarily from the standpoint of its common ion effect. The pH controls which phosphate ion will be dominant, and in calcareous soils it controls the solubility of $CaCO_3$ and hence the Ca^{2+} concentration.

Most investigators have found that the minimum solubility of the calcium phosphates in soils occurs between pH 7 and 8. Below pH 7.0 the HPO_4^{2-} concentration decreases rapidly. Above pH 8.0 the increasing alkalinity is usually the result of increasing amounts of alkali carbonates. These carbonates supply large amounts of CO_3^{2-} which decreases the solubility of the $CaCO_3$ and hence the concentration of Ca^{2+}.

Organic Matter and Phosphorus Fixation

No discussion of the phosphorus fixation problem would be complete without a consideration of the part played by organic compounds in the various reactions. Because of the complexity of the organic phase in soil, the effects are discussed briefly and only with respect to their bearing on the three classes of reactions previously described.

(1) In adsorption reactions the organic compounds, being dominantly

anionic in character, would be expected to compete with the phosphate anion in polar adsorption phenomena and, hence, decrease fixation.

(2) With respect to isomorphous replacement reactions, it is highly improbable that the organic anions would fit into the clay mineral lattices, but to any extent that they did do so, they would be competitive with phosphate anions and, hence, decrease fixation.

(3) In double-decomposition reactions involving Fe and Al, the acids produced in organic matter transformations could decrease the pH and thus increase fixation by solubilizing larger amounts of Fe and Al. On the other hand, hydroxy acids, such as tartaric, citric, malonic, and malic, are known to be formed in soils, and such acids can chelate Fe and Al and prevent them from reacting with phosphate to form insoluble precipitates. In Ca systems the organic and carbonic acids from the organic matter would tend to decrease phosphate fixation in noncalcareous soils, but in calcareous soils the effect would not be important so long as solid-phase $CaCO_3$ remained to control the pH and Ca^{2+} concentration.

In general, the over-all effect of the organic phase in soils has been found to be such as to decrease phosphorus fixation.

Summary: Phosphorus Fixation

From the preceding discussion one must conclude that no one type of reaction can account for all the phosphorus fixation in a complex chemical medium like the soil. In acid soils the initial or fast reaction appears to be dominated by polar adsorption and by simple precipitation according to solubility product principles. The continuing or slow reaction, on the other hand, seems to be more conclusively explained by an isomorphous replacement reaction and by a continuing precipitation reaction as the aluminosilicates and hydrous oxides try to maintain their "solubility product" concentrations.

All the facts are not incontrovertible, and the various mechanisms of fixation are aptly described in the literature as theoretical rather than dogmatic. There is a particular need for more information on the limiting conditions, both energetic and geometric, in isomorphous replacement reactions. The "solubility product constants" of the silicate clay minerals, the hydrous oxides, and the phosphates, in pure and in mixed systems, also need further study.

NUTRIENT CATION FIXATION

Cation fixation in soils occurs when exchangeable or water-soluble cations are converted to a form that cannot be readily extracted with a neutral solution of a salt. Potassium and ammonium ions have been primarily involved in such reactions, but most of the work has been directed at the potassium problem.

It has been known for some time that potassium added to soils over a period of years could not be completely accounted for without an assumption that the soil had in some manner converted a considerable amount of the element into some difficultly soluble form. But the first detailed study of this fixation was that of Volk (54, 55). His results pointed out that: (*a*) drying the soil was extremely important in the reaction; (*b*) the clay-size fraction was primarily responsible, but the quality of the clay was also important, since a kaolin clay sample and two laterite soils with large amounts of clay-size material did not fix potassium; (*c*) HCl treatment of a soil decreased its fixing power, whereas treatment with Na_2CO_3 or $Ca(OH)_2$ increased it; (*d*) long-continued K-fertilization resulted in decreasing the K-fixing power and increasing the muscovite or muscovite-like component of the clay-size fraction, and (*e*) the greater the amount of potassium added, the greater was the amount fixed upon drying but the smaller was the percentage fixed. Table 9.3 and Figure 9.6 indicate the order of magnitude of some of these effects.

From his examination, by x-ray diffraction methods, of the clay fraction (2-0.3 μ) of specific gravity greater than 2.667, Volk found that the sample from the KCl plot of the long-time Pennsylvania Soil Fertility Plats gave stronger second-order diffraction lines of muscovite than did that from the check plat which received no potassium. He concluded, therefore, that "a portion of the added potassium had reacted with colloidal silicates and become fixed in the form of muscovite."

Volk's conclusion is weakened by apparent dissimilarities in the mineral-

TABLE 9.3. POTASSIUM-FIXING CAPACITY OF HAGERSTOWN SILT LOAM

Treatment of soil*	Fixing† capacity (lb/A)
A. Kept moist	0
B. Air-dried once	252
C. Dried once at 70°C	320
D. Wetted and dried 10 times at 70°C	410
E. Wetted and dried 32 times at 70°C	428
F. Limed, then treated same as D	488
G. Leached with HCl, then same as D	280
H. Fertilizer with 5000 lb of K_2O over a period of 50 yr (KCl plat); treated same as D	240

* Surface soil from check plat of Pennsylvania Soil Fertility Plats [selected data from Volk (55)].
† From KCl added in solution at a rate equivalent to 1000 lb K_2O per 2 million lb of soil.

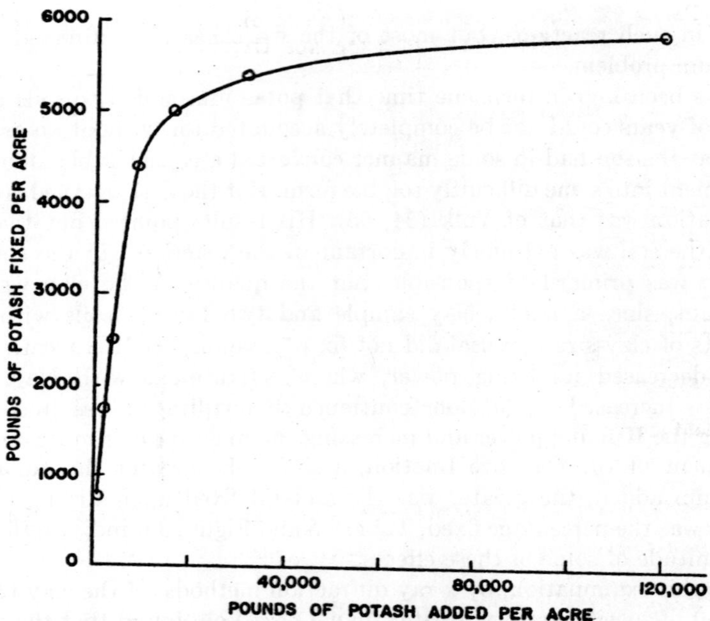

Figure 9.6. Effect of concentration of soluble potash on amount fixed by Miami silt loam when treated with varying quantities of potassium bicarbonate and then alternately wetted and dried 10 times at 70°C (55).

ogical composition of the coarser fractions of the two samples and by lack of data on the original muscovite content of the two plats. But it is quite possible, in light of our present knowledge of the clay minerals in Pennsylvania Hagerstown soil, that potassication of hydrous mica lattices had occurred, causing a sharpening of the x-ray diffraction pattern of the hydrous mica and emphasizing its resemblance to that of muscovite.

Other investigators continued the work of Volk. The next important contribution was that of Chaminade (11), who related the fixation to the exchange capacity of the soil. He found that maximum potassium fixation resulted if more than 4 per cent of the exchange capacity was occupied by potassium. Chaminade and Drouineau (12) indicated that NH_4^+ and Mg^{2+} could also be fixed but that Ca^{2+} and Na^+ could not. They suggested that the fixation resulted from a diffusion of the cations into the crystal nuclei of the soil colloids but offered no adequate explanation for the differences among the cations.

The next important contributions were those of Truog and Jones (53)

TABLE 9.4. EQUIVALENCE OF MAGNITUDE OF K-FIXATION AND THE REDUCTION IN EXCHANGE CAPACITY*

Material and treatment		K_2O Fixed (meq/100 g)	Reduction in exchange capacity after K-fixation (meq/100 g)
Bentonite:			
Wetted and dried 20 times at 80°C		33.9	33.0
Dried at 110°C for 75 hours		27.0	28.0
Dried at 145°C for 48 hours		31.5	34.0
Soils†	Horizon		
Miami silt loam	A	1.5	1.9
	B	4.7	4.8
Carrington silt loam	A	2.7	2.7
	B	3.4	3.2
Richfield clay	A	4.0	3.8
	B	5.0	5.1

* Courtesy E. Truog and R. J. Jones and American Chemical Society, Washington, D. C., publishers of Ind. Eng. Chem.

† Soils freed of organic matter, saturated with K, wetted and dried 20 times at 80°C.

and of Kolodny (30), who independently pointed out the relation between the amount of potassium fixed and the amount of reduction in exchange capacity of bentonite clay when the K-saturated clay was wetted and dried or when KCl was added to a bentonite sample. Truog and Jones also found that soils behaved similarly (Table 9.4). There was a strong indication that the exchangeable ions were the seat of the fixation reaction. Since such ions had been described as located in positions between the layers of expanding lattice minerals, such as Wyoming bentonite (montmorillonite) and Putnam clay (beidellite), Joffe and Kolodny (28) suggested that a part of the exchangeable K ions is converted to a fixed position when one-dimensional lattice contraction takes place as a result of heating.

A clue to the specificity of the fixation reaction for particular cations was provided by the work of Page and Baver (45). They studied the fixation of cations by the colloids from a Wyoming bentonite (montmorillonite) and from a Miami silt loam soil and related it to the size of the unhydrated ions (Figure 9.7). Although their equilibrium concentration procedure as a measure of fixation was not so severe as the leaching procedure used by other workers, it indicated that not only K but NH_4, Rb, Cs, and Ba could be fixed.

A theory was proposed that gave consideration to the ionic size of the unhydrated ions, the expanding-contracting nature of the montmorillonite type of minerals, and the geometric arrangement of the oxygen ions at the

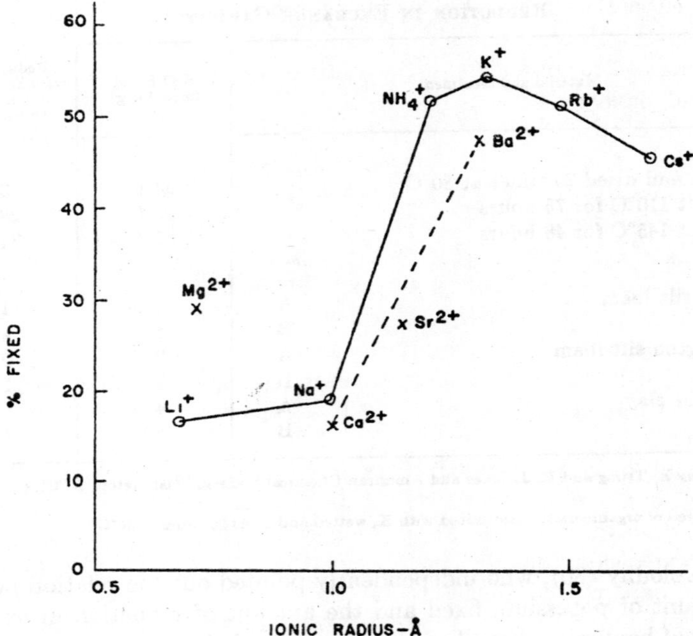

Figure 9.7. The relation of ionic size to cation fixation by Miami colloidal clay saturated by various cations (45).

surface of the crystal layers. The essential elements of their theory were as follows:

1. The exposed surface between the layers of the 2:1 expanding-lattice clays consists of a sheet of oxygen ions arranged hexagonally, the opening within the hexagon being 2.8 Å in diameter.

2. As the clay is dehydrated, the layers contract and the ions lose their hulls of oriented water molecules, approaching the unhydrated ionic diameter in size. These diameters are respectively: Li, 1.20 Å; Na, 1.90 Å, K, 2.66 Å; NH_4, 2.96 Å; Rb, 2.96 Å; Cs, 3.38 Å; Mg, 1.30 Å; Ca, 1.98 Å; and Ba, 2.70 Å.

3. Ions whose diameters allow them to fit snugly into the lattice "holes" should be held very tightly, because they are closer to the negative electrical charges within the crystal, and by fitting into the "hole" they would allow the layers to come closer together and be locked against rehydration and re-expansion. Larger cations, which could not enter the "holes," would remain more loosely held between the layers rather than within the layer, and would be more accessible for rehydration. Smaller cations would be

able to enter the "holes" but would be too small to contact and bind the two layers together.

It remained for other investigators to refine the "lattice-hole theory" of Page and Baver.

Hendricks, Nelson, and Alexander (24), using controlled vapor hydration of montmorillonite, found a maximum basal spacing of 12 Å for K-saturated material, whereas Ca^{2+}-, Mg^{2+}-, Na^{+}-, and H^{+}-saturated samples had a value of about 16 Å. The 12 Å spacing corresponded to a single layer of water molecules, whereas the 16 Å spacing was equivalent to a double layer. The thermal water-loss curves were regarded as indicative of the unhydrated character of the K ion.

Jackson and Hellman (27) found that when a K-saturated montmorillonite was dried at 30°C it showed a basal spacing of 12.3 Å, whereas the Na- and Ca-saturated samples had basal spacings of 14.6 Å and 15.1 Å, respectively.

Barshad (5) investigated the basal spacings of montmorillonite saturated with various cations under air-dried conditions and when subsequently immersed in water. Some of his data are shown in Table 9.5.

The expansion that occurred when the Mg, Ca, Ba, H, Li, and Na samples were wetted suggests that these ions would not be fixed. With the K, NH_4, and Rb, the lattice remained contracted and hence these ions would be expected to be less accessible to the displacing cation.

Under no circumstances has fixation of K by montmorillonite occurred in absence of drying. Could it, then, be expected that, if a Ca- or Mg-saturated montmorillonite were kept moist while being converted to a K-saturated condition, the lattice would remain expanded? Further evidence is needed to determine whether a critical distance of contraction is necessary to fix the K.

Indirect evidence of the critical nature of the distance of contraction is given by an experiment of Page and Baver (45), who placed large organic

TABLE 9.5. BASAL SPACINGS (Å) OF MONTMORILLONITE (OTAYLITE) AS AFFECTED BY TYPE OF EXCHANGE CATION AND DEGREE OF HYDRATION*

Treatment	Saturating cation								
	Mg	Ca	Ba	H	Li	Na	K	NH_4	Rb
Air-dry	14.8	15.1	12.9	14.5	13.4	11.9	12.0	12.1	12.3
Immersed in H_2O†	20.7	17.9	18.5	19.3	23–18	24–18	13.2	13.2	12.3

* Courtesy I. Barshad and Mineralogical Society of America, publishers of Am. Minerologist.

† Thick pastes were prepared and packed into capillary tubes, which were then sealed with deKhotinsky cement to prevent loss during exposure to x-rays

cations with the K between the plates. A decrease in K fixation was ascribed to the layers being kept apart by the large cations, thus making the K ions accessible for exchange. They did not measure the basal spacing of the dried clay or the remoistened clay, but Gieseking (21) showed that as increasing amounts of a large organic cation (tributyl monoheptyl ammonium ion) were added to a H montmorillonite, the H ion was displaced and the basal (001) spacing increased from 13.7 Å, in the air-dried condition, to 17.2 Å. The evidence is suggestive, at least, that, although K ions may be dehydrated, their fixation is decreased when the sheets of oxygen ions are prevented from approaching each other.

Additional clues to the problem are furnished by the work of Barshad (4, 5) with vermiculites, which fix K against extraction by NH_4, and NH_4 against extraction by K, both ions being slowly but completely extractable by Mg, Ca, and Na ions.

From an examination of the changes in basal spacings, differential thermal curves, and water-loss data for vermiculite saturated with various cations, it was concluded:

1. In air-dry condition the Mg- and Ca-saturated samples expanded a distance equivalent to about two layers of water molecules; the Ba-, Li-, and Na-saturated samples expanded a distance of about one layer; but the K-, NH_4-, Rb-, and Cs-saturated samples were almost completely contracted.

2. Upon immersion in water, the Mg and Ca forms did not expand any further. The Li, Na, and Ba forms expanded the equivalent of a monomolecular layer of water, thus attaining, but not exceeding, the total interlayer expansion of two monomolecular layers. In sharp contrast, the K, NH_4, Rb, and Cs forms did not change their interlayer spacing when immersed in water. Thus the K, NH_4, Rb, and Cs ions appear capable of keeping the lattice completely contracted when saturating the vermiculite, but the force of contraction can be overcome by subjecting the material to a cation-exchange reaction with NaCl and $MgCl_2$ and displacing the contraction stabilizing ion.

It would be interesting to know at what level of K depletion or Mg saturation the forces of contraction would be overcome, and whether the energy relations can be approached from an analysis of the hydration energies of the respective cations in relation to the bonding energy of the K ion. Dennis and Ellis (17) indicated that vermiculite collapsed, in the moist state, from 15.9 to 13.2 Å when 40 per cent of the total exchange capacity was occupied by K if the complementary cation was Na. With Ca as the complementary cation, collapse in the moist state occurred when as little as 20 per cent of the exchange sites were occupied by K.

Latimer (32) has shown that the entropy of hydration of the K ions is much greater than that of the other common soil cations, such as Mg, Ca, and Na, and is of about the same order of magnitude as that of Rb. This relationship is more than fortuitous and suggests that a study of the energy relations involved in the hydration of the ions should be examined more thoroughly.

Bailey (3) and Wear and White (57) have emphasized the greater importance of tetrahedral-layer charges resulting from isomorphous substitution of Al for Si in the tetrahedral layers, in comparison with the octahedral charges in fixation of the interlayer cations, because the force of attraction between the negative charges in the lattice and the interlayer cations would be greater. Wear and White indicated the distance between the K ion and negative lattice charges to be 2.19 and 4.99 Å, respectively, for the tetrahedral and octahedral negative sites. Since the force of attraction is inversely proportional to the square of the distance, the force of attraction to the tetrahedral charge would be more than four times as great. Beidellite, hydrous mica (illite), and vermiculite clays have relatively more tetrahedral charges than the Mg montmorillonites and should therefore be able to hold more potash in a fixed condition. Data in Table 9.6 supports this relationship.

In a study of ion activities in clay suspensions, using clay membrane electrodes, Marshall and McLean (41) examined the effect of clay type and drying on the activity of the K ion in K-saturated clays. In the clay systems kept continually moist after K saturation, it was found that the fractions of the total K added which was active were, respectively, 26.9,

TABLE 9.6. RELATIONSHIP BETWEEN TOTAL POTASSIUM FIXED IN CRYSTAL STRUCTURE AND NEGATIVE CHARGES ORIGINATING IN THE TETRAHEDRAL POSITIONS*

Clay mineral	Potassium prefixed in original crystal structure	Potassium in crystal structure after fixation, dried from benzene	Calculated tetrahedral charge†
	meq/100 g Clay		
California bentonite	6.0	22.7	18
Wyoming bentonite	1.5	27.1	47
Putnam clay	27.1	38.9	73
Illite	128.5	130.9	96

* Courtesy J. I. Wear and J. L. White and The Williams & Wilkins Co., Baltimore, Md., publishers of Soil Sci.

† Isomorphous substitution in tetrahedral layer calculated from total analysis. Theoretically, 25 per cent replacement of Al for Si in the tetrahedral positions provides a negative charge of 251 meq per 100 g of clay material.

7.3, and 11.9 per cent for montmorillonite, beidellite, and illite. When the K clays were dried at 35°C and 35 per cent relative humidity and resuspended, the values became 20.0, 4.6, and 10.9 per cent, respectively.

Obviously, the two clays having the higher degree of tetrahedral substitution, beidellite and illite, had the smaller fraction of their potassium remaining active. At the same time, the beidellite, which has about the same amount of tetrahedral substitution as illite but less fixed potassium and, hence, more active tetrahedral charges, rendered more of the added potassium inactive on drying.

Another approach to the problem of the energetics of the cations in the interlayer positions is the study by Barshad (7) of the influence, on the interlayer spacing, of various organic solvents and solutions of varying dielectric properties and containing molecules of varying dipole moments. Expansion of Mg-saturated Macon vermiculite was essentially unchanged as the dielectric constant increased from that of pure water (78.5) to a value of 115.6. From this value and up to 289.6, expansion increased to a point where the mineral gelatinized. On the other hand, the spacing of the K vermiculite remained unchanged over the entire range of variation in dielectric constant. The effect of the dielectric constant appeared to be of greater importance than that of the dipole moment.

Although the media of extremely high dielectric constant were incapable of effecting a separation of the oxygen planes by hydration of the K ions in vermiculite, media of relatively low dielectric constant (butanol, 17.7, and ethanol, 25.0) could do this in a K montmorillonite. Since the K ion is the same, the difference must reside in the character of the negative charges on the lattice, particularly in their intensity as related to the octahedral or tetrahedral origin.

Although most of the early work indicated that appreciable cationic fixation occurred only on drying, it soon became apparent that in some instances fixation occurred even though the soil was kept continuously moist. A rather thorough study of this relation was made by Illinois workers (18). They observed that when potassium salts were added to field soils and stored under moist conditions, fixation occurred and generally increased in magnitude up to about 25 weeks. When no potassium was added, the soils either fixed K or released it, according to the relative levels of fixed and replaceable K present. A reversible equilibrium of the following nature seemed to occur:

$$\text{fixed K} \overset{A}{\rightleftharpoons} \text{replaceable K} \overset{B}{\rightleftharpoons} \text{water-soluble K},$$

equilibrium A being strongly hysteretic.

Stanford and Pierre (49) reported both K and NH_4 fixation under moist conditions in several calcareous Iowa soils and indicated that the mechanism of fixation of these two ions appeared to be similar. Later Stanford (50) showed that under moist conditions illite fixed K whereas montmorillonite did not; when dried, both clays fixed K. He further indicated that both the moist and the dry fixation of K by illite were greatly increased by increasing the pH with either NaOH or $Ca(OH)_2$. The fixing power of the montmorillonite, on the other hand, decreased with increasing pH up to 6.6 and then remained constant.

Results by Allison, Kefauver, and Roller (2) emphasized the importance of illite and vermiculite in the moist fixation of NH_4. Montmorillonitic soils, however, fixed little or no NH_4 under moist conditions. The ease of nitrification of the fixed NH_4 was least in the vermiculitic soils and greatest in the montmorillonitic soils (Table 9.7).

Young and Cattani (58) investigated 17 soil profiles from Oregon and Washington and found that air-dry soil samples fixed anhydrous NH_3 in amounts ranging from 6 to 1015 ppm N, and that this fixation from the gaseous phase generally exceeded by several-fold the fixation of N from aqua ammonia. In some samples where large amounts of anhydrous NH_3 were fixed by air-dry soil it was noted that clay lattice collapse from 14 to 10 A had occurred.

Brown and Bartholomew (9) postulated that anhydrous NH_3 sorbed by dry clays and retained against evacuation at room temperature for 10 hours was chemisorbed, in other words "fixed." They suggested that the mechanisms involved include (*a*) reaction of NH_3 with H^+ on exchange sites to form NH_4^+; (*b*) reaction with lattice hydroxyls at the broken edges of the crystal lattice in association with silicon but non-reaction with

TABLE 9.7. AMMONIUM FIXATION AND NITRIFICATION OF FIXED AMMONIUM IN SOILS CONTAINING DIFFERENT CLAY MINERALS*

Soil	Predominant clay mineral	Exchange capacity (meq/100 g)	pH	Fixation treatment	NH_4-fixed (meq/100 g)	Nitrification† (%)
Miami silt loam, 9–13 in.	Illite	17.1	5.7	Moist	1.7	
				Air-dried	3.3	4.7
Traver fine sandy loam, 24–60 in.	Vermiculite	7.6	9.9	Moist	3.4	
				Air-dried	3.4	2.1
Yolo silt loam, 10–21 in.	Montmorillonite	22.4	7.1	Moist	1.1	
				Air-dried	1.6	14.0

* Courtesy F. E. Allison, M. Kefauver, and E. M. Roller and Soil Science Society of America.

† Nitrification at end of 8 weeks, per cent of NH_4-N fixed by air drying.

gibbsitic hydroxyls; and (*c*) reaction with exchangeable ions such as aluminum hexahydronium ion [$Al(H_2O)_6^{3+}$] to produce $Al(OH)_3$ and H_2O and with $Ca(H_2O)_6^{2+}$ to produce $Ca(OH)_2$ and water and non-reaction with nonhydrated K^+ or NH_4^+.

Unwittingly Brown and Bartholomew (9) cited evidence concerning the third mechanism when they reported, "In most instances a number of 'zero' pressure readings were recorded before any apparent pressure could be detected in the reaction flask." No "zero" pressures should have been observed if the third mechanism were operative, since in a reaction such as:

$$Al(H_2O)_6^{3+} + 3NH_3 \leftrightharpoons 3NH_4^+ + Al(OH)_3 + 3H_2O$$

one mole of water would be released for each 3 moles of NH_3 adsorbed and ΔP would be zero. In the case of the suggested reactions:

$$Ca(H_2O)_6^{2+} + 2NH_3 \leftrightharpoons 2NH_4^+ + CaO + 5H_2O$$

or

$$Ca(H_2O)_6^{2+} + 2NH_3 \leftrightharpoons 2NH_4^+ + Ca(OH)_2 + 4H_2O$$

release of 2 or $2\frac{1}{2}$ moles of H_2O for each mole of NH_3 adsorbed should have resulted in positive ΔP values during the sorption measurements. In neither case, therefore, did the data of Brown and Bartholomew (9) support the third mechanism as proposed in their equations.

Mortland (44) suggested that the NH_3 which was irreversibly adsorbed by Ca bentonite may have reacted with traces of water to form NH_4OH and may then have entered into cation-exchange reactions. It is highly improbable that this could account for the large amounts of irreversibly adsorbed NH_3, and further research will be necessary to secure an adequate explanation.

From the results obtained with the pure clay minerals it appears that montmorillonitic clays are not responsible for cationic fixation in the moist state. Furthermore, the order of magnitude of K fixation by illite in the moist state (1.39 meq/100 g of clay) which was obtained by Stanford (50) would seem to preclude the possibility of such a material's being responsible for the large fixations (4.0 meq/100 g of soil) that have been observed. Other hydrous micas, lower in potassium and sufficiently Ca-saturated to be slightly expanded, may have higher cation-fixing powers under moist conditions.

The best explanation seems to lie in the evidence presented by Barshad (4, 5) on the fixation of NH_4 and K by vermiculites and biotites that may be present in the coarse fraction as well as in the clay fraction of soils. He found, for example, that the 100–250 μ fraction of a Sweeney subsoil layer (30 to 32-inch layer) fixed more NH_4 than did the $<1\ \mu$ fraction, notably, 17.8 meq as against 14.5 meq/100 g.

Although the chemistry of the cationic (K and NH_4) fixation reaction appears to be fairly well worked out, many critical details still deserve further attention. It is hoped that this review will continue to excite the interest of theoretical physical chemists, other than those in the field of soil chemistry, to take up the challenge and tackle some of the unsolved problems of the hydration phenomena of charged ions in the interlayer region. More study is also needed on the geometry of the distribution of the charges, both positive and negative, on and in the interlayer surfaces of the oxygen ions. For example, additional information is needed on the possible geometric distribution of aluminum tetrahedrons and the influence of various distributions of the charge density in the lattice. Other ions having entropies of hydration similar to that of K, and similar or dissimilar in size, should be studied in an effort to segregate further the qualitative characteristics of the ions that make them susceptible to fixation.

REFERENCES

1. Alexander, L. T., Hendricks, S. B., and Nelson, R. A., *Soil Sci.* **48**, 273–279 (1939).
2. Allison, F. E., Kefauver, M., and Roller, E. M., *Soil Sci. Soc. Am. Proc.* **17**, 107–110 (1953).
3. Bailey, T. A., Ph.D. thesis, Madison, Wisconsin, University of Wisconsin, 1942.
4. Barshad, I., *Am. Mineralogist* **33**, 655–678 (1948).
5. Barshad, I., *Am. Mineralogist* **35**, 225–238 (1950).
6. Barshad, I., *Soil Sci.* **72**, 361–371 (1951).
7. Barshad, I., *Soil Sci. Soc. Am. Proc.* **16**, 176–182 (1952).
8. Black, C. A., *Soil Sci. Soc. Am. Proc.* **7**, 123–133 (1943).
9. Brown, J. M., and Bartholomew, W. V., *Soil Sci. Soc. Am. Proc.* **26**, 258–262 (1962).
10. Buehrer, T. F., "Arizona Agr. Expt. Sta. Tech. Bull. **42**," pp. 154–212, 1932.
11. Chaminade, R., *Compt. Rend.* **203**, 682–684 (1936).
12. Chaminade, R., and Drouineau, G., *Ann. Agron.* **6**, 677–690 (1936).
13. Cole, C. V., and Jackson, M. L., *Soil Sci. Soc. Am. Proc.* **15**, 84–89 (1951).
14. Cole, C. V., Olsen, S. R., and Scott, C. O., *Soil Sci. Soc. Am. Proc.* **17**, 352–356 (1953).
15. Coleman, R., *Soil Sci. Soc. Am. Proc.* **9**, 72–78 (1945).
16. Davis, L. E., *Soil Sci.* **40**, 129–158 (1935).
17. Dennis, E. J., and Ellis, R., Jr., *Soil Sci. Soc. Am. Proc.* **26**, 230–233 (1962).
18. DeTurk, E. E., Wood, L. K., and Bray, R. H., *Soil Sci.* **55**, 1–12 (1943).
19. Dyer, B., *J. Chem. Soc.* **65**, 115–167 (1894).
20. Frink, C. R., and Peech, M., *Soil Sci. Soc. Am. Proc.* **26**, 346–347 (1962).
21. Gieseking, J. E., *Soil Sci.* **47**, 1–13 (1939).
22. Haseman, J. F., Brown, E. H., and Whitt, C. D., *Soil Sci.* **70**, 257–271 (1950).
23. Haseman, J. F., Lehr, J. R., and Smith, J. P., *Soil Sci. Soc. Am. Proc.* **15**, 76–84 (1951).
24. Hendricks, S. B., Nelson, R. A., and Alexander, L. T., *J. Am. Chem. Soc.* **62**, 1457–1464 (1940).

25. Huffman, E. O., Cate, W. E., Deming, M. E., and Elmore, K. L., *Trans. Intern. Congr. Soil Sci. 7th Congr. Madison* **2**, 404–412 (1960).
26. Jackson, M. L., *Trans. Intern. Congr. Soil Sci. 7th Congr. Madison* **2**, 445–455 (1960).
27. Jackson, M. L., and Hellman, N. N., *Soil Sci. Soc. Am. Proc.* **6**, 133–145 (1942).
28. Joffe, J. S., and Kolodny, L., *Soil Sci. Soc. Am. Proc.* **3**, 107–111 (1939).
29. Kelly, J. B., and Midgley, A. R., *Soil Sci.* **55**, 167–176 (1943).
30. Kolodny, L., Ph.D. thesis, New Brunswick, N. J., Rutgers University, 1938.
31. Kolthoff, J. M., *Kolloid-Z.* **30**, 35–44 (1922).
32. Latimer, W. M., *Chem. Rev.* **18**, 349–358 (1936).
33. Latimer, W. M., "Oxidation Potentials," 2nd ed., New York, Prentice-Hall, Inc., 1952.
34. Lindsay, W. L., and Moreno, E. C., *Soil Sci. Soc. Am. Proc.* **24**, 177–182 (1960).
35. Lindsay, W. L., Peech, M., and Clark, J. S., *Soil Sci. Soc. Am. Proc.* **23**, 266–269 (1959).
36. Low, P. F., and Black, C. A., *Soil Sci. Soc. Am. Proc.* **12**, 180–184 (1948).
37. Low, P. F., and Black, C. A., *Soil Sci.* **70**, 273–290 (1950).
38. McAuliffe, C. A. *et al.*, *Soil Sci. Soc. Am. Proc.* **12**, 119–123 (1948).
39. Magistad, O. C., *Soil Sci.* **20**, 181–225 (1925).
40. Marshall, C. E., *Z. Krist.* **91(A)**, 433–449 (1935).
41. Marshall, C. E., and McLean, E. O., *Soil Sci. Soc. Am. Proc.* **12**, 172–175 (1948).
42. Mattson, S., *Soil Sci.* **30**, 459–495 (1930).
43. Mattson, S., *Soil Sci.* **31**, 57–77 (1931).
44. Mortland, M. M., *Soil Sci.* **80**, 11–18 (1955).
45. Page, J. B., and Baver, L. D., *Soil Sci. Soc. Am. Proc.* **4**, 150–155 (1940).
46. Pauling, L., "The Nature of the Chemical Bond," 2nd ed., Ithaca, N. Y., Cornell University Press, 1945.
47. Pierre, W. H., and Norman, W. G., "Soil and Fertilizer," Agronomy Monograph Series, Vol. 4, p. 93, New York, Academic Press, Inc., 1953.
48. Ross, C. S., and Hendricks, S. B., "U. S. Geol. Survey Profess. Paper 205-B," 1945.
49. Stanford, G., and Pierre, W. H., *Soil Sci. Soc. Am. Proc.* **11**, 155–160 (1947).
50. Stanford, G., *Soil Sci. Soc. Am. Proc.* **12**, 167–171 (1948).
51. Stout, P. R., *Soil Sci. Soc. Am. Proc.* **4**, 177–182 (1940).
52. Toth, S. J., *Soil Sci.* **44**, 299–314 (1937).
53. Truog, E., and Jones, R. J., *Ind. Eng. Chem.* **30**, 882–885 (1938).
54. Volk, N. J., *Am. J. Sci.* **26**, 114–129 (1933).
55. Volk, N. J., *Soil Sci.* **37**, 267–287 (1934).
56. Wada, K., *Soil Sci.* **87**, 325–330 (1959).
57. Wear, J. I., and White, J. L., *Soil Sci.* **71**, 1–14 (1951).
58. Young, J. L., and Cattani, R. A., *Soil Sci. Soc. Am. Proc.* **26**, 147–152 (1962).

CHAPTER 10

Soil Chemistry and Plant Nutrition

MACK DRAKE

Department of Agronomy
University of Massachusetts, Amherst

"A rational system of agriculture cannot be formed without the application of scientific principles, for such a system must be based on an exact acquaintance with the means of nutrition of vegetables, and with the influence of soils and action of manure upon them." Such were the thoughts expressed by Justus von Liebig (47, 81) more than a hundred years ago, thoughts which led him to examine the then-existing theories of plant nutrition, embark on theories of his own, and stress the significance of plant analysis as a means of measuring the needs of plants.

A more important contribution by Liebig was his tremendous stimulation of scholars to apply chemistry to the study of plant nutrition and to the examination of the nature of the soil. This stimulation carried on to America and was a central theme in, and guide to, the infant land-grant colleges. Liebig's influence was especially important to the development of soil and plant science and to the training of graduate students and young scientists in the laboratories of Truog, Albrecht, Baver, Bear, Bradfield, Funchess, Goessmann, Jenny, Mattson, Marshall, Pierre, Scarseth, and others. Similar stimulation has occurred in other countries, notably that by Dr. W. S. Clark of the Massachusetts Agricultural College, who in 1876 helped establish the Sapporo Agricultural College of Hokkaido, Japan. There many soil and plant scientists have been trained in the laboratories of Drs. Miyake and Ishizuka.

Early in the twentieth century, teams of research scientists, such as Hoffer and Carr at the Indiana Agricultural Experiment Station, cooperated to unravel some of the mysteries of soil-plant nutrition. The application of basic plant physiology and chemistry enabled Hoffer and Carr (36), in a study of the mineral nutrition of the corn plant, to demonstrate effectively the important function of K in translocation of carbohydrates. This work with K, including their explanation of the role of superphosphate in the overcoming of aluminum toxicity to corn (14, 36) on the grey poorly drained soils of Indiana, is, though almost forgotten, a classic. It led to the development and interpretation of the Hoffer, Thornton, and Scarseth (79, 85) diagnostic approach to plant nutrition using rapid qualitative chemical tests on plant tissue. This has been used as an effective means to study soil-plant relationships and to demonstrate the importance of balanced nutrient level in the growing plant.

The French scientist Devaux (19) reported on cation-exchange properties of plant roots, suggesting that the pectin in the walls of the root hairs contributed this important chemical property. Mattson's group (27, 58, 59, 90) further developed and applied this root cation-exchange theory to help explain ion uptake by plants, the different abilities of plants to obtain nutrients, and the widely varying composition of plants. Jenny's theory (42, 44) of contact exchange aided in visualizing cation transfer from primary and secondary minerals to the soil clay colloids and, also, the transfer or exchange of H^+ from the plant root for cations adsorbed by the clay colloids. Hoagland's group (35) developed new concepts of ion uptake and transport in the plant. The chemistry of fixation and release of phosphates in the soil developed by Sieling and his associates (7, 17, 83, 84) has provided new concepts of the value of organic matter, lime, and precision placement of phosphates in the soil. Ohlrogge's group (25, 67, 72, 73) has provided valuable information on the placement of phosphates, and on the increased uptake of applied phosphate produced by placing fertilizer nitrogen with the phosphate. His demonstration of the ability of a single corn root to obtain adequate nitrogen and phosphate for complete growth is outstanding, as is his research on plant root absorption of fertilizer nutrients from highly concentrated fertilizer bands. There are many other outstanding examples of research men and teams who have advanced our knowledge of soil-plant nutrition relationships by applying fundamental principles as first set forth by Liebig (47). These research concepts are valuable resources, and their demonstration and application by trained agricultural leaders can result in more abundant food production in many underdeveloped countries. And these are the types of fundamental concepts of chemistry and plant physiology as applied to soil and plant nutrition which will be sought, reviewed, and eagerly applied in soil management in the next quarter century, when increased population has consumed the so-called food surplus in North America.

Before Liebig, the natural philosophers considered that minerals in plants were nonessential; that they were, perhaps, stimulants; and that they were accidental plant constituents or skeleton, similar to animal bones. The "humus theory," that carbon was derived from humus, prevailed and was supported by the favorable effect of animal manures and plant residues on plant growth. In 1804, De Saussure (78) established that green plants assimilate carbon from the carbon dioxide of the atmosphere, and he believed that the soil supplied a small but essential part of the plant nutrients, including nitrogen and mineral elements. Sprengel (81) concluded that plants obtain their minerals directly from the soil and there was no basis for the humus theory. In 1840, when Liebig explained that, since humus arose from decay of plants, plants must have preceded the humus, the humus theory ended.

The importance of organic matter in plant nutrition was developed later by soil microbiologists and chemists, who showed that the decay of organic matter releases nitrogen and mineral nutrients in forms available to plants. Decomposition products of organic matter have a beneficial effect on soil aggregation, increased water-holding capacity, cation-exchange capacity, and availability of nutrients, especially phosphorus. Carbon dioxide liberated by organic matter decomposition, when diffused into the atmosphere, serves as a source of carbon for the plant.

During the last century, soil scientists have sought the key to a correlation between chemical properties of soil and the growth response and nutrient content of plants. In seeking these correlations, total analysis of soils, fractional analysis, including strong and weak acid extractions, and analysis of soil solution, exchangeable cations, and exchangeable anions were determined. Analysis of the plant was also used to study the effect of soil and soil treatment upon nutrition and plant growth. In addition to analysis of the entire plant, some investigators used leaves or other special parts and some analyzed only weak-acid-extractable or water-soluble constituents.

Interpretation of results from soil-plant studies often leads to contradictory conclusions, and today many of the interrelationships involving nutrients in the soil and in the plant remain subjects of controversy. Fortunately as a result of a more comprehensive characterization of the soil-root ionic environment, improved appreciation and control of environmental conditions for plant growth, and increased analytical precision, these soil-plant interrelationships are better understood. The use of radioactive isotopes, chelating agents, nutrient solutions, both intact and excised root techniques, and the energy relationship concept of both the ionic soil and plant colloid systems are highly useful in increasing our understanding of soil-plant relationships.

NUTRIENTS ESSENTIAL TO PLANTS

Sixteen elements are known to be essential to plant growth. Although all sixteen are required, these may arbitrarily be divided into two groups, macronutrients and micronutrients, based on the relative quantities required for normal plant growth:

Essential macronutrients. C, H, O, N, P, K, Ca, Mg, and S.
Essential micronutrients. Fe, Mn, Cu, Zn, B, Mo, and Cl.
Elements beneficial to some plants under some conditions. Na, Si, Co, and Al.

The essential nutrient elements may be classified as nutritive or regulatory—some serve both roles. Carbon, hydrogen, oxygen, nitrogen, phosphorus, and sulfur are nutritive. Sulfur, phosphorus, potassium, calcium, magnesium, and the other essential nutrient elements are primarily regulatory. For the functions of these essential nutrient elements, see "Soil Conditions and Plant Growth" by Russell (77), "The Nature and Property of Soils" by Buckman and Brady (12), and "Plant Physiology" by Meyer, Anderson, and Bohning (66).

MECHANISM OF ION ADSORPTION BY PLANT ROOTS

Plants absorb ions primarily through the roots, but many plants are capable of absorbing a variety of nutrient ions through the leaves, stems, or other organs*. Foliar sprays to supply nitrogen, iron, zinc, copper, and molybdenum are examples of the latter. The salient features of ion adsorption through roots have been summarized by Truog, Overstreet and Dean, Marshall, Jenny, Broyer, and Burstrom in "Mineral Nutrition of Plants" (87). Ion intake is, in general, considered to be an exchange process. Hydrogen ions are released to the culture medium in exchange for metal cations, and OH^- and HCO_3^- are released in exchange for anions.

Various compounds and absorption mechanisms are involved in this process. The compounds responsible for cation and anion absorption are generally given the designation R·H and R·OH, respectively. Osterhout postulated that the plant substance R·H may be similar in its properties to some of the aromatic alcohols, whereas Brooks advanced the theory that properties of both R·H and R·OH were inherent in the amino acid molecule, the H^+ of the —COOH being exchangeable for cations, and the OH^- of the $—NH_3OH$ groups being exchangeable for anions.

According to Lundegardh (50), R·H in plants designates the protoplasm, which, as a whole, is negatively charged and contains appreciable qualities of substances with comparatively strong acid properties. Cations in the culture medium exchange for H^+ in the plasma membrane and proceed

* See *Soils and Fertilizer* **16, 246–262 (1953).**

inward through the protoplasm by paths of substances of acid dissociation. Supposedly in this process no energy is required. Moore, Jacobson, and Overstreet (69) concluded that nearly all the Ca absorbed in a 3-hour period by excised roots of 6-day-old barley plants, was not metabolic because absorption was not reduced either by low temperature or by 2-4 dinitrophenol. They postulated that most of the Ca in young barley roots is associated with the cell surface region and that it is this surface Ca which is active in influencing the absorption of other ions. Steward and Street believed ion accumulation to be associated with protein synthesis and suggested that R·H and R·OH may correspond to the acidic and basic groups of certain phosphorylated energy-rich nitrogen compounds in the protoplasm.

Jacobson and Overstreet suggested that cations in plants may be bound in the form of chelated complexes, since many plant substances, such as amino acids, proteins, and organic acids, are known to form chelated compounds, notably with polyvalent cations. For example, Stewart and Leonard (87) observed that iron supplied as Fe EDTA entered the roots and was translocated into plant leaves as that iron chelate. The chlorophyll molecule may be considered as a magnesium chelate. In this class of structure, the metallic ion is attached at two or more points in the same molecule, and one of the bonds is frequently coordinate in nature.

The presence, in excised barley roots, of ion-binding compounds generated by metabolic processes, have been postulated by Epstein and Hagen (28). Their experimental results support the hypothesis "that the adsorption process entails the formation and breakdown of an intermediate labile complex, M·R of the metal ion M, with a metabolically produced binding compound or carrier, R." These binding compounds are specific for certain ions. The ions K and Cs were found to interfere competitively with Rb absorption, and all three ions are considered to be bound by the same reactive centers or binding sites. Sodium did not interfere competitively with Rb except at high concentrations of both, indicating that Na is not bound by the same sites. Jacobson and co-workers (40) reported that the presence of Ca altered the ratio of Na and K absorbed. However, the essentially constant sum of Na and K absorbed in spite of large changes in this ratio suggested a common metabolic carrier of Na and K.

Though, according to Lundegardh (50), no special energy is required to transport cations into the plant, energy is essential for anion assimilation. To account for anion entry, he made use of the system Fe^{2+}-Fe^{3+} in the cytochrome-cytochrome oxidase system in the root cell membrane. According to this scheme, the Fe^{3+} ion in the hemic group of a respiratory enzyme attracts one more anion than the Fe^{2+}. The anion then can move in a structural unit in which electrons move from one atom to another in the next molecule. The transference of an electron between two Fe atoms moves one anion from the oxidized to the reduced stage, hence in opposite direction to the electron movement.

Anion entry into plant roots has been observed only with metabolically active roots. Roots rendered inactive by treatment with ether drastically reduced their intake of both cations and anions from neutral salts. Therefore, whether the metal cation associated with the anion also requires energy for its entry, remains an open question. The relative absorption of ions depends greatly, however, upon the nature of the cation and the anion.

Cation Movement from Nutrient Medium to Root Surface

Consider the movement of cations from the growth medium to the root surface. Although several mechanisms have been proposed to explain the uptake of cations by plants, the first step in this uptake process is that of movement of the cation from the growth medium to the root surface. Random migration of cations can possibly account for cation accumulation at the root surface in either natural (lake or sea water) or artificial (hydroponics) nutrient solutions. Thus it is reasoned that in nutrient solutions, an attraction mechanism is not required to explain the movement of the cation to the root surface. The root-soil system, however, is greatly different from the root-nutrient solution system, and an attraction property of the root itself is of much greater importance than the simple random migration in accounting for this accumulation of cations at the root surface. Dissociation and random movement of cations from the soil colloids to the root surface does not adequately explain the magnitude of cation accumulation by roots in the soil.

Mattson (58, 59), Albrecht (1, 2, 11), Jenny (43, 44), Marshall (55), McLean (53), and others have demonstrated that in normal soils, both organic and inorganic (clay) colloids attract or bond cations with considerable force, and that a cation-exchange reaction is required to remove appreciable amounts of those cations adsorbed by the soil colloids. Thus it has been proposed by these investigators that, in soil-root systems, a chemical attraction for cations by the plant root and a cation-exchange reaction between the root colloid and the soil colloid is a most important factor in the accumulation of cations at the root surface. This brings into play Mattson's concept (58) of active competition between the soil colloid and the root colloid for cations.

These investigators have established that the clay colloid bonding energy for cations depends on factors such as the nature of the clay, the kind of cation, the per cent saturation, and kinds and amounts of associated cations present. Mattson (59) and Wiklander (27) demonstrated that root colloids of different plant species vary widely in the attraction and adsorption of different cations. One can visualize many important possibilities for improved production and quality of crops by considering the characteristics of plant root colloids, as well as the properties of the soil colloids, as a guide to studying kinds and rates of fertilizer nutrients and limestone and in the selection of crops and cropping sequence. And where two or more species of plants are grown in association, such as legume-grass, orchard-tree–grass-sod-mulch, or cultivated-crop–weeds, the competition of the roots of these associated species for cations and anions also must be considered.

Cation-exchange Properties of Plant Roots

The sphere of concentration of the soil chemist interested in problems of plant nutrition is in the characterization of the plant root-soil ionic environment. The properties of roots and soil colloid, notably ion exchange, are his predominant concern. The existence of cation-exchange properties of roots was reported in 1916 by the French chemist Devaux (19), who found this property to arise from the pectose substances in the walls of the root hairs. These roots "being in intimate contact with the soil particle, form a colloidal system—the soil and the cellular walls." Though a pectic substance had been recognized as a constituent of plant roots, its role as a cation exchanger and accumulator had not been properly appreciated.

Jenny (43) noted that many experiments involving cation accumulation or cation depletion could be explained on the basis of cation exchange between the root surface and the soil colloid. Concerning contact exchange, Jenny said, "It assigns to the root surface the role of active individualistic, genetic, and physiological conditioned participation in the liberation of adsorbed nutrient ions. It enables the plant to feed directly upon the solid phase. It insists upon the interplay of soil colloids and plant colloids." Recently, Jenny has studied highly magnified pictures of thin sections made from resin-embedded root-soil systems. From these pictures, it appears that a mucilaginous substance of the root contacts, and indeed surrounds, clay and other small soil particles.

Contact exchange does not require the presence of carbon dioxide. Mattson (59) wrote: "The great ability of plants to absorb nutrient ions from the soil, generally credited to the excretion of carbonic acid, is more easily accounted for by the continuous formation of a strong acidoid (H-colloid) in the surface of the root. Consider the growing root penetrating the soil, always establishing new contacts and always forming new acidoids, originally unsaturated, and of a strength and concentration much greater than that of carbonic acid. The result must be a great solvent action and a rapid exchange due to the intimate contact with the soil particles which is established by virtue of the root pressure."

More recently cation-exchange capacities (CEC) of plant roots have been measured by means of different techniques*. The presence of H ion swarms at the root surface was demonstrated by the suspension effect [Williams and Coleman (91)]. When the electrodes of the glass-electrode–calomel-electrode assembly were pressed against the plant roots that had been thoroughly washed with CO_2-saturated water, pH values between 3 and 4, often between 2 and 3, were recorded for the roots, as compared to pH 3.95 to 4.10 for the supernatant liquid (CO_2-saturated H_2O). In addition, titratable amounts of root H were recorded by displacement with KCl.

For the determination of CEC, fresh excised roots are saturated with H^+ by dipping in acid (15) or by electrodialysis (21), and are titrated with 0.01 *N* KOH to pH 7 in 5 minutes in a 1 *N* KCl solution. The CEC of different plant species are recorded in Table 10.1. Differences ranging from 9.0 meq/100 g dry root, for wheat, to 94 meq/100 g, for larkspur, are recorded. In general, the cation-exchange values are much higher for dicotyledons than for monocotyledons. Roots of several plant species with high CEC are coarse, thick, or somewhat gelatinous (lupine, pea, ragweed). In contrast, those of several species with low cation exchange (as redtop and timothy) have fine fibrous roots. On a surface-area basis, the CEC of the thick coarse roots is probably ten to a hundred times that of the fine roots.

The pH of electrodialyzed colloids in 1 *N* KCl was defined by Mattson as the ultimate pH, and is used as a measure of the "acidoid" strength of soil colloids. The results in Table 10.1 show this value to range between 3.1 and 4.7. For a given weight of H^+ roots, there is a tendency for the pH to be lower when the CEC is high and to increase with decreasing CEC.

The exchangeable hydrogen of live-intact roots may vary between 8 and 65 per cent; it increases with temperature, age of plant, and prolonged withholding of nutrients. Crooke's group (15) reported much higher CEC values near the apex of leek roots. Although the cation content was, by difference, highest near the apex, the H^+ concentration was much higher near the root apex and decreased with distance from the apex.

Pectic substances and the cation-exchange capacity of plants. By using methylene-blue stain, Mattson's group (59) showed that the most active acidoids (colloids) of the root were pectic substances located in the surface layer localized near the root tip. Crooke, Knight, and Macdonald

* See references 15, 21, 37, 51, 58, 68, and 91.

TABLE 10.1. CATION-EXCHANGE CAPACITY OF PLANT ROOTS

Plant species	pH_u[b]	CEC[a] (meq/100 g)
DICOTYLEDONS		
Larkspur, *Delphinium ajacis*	3.10	94.0
Lettuce (Penn Lake)	3.20	65.1
Soybeans (Lincoln), *Glycine soja*	3.26	65.1
Cucumber (National Pickling)	3.26	53.9
Blue lupine, *Lupinus angustifolius*	3.23	53.3
Carrot (Supreme half-long)	3.27	51.7
Sugar beet		51.3
Canadian field peas, *Pisum sativum arvense*	3.35	49.6
Alfalfa, *Medicago sativa*		
Atlantic	3.42	48.0
Kansas common	3.49	40.0
Yellow lupine, *Lupinus luteus*	3.28	47.7
Red clover, *Trifolium pratense*	3.37	47.5
Turnip		47.5
Rutabaga		47.5
Sunflower, various growth stages		47.2
Celery (Summer pascal)	3.40	47.0
Chickweed, *Stellaria media*	3.39	44.7
Hairy vetch, *Vicia villosa*	3.41	44.1
Ladino clover, *Trifolium repens* var.	3.43	43.4
Crimson clover, *Trifolium incarnatum*	3.36	41.7
White lupine, *Lupinus albus*	3.31	41.3
Squash (Golden delicious)	3.45	40.1
Kale		40.0
Buckwheat, *Fagopurum esculentum*	3.39	39.6
Irish potato, *Solanum tuberosum*	3.62	38.1
Peanut, *Arachis hypogaea*	3.42	36.5
Cotton, *Gossypium herbaceum*	3.47	36.1
Spinach (L.S. Bloomsdale)	3.61	35.1
Snapbean (Top crop)	3.67	34.8
Tomato, *Lycopersicum esculentum*	3.67	34.6
Cauliflower, edible portion		34.6
Cabbage, edible portion		34.0
Rape, *Brassica napus*	3.37	33.2
Lespedeza, all stages maturity		32.4
Sweet potatoes (Puerto Rican)		31.4
Birdsfoot trefoil, *Lotus corniculatus*	3.54	23.9
Leek		36.0
Onion (Brigham yellow globe)	3.63	29.5

TABLE 10.1—*Continued*

Plant species	pH$_u$[b]	CEC* (meq/100 g)
MONOCOTYLEDONS		
Grasses		
Alta fescue, *Festuca elatier arundinacea*	3.60	30.4
Orchard grass, *Dactylis glomerata*	3.72	24.9
Smooth brome, *Bromus inermis*	3.70	24.6
Canada bluegrass		24.1
Timothy, *Phleum pratense*	3.78	22.6
Rye grass, *Lolium perenne*	3.53	22.5
Tall meadow oat grass, *Arrhenatherum elatius*	3.67	22.5
Kentucky bluegrass, *Poa pratensis*	3.83	21.6
Field bromegrass, *Bromus arvensis*	3.75	18.0
Reed canary, *Phalaris arundinacea*	3.80	17.4
Red top, *Agrostis alba*	3.92	17.3
Colonial bentgrass, *Agrostis tenuis*	3.95	16.3
Rice paddy		15.2
Bermuda grass, *Cynodon dactylon*		10.5
Cereals		
Spring oats, *Avena sativa*	3.78	22.8
Corn (Yellow Dent), *Zea mays indentata*		
Ind. Wf9	3.80	17.0
Ohio 40B	3.94	13.5
Rosen rye, *Secale cereale*	4.12	15.1
Sudan grass, *Sorghum vulgare*	4.06	13.5
Barley, *Hordeum vulgare*	4.25	12.3
Millet, *Panicum miliaceum*	4.28	12.2
Winter wheat, *Triticum vulgare*	4.70	9.0
Weeds		
Nutgrass, *Cyperus esculentus*	3.51	28.7
Quackgrass, *Agropyron repens*	3.78	19.8
Foxtail, Setaria glauca	4.65	11.4
WEEDS—DICOTYLEDONS		
Ragweed, *Ambrosia artemisiifolia*	3.27	58.9
Chickweed, *Stellaria media*	3.39	44.7
Pigweed, *Amaranthus retroflexus*	3.35	42.3
Smartweed, *Polygonum pensylvanicum*	3.48	41.1
Purslane, *Portulaca oleracea*	3.53	40.7
Lamb's-quarters, *Chemopodium album*	3.94	25.0

* meq/100 g oven-dry weight.

TABLE 10.2. CATION-EXCHANGE CAPACITY, PECTIN CONTENT, RESPIRATION, AND MINERAL CONTENT OF LEEK ROOTS*

Distance from apex (mm)	CEC		Pectin (meq/100 g dry roots)	Respiration (mlO_2/100 g dry wt/hour)	Mineral content (% in dry matter)				
	(meq/100 g dry roots)	(meq/sq m)			N	K	Ca	Mg	Na
0–2.5	134	11.9	63	212	2.17	3.10	0.33	0.26	0.50
2.5–5	120	10.1	56	201	2.15	3.11	0.28	0.27	0.52
5–10	57	5.8	53	152	2.06	3.66	0.24	0.27	0.48
10–20	65	6.0	51	160	2.06	5.00	0.24	0.28	0.66
20–40	39	3.6	50	130	1.96	4.66	0.21	0.28	0.55
40–60	28	3.2	48	119	1.92	4.17	0.21	0.29	0.48
60–80	26	3.4	48	116	—	4.42	0.21	0.28	0.48
80–100	30	4.1	45	110	1.76	4.50	0.24	0.28	0.46
100–120	33	5.1	40	94	1.65	4.50	0.23	—	0.44
120–140	46	7.3	36	80	1.64	4.41	0.21	0.23	0.40
Whole roots	26	5.0	36	—	—	—	—	—	—

* After Crooke, Knight and MacDonald (15).

(15), who measured CEC of different sections of leek roots, found highest CEC and highest per cent N and Ca in the root section 0 to 2.5 mm from the apex. Root CEC values of 134, 120, 57, 39, and 28 meq/100 g dry tissue were reported, respectively, for root sections 0 to 2.5, 2.5 to 5, 5 to 10, 20 to 40, and 40 to 60 mm from the apex (Table 10.2). Per cent N and Ca decreased, and per cent K increased, as root CEC decreased. Not only was the newest (tip) 5 mm of root material unusually high in CEC, but the exchange sites must have been predominately saturated with exchangeable H. While the CEC of the root tip was about four times that of the section 40 to 80 mm from the apex, the cation contents were similar (K + Na + Ca + Mg). One must inquire why the CEC value decreases with increased distance from the root apex. Is it because, as the root cell wall develops and thickens, pectic substances and other carbohydrate materials are built into layers which become less exposed and are reduced in exchange reactions by physical reasons?

McLean (51), Smith and Wallace (80), and others have shown that the root CEC of a given plant species can be increased 10 to 30 per cent by increasing the level of nitrogen applied to the soil or supplied in a nutrient solution. This increase in root CEC was paralleled by a corresponding increase in nitrogen content of the root tissue. What is the nature of this nitrogen-induced increase in the CEC of a given species? Does this indicate that the increased number or greater density of exchange sites arise in part from protein amino groups? Crooke's results (15) indicated that nearly 100 per cent of the exchange sites could be accounted for by pectin groups. The zone immediately back of the root cap or apex has been described as a region of rapidly dividing cells. In addition to the proposed role of increased N in increasing the CEC by additional protein-amino groups, the higher nitrogen level may raise root CEC primarily by stimulating greater cell division and by reducing non-pectic carbohydrate deposition in the root cell walls. This effect may be to extend the highly reactive 120 to 130 meq CEC, 0 to 5 mm zone up to 0 to 20 or 0 to 40 mm from the apex. Observation of oat and barley roots under high magnification indicates the possibility of N-induced CEC increases by increased production of root hairs and by increased lateral root-tip development.

Cation Bonding Energies of Root Colloids

Mean free bonding energies* of cations calculated from activity measurements of soil colloids have been calculated by Marshall (55). Similar calculations were made by McLean and Baker (53) from activity measurements of Na, K, and Ca on roots of several plant species (Table 10.3). They reported root cation-exchange values of 42.4 meq/100 g for alfalfa, 41.1 for soybeans, 14.1 for red top, and 11.8 for Reed canary grass. The alfalfa and soybean roots, with relatively high cation exchange, have bonding energies for calcium which are more than double the bonding energies for K. For Ca, on redtop roots with low cation exchange, the bonding energy was about 50 per cent of the value with the legumes, whereas K was held with about 80 per cent of the value for legumes. Reed canary grass bonded K with about 90 per cent of the value for Ca. Buckwheat roots bond Ca with almost three times as much energy as K. Bonding energy values of clays usually increase as percentage saturation decreases. These bonding energies were determined on roots saturated with K, Ca or Na. Thus, at lower saturation values it would be reasonable to expect that the bonding energies of the grasses for K would exceed those for Ca. Unusually high Ca-bonding values would result if the roots contained materials such as oxalate which combine with Ca to form relatively insoluble compounds. Plant roots containing oxalate and citrate would be highly effective in releasing Ca from minerals such as limestone and rock phosphate.

* Mean free bonding energy in calories per mole. Derived from (ΔF) cation $= RT \ln 1/f$, where $f = a/c$, a being the measured cation activity and c the total concentration of the cation in the system. Complete dissociation of the cation from the surface would correspond to $f = 1$ and (Δf) cation $= 0$.

TABLE 10.3. ACTIVITIES, FRACTIONS ACTIVE, AND MEAN FREE BONDING ENERGIES OF CATIONS IN HOMOIONIC PLANT ROOT SYSTEMS*

CEC (meq/100 g)	Saturated cation	Activity (× 10⁸)	Fraction Active (%)	ΔF† (cal)
		Alfalfa		
42.4	Na	4.85	36.4	599
42.4	K	4.26	32.0	672
42.4	Ca	0.49	7.3	1549
		Soybean		
41.1	Na	3.05	24.1	843
41.1	K	3.81	30.1	711
41.1	Ca	0.38	6.1	1654
		Redtop		
14.1	Na	2.86	59.6	305
14.1	K	1.79	37.3	583
14.1	Ca	0.54	22.4	885
		Reed Canary		
11.8	Na	3.40	91.0	55
11.8	K	2.36	63.2	272
11.8	Ca	1.10	58.9	314

* From McLean and Baker (53).
† Mean free bonding energy in cal/mole.

Cation-exchange Capacity and Selective Adsorption of Ca and K

The ability of plants to absorb or exclude ions selectively has been the subject of extensive study and speculation. The existence of compounds in the plasma membrane having adsorption sites that can be occupied by some monovalent cation species and not by others has been referred to previously.

Mattson (58) has used the Donnan principle to help explain the differential adsorption of monovalent and divalent cations by colloids of high and low CEC (see the following section, "Cations from Soil to Plant"). When this principle is applied to plant species having different root CEC, Mattson and co-workers stipulated a cation distribution in which the outer layers or outside solutions of a high-CEC colloid would be of a lower concentration than that of a low-CEC colloid. This greater dilution of the outside solution requires relatively greater adsorption of divalent, than of monovalent, cation, since activity of divalent cation is inserted as the square root in the Donnan distribution ($K:\sqrt{Ca}$). Accordingly, the higher the cation-exchange capacity of roots, the greater is the relative adsorption of Ca over K, when nearly all the cations are held in the exchangeable state.

Differential uptake of monovalent and divalent cations by plant roots with different cation-exchange capacities supports Mattson's theory. The root colloid and the soil colloid compete for cations, and the cation uptake by the plant depends in part upon the relative exchange capacities of the root and soil as demonstrated by Elgabaly and Wiklander (27). Their concept of the valence effect was used to predict differential monovalent and divalent cation uptake by roots of the same plant from different clays, and by the roots of different plants from the same clay. Pea roots (71 meq/100 g) and barley roots (22.7 meq/100 g) were placed in Na-Ca bentonite systems for 10 hours. After equilibrium the pea roots contained two to three times as much Ca as did barley roots, while barley roots contained four to five times as much Na as did pea roots. These differences might have been even greater if beidellite, which bonds Ca with 2670 calories, had been used instead of montmorillonite, which bonds Ca with 1396 calories per mole (53). In a similar experiment, tops of both peas and barley contained more Ca when these plants were grown on kaolin than when grown in montmorillonite. Peas contained 10 per cent more K when grown on 10 per cent K–90 per cent Ca-saturated montmorillonite as compared to 10 per cent K–90 per cent kaolin (Mattson).

Potassium uptake (20) by three cereals of similar physical growth characteristics is shown in Table 10.4. At each of the three lower levels of soil potassium (0.095, 0.201, and 0.307 meq/100 g), the uptake of K by wheat (9.0 meq), barley (12.3 meq), and oats (22.8 meq) correlated well with root cation-exchange capacity. At the lowest level of exchangeable soil K, wheat plants contained amounts of K equal to 155 per cent of exchangeable K (0.095 meq/100 g); barley, 90 per cent; and oats, 55 per cent. It is important to note that at higher increments of soil K, the effect of root cation-exchange capacity on K uptake vanished. This illustrates the principle that the valence effect can be destroyed by greatly increasing the outside cation concentration.

Similar reduction in relative differences in K uptake (33) by roots of bentgrass, Kentucky bluegrass, smooth brome grass, and Ladino clover (16, 21, 24, and 43 meq/100 g, respectively) were reported (Table 10.5).

It is only when nearly all of the cations exist in an exchangeable state and the plant root colloids must compete with the soil colloids for these cations by

TABLE 10.4. POTASSIUM UPTAKE BY PLANTS FROM GREENHOUSE POTS OF SOIL CONTAINING EIGHT LEVELS OF POTASSIUM (EXCHANGEABLE K PLUS SEVEN INCREMENTS OF KCl)*

K_2O Content of soil†		K_2O Content of plants‡		
(g/pot)	(meq/100 g)	Oats (g/pot)	Barley (g/pot)	Wheat (g/pot)
0.267	0.095	0.15	0.24	0.42
0.567	0.210	0.33	0.36	0.50
0.867	0.307	0.50	0.56	0.77
1.167	0.414	0.81	0.81	1.09
1.467	0.520	1.12	0.96	1.02
1.767	0.627	1.35	1.24	1.37
2.517	0.893	1.70	1.56	1.83
3.267	1.158	2.04	2.05	1.88

* After Drake and Scarseth (20).
† kg soil/pot.
‡ Roots excluded.

TABLE 10.5. RELATIVE UPTAKE OF K BY THREE CUTTINGS AT TWO LEVELS OF SOIL K*

Plant species	Exchangeable K (100 lb K_2O/acre)	Exchangeable K +(120 lb K_2O/acre)
Bentgrass	100	100
Kentucky bluegrass	72	93
Ladino clover	40	65
Smooth Brome	27	58

* After Gray, Drake, and Colby (33).

exchange that the Donnan distribution will be reflected in the composition of plants (Mattson). With an increase of free electrolytes, the inequalities of the Donnan distribution of ions will be evened out. Many dicots with relatively high CEC roots cannot obtain the large amounts of K required for normal growth from soil colloids, especially at low per cent K saturation, whereas many monocots can. By increasing the outside ionic concentration, the "valence effect" is reduced. For example, by liberal applications of potassium fertilizer salts to a soil with low K-supplying power, the "valence effect" is greatly reduced, so that roots with high CEC, such as alfalfa, Ladino clover, and head lettuce, adsorb adequate amounts of K (22). Cation uptake by plants from nutrient solution omits this important cation competition of the soil colloid with plant root colloid. For example, Newton's studies (71) with barley and peas also showed little differences in the

Ca and K content when these plants were grown in culture solution (high electrolyte concentration); when these two plant species were grown in soil, however, the Ca in peas was 2.3 times higher than that in barley.

It has been suggested that the great Ca bonding energy of high CEC roots should detract from, instead of aiding in, the uptake of Ca, in that the calcium ions once adsorbed by the root colloid would be difficult to translocate to stem and leaf tissue. In a limited sense this may be true and may explain why tomatoes and many legumes require soils either with a high reserve of exchangeable Ca or well supplied with limestone. Unless there is an abundance of Ca present, the high-CEC root colloid may seriously compete for Ca with the stem and leaves, resulting in poor growth and calcium-deficient foliage or fruit.

Truog suggested that plants high in Ca require a large portion of this element in the form of carbonates and bicarbonates for neutralizing and precipitating the acids in plant sap. He assumed that relatively small amounts of Ca needed as an indispensable constituent of plant substances can be furnished by Ca salts of different kinds. There is some question whether the bicarbonate anion enters the plasma membrane to any great extent. Such a process does not appear to be essential. But for continued root metabolic activity, removal of the respiration by-product CO_2, is essential. In culture solution this is accomplished by aeration, together with use of salts that buffer the solution against great fluctuations in acidity. In soils, the colloids and the bicarbonate, carbonate, and possibly phosphate anions serve as buffers through the exchange adsorption or root-produced H ions.

Mode of Ion Entry

The path of entry of ions into the plasma membrane may be postulated with the aid of the schematic illustration of the cross section of a root hair in Figure 10.1 (29). The thickness of the cell wall is given as 0.5 μ, although thicknesses ranging between 0.1 and 3 μ have been reported. The diameters of soil colloids are 0.2 μ and less. The cell wall is made up of a network of cellulose strands about 0.0005 μ thick. Under optimum conditions of growth, films of H_2O surround cell walls and soil colloids. This H_2O contains diffusible ions as well as CO_2 produced from the respiratory reactions of the roots. The gradient for CO_2 and H^+ is away from the plasma membrane, and for the nutrient cations it is toward the membrane. Movement of nutrient cations in that direction depends on their concentration and the nature of their associated anions. Under certain conditions, therefore, root-absorbed cations may be lost to the outer medium.

It seems fairly certain that ion absorption involves physiochemical and

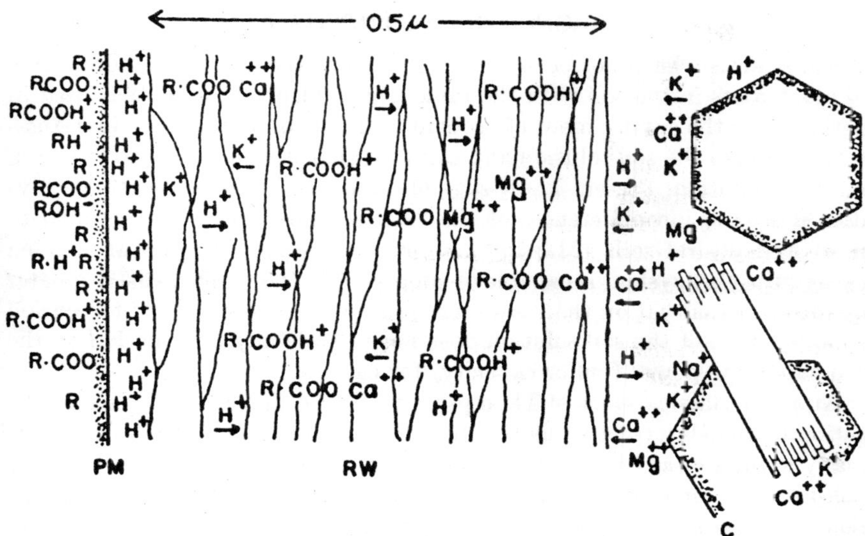

Figure 10.1. Schematic illustration of the cross section of root hair and relative concentration and direction of movement of ions. *PM* = plasma membrane; *RW* = root-cell wall; *C* = soil colloid. (*Modified from Frey-Wyssling*)

metabolic-physiological processes. To distinguish the acid systems of these two processes, identification of these systems by the generalized designation R·COOH and R·H, respectively, is suggested. As illustrated in Figure 10.1, R·COOH is distributed throughout the cell wall and in the plasma membrane, while R·H is restricted to the membrane. Pectic acid in cell-wall material probably constitutes the R·COOH groups. In the whole roots, malic, citric, oxalic, and other acids have been found. According to the investigation by Jacobson and Ordin (39), the concentrations of some of these acids, notably malic acid, change considerably during the process of ion adsorption, while acetic, citric, and oxalic acid remain relatively more constant. These changes occur as a result of unequal cation and anion absorption and thereby contribute to the maintenance of ion equivalency. When under suitable conditions the external medium contained KBr, excised barley roots absorbed K and Br in nearly equal amounts, from which little change in the organic acid concentrations occurred. With $KHCO_3$ in the medium, preferential absorption of K was compensated by an increase in malic acid, whereas with $CaBr_2$, preferential absorption of Br was compensated by a decrease in malic acid. In addition to changes in organic acids as a result of unequal ion absorption, equivalency was further established by displacement of ions present within the roots.

CATIONS FROM SOIL TO PLANT

Soil scientists have not been able to determine the chemical environment of plant roots in the soil with the same degree of precision as that used in determining the environment of roots in nutrient solutions. Some investigators extracted the soil solution and assumed that it was similar to that of a nutrient solution. Present knowledge of cation exchange and cation activities show the incompleteness of the "soil solution theory" as applied to most agricultural soils (Marshall, Jenny). At equal activities of a given cation, the clay system is capable of supplying to the plant a vastly greater quantity of that cation than is a true solution. Measurement of the cation concentration of the soil solution does not provide any information on the soil-supplying power (buffer capacity) for cations.

After the classical work of Thompson, Way, and Forschamer, the identification of the clay fraction as the inorganic source of soil cation exchange, many studies of soil-clay cation exchange were initiated [Bradfield (8)]. It has been demonstrated that plants obtain adsorbed cations by exchange reactions. Jarusov (41) and Jenny and Ayers (44) set forth certain general relationships on the effect of one cation on the replacement of another, based on the concept of different adsorption or bonding energies of the clay for the various cations. They reported that the energy of cation attraction decreased with increased per cent cation saturation. The nature of the clay colloid itself must be considered, since montmorillonite was reported to hold Ca in preference to K and NH_4, whereas mica-type clays were reported to hold K and NH_4 in preference to Ca.

Cation Activities

Marshall and co-workers (51, 55) have used the clay-membrane electrodes to measure cation activities of homoionic clays and Ca-K clay systems. Bonding energy and activity values show that:

1. A single cation on a given clay mineral can be held with a wide range of bonding energies. Although the activity of Ca in an H-Ca montmorillonite system changes very little from 30 to 70 per cent Ca saturation (little change in bonding energy), the activity of Ca almost triples as the Ca saturation increases from 70 to 90 per cent.

2. A given clay such as montmorillonite has different bonding energies for different cations; for example, K = 710 cal/mole as compared to 1398 for Ca.

3. Different clays have widely different bonding energies for a given cation; for example, K montmorillonite = 710 calories/mole and K beidellite = 1544; Ca montmorillonite = 1396 and Ca beidellite = 2670.

4. The complementary ion (H-Ca in comparison with K-Ca) has im-

portant effects on bonding energies. In a comparison of a H-K montmorillonite system with a Ca-K montmorillonite system, the substitution of Ca for H greatly increases K activity (reduced K-bonding energy). In contrast, substituting K for H depresses Ca activity, since for montmorillonite the bonding energy for H is greater than that for K. The addition of as little as 10 per cent exchangeable K, for example, can reduce Ca activity by 50 per cent.

Activity measurements indicating K to be bonded by about the same energy by both kaolin and montmorillonite have been reported. But theoretical considerations by Wiklander (90) and studies of K uptake by plants indicate that K is bonded with relatively much greater energy by kaolin than by montmorillonite, particularly at low per cent K saturation (less than 10 per cent).

Cation-exchange Capacity and Valence Effect

Mattson (58), Wiklander (90), and co-workers have made applications of the Donnan equation to develop the theory that colloids with high cation-exchange capacity (strong acidoids) adsorb divalent cations with relatively greater attraction or bonding energy than they do monovalent cations. The distribution of cations of different valence in a Donnan system is governed by the relative activity of both inside and outside solutions. The distribution of monovalent and divalent cations between the colloid and the outside solution is generally given by the equation

$$\frac{(M^+)_i}{(M^+)_o} = \frac{\sqrt{(M^{2+})_i}}{\sqrt{(M^{2+})_o}}$$

where $(M^+)_i$ and $(M^+)_o$ represent the activity of the inside and outside solutions, respectively, of the cation M^+.

According to the Donnan distribution, a volume increase or a dilution of the outside solution favors adsorption of the divalent cation over the monovalent. *This is referred to as the valence effect.* How does this dilution of the outside solution occur? Visualize a solution of monovalent and divalent cations with activities of 1×10^{-4}. On the addition of a colloidal particle or micelle with cation-exchange capacity to this solution, both divalent and monovalent cations will tend to be adsorbed by the colloidal micelle, with the resulting dilution of the outside solution (that solution around the micelle). From the Donnan equation

$$\frac{K_i^+}{K_o^+} = \frac{\sqrt{(Ca^{2+})_i}}{\sqrt{(Ca^{2+})_o}}$$

if activities of K_o^+ and $Ca_o^{2+} = 0.01$ and $K_i^+ = 0.1$, the Ca_i^{2+} activity $= 1$.

or the Ca/K adsorbed ratio = 10 to 1. If by dilution the outside activities of K_o^+ and Ca_o^{2+} = 0.0001, and if the inside activity of K_i^+ = 0.01, then Ca_i^{2+} = 1 and the adsorbed Ca/K ratio has increased to 100 to 1.

From the square-root relationship, the divalent cations will accumulate to a greater degree than monovalent cations in that phase with the higher activity. Colloids with higher exchange capacity have higher micellar adsorption, and the higher exchange capacity colloids have inside solutions with higher activity. Thus, the higher the exchange capacity of the colloid, the higher will be the ratio of divalent to monovalent cations adsorbed from equal divalent/monovalent cation activities in a solution. Conversely, among cation exchangers with different exchange capacities and with adsorbed monovalent and divalent cations present in equivalent amounts, the higher exchange material will release its monovalent cations relatively more readily, and its divalent cations less readily, than the lower capacity exchanger.

Montmorillonite, a colloid of high cation-exchange capacity would be expected to bond Ca with much greater energy than it would bond K. Kaolinite, with low cation-exchange capacity, would attract K with relatively greater bonding energy than Ca. Also, montmorillonite would be expected to adsorb Ca with higher, and K with relatively lower, bonding energy, than would kaolinite. The magnitude of the valence effect produced by the dilution of the outside solution was demonstrated with a high cation-exchange resin Wofatit K(200 meq/100 g). With a 1 *N* NH_4 Cl-$CaCl_2$ solution, the adsorbed cations were 42 per cent NH_4^+ and 58 per cent Ca^{2+}. The corresponding adsorption values for a 0.0001 *N* solution were 1.1 per cent NH_4 and 98.9 per cent Ca (90). From these results one can appreciate that cation adsorption by plant roots from the soil solution would lower the cation activity of the outside micellar solution and greatly increase the magnitude of the valence effect.

Crooke and Knight (16) have reported on the relationship of the root CEC and mineral composition of plants. They cited the results of Mattson and his co-workers, who stated that the CEC of plant roots will determine, according to Donnan principles, the relative amounts of divalent to monovalent cations adsorbed by the roots, and who demonstrated this experimentally, using excised roots of pea (high CEC) and barley (low CEC). As predicted by the Donnan distribution, pea roots adsorbed more Ca in relation to Na than did barley. In another experiment, pea roots adsorbed relatively more calcium and less potassium K than barley roots. These results were in agreement with the cation composition differences predicted between monocots and dicots.

Drake, Vengris, and Colby (21) reasoned that, if divalent cations are adsorbed in relatively greater amounts by those root colloids with higher

CEC, there must be a corresponding decrease in adsorption of monovalent cations, particularly potassium, since the sum of the cations in the plant tends to have a constant value [Bear and Prince (6); Lucas, Scarseth (49), Sieling (48), Itallie (38)]. On soils with low levels of exchangeable K, legume and other plant roots characterized by high CEC may adsorb relatively large amounts of divalent cations (Ca, Mg) but may be unable to obtain sufficient monovalent K to maintain normal growth. Conversely, those plant species with relatively low-CEC roots, such as the grasses, absorb relatively small amounts of divalent Ca and Mg but relatively large amounts of monovalent K, even from soils low in exchangeable K.

The mineral composition of plants is largely determined by physical-chemical relations between the colloidal system of both the plant root and the soil. Plant material of legumes and many other dicotyledonous plants contains relatively larger amounts of Ca and Mg than grasses and other monocotyledonous plant material, because the higher-CEC root colloid of the dicotyledonous plants attracts and adsorbs a larger proportion of divalent than monovalent cations. Because of this relatively greater attraction and adsorption of divalent cations, legumes and other dicots require a high activity of K in the soil solution to overcome this valence effect of the Donnan distribution if adequate amounts of K are to enter the plant.

Crooke and Knight (16) stated that experimental proof of the valence effect in the cation composition of the growing plant appears, when compared to Mattson's monovalent-divalent cation adsorption by roots, to be more difficult to demonstrate for two reasons: (*a*) the *maximum valence effect* occurs in very dilute solutions—conditions not met in nutrient solutions and in well-fertilized soils; and (*b*) while cations may be adsorbed on the roots in the ratios predicted by the Donnan distribution, once translocation occurs, the composition of the plant top may not reflect those initial adsorption ratios. It is assumed that cation translocation involves several mechanisms. While the large volume of plant analysis published by Beeson (6a) showed a general positive trend between root CEC and the ratio of divalent to monovalent cations in the tops, Crooke and Knight concluded that data which consisted of analysis of tops only were not adequate to test for the effect of root CEC on the differential cation composition of the entire plant. In summary, Crooke and Knight stated that root CEC was positively correlated with the content of the tops of total cations, ash, excess base, and total trace elements, and that the listing of plant species according to ascending root CEC was interpreted in terms of a change from a carbonaceous to a proteinaceous type of plant. They cited Albrecht's 1943 statement (2) that carbonaceous plants would have a low Ca/K ratio and proteinaceous plants a high ratio.

Knight, Crooke, and Inkson (45) collected five monocots and seven dicotyledon plants from four soils (sand, peat, acid soil, calcareous soil) which varied widely in drainage, pH, and calcium supply. Tops were separated into leaves and stems, and the roots were washed and dried at 80°C. The CEC of root material was measured by an acid-washing technique followed by titration to pH 7.0 with 0.01 *N* KOH. Uronic acids were estimated by a micro-decarboxylation method and the results were expressed in meq/100 g, assuming all CO_2 was derived from carboxyl groups. Their data showed that root CEC is, in general, accounted for by uronic acid content, and that this CEC–uronic-acid relationship held for different parts of the same plant. For *Heradeum sp.*, the values (meq/g dry matter) for CEC and uronic acid were, respectively: roots 25.8, 49.5; tap root, 26.0, 52.0; stem, 26.7, 52.5; and leaves 26.3, 70.7. Apparently leaves contain a high proportion of substances which yield CO_2 on decarboxylation but do not take part in cation exchange.

The most important findings, however, were that the CEC of the plant was largely independent of soil type and that the characteristic CEC is also shown by other organs of the plant-rhizomes, stems, and leaves, as well as by roots. This suggests the possibility that the colloids of different parts of a given plant—roots, stems, leaves, etc.—compete for cations.

Asher and Ozanne (4) stated that root CEC may exert a qualitative influence on nutrient uptake by: (*a*) raising the concentration of cations in the Donnan free space (DFS) above the level of the surrounding aqueous solutions, and (*b*) influencing the relative proportions of ions of different valency in the DFS as a consequence of the Donnan valence effect. They stated that these two effects should be most marked where a plant is growing in a medium of low ion activity. They electrodialized roots in nylon bags at 120 volts for 25 minutes and reported important differences between the root CEC of various species. Although their CEC values were lower than those reported by Crooke, Drake, McLean, Wallace, and others, the relative values were in the same order: legumes > herbs > grasses.

They reported a highly significant correlation between root CEC and the percentage of both calcium and phosphorus in the tops of plants growing with relatively insoluble ground rock phosphorus as the only source of calcium and phosphorus. This supports the theory of Graham (32) and of Drake and Steckel (23) that amounts of phosphorus released from rock phosphate should increase with higher root CEC values. The increased bonding of calcium by the higher-CEC roots should dissolve the rock phosphate crystal. This was supported also by results of field trials reported by Rossiter and Ozanne (76), which showed subterranean clover (CEC = 27) to be clearly superior to cape-weed (CEC = 19) in its ability to utilize the phosphorus of rock phosphate.

TABLE 10.6. EXCHANGEABLE K IN A MONTMORILLONITIC AND A KAOLINITIC SOIL AFTER SUCCESSIVE CROPS OF LADINO CLOVER*

Soil	Initial K_2O in Soil (lb/A)	K_2O in Soil after successive harvests (lb/A)							Total K_2O removed in 15 crops (lb/A)
		No. 2	No. 4	No. 6	No. 8	No. 10	No. 12	No. 15	
Wooster†	200	90	88	76	84	76	89	84	365
Decatur‡	263	148	112	116	120	138	129	114	318

* After Reitemeier *et al.* (75).
† Wooster silt loam, gray-brown podzolic, 14.6 per cent clay, of hydrous mica and montmorillonitic types.
‡ Decatur clay loam, red podzolic, 40.9 per cent clay, predominately of the kaolinitic type.

Clay Bonding Energy and Relative K Uptake

Cation adsorption or bonding values and an appreciation of the valence effect should be useful in predicting and interpreting the relative availability of exchangeable cations in a given soil. Potassium uptake by 15 successive crops of Ladino clover grown on two soils, one of the hydrous-mica-montmorillonite-type clay (high exchange-low K bonding energy), and the other of the kaolinitic-type clay (low exchange-high K bonding energy) is listed in Table 10.6.

Ladino clover obtained 15 per cent more K from the Wooster soil than from the Decatur (75). The Wooster soil, derived from glacial till, was much less weathered than the Decatur, a red podzolic weathered from limestone. The Wooster soil would be expected to supply more K from nonexchangeable K-bearing minerals. It is important to note, however, that the level of exchangeable K after successive crops was always much higher for the Decatur soil (low exchange capacity-high K bond adsorption) than for the Wooster (high exchange-low K bond adsorption). These data are presented to illustrate the principle that at equal levels of exchangeable soil K, a given plant species will obtain relatively more exchangeable K from the soil that has the lower bonding energy for K. In estimating the cation-supplying power of the soil, the nature of the plant root and the bonding energy of the clay mineral for a specific cation, as well as the total amount of that cation in the exchangeable and nonexchangeable forms, must be considered. This consideration is of special importance in the interpretation of chemical soil tests.

Per Cent Base Saturation and Type of Colloid

Allaway (3) measured calcium uptake from H-Ca systems by soybeans grown on 40, 60, and 80 per cent Ca-saturated peat, kaolin, illite, Wyoming bentonite, and Mississippi bentonite. Calcium uptake decreased in the following order: peat—kaolin—illite—Wyoming bentonite—Mississippi bentonite. Calcium uptake also decreased as the per cent Ca saturation decreased. Earlier Albrecht and McCalla had reported Ca uptake by soybeans to decrease with decrease in per cent Ca saturation. In an investigation by Mehlich and Colwell (60), two soils, one with montmorillonitic and the other containing kaolinitic clay, were 20, 40, 60, and 80 per cent Ca-saturated. The uptake of Ca by soybeans and cotton increased with increase in per cent Ca saturation on both the kaolinitic and montmorillonitic soils. No increased uptake of Ca by cotton occurred above 40 per cent Ca saturation of the kaolin. It has been suggested that the amount of Ca in the system may have been such that 40 per cent Ca saturation supplied adequate Ca. In Allaway's experiment about 1/100 as much Ca was offered to each plant, and Ca uptake from kaolin increased with each increase in per cent Ca saturation.

Calcium uptake by tomato plants was studied by Eck *et al.* (26) in an environment where no other nutrient cation was present (Figure 10.2). A split-root system permitted half the plant roots to be placed in either a Ca-H kaolinite or a Ca-H-bentonite-and-sand system, and the other half

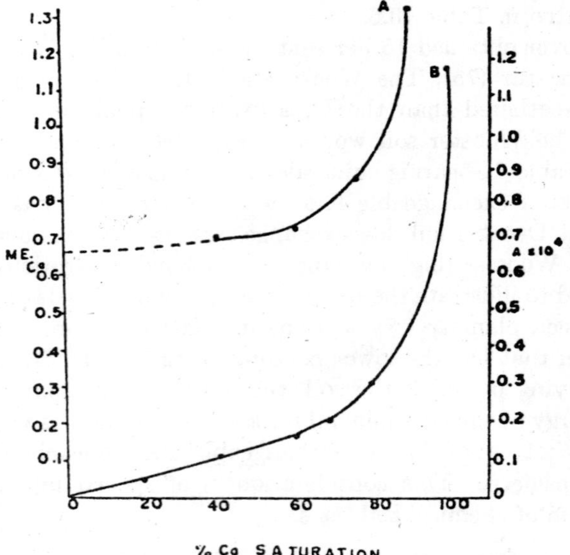

Figure 10.2. (*A*) Me. of Ca in tomato plants plotted against per cent Ca saturation of Ca-H bentonite; (*B*) activities of the Ca ion plotted against per cent Ca saturation of Ca-H bentonite systems from Marshall (87).

in sand containing a nutrient solution without Ca. No increase in yield of tops or in Ca content of the tomato plants occurred when the Ca saturation of bentonite was increased from 45 to 60 to 75 per cent. Growth of roots was doubled, however, by increasing from 75 to 90 per cent. Increasing the per cent Ca saturation of bentonite from 75 to 90 also resulted in greater top growth and twice as great uptake of Ca. The 45 per cent Ca-saturated Ca-H-kaolinite supplied as much Ca as the 90 per cent, and each produced greater top growth and greater Ca uptake than the 90 per cent Ca-bentonite, demonstrating that bentonite has a much greater bonding for Ca than kaolin. These results represent a biological confirmation of the implications of Marshall and McLean's Ca activity values (55) for different per cent Ca saturations of a Ca-H bentonite system (Figure 10.2). Similar K-H-kaolin and bentonite systems were designed to measure K uptake, but the roots died. Analysis of a normal ammonium acetate extraction showed that an appreciable amount of Ca had migrated from the transplanted root onto the K-H-bentonite system—ten times as much as onto the K-H-kaolin.

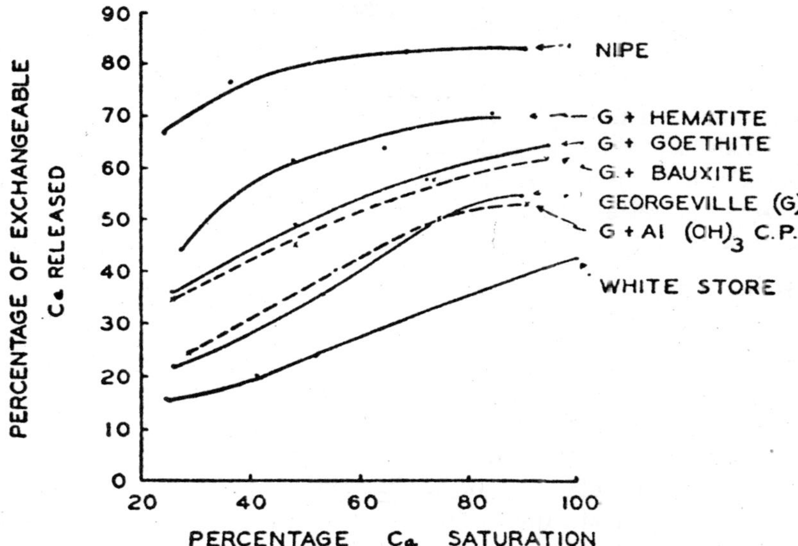

Figure 10.3. Percentage release of Ca from Nipe (containing 85% iron and aluminum oxide and 15% kaolinite in colloid fraction); White Store (montmorillonite-kaolinite colloid); Georgeville (kaolinite colloid); and Georgeville plus sesquioxide mixtures. HCl added in amounts equivalent to the exchangeable Ca. (*After Mehlich. Courtesy The Williams & Wilkins Co., Baltimore, Md., publishers of Soil Science*).

Peanuts were grown in kaolin and montmorillonite clay systems with different degrees of Ca saturation. For comparable Ca levels, more Ca invariably was absorbed by the peanut shell from the kaolinitic system than from the montmorillonitic. The Ca uptake increased greatly with additional increments of Ca in kaolin, but only moderately with montmorillonitic colloid. In the montmorillonite, per cent Ca saturation was more important than the total Ca supplied. For example, 80 per cent saturation in the 0.8-meq series (0.64 meq Ca) produced shells containing more than 7 meq Ca per 100 g, whereas systems with the same amount of Ca in the soil, but only 27 per cent Ca saturation in the 2.4-meq series (0.64 meq Ca), produced shells containing only half as much Ca. With kaolin systems, however, the Ca content of peanut shells was about the same for either 79 per cent Ca saturation in the 0.8-meq series or 26 per cent Ca saturation in the 2.4-meq series. This indicates that kaolinite bonds Ca very weakly, even at only 26 per cent Ca saturation. In contrast to the weak kaolin bond for Ca, only the 99 per cent Ca-saturated montmorillonitic clay system was as good a peanut fruit producer as the kaolin system.

In a subsequent experiment, Mehlich and Reed (61, 62) found Ca uptake by peanut shells from a predominately organic soil to be more like that from montmorillonite than from the kaolinite colloid. Calcium uptake by peanut roots in relation to type of colloid was, however, in the same order as that reported for soybeans by Allaway and by Mehlich and Colwell (60). Analysis of the three colloids for Ca in solution showed relationships to the Ca in the peanut shells; hence, it was concluded that the peanut fruit was but weakly endowed to engender H through the peanut pegs (shells) in exchange for Ca on the soil colloid.

Nitrogen Effect on Cation-exchange Capacity and Cation Uptake

McLean, Adams, and Franklin (52) reported that the CEC of roots of several plant species were increased by increasing the level of N in nutrient solutions (Table 10.7); the per cent nitrogen of the root material increased with the increase in root CEC of each species studied. Smith and Wallace (80) reported on the effects of nitrogen fertilizer on root CEC and K/Ca ratios of leaves of Valencia oranges. In an orchard experiment, two levels, 0 and 4 pounds of nitrogen per tree per year, had been supplied over a 10-year period. Roots from trees in this experiment were carefully washed free of soil and CEC was determined. Cation-exchange values of 19.7 and 25.0, respectively, for low and high N were measured (Table 10.8). The leaf analysis for Ca and K reflected the expected Donnan distribution of cations. Leaves from trees with the lower CEC roots contained 5.74 per cent Ca and 1.18 per cent K, in comparison to 6.24 per cent Ca and 0.94 per cent K with the induced higher CEC roots.

TABLE 10.7. CATION-EXCHANGE CAPACITIES AND PERCENTAGE OF NITROGEN IN ROOTS AND TOPS OF SEVERAL CROPS (SEEDLINGS) GROWN IN SOLUTIONS WITH THREE NITROGEN LEVELS*

Crop	Age (days)	Nitrogen levels†	CEC‡	% Nitrogen	
				Roots	Tops
Oats	15	low	17.7	1.29	3.50
		medium	21.0	1.51	4.15
		high	24.4	2.14	5.08
Corn	15	low	21.6	1.61	3.16
		medium	24.6	1.75	3.21
		high	27.4	1.96	3.84
Buckwheat	15	low	41.6	2.00	1.98
		medium	42.4	2.05	2.41
		high	44.0	2.25	3.00
Buckwheat	38	low	46.8	2.13	1.79
		medium	49.6	2.50	2.41
		high	51.8	2.77	3.32
Cotton	25	low	44.7	2.35	4.84
		medium	45.9	2.65	5.39
		high	50.0	3.03	6.40
Vetch	25	low	52.4	4.72	6.57
		medium	48.2	4.69	6.60
		high	50.3	5.24	7.33
Soybeans	25	low	54.6	3.93	7.18
		medium	56.3	4.24	7.10
		high	60.9	4.21	7.21

* After McLean, Adams, and Franklin (52).
† One-third normal, normal, and twice normal nitrogen (nitrate) levels of one-fifth Hoagland solution.
‡ Cation-exchange capacity in meq/100 g dry roots.

TABLE 10.8. K AND CA CONTENTS OF LEAVES OF 22-YEAR-OLD VALENCIA ORANGE TREES MAINTAINED AT TWO DIFFERENT N LEVELS FOR 10 YEARS BEFORE ANALYSES WERE MADE, AND CATION RATIOS FOR THE TWO N LEVELS*

Age of leaves (months)	Potassium content†		Calcium content†		Cation ratios‡		
	No N (%)	4 N (%)	No N (%)	4 N (%)	K_0/K_4	Ca_4/Ca_0	$[Ca_4/Ca_0]^2$
6	1.49	1.24	5.04	5.30	1.20	1.05	1.10
13	1.12	0.88	5.58	6.14	1.27	1.10	1.21
15	1.10	0.83	5.90	6.58	1.32	1.11	1.23
18	1.01	0.81	6.45	6.97	1.24	1.08	1.17
Means	1.18	0.94	5.74	6.24	1.26	1.09	1.19

* After Smith and Wallace (80).
† No N = trees not receiving N; and 4N = trees receiving 4 lb N/tree/year.
‡ $\frac{\text{Root CEC 4N}}{\text{Root CEC 0N}} = \frac{25.0 \text{ meq/100 g}}{19.7 \text{ meq/100 g}} = 1.27.$

Drake and White (24) grew buckwheat, tomato, and oat plants in soil low in exchangeable and total Ca. Calcium was supplied at the rate of 4000 pounds per acre as 30- to 50-mesh limestone placed in a 2-inch soil zone, 1 inch below the surface. Nitrogen was supplied at four rates, 170, 340, 510, and 680 pounds per acre, in several applications, as NH_4NO_3. Per cent Ca in the plant tissue was doubled, and total Ca uptake was doubled for tomato and tripled for buckwheat by rates of N. It was assumed that the nitrogen levels used induced increases in root CEC and that these induced increases resulted in the increased uptake of calcium from the relatively insoluble, coarse limestone particles.

Other Factors Influencing Calcium Availability to Plants

Calcium release is influenced not only by the nature of the organic and inorganic mineral silicate soil colloids but also by aluminum and iron oxides (63). Some of these oxides may increase the Ca release, as indicated in Figure 10.3. While, in general, the availability of a given cation increases with increased per cent saturation, the kind and per cent saturation of associated cation or combination of cations also influence this cation availability. Marshall demonstrated that Ca activity could be reduced 50 per cent by adding 10 per cent exchangeable K to an H-Ca montmorillonite system.

For New Jersey soils, Bear *et al.* (5) reported optimum growing conditions for alfalfa when the cation-exchange complex contained 65 per cent Ca, 10 per cent Mg, 5 per cent K, and 20 per cent H. The Ca/Mg and Ca/K ratios were 6.5 and 13, respectively. Plant growth was not substantially affected by a considerable deviation from these ratios, provided the per cent saturation of an individual cation or the sum of the nutrient cations was not limiting. These limiting values depended on the nutrient requirements of different plant species.

Mehlich stated that the exchange of H between R·COOH and the cations on the exchange complex has been conceived to be primarily a neutralization reaction. For this reaction to proceed, the H dissociation of the soil exchange complex must be lower than that of R·COOH. This is usually achieved when the per cent base saturation is sufficiently high. Below a critical per cent base saturation, nutrient cations are then lost from the plant roots to the external medium. Attempts to overcome this cation loss by adding neutral salts such as $CaSO_4$ or $MgSO_4$ to soils have usually failed. This was to be expected, since such a treatment did not alter the per cent base saturation, because, to the extent that Ca or Mg has exchanged for H, the latter appears in solution as H_2SO_4. If the soil also contains exchangeable Al and Mn in toxic amounts, their concentrations may further depress Ca uptake. It is only after the excess H, Al, and Mn associated with SO_4 have been removed by leaching, that the degree of base saturation will be increased by $CaSO_4$. This explains the ineffectiveness of Mg SO_4 to correct magnesium deficiency of apple trees when the soil was low in base saturation. Since acids and salts are more easily leached from sandy soils, $CaSO_4$ has proved to be of some value for these lighter-textured soils as an indirect supply of Ca. But even with sandy soils the per cent base saturation is more effectively increased, and the plant's Ca requirements are more adequately met, when liming materials such as CaO or $CaCO_3$ are added.

Root-soil Ionic Environment–Macronutrient Cations

Mehlich and Coleman (64) summarized results of many studies of the ionic environment of plant roots in the soil. Cation-exchange and cation-activity measurements were the principal approaches used. The reciprocal relationships of exchangeable Ca, Mg, and K in the soil to cation composition in the plants was reported by Lucas and Scarseth (49). Marshall's group has emphasized the importance of cation activity as a measure to characterize the soil-root ionic environment. He proposed that the proportions of cations which arrive at the root surface will depend on the rate-of-delivery processes. If the rate of removal of cations from colloidal particles adjacent to the root is low in comparison to the rate of renewal, then over a short interval the exchange for root H will be equal to the activity measurements. However, if the removal rate is rapid and the replacement rate is low, the proportions exchanged on the root surface will approach the value for exchangeable cations. It would appear from this that a measure of the delivery rate, together with a measure of the cation activity, would adequately describe the soil-root ionic environment.

Bray (9, 10) described the ionic environment of the soil solution as an exchange of part of the total exchangeable cations for hydrogen. Marshall questioned this approach on the basis that under field conditions cation exchange is governed by rate processes which are likely to vary from one plant to another. As a measure of these rate processes it would appear preferable to determine a series of points, thus describing a curve connecting the proportions of cations released with the per cent of exchange. For smaller release by exchange, these proportions will approach more closely those determined by activity measurements. Brown and Albrecht (11) studied this concept with a H colloid and soil clays separated by a colloidal membrane. The results were essentially the same as those obtained by the release with varying amounts of H from HCl.

Micronutrient Cations

The characterization of the soil-root-ionic environment as related to the micronutrient cations is more difficult than for the macronutrient cations. Epstein and Stout reported that the uptake of Fe, Mn, Zn, and Cu by tomato plants increased with increasing saturation of these ions when present on the colloids in amounts of less than 0.1 per cent. At higher levels, the rate of intake as related to degree of saturation decreased. With increasing Ca saturation, Fe and Mn in the plant increased while Zn decreased, and each of the three cations decreased in the supernatant liquid with increased Ca saturation. Gilbert (31) reported Cu availability decreased with increasing acidity, especially for plants grown on organic soils. According to Truog (86), Cu, Zn, and Mn availability is at an optimum in the range of pH 5 to 7. The uptake of micronutrient cations is influenced not only by the reaction of the soil and the concentration of each respective element, but also by the nature and activity of the complementary cations. Thus an increased supply of Fe may depress plant uptake of Mn, and in some cases increased Mn may depress Fe. The addition of Mn increased both Mn and Fe in tomato plants. Additions of Cu may cause Mn deficiency, whereas the addition of ammonium sulfate and phosphate may benefit Mn uptake. Hoffer* has reported that Zn and B have increased yields of heavily fertilized corn. Large applications of fertilizer phosphate placed in bands in combination with nitrogen produces high uptake of phosphorus by the plant. This high level of phosphorus in the conducting tissue may reduce the activity of Zn. It is reasoned that in those plants with very high levels of P in the conducting tissue, the application of Zn may be necessary to maintain a sufficient activity of Zn in the plant.

Interactions between Cu and N, Cu and Zn, Cu and Mo, and Cu and Mn have been reported. Cu reduces Zn and Mo injury and regulates uptake of Fe and Mn.

PHOSPHORUS FROM SOIL TO PLANT

The element phosphorus has several essential functions in plant growth. Because most of the total phosphorus content of soils is relatively unavailable to plants, fertilizers high in phosphorus (as compared to nitrogen or potassium) usually produce the greatest yield responses on new agricultural soils. In the United States the total tonnage of fertilizer phosphorus, expressed as P_2O_5, exceeds that of nitrogen or potassium. In recent years, increased knowledge of the chemistry of phosphates in the soil has made it possible to increase the effectiveness of applied fertilizer phosphorus and

* Private communication; G. N. Hoffer, Lafayette, Indiana.

has pointed the way to more efficient uses of the world's limited reserves of high-grade phosphate ore.

Annual crops planted immediately after the application of soluble phosphate fertilizer often recover only 1 to 20 per cent of the phosphorus applied. A large part of the applied soluble phosphate quickly becomes relatively insoluble and unavailable to plants. This rapid reduction in availability of soluble phosphates is known as phosphate "fixation" and is due to the chemical precipitation of phosphorus by hydrated aluminum, iron, and/or calcium in the soil.

It was proposed by Connor (14) that much of the harmful acidity arising from acid soils is due to toxic acid salts of aluminum and iron, and that the reduction of soil acidity by fertilizer phosphate is due to a combination of the soluble phosphoric acid with the acid salts of aluminum and the consequent formation of insoluble nonacid compounds. Hoffer's group (36) reported that high amounts of soluble aluminum both in the soil and in the corn plant tissue were responsible for the stunted plant growth on grey, acid, poorly drained, middle western soils, particularly in cool, wet seasons. They showed that either superphosphate or barnyard manure corrected this stunted growth in the early spring. While the combined treatment with superphosphate and calcium carbonate was highly beneficial, calcium carbonate applied alone was harmful and caused greater accumulation of iron and aluminum in the nodal tissue of corn plants than in plants on untreated or superphosphate-treated plots. Duncan and Ohlrogge (25) have demonstrated that aluminum sulfate completely prevented the penetration of corn roots into the fertilizer band unless monocalcium phosphate was mixed in the soil with the aluminum sulfate.

This stunted growth of plants, which could be prevented by the application of fertilizer phosphorus, became recognized as phosphorus deficiency. This condition resulted from the small amount of native, soluble, soil phosphorus reacting with active aluminum and iron to form insoluble aluminum and iron phosphates. Although some investigators speak of "aluminum toxicity," those problems with active aluminum that can be corrected by supplying adequate fertilizer phosphorus, are more logically termed "phosphorus deficiency." If there is active aluminum within the plant cell or conducting tissue, the phosphate activity must be reduced, and plant growth and quality will be limited by varying degrees of this phosphorus deficiency.

Available Phosphorus and Development of Plant Seedlings

Numerous phosphate fertilizer studies have demonstrated the high available-phosphate requirement of plants during the seedling period of growth. Studies with radioactive phosphorus have shown that, as the plant develops, the percentage of phosphorus in the plant from applied fertilizer phosphorus rapidly decreases. According to Nelson *et al.* (70) applied fertilizer phosphorus constituted 64, 26, and 16 per cent, respectively, of the total phosphorus in corn plants sampled 1, 2, and $3\frac{1}{2}$ months after planting. In regions having a long growing season, plants which are stunted and deficient in phosphorus during the first 30 days of growth, if they survive at all, often recover, mature somewhat later, but have almost as good a yield as plants initially supplied with adequate phosphate. Although under otherwise favorable conditions corn and certain other plants have shown a remarkable ability to obtain phosphorus from a low level of available soil phosphorus, a high level of available phosphorus, especially during the initial development, is essential for rapid plant development.

Concentrating the phosphate fertilizer in bands near the seed reduces phosphate fixation and insures a high concentration of soluble phosphorus for the plant. The practice of thoroughly mixing the soluble fertilizer phosphorus with the soil should be discouraged. This mixing increases the contact between soil and fertilizer particles and increases the amount of chemical reaction between the soluble phosphate and the active aluminum and iron in the soil. This chemical precipitation or "fixation" of soluble fertilizer phosphates rapidly decreases its availability to plants. In contrast, banding the soluble phosphates in a zone near the seed provides minimum contact with soil particles, thereby reducing fixation and maintaining a high level of available phosphorus. The band of fertilizer phosphorus must be placed below and preferably to one side of the seed where the band will intercept the roots of the germinating seed. Phosphate fixation may be reduced also by granulating, pelleting, or coating the fertilizer particles with plastic to reduce soil-fertilizer contact.

Ohlrogge (73) demonstrated that a single root of corn could obtain an adequate amount of phosphorus and nitrogen for the normal development of the plant. This would indicate that although the roots of the developing plant extend far beyond the phosphate banded near the seed, this banded phosphate remains potentially available to the plant.

Development of a sizable root system seems to influence phosphorus uptake. For perennials (74) the seedling period of growth is the critical time when available phosphate is most needed. If at this period the available phosphate is inadequate, the stand will be poor, and many surviving plants will be stunted. Failure to establish a stand helps explain why rock phosphate has shown up poorly in some field tests but well in others. In contrast, it is reasoned that adequate available phosphate to establish the perennial plant ensures an abundant production of carbohydrates for root development of the established plant and results in a valuable supply of chelating materials to increase the availability of soil phosphates. This may explain the observed value of using a combination of soluble fertilizer phosphate and rock phosphate for perennial forage crops. It is important to note, however, that for many annual plants there is insufficient time for the production of a root system to adsorb calcium or to form the chelating agents required to solubilize rock phosphate.

Nitrogen Stimulation of Plant Uptake of Fertilizer Phosphorus

Ohlrogge's group showed that nitrogen increased the uptake of fertilizer phosphorus throughout a wide range from low to high levels of available soil phosphorus when the nitrogen and phosphorus were mixed together in the band. The mechanism of nitrogen-induced increased feeding power did not seem to arise from a chemically increased availability of the fertilizer phosphorus, since nitrogen placed 3 to 4 inches from the phosphorus band increased the uptake of fertilizer phosphorus at low levels of available soil phosphorus (100 pounds P_2O_5/A) but nitrogen placed separately had no effect at higher levels of available soil phosphorus. Weight of roots per unit area of soil was increased by a factor of 1.5 by placement of nitrogen with, rather than separated from, the phosphorus band. Duncan and Ohlrogge (25) reported that nitrogen, as either ammonium nitrate or ammonium sulfate, when combined with the phosphate fertilizer band, produced a mass of fine silky roots. There was much less root development when the phosphate was used alone or when the nitrogen band was separated from the phosphate band. There was little or no development of roots in either the ammonium sulfate or ammonium nitrate zone when these materials were banded alone.

They stated that, if it is assumed that both nitrogen and phosphorus are required for maximum proliferation of a root mass in a specified area, and (for the condition of nitrogen and phosphorus banded separately) if part of the nitrogen was transfered through either the soil or the plant to become available to the root system in general, the greatest stimulation would be expected to occur in the area of the higher phosphorus concentration. With very low levels of available soil phosphorus, this stimulation would occur in the phosphate fertilizer band, resulting in a relatively greater feeding power in this zone. But at higher levels of available soil phosphate, a large differential in phosphorus concentration no longer exists. Thus a more general root stimulation would result, giving little or no change in the relative feeding power of that part of the root system in the phosphate fertilizer band. In contrast, when the nitrogen was mixed with phosphorus in the band, both the nitrogen and phosphorus concentrations would be higher in this zone than in other areas of the soil, resulting in root proliferation in the phosphorus band at all levels of available soil phosphorus. They reasoned that this would account for the greater feeding power of the system on the phosphorus band, even at high levels of available soil phosphorus. Their conclusion was that although root development appeared to be the primary factor responsible for increased uptake of fertilizer phosphorus, chemical and physical factors also may be important.

Root Development Within the Band of Phosphate Fertilizer

In a review of soil-root-plant relationships, Ohlrogge considered the uptake of plant nutrients from concentrated solutions and banded fertilizers in the soil. He emphasized that most of the basic knowledge of root growth processes has been gained from work with segmented and excised roots in nutrient solutions, and that since nutrient solutions are not a natural environment for roots, the application of root-nutrient solution information requires careful consideration and adjustment: "It would thus seem worthwhile to examine some of the possible characteristics that roots of intact plants growing in soil might possess which would necessitate some adjustment."

Duncan and Ohlrogge (25) reported on the development in the soil of corn roots in a band containing a mixture of nitrogen and phosphorus salts and described the large seminal root, or in some cases one of the secondary roots originally next above the seminal root, as the first-order root, with the branches from it as second-order, branches from second-order as third-order, etc. They reported that the tip of the first-order root was sensitive to salt concentration of the N-P band, which killed the meristem, and they suggested that this rapidly extending root tip penetrated into the fertilizer salt too rapidly to permit intercellular adjustments in osmotic pressure. The resulting plasmolysis killed the tip meristem. In contrast, however, the second-order roots were not injured by the same N-P fertilizer band, possibly because, they suggested, a slower growth rate permitted root cell adjustment to the salt concentration. (More recent studies indicate that the "wall effect" and other restricted area dilution factors must be considered.) Where second-order roots grew into fertilized soil, these developed long third-, fourth-, and fifth-order roots. In contrast, the second-order roots found in the unfertilized soil were often short and unbranched or with short third-order branches. Since the tips of first-order roots apparently were killed by plasmolysis on entering the N-P fertilizer band, the question arises, did not a somewhat lesser degree of plasmolysis occur when the second-order roots entered this zone? Assume that some lesser degree of plasmolysis of the second-order root occurred. If one considers a greatly restricted zone at the root surface, is it not possible that the osmotic pressure was reduced enough to allow rapid uptake of the nitrogen and phosphate salts by the actively dividing root cells? This is oversimplified, but one must not ignore the possibilities of large differences between the concentration of nutrient ions at the root surface and in the solution only a few mm distant, provided rapid uptake of ions occurs.

The Purdue group has observed many cases where individual corn roots proliferated in a zone of fertilizer where the concentration of salts would be expected to plasmolyze the cells. They reported that the level of available soil water into which the root emerges from the fertilizer band has a great influence on the uptake of phosphorus from the band. Also, the level of available soil water in which the other roots of the plant develop has much less effect than the soil water level into which the root emerges from the band. This effect of the level of soil water into which the root emerges suggests that the immediate root system itself may transfer water to the fertilizer band, as well as supply water to those cells of roots developing in the band. This effect of level of adjacent soil water may be related to the root cell adjustment to the high salt concentration, as suggested by Duncan and Ohlrogge.

Grunes (34) has reviewed and discussed research results on the biological and chemical effects of nitrogen on the availability of soil and fertilizer phosphorus to plants. This included the nitrogen effect on increased uptake of phosphorus by plants by increasing root efficiency, root area, and absorbing capacity; the effect on plant metabolism and root properties; the effects of nitrogen carriers on pH; and salt effects.

Effect of Lime on Phosphorus Availability

From pH 3 to 7 the effectiveness of hydrated iron to precipitate soluble phosphate remains very high, but it decreases rapidly from pH 7 to 7.5. Hydrated aluminum is, however, highly effective in precipitating phosphate from pH 3.5 to 9.0. Approximately 90 per cent of the phosphate will be fixed by aluminum at pH 6.5, and 70 per cent at pH 9.0. This indicates that even with alkaline and calcareous soils, aluminum is a serious fixer of soluble phosphates. Less than 10 per cent of the phosphate precipitated by iron or aluminum at pH 4.0 would be solubilized by increasing the pH to 6.0. But Struthers (83) determined that no phosphate would be precipitated in this pH zone in the presence of citrate and certain other organic materials.

Thus it would appear that the direct effect on phosphate availability produced by liming acid soils is probably less than the indirect effect produced by creating more effective conditions for increased production of plant residues and improved microorganic activity.

Anion Solubilization of Soil Phosphates

An increase in hydroxyl ion concentration has been found by Cole and Jackson to release phosphorus from variscite by decreasing the activity of the aluminum ion. Many anions decrease the activity of aluminum, iron, and calcium in the soil by "chelation," forming stable complex ions, or by forming relatively insoluble precipitation products. Thus, the availability of phosphorus in the soil will be increased by those chemical reactions that lower the effective concentration of the phosphate fixers—aluminum, iron, and calcium. These are anion solubilization reactions, not anion-exchange reactions. The ability of an anion to prevent phosphate fixation or to release fixed phosphate depends upon the stability of the complex or precipitate formed by the anion and the specific cation involved, that is, aluminum, iron, or calcium. The anions sulfate, nitrate, chloride, borate, and acetate do not form stable compounds with aluminum and iron and therefore are not effective in the release of phosphates. Only "saloid-bound" phosphates are released by sulfate, nitrate, chloride, borate, and acetate anions.

Dean and Rubins (18) saturated eight soils with phosphate and found the release of fixed phosphates by anions to be in the order hydroxyl > citrate > fluoride > tartrate > arsenate > acetate. The great effectiveness of the citrate solution in removing fixed phosphate, they concluded, involved the formulation of complex ions.

Many soils contain relatively large amounts of phosphorus in the organic fraction. Phosphorus from these organic forms cannot be utilized unless it has been released or mineralized. Some believe that increased phosphorus availability to plants effected by organic matter can be explained by the mineralization of organic phosphates through bacterial decomposition. However, the decomposition of this organic matter is accomplished by soil microorganisms which themselves use large amounts of phosphorus and compete directly with the plant root for it. Mineralization of organic phosphates by the decomposition of organic matter in itself does not appear to be a sizeable source of phosphorus for plants. A more fundamental explanation of the role of soil organic matter in phosphorus uptake is that the organic anions produced by microorganisms, complex the aluminum, calcium, and iron and thereby solubilize the relatively insoluble soil phosphates. The important solubilization effect of organic anions on relatively insoluble forms of soil phosphorus helps explain the frequent inability of the chemical soil test to predict the phosphorus-supplying power of a soil. The factor of soil management and its effect on microorganic activity as well as the results of chemical extraction must be considered. The effect of fertilizer nitrogen in stimulating organic anion production from soil organic matter must be considered. The large yield response of grain and grass to high applications of nitrogen fertilizers probably results indirectly from an increased availability of soil phosphates as well as from the available nitrogen.

Organic Anions Prevent Fixation and Release Fixed Phosphorus

In a system of 1 mmole iron or aluminum, 1 mmole phosphate, and 1.5 mmoles of citrate anion, no precipitation of phosphate occurs (84). In fact, Swenson *et al.* (84) reported that 1 mmole of citrate is much more effective than 1 mmole of hydroxyl in keeping the phosphate in solution. Struthers and Sieling (83) reported the effects of nine organic anions in preventing phosphate precipitation by aluminum and iron at different pH values. In addition, the specific effect of each anion with iron was different, although in some cases there was a general similarity. Some of the anions were most effective at low pH levels, some in moderately acid solutions, and others at or above pH 7.0. The most effective anions were citrate, oxalate, and tartrate (Figure 10.4). Citrate anion was found to be effective in dissolving various fixed phosphates, that is, iron and aluminum phosphates.

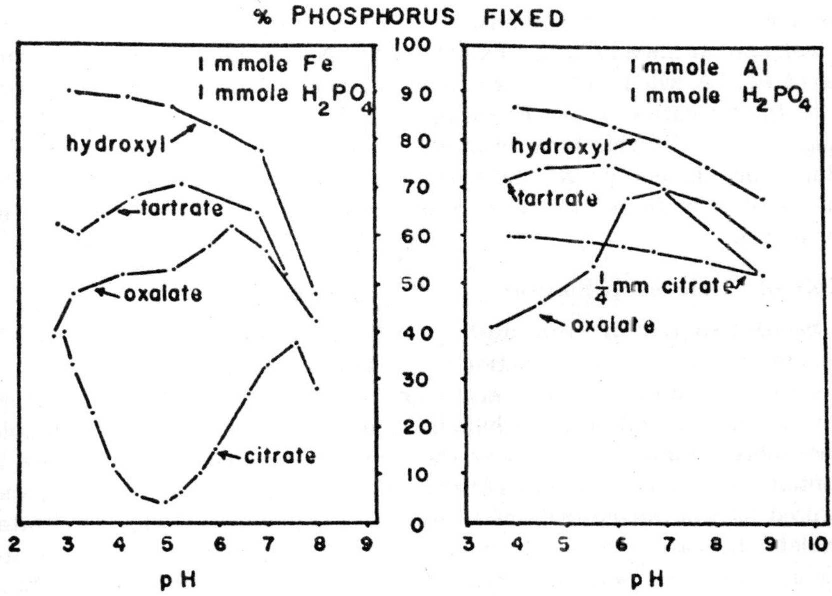

Figure 10.4. Influence of hydroxyl, tartrate, oxylate, and citrate anions in preventing fixation of phosphorus by iron (*left*) and by aluminum (*right*) (83).

All of these organic anions occur in soils as a result of microorganic activity on organic matter. When present, they are highly effective either in preventing phosphate fixation or in solubilizing aluminum and iron phosphates, since they inactivate aluminum and iron in a pH range in which these elements are active phosphate fixers.

In general, inorganic anions are less effective than organic anions in releasing fixed iron and aluminum phosphates. At pH 7.0 the relative effect of inorganic anions in releasing fixed phosphates from high to low effectiveness is in the order: hydroxyl > fluoride > arsenate > borate > thiocyanate > sulfate > chloride > nitrate. The relative position of *fluoride in this series would depend on, among other things, the ratio of aluminum to iron phosphates, since Swenson et al. and Turner and Rice* (88) *demonstrated that fluoride ion does not release phosphorus from iron phosphate but is highly effective in releasing phosphate from aluminum phosphate.*

CHELATION AND PLANT NUTRITION

By the process of "chelation," many organic anions form highly stable Werner-type complex ions with such cations as Ca^{2+}, Zn^{2+}, Mn^{2+}, Al^{3+}, and Fe^{3+}. The solution then behaves as though these cations were not present. For example, the addition of the chelating agent ethylenediaminetetraacetic acid (EDTA) to a solution of $CaCl_2$ prevents the precipitation of calcium on the addition of oxalate anion. And in contrast, this chelating agent EDTA when added to calcium oxalate will dissolve this precipitate with the formation of oxalic acid and calcium EDTA. This is a salt-splitting reaction. By this same reaction, chelating agents inactivate iron and aluminum, thereby preventing phosphate fixation. They also split off iron and aluminum from the precipitated phosphates, leaving the phosphates in solution.

Role of Chelates in Solubility of Phosphates

Solubilization of iron and aluminum phosphates. Dalton (17) demonstrated that decomposition products of organic materials are highly effective in increasing the availability of fixed soil phosphate to Ladino clover. The acid soil used was high in fixed phosphorus but low in available phosphorus. Starch was used as the source of organic matter because it contains insignificant amounts of phosphorus, occurs in plant residues, and is used by soil microorganisms to produce organic anions such as citrate, oxalate, tartrate, malate, malonate, and galacturonate, all of which are highly effective in solubilizing fixed iron and aluminum phosphates. Plant uptake of phosphorus from treatments was as follows: no soluble fertilizer P_2O_5 and no starch, 790 mg P; 80 pounds soluble P_2O_5 per acre, 885 mg P; 160 pounds soluble P_2O_5 per acre, 960 mg P; and 4 tons starch per acre but no soluble P_2O_5, 1000 mg P. Addition of sucrose, glucose, and citrus pectin also increased the plant uptake of phosphorus.

Protection of applied soluble phosphates. Mixing certain chelating materials with soluble phosphate fertilizers increased the uptake of the applied phosphorus. Oxalate anion, an effective chelator of iron, aluminum, and calcium, was mixed with superphosphate, and this in turn was thoroughly mixed into an acid soil with high phosphorus-fixing capacity*. Phosphorus uptake by wheat was increased by superphosphate applications of 300 and 600 pounds per acre. At the 300-pound application, the addition of 240 pounds of oxalic acid produced a 13 per cent increase in phosphorus uptake, and 480 pounds of oxalic acid produced an increase of 32 per cent. Similar increases in phosphorus uptake were produced by oxalic acid at the 600-pound rate of superphosphate application. This protective action, which resulted in greater efficiency in plant uptake of applied phosphorus, can have a great effect on more economical use of our phosphate resources. In theory, one of the important effects of adding detergents to commercial fertilizers results from the chelating properties of these detergents in protecting the applied soluble phosphate from fixation in the soil.

Solubilization of Rock Phosphate by Calcium Chelation

Chelation of calcium is a highly important natural process in solubilizing relatively insoluble rock phosphate. The concentration of phosphate in solution will be increased in a system containing calcium phosphate by those reactions that lower the calcium activity. Substances such as ethylene-diamine-tetraacetate that react with calcium to form complexes of greater stability than the rock phosphate crystal have been shown to be highly effective in dissolving the rock phosphate crystal†. Such organic anions, as citrate and oxalate, liberated appreciable amounts of phosphorus from rock phosphate.

In theory, any colloidal material that adsorbs calcium with high bonding energy will lower the calcium activity in the rock phosphate system and will dissolve the rock phosphate. Graham (32) used H-"Amberlite", H-Putnam, and H-humus to weather rock phosphate. The amount of phosphorus weathered, as well as the amounts of phosphorus taken up by plants, was related to the initial pH and to the bonding energy of the colloidal system for calcium. Of these materials, H-"Amberlite" with the highest bonding energy for Ca was the most effective in making the phosphorus available.

Plant roots that possess high bonding energies for calcium, high cation-

* Unpublished: J. E. Steckel, University of Massachusetts, Amherst.

† M. S. Angelini, M.S. thesis (copy filed at University of Massachusetts Library, Amherst, 1953).

exchange capacity, and high acid dissociation would be expected to obtain more phosphorus from rock phosphate than would those low in these properties. Fried (30) and others have reported that the phosphorus of rock phosphate was more available to buckwheat and legumes, such as alfalfa, crotalaria, and Ladino clover, than to the grasses (orchardgrass, bromegrass, perennial ryegrass, millet). Asher and Ozanne (4) reported a high correlation between the CEC of plant roots and the uptake of both Ca and P where rock phosphate was the only source of Ca and P.

Plants of high, medium, and low cation-exchange capacity were grown in greenhouse pots on a soil with less than 5 lb acre of (Truog) phosphorus (900 pounds total P per acre) (23). Ground rock phosphate was applied at the rate of 1000 lb/acre (230 mg P to each pot, limed to pH 6.5). The results in Table 10.9 show that the plants with high-cation-exchange roots were more effective in obtaining phosphorus from soil and rock phosphate for themselves, and were two or three times as effective as the plants with lower-cation-exchange roots in solubilizing rock phosphate for the following crop of sudan grass. In this rock phosphate series, oat plants grown in association with vigorous red clover contained 62 per cent more phosphorus than did oats grown alone (adequate nitrogen was supplied). This demonstrated that the relatively insoluble soil and rock phosphate was solubilized by red clover roots not only for the red clover itself but for the associated cereal roots as well.

One must recognize that at least two distinct aspects of calcium release are involved in the plant-root–rock-phosphate system. One is the exchange adsorption of calcium from the rock phosphate crystal by the root colloid. The effect of this exchange adsorption increases with increased root bonding energy for calcium. In the early growth of the seedling root, this is probably the most important mechanism for phosphate release. As the

TABLE 10.9. PHOSPHORUS UPTAKE FROM SOIL AND ROCK PHOSPHATE BY FIRST CROP AND BY SECOND CROP OF SUDAN GRASS*

First crop	Root exchange (meq/100 g)	P uptake (mg)			P in applied rock (%)
		1st crop	Sudan†	Total	
No crop	—	—	1.62	1.62	—
Wheat	9	2.95	2.75	5.70	2.5
Lamb's quarters	25	8.29	1.83	10.12	4.4
Smartweed	41	13.77	6.43	20.20	8.8
Ragweed	59	8.76	9.07	17.83	7.7

* Unpublished data: M. Drake, University of Massachusetts, Amherst.

† Sudan grass root exchange 13 meq/100 g.

root becomes older a second effect grows in importance. The second effect is produced by organic matter supplied from the sloughing off of root walls and root hairs as well as from dead roots. This material is acted upon by soil microorganisms to produce a series of important chelating anions. This second effect of chelation by organic anions probably accounted for most of the difference in phosphorus uptake between sudan grass grown after wheat (2.75 mg) and with no preceding crop (1.62 mg).

Release of the phosphorus by calcium adsorption by the colloid of the living plant root and by organic anions from root material must be considered as well as the release of phosphorus from relatively insoluble iron and aluminum phosphates of the soil. In addition, some variation between plant species in total phosphorus uptake would be expected, because the growth rate and the volume of soil occupied by the roots differ greatly.

Chelation of Iron and Other Metals

Wallace (89) in his "A Decade of Synthetic Chelating Agents in Inorganic Plant Nutrition," discussed the chemical nature of chelation of the nutrient cations as well as some new concepts of chelation as related to biological systems and plant nutrition. In this volume, in addition to research papers by Wallace and his associates, there are sections on a decade of literature references on chelating agents in plant nutrition, questions and answers on the use of metal chelates in plant nutrition, and 39 general conclusions. Wallace's group suggest that contact chelation may more effectively describe that mechanism of cation accumulation by plant roots in contact with the clay fraction which Jenny called "contact exchange." Some of the conclusions are as follows:

> In soils, the important chemical reactions of metal chelates include hydrolysis of metal chelates, fixation of metal chelates on the clay surfaces, replacement of chelated metals by other metals, extraction of metals from the soil by salt or acid forms of chelating agents, and the apparent failure of soil microorganisms to decompose certain chelating agents.
>
> The effectiveness of some metal chelating agents is decreased by clay fixation. Either the chelating agents can chelate metals on the edge of the clay lattice or chelated metals can become attached to oxygen groups on the edge of the clay lattice.
>
> The application of chelating agents without metals to soil has not proved to be a satisfactory means of supplying micronutrients to plants. Direct application to plants by injection, or by foliar sprays of synthetic chelating agents without iron, have not satisfactorily corrected lime-induced chlorosis.
>
> EDDHA has proved to be the most effective chelating agent for use with iron on calcareous soils. Fe EDDHA is highly stable, not easily exchanged with calcium, not toxic, and does not fix on clay. Some iron chelates failed to satisfactorily supply iron to plants in the presence of bicarbonate. The apparent reason is the competition of HCO_3^- and OH^- from the nutrient solution with donor groups of the chelating agents

to supply electrons to available orbitals of the iron atoms. EDDHA is more successful than other chelating agents in the presence of bicarbonate, but even competition with it results in decreased iron uptake.

EDDHA appears to be the best source of iron in nutrient solutions (hydroponics). It is stable over a wide pH range, and iron is not as subject to replacement from it by copper, zinc, or calcium as with other chelating agents.

Roots have sites that can remove metals from chelating agents and chelating agents can diffuse from roots and collect in the external solution.

There appear to be quantitative and qualitative differences among plant species in natural chelating agents associated with roots that participate in cation uptake. An excess of chelating agents in a nutrient solution can inhibit the uptake of all cations by plant roots and can be toxic. The effect seems to be either one of competition between the synthetic chelating agent and the binding site (carrier or chelate) on the root for the metal, or one of removing calcium from solution.

Iron chelates can compete with or inhibit uptake of manganese by plant roots and thereby may control manganese toxicity.

The chelating agents can, to an extent, be taken into plant roots and transported to the leaves of plants. They seem to enter passively in part, through broken or injured roots, and seem to be transported in the xylem. There is considerable evidence that synthetic chelating agents do not penetrate plant cells. Chelating agents are retranslocated poorly in plants—they do not seem to move from tissues of original deposition.

Chelating agents appear to be taken into plants in direct proportion to that supplied in the external solution, that is, a tenfold increase in that supplied results in a tenfold increase in plant uptake.

Chelating agents can increase the yield of plants by some 20 to 30 per cent independently of supplying deficient micronutrients, possibly by providing a more favorable balance of the micronutrients.

Some fungicides are metal chelating agents, and chelation seems to be their mode of action.

Other research has shown that many organic anions chelate or combine with such metals as iron to form stable Werner-type complex ions (57). These chelated or complexed metallic ions are relatively free to migrate in the soil solution. Relatively insoluble compounds in the soil containing essential iron, manganese, copper, and zinc are solubilized by chelating agents such as the citrate anion. In most soils, several of these effective chelating agents are produced naturally from soil organic matter by soil microorganisms. An understanding of the chemistry of metal chelates [Martell and Calvin (57)] and of the factors that influence the natural production of chelating agents in the soil helps to explain the following problems and observations.

In nutrient solution studies, particularly above pH 7.0, iron chlorosis is prevented by the addition of ferric citrate or citrate anion. All the iron remained in solution, even in basic solutions with a citrate to iron ratio greater than 1:1, but at a 16:1 ratio growth was retarded. In 1951 Bitcover and Sieling (7) postulated that iron entered the plant root as ferric organic complex ion or was transported to the plant root as the complex ion. Natural chelating agents in the plant root could effect exchange of iron from the complex at the root surface. Recent studies using iron EDTA tagged with C^{14} indicated, however, that the complexed iron enters the plant root and is translocated in the plant as the iron complex [Stewart and Leonard (82)].

In the treatment of iron chlorosis of citrus, placing the roots of chlorotic grapefruit trees in a system containing one part citric acid and ten parts ferrous sulfate increased the iron uptake and produced green leaves. Soil applications of the citric-acid–ferrous-sulfate combination did not correct the chlorosis. But a ratio of 1 mole of citrate to 1 mole of iron is required to chelate all the iron. A soil application of chelated iron produced greener orange leaves containing more than twice as much iron as the chlorotic untreated tree leaves (82). Iron chlorosis in blueberries can be corrected by an increase in iron activity, either by increased soil acidity or by proper use of organic matter to complex the iron in neutral soils.

Zinc-deficiency symptoms were particularly acute where topsoil had been removed in leveling for irrigation. This could be explained by insufficient chelating agents in the subsoil, which was low in organic matter. Corn and apricot seedlings when grown between 1-year established alfalfa plants were almost free of white-bud and little-leaf symptoms of zinc deficiency, whereas without alfalfa they were seriously affected. One of several possible explanations is that zinc was chelated by alfalfa roots or by the chelating effects from these roots. Decomposition of tomato roots in the soil allowed a second crop of corn to develop free of zinc-deficiency symptoms (Hoagland). Yield increases of sweet corn were produced by addition of copper sulfate and of zinc sulfate on Miami silt loam soils in Wisconsin. But no yield increase was produced by copper or zinc on Carrington silt loam, a prairie soil of the same area that was high in organic matter. It is suggested that greater chelation of copper and zinc may have occurred in the Carrington soil.

Zinc-deficiency symptoms of young corn plants in the southeastern United States frequently disappear as the growing season advances. Similarly, early in cold, wet, growing seasons in northern Indiana, oats and soybeans are often deficient in manganese, but with the warm weather of July, they usually recover. Low microorganic activity during cool, wet soil conditions apparently did not produce enough chelating agents to solubilize sufficient zinc in the southeast or manganese in northern Indiana for normal plant requirements. With warmer weather, microorganic activity increased and probably produced enough chelating agents, which, in turn, solubilized adequate amounts of zinc and manganese.

PLANT TISSUE ANALYSIS—A DIAGNOSTIC TOOL

For well over a century scientists have been seeking ways to define a well-nourished plant. In an attempt to discover factors limiting plant growth and quality, to evaluate how well we are succeeding in this complex area of soil-plant nutrition, or to learn when plants grow poorly, [for example, the corn on the unphosphated plots of Hoffer and Carr (36)], scientists have turned to chemical analysis of the plant. Comparing chemical analysis of normal and deficient plants or of plants supplied with different treatments of mineral nutrients often provides a clue. While total chemical analysis of plant material has been widely used, future use will probably concentrate on total content of the several separate organs of functioning tissue of the plant at different stages of physiological development. The importance of sampling the different parts of the plant at different physiological stages of development is apparent when one recognizes that such nutrients as potassium are mobile. These may be translocated to more active tissue, especially when approaching a condition of deficiency. Under conditions of unbalanced or so-called luxury consumption, the nutrient may accumulate in older tissue. Determination of relative amounts of nutrients in old as well as new tissue often will reflect seasonal changes of nutrient status of such immobile nutrients as calcium.

The rapid chemical plant-tissue test has great value, especially when used by an experienced person with an eye trained to look for plant hunger signs and other limiting environmental factors (79). Some of these relatively simple, inexpensive, qualitative plant-tissue tests, such as that developed by Thornton (85), have been used with excellent results, especially as a diagnostic tool [Scarseth (79), Krantz (46), *et al*]. While the absolute precision is low, the rapid tissue test provides a qualitative measure of the active fraction of major macronutrient ions in a specific part of the plant: conducting, synthesizing leaf, or storage tissue. In the field, this test reveals the relative quantity and balance of active nutrient ions in each specific area of plant tissue analyzed, whereas the total ash analysis cannot separate the active from the accumulated inactive nutrients.

Interpretation of the rapid plant-tissue test is of the greatest importance (79). Of initial interest is the determination, if possible, of the first limiting factor or deficient nutrient element. The rapid plant-tissue test reveals whether there is an abundance, just a little, or only a trace of the nutrient element present. If there is only a trace of the major nutrient (or macronutrient) present, this is limiting growth; if there is only a small or medium amount present, it may become limiting. In case one nutrient is limiting, the other nutrients often accumulate and produce high tests which are misleading. Often when the first limiting nutrient has been supplied by fertilizer as a sidedressing, one or more of the nutrients that had tested high, may decrease rapidly and may become limiting.

One of the chief values of the rapid plant-tissue test is that it provides information on the relative balance of the active fraction of major nutrients in the plant tissue, and a second is that the simplicity and rapidity of analysis permits frequent evaluation of the nutrient status throughout the growing period. This rapid on-the-spot detection of a nutrient deficiency often makes it possible to correct this and save the growing plants by the sidedress application of an appropriate fertilizer material.

PRACTICAL APPLICATIONS

Results of studies designed to evaluate the factors affecting the soil-root-ionic environment have developed a number of principles that are important in crop production. Ion absorption involves physiochemical and metabolic-physiological processes. Both processes are encountered in cation absorption, whereas anion absorption is predominately of a metabolic-physiological nature. The following designations are suggested as representative of the three processes: R·COOH, physiochemical; R·H and R·OH, metabolic-physiological for cations and anions, respectively.

Characterization of root properties revealed large variations between plant species with respect to their cation-exchange capacities, and possibly with respect to kinds, and H dissociation of organic acids. These root CEC properties are believed to be important in the attraction of cations to the root and in adsorption at the root surface. Pectic substances located in the root surface account for most of the root cation-exchange properties. Root CEC values are highest near the apex and decrease rapidly with distance from the apex. The CEC of plant roots was shown to be largely independent of soil type. Characteristic CEC of roots is also shown by other organs of the plant, rhizomes, stems, and leaves as well as roots, suggesting that cells of these different organs may compete for cations. One type of cation movement in the plant may involve migration from cell surface to cell surface across adjacent exchange sites. Experimental proof of the valence effect in the cation composition of the plant grown in soil as compared to Mattson's monovalent-divalent cation adsorption by roots from clay systems appears difficult to demonstrate in some instances for two reasons: (*a*) the maximum valence effect occurs in very dilute solutions—conditions not met in nutrient solutions and in well-fertilized soils; and (*b*) while cations may be adsorbed on roots in ratios predicted by the Donnan distribution, once translocation occurs the composition of the plant top may not reflect those initial adsorption ratios. When rock phosphate was the only source of Ca and P, there was a high correlation between CEC of plant roots and uptake of Ca and P.

Root CEC may exert a qualitative influence on nutrient uptake by: (*a*) raising the concentration of cations in the Donnan free space (DFS) above the level of the surrounding aqueous solution; and (*b*) influencing the relative proportions of ions of different valency in the DFS as a consequence of the Donnan valence effect.

Principles of cation-exchange reactions support the following practical applications and considerations:

1. Colloids of the plant root compete with soil colloids for cations.

2. Plants with high CEC roots (high bonding of Ca), as alfalfa and Ladino clover, absorb Ca strongly and in preference to K. It is necessary to maintain relatively high levels of K in the soil solution to effect adequate K uptake by these plants. Plants with low CEC roots, as wheat and bent grass, bond Ca with relatively low energy and are highly effective in obtaining K from soils, even at low levels of exchangeable K.

3. Grasses grown in association with alfalfa and the clovers seriously compete for K, thereby increasing the difficulty of maintaining the high level of available soil K required by the legume. It is for this reason that on soils low in exchangeable K, frequent applications of fertilizer K to legume-grass associations greatly increase the longevity of the legume.

4. Plant species having roots with high CEC and high bonding energy for Ca are more capable than those of lower exchange capacities in obtaining Ca from relatively insoluble sources such as rock phosphate. In this process, phosphates become available not only to the roots feeding upon them but also to roots of associated plant species having lower CEC.

5. The mineral composition of plants is largely determined by physical-chemical relations between the colloidal systems of both the plant root and the soil. Plant material of legumes and many other dicotyledons contains larger amounts of Ca and Mg than do grasses and other monocots, not because these dicot plants require larger amounts of Ca and Mg, but because the nature of the root colloid causes the roots to attract and adsorb larger amounts of Ca and Mg. These adsorbed cations are in turn translocated into the plant. Legumes and other dicots with roots of high cation-exchange capacity must be liberally fertilized with K, not because these plants have a higher K requirement, but because the nature of their root colloid attracts relatively more Ca than K, and thus a higher K activity is required in the soil to overcome the higher bonding energy for Ca.

Identification of various soil colloids and investigation of their physicochemical surface properties have led to the following applications:

1. Below a critical per cent base saturation of the soil colloid, nutrient cations may be lost from the root to the external solution. On soils with low per cent base saturation, $CaSO_4$ or $MgSO_4$ are far less effective suppliers of cations than are the less-soluble carbonates. Frequently the application of $MgSO_4$ has not corrected Mg deficiency in apple trees growing on soils with low per cent base saturation.

2. Because of the high energy of retention of Ca by montmorillonite and beidellite clays in contrast to kaolinite clays, soils in which the former predominate should be limed above 70 per cent Ca saturation, whereas those in which the latter predominate require lower rates but more frequent applications of lime. Because of the lower bonding of Ca by kaolinitic clays, one would expect such overliming injury as boron deficiency induced by increased Ca uptake of the plant to occur more frequently on kaolinitic than on montmorillonitic soils.

3. Since Ca is more strongly retained by montmorillonite-beidellite clays than by kaolinite clays, the availability of Na and K to plants (for a given percentage Ca saturation) will be greater from the montmorillonite-beidellite soil colloid association. Conversely, applications of fertilizer K will depress the activity or plant availability of Ca to a greater degree in montmorillonite than in kaolinite soils.

4. As a general guide under New Jersey conditions, the cation-exchange complex should contain 65 per cent Ca, 15 per cent Mg, 5 per cent K, and 20 per cent H.

Placing soluble fertilizer phosphate in bands near the seed reduces phosphorus fixation by active Al and Fe and results in greater uptake of fertilizer P. Placing ammonium or nitrate nitrogen salts with the band of soluble fertilizer phosphate, increases root proliferation and further increases uptake of P from the fertilizer P.

A single root can obtain all the N and P required for the development of normal plants.

Demonstrated properties of chelating agents in rendering and keeping ions stable in solution (Werner-type complex ions) lead to new evaluations of the significance of organic matter in affecting nutrient availability. Natural chelating agents are produced in the soil by microbial breakdown of organic matter. One important effect of these chelating organic ions is to inactivate iron, aluminum, and calcium and thereby reduce phosphate fixation by these means. Release of some of the fixed phosphates likewise may be anticipated. The use of Fe, Mn, Zn, and Cu chelating compounds as a means of controlling plant deficiencies of these nutrients is assuming increasing importance.

Several fundamental reasons are at hand to explain the permanency and soundness of Cyril G. Hopkins' system of ground rock phosphate, organic matter, manure, limestone, and legumes. The two weak points were potassium and nitrogen. In our modern agriculture, the systems of handling manure fail to return the potassium removed by crops. Fertilizer K must be applied to avoid soil depletion. Maximum yields of several crops require fertilizer nitrogen to add to, as well as to replace, that supplied by legumes. It is reasoned that the legume roots and chelating agents of organic matter split Ca from rock phosphate, leaving the phosphate soluble. The legume roots and chelating agents also split Al and Fe from soil phosphates, leaving the phosphate soluble. In addition, the chelating agents protect the soluble phosphate in the inactivation of Al, Ca, and Fe. The relationship of soil organic matter to the availability of phosphorus and other elements is the keystone to the adage:

Lime and lime without manure
Make both farm and farmer poorer
But lime, manure, and vigorous clover
Make the old farm rich all over.

REFERENCES

1. Albrecht, W. A., and McCalla, T. M., *Am. J. Botany* **25**, 403–407 (1938).
2. Albrecht, W. A., *Soil Sci.* **55**, 13–21 (1943).
3. Allaway, W. H., *Soil Sci.* **59**, 207-217 (1945).
4. Asher, C. J., and Ozanne, P. G., *Australian J. Agri. Res.* **12**, 755–766 (1961).
5. Bear, F. E., *et al.* "New Jersey Agr. Expt. Sta. Bull. 721," 1945.
6. Bear, F. E., and Prince, A. L., *J. Am. Soc. Agron.* **37**, 217–222 (1945).
6a. Beeson, K. C., "U.S.D.A. Miscellaneous Publication 369," 1941.
7. Bitcover, E. H., and Sieling, D. H., *Plant Physiol.* **26**, 290–303 (1951).
8. Bradfield, R., *J. Am. Chem. Soc.* **45**, 2669 (1923).
9. Bray, R. H., *J. Am. Chem. Soc.* **64**, 954–963 (1942).
10. Bray, R. H., *Soil Sci.* **58**, 305–324 (1944).
11. Brown, D. A., and Albrecht, Wm. A., "Missouri Agr. Expt. Sta. Research Bull. 477," 1951.
12. Buckman, H. O., and Brady, N. C., "The Nature and Properties of Soils," New York, The Macmillan Co., 1960.
13. Cole, C. V., and Jackson, M. L., *J. Phys. Colloid Chem.* **54**, 128–142 (1950).
14. Conner, S. D., *J. Ind. Eng. Chem.* **8**, 35–40 (1916).
15. Crooke, W. A., Knight, A. H., and Macdonald, I. R., *Plant Soil* **13**, 123–127 (1960)
16. Crooke, W. M., and Knight, A. H., *Soil Sci.* **93**, 365–373 (1962).
17. Dalton, J. D., Russell, G. C., and Sieling, D. H., *Soil Sci.* **73**, 173–181 (1952).
18. Dean, L. A., and Rubins, E. J., *Soil Sci.* **63**, 377–387 (1947).
19. Devaux, H., *Compt. Rend. Acad. Sci. (Paris)* **162**, 561 (1916).
20. Drake, M., and Scarseth, G. D., *Soil Sci. Soc. Am. Proc.* **4**, 201–204 (1940).
21. Drake, M., Vengris, J., and Colby, W. G., *Soil Sci.* **72**, 139–147 (1951).
22. Drake, M., Colby, W. G., and Vengris, J., *Better Crops* **36**, 13 (1952).
23. Drake, M., and Steckel, J. E., *Soil Sci. Soc. Am. Proc.* **19**, 449–450 (1955).

24. Drake, M., and White, J. M., *Soil Sci.* **91**, 66–69 (1961).
25. Duncan, W. G., and Ohlrogge, A. J., *Agron. J.* **50**, 605–608 (1958).
26. Eck, P., Drake, M., and Steckel, J. E., *Soil Sci.* **84**, 145–154 (1957).
27. Elgabaly, M. M., and Wiklander, L., *Soil Sci.* **67**, 419–424 (1949).
28. Epstein, E., and Hagen, C. E., *Plant Physiol.* **27**, 457–474 (1952).
29. Frey-Wyssling, A., "Ernährung und Stoffwechsel der Pflanzen," Zurich, Buchergilde Gutenberg, 1945.
30. Fried, M., *Soil Sci. Soc. Am. Proc.* **17**, 357–359 (1953).
31. Gilbert, F. A., *Advan. Agron.* **4**, 147–177 (1952).
32. Graham, E. R., *Soil Sci. Soc. Am. Proc.* **19**, 26–29 (1955).
33. Gray, B., Drake, M., and Colby, W. G., *Soil Sci. Soc. Am. Proc.* **17**, 235–239 (1953).
34. Grunes, D. L., *Advan. Agron.* **11**, 369–396 (1959).
35. Hoagland, D. R., "Inorganic Nutrition of Plants," Waltham, Mass., Chronica Botanica Co., 1944.
36. Hoffer, G. N., and Carr, R. H., *J. Agr. Research* **13**, 801–823 (1923).
37. Inden, T., Hori, S., and Okuda, A., *J. Sci. Soil Manure* Japan **29**, 259–262 (1958).
38. Itallic, Th.B. van, *Soil Sci.* **46**, 175–186 (1938).
39. Jacobson, L., and Ordin, L., *Plant Physiol.* **29**, 70–75 (1954).
40. Jacobson, L., Hannapel, R. J., Moore, D. P., and Schaedle, M., *Plant Physiol.* **36**, 58–61 (1961).
41. Jarusov, S. S., *Soil Sci.* **43**, 285–303 (1937).
42. Jenny, H., *J. Phys. Chem.* **36**, 2217–2258 (1932).
43. Jenny, H., and Overstreet, R., *Soil Sci.* **47**, 257–272 (1939).
44. Jenny, H., and Ayers, A. D., *Soil Sci.* **48**, 443–459 (1939).
45. Knight, A. H., Crooke, W. M., and Inkson, R. H. E., *Nature* **192** (No. 4798), 142–143 (1961).
46. Krantz, B. A., Nelson, W. L., and Burkhart, L. F., "Diagnostic Techniques for Soils and Crops," pp. 137–155, The American Potash Institute, 1948.
47. Liebig, J. seen in: "Liebig and After Liebig," Washington, D. C., American Association for the Advancement of Science, 1942.
48. Lucas, R. E., *et al.*, "Indiana Agr. Expt. Sta. Bull. 468," 1940.
49. Lucas, R. E., and Scarseth, G. D., *J. Am. Soc. Agron.* **39**, 887–896 (1947).
50. Lundegardh, H., *Soil Sci.* **54**, 177–189 (1942).
51. McLean, E. O., and Marshall, C. E., *Soil Sci. Soc. Am. Proc.* **13**, 179–182 (1948).
52. McLean, E. O., Adams, D., Franklin, R. E., Jr., *Soil Sci. Soc. Am. Proc.* **20**, 345–347 (1956).
53. McLean, E. O., and Baker, F. E., *Soil Sci. Soc. Am. Proc.* (1952) **17**, 100–102 (1953).
54. Marshall, C. E., "Missouri Agr. Expt. Sta. Research Bull. 385," pp. 1–60, 1944.
55. Marshall, C. E., "The Colloid Chemistry of the Silicate Minerals," New York, Academic Press, Inc., 1949.
56. Marshall, C. E., and Upchurch, W. J., *Soil Sci. Soc. Am. Proc.* (1952) **17**, 222–227 (1953).
57. Martell, A. E., and Calvin, M., "Chemistry of the Metal Chelate Compounds," New York, Prentice-Hall, Inc., 1952.
58. Mattson, S., *Ann. Agr. Coll. Swed.* **15**, 308–316 (1948).
59. Mattson, S., *et al.*, *Ann. Agr. Coll. Swed.* **16**, 457–484 (1949).
60. Mehlich, A., and Colwell, W. E., *Soil Sci. Soc. Am. Proc.* **8**, 179–184 (1944).
61. Mehlich, A., and Reed, J. F., *Soil Sci. Soc. Am. Proc.* **11**, 201–205 (1947).
62. Mehlich, A., and Reed, J. F., *Soil Sci. Soc. Am. Proc.* **13**, 399–401 (1949).
63. Mehlich, A., *Soil Sci.* **73**, 361–374 (1952).

64. Mehlich, A., and Coleman, N. T., *Advan. Agron.* **4**, 67–99 (1952).
65. Mehlich, A., *Soil Sci. Soc. Am. Proc.* (1952) **17**, 231–234 (1953).
66. Meyer, B. S., Anderson, D. A., Bohning, R. H., "Plant Physiology," New York, D. Van Nostrand Co., Inc., 1960.
67. Miller, M. H., and Ohlrogge, A. J., *Agron. J.* **50**, 95–97 (1958).
68. Mitsui, S., Nakagawa, M., Baba, A., Tensho, K., and Kumazawa, K., *J. Sci. Soil Tokyo* **26**, 497. [*Chemical Abstr.* 50, 12380e (1956)].
69. Moore, D. P., Jacobson, L., Overstreet, R., *Plant Physiol.* **36**, 53–57 (1961).
70. Nelson, W. L., *et al.*, *Soil Sci. Soc. Am. Proc.* **12**, 113–118 (1948).
71. Newton, J. D., *Soil Sci.* **15**, 181–204 (1923).
72. Ohlrogge, A. J., Miller, M. H., and Duncan, W. G., *Better Crops* **41**, 26–30 (1957).
73. Ohlrogge, A. J., "Proc. 12th Hybrid Corn Industry Research Conf.," pp. 55–62, 1957.
74. Parsons, J. L., *Better Crops* **42** (1958).
75. Reitemeier, R. F., *et al.*, "U.S.D.A. Tech. Bull. 1049," 1951.
76. Rosseter, R. C., and Ozanne, P. G., *Australian J. Agri. Res.* **6**, 553–64 (1955).
77. Russell, E. J., and Russell, E. W., "Soil Conditions and Plant Growth," New York and London, Longmans, Green, and Co., 1950.
78. Saussure, Th. de. seen in: "Liebig and After Liebig," Washington, D. C., American Association for the Advancement of Science, 1942.
79. Scarseth, G. D., *Soil Sci.* **55**, 113–120 (1943).
80. Smith, R. L., and Wallace, A., *Soil Sci.* **82**, 165–172 (1956).
81. Sprengel, C. seen in: "Liebig and After Liebig," Washington, D. C., American Association for the Advancement of Science, 1942.
82. Stewart, I., and Leonard, C. D., *Science* **116**, 546–566 (1952).
83. Struthers, P. H., and Sieling, D. H., *Soil Sci.* **69**, 205–213 (1950).
84. Swenson, R. M., Cole, C. V., and Sieling, D. H., *Soil Sci.* **67**, 3–22 (1949).
85. Thornton, S. F., Conner, S. D., and Fraser, R. R., "Purdue Univ. Agr. Expt. Sta. Circ. 204," (Rev.) (1939).
86. Truog, E., *Soil Sci.* **5**, 169–195 (1918).
87. Truog, E., "Mineral Nutrition of Plants," Madison, Wisconsin, The University of Wisconsin Press, 1951.
88. Turner, R. C., and Rice, H. M., *Soil Sci.* **74**, 141–148 (1952).
89. Wallace, A. "A Decade of Synthetic Chelating Agents in Inorganic Plant Nutrition." Copyright 1962 by Arthur Wallace. Lithographed by Edwards Brothers, Inc. Ann Arbor, Michigan, 1962.
90. Wiklander, L., *Ann. Agr. Coll. Swed.* **14**, 1–171 (1946).
91. Williams, D. E., and Coleman, N. T., *Plant Soil* **2**, 243–256 (1950).

CHAPTER 11

Radioisotopes and Soils

E. R. GRAHAM

Department of Soils
University of Missouri, Columbia

Many isotopes, both stable and radioactive, of natural and man-made origin may be found in soils. Since the minerals that make up the soil bed may have come from air, land, or sea, the kind and amount of any one isotope might well be associated with a given landscape and be characteristic of a given soil type. It will be the purpose of this review to indicate what isotopes of interest occur in soils, along with their amounts, properties, areas of distribution throughout the earth's surface, and a brief discussion of the use of man-made isotopes in soil studies.

In considering the natural radioactivity of the earth's crust, one thinks first of the radioactive series dominated by the two elements, uranium and thorium, because these change to many interesting isotopes such as radium and radon along with radioactive lead. The existence and distribution of long-life radioactive nuclides among the lighter elements in soils is well known, but the extensive concentration and distribution in the earth's surface is sometimes overlooked. It might well be anticipated that the concentration of the less-dense elements in the earth's crust could, in some instances, be appreciable. Largely due to fundamental discoveries made by Ernest Rutherford and the cooperation of the English chemist, Frederick Soddy, 20 elements possessing radioactive properties had been described by the end of the year 1904. By 1912 this number was extended to more than 30. At the present time over 40 radioactive elements, mostly of high atomic weight, are known in nature. Those elements that are of light atomic weight, namely, potassium, rubidium, samarium, lutetium, rhenium, and a few others, possess feeble radioactive properties in their normal states.

These elements, however, occur naturally and should be considered as part of the soil. Of the many radioactive elements found in nature, the most common (see Tables 11.1 to 11.4) would probably be K^{40}, Th^{232}, U^{238} and U^{235}, Rb^{87}, Nd^{150}, Sm^{152}, and Lu^{176}. Since, in most cases in nature the radioactive nuclide occurs together with one or more stable nuclides of the same element, it is important to consider the fractional occurrence of the radionuclide before conclusions can be drawn about the element as a whole.

Since isotopes of an element may be either stable or radioactive, some comments on the nature of radioactivity are in order. Becquerel showed that beta rays emitted by radioactive substances are deflected by electric and magnetic fields in the same manner as cathode rays or high-speed electrons. More accurate measurements made later definitely identified the beta particles as high-speed electrons. Alpha rays may readily be shown to be high-speed particles carrying a positive charge.

Rutherford and R. Royds collected enough of these particles to identify them as helium gas. Gamma rays have neither mass nor charge, and travel

TABLE 11.1. RADIOACTIVE DISINTEGRATION SERIES FOR URANIUM 238

Element	Half-life	Type of decay and energy of radiation (Mev)
$_{92}U^{238}$ ↘	4.50×10^9 yr	Alpha 4.21
$_{90}Th^{234}$ ↙	24.1 days	Beta⁻ 0.13, gamma 0.09
↓ $_{91}Pa^{234}$	1.14 min.	Beta⁻ 2.32, gamma 0.80
$_{92}U^{234}$ ↘	2.35×10^5 yr	Alpha 4.76
$_{90}Th^{230}$	8.0×10^4 yr	Alpha 4.66, gamma 0.68, 0.14–0.25
↓ $_{88}Ra^{226}$	1.6×10^3 yr	Alpha 4.79, gamma 0.19
↓ $_{86}Rn^{222}$	3.825 days	Alpha 5.49
↓ $_{84}Po^{218}$	3.05 min.	Alpha 5.99
↓ $_{82}Pb^{214}$ ↙	26.8 min.	Beta⁻ 0.65, gamma 0.29
$_{83}Bi^{214}$ ↘	19.7 min.	Alpha 5.50, beta⁻ 3.15, gamma 1.8
$_{81}Tl^{210}$	1.32 min.	Beta⁻ 1.80
↓ $_{84}Po^{214}$ ↘	1.5×10^{-4} sec.	Alpha 7.68
$_{82}Pb^{210}$ ↙	22 yr	Beta⁻ 0.026, gamma 0.047
↓ $_{83}Bi^{210}$	4.97 days	Beta⁻ 1.17
$_{84}Po^{210}$ ↘	138 days	Alpha 5.30, gamma 0.80
$_{82}Pb^{206}$	S	

Note: ↘ denotes a parent-daughter relationship (i.e., Sr^{90} decays by beta decay to Y^{90} which is radioactive and decays by beta decay to stable Zr^{90}).

TABLE 11.2. RADIOACTIVE DISINTEGRATION SERIES FOR THORIUM 232

Element*	Half-life	Type of decay and energy of radiation (Mev)
$_{90}Th^{232}$ ↘	1.39×10^{10} yr	Alpha 4.20
$_{88}Ra^{228}$ ↙	6.7 yr	Beta⁻ 0.053
↓ $_{89}Ac^{228}$	6.13 hr	Beta⁻ 1.55, gamma 0.96
$_{90}Th^{228}$ ↘	1.90 hr	Alpha 5.42
$_{88}Ra^{224}$	3.64 days	Alpha 5.68
↓ $_{86}Rn^{220}$	54.5 sec	Alpha 6.28
↓ $_{84}Po^{216}$	.158 sec	Alpha 6.77
↓ $_{82}Pb^{212}$ ↙	10.6 hr	Beta⁻ 0.36
$_{83}Bi^{212}$ ↓ ↘	60.5 min.	Alpha 6.05, beta⁻ 2.20, gamma 0.040–2.20
$_{81}Tl^{208}$	3.1 min.	Beta 1.82, gamma 2.62
$_{84}Po^{212}$	3×10^{-7} sec	Alpha 8.95
$_{82}Pb^{208}$ $_{82}Pb^{208}$	S	

* See footnote to Table 11.1.

TABLE 11.3. RADIOACTIVE DISINTEGRATION SERIES FOR URANIUM 235

Element*	Half-life	Type of decay and energy of radiation (Mev)
$_{92}U^{235}$ ↘	7.07×10^{8} yr	Alpha 4.52
$_{90}Th^{231}$ ↙	24.6 hr	Beta⁻ 0.2, gamma 0.03
$_{91}Pa^{231}$ ↘	3.2×10^{4} yr	Alpha 5.05, gamma 0.32
$_{89}Ac^{227}$ ↙ ↓	13.5 yr	Alpha 5.00, beta⁻ 0.22
$_{90}Th^{227}$	18.2 days	Alpha 6.05, gamma 0.05
$_{87}Fr^{223}$	22 min.	Beta⁻ 1.15, gamma 0.05
$_{88}Ra^{223}$ $_{88}Ra^{223}$	11.7 days	Alpha 5.72, gamma 0.03
↓ $_{86}Rn^{219}$	3.92 sec	Alpha 6.82, gamma 0.27
↓ $_{84}Po^{215}$	1.83×10^{-3} sec.	Alpha 7.36
↓ $_{82}Pb^{211}$ ↙	36.1 min.	Beta⁻ 0.5, gamma 0.8
$_{83}Bi^{211}$ ↘	2.16 min.	Alpha 6.62, beta⁻ 1.4, gamma 0.35
$_{84}Po^{211}$	0.5 sec	Alpha 7.43
$_{81}Tl^{207}$	4.76 min.	Beta⁻ 1.47, gamma 0.89
$_{82}Pb^{207}$ $_{82}Pb^{207}$	S	

* See footnote to Table 1.11.

TABLE 11.4. NON-SERIES, NATURAL RADIO ISOTOPES

Element	% Abundance	Half-life	Emission
$_{19}K^{40}$	0.012	1.3×10^{9} yr	Beta, gamma, K
$_{37}Rb^{87}$	27.2	6.0×10^{10} yr	Beta, gamma
$_{62}Sm^{147}$	15.1	1.3×10^{11} yr	Alpha
$_{71}Lu^{176}$	2.6	2.4×10^{10} yr	Beta, gamma, K
$_{75}Re^{187}$	62.9	5.0×10^{10} yr	Beta
$_{83}Bi^{209}$	100.0	2.7×10^{17} yr	Alpha

at the velocity of light. They are, therefore, electromagnetic waves. Their behaviors are identical to x-rays, except that the energies of the gamma rays are usually higher than those associated with x-rays. Gamma rays originate along with changes in nuclear energy levels, while x-rays originate because of energy changes of the orbital electrons, or with the sudden slowing of fast-moving electrons. Gamma rays, emitted by certain radioactive substances, will expose a photoplate, similar to x-ray pictures, of thick sections of metal castings, weldings, pipes, structural sections, and armor plate. Radioactive substances may often be used photographically in places inaccessible to x-ray generators, so that pictures may be taken that otherwise would be impossible.

Many radioactive substances have been produced by means of particle accelerators and nuclear reactors. These are often called artificial radioactive substances, since they do not occur naturally on the earth. The artificial radioactive isotopes emit either alpha particles, beta particles, or gamma rays, just the same as the naturally occurring radioactive elements. Some of the radioactive elements decay through the capture of electrons by the nucleus. Since the captured electrons are usually from the K orbit, this type of decay is called K-capture. The energy resulting from the decay is released as gamma ray.

A fourth type of radioactive element is found to emit a particle carrying a positive electric charge identical to the charge on a proton, but with the mass of an electron. These particles are called positive electrons, or positrons. Positrons have a very short life within the vicinity of other matter. These positive charges cause them to be attracted by electrons. Upon meeting, the charges on the two particles (positron and electron) are neutralized, and their masses change completely into energy in the form of two gamma rays. This was the first example discovered of annihilation of mass, or of the conversion of mass into energy.

The radioactive decay of any atom is associated with the emission of a charged particle of positive or negative beta, or from the capture of an electron by the nucleus. The nuclear charge, and hence the atomic number, always changes in the decay process. Radioactive decay is therefore characterized by change of the atoms of one element into a different element. The atomic number decreases by two units with the emission of an alpha particle, and decreases one unit with the emission of a positron or the capture of an electron. It goes up one with the emission of a negative electron. Energy is released in each of such transformations.

Alpha Decay

$$_{88}Ra^{226} \rightarrow {}_{86}Rn^{222} + {}_{2}He^{4} + Q$$

Beta(+) Decay

$$_{7}N^{13} \rightarrow {}_{6}C^{12} + 1^{e0} + \gamma + Q$$

Beta(−) Decay

$$_{82}Pb^{214} \rightarrow {}_{83}Bi^{214} - 1^{e0} + Q$$

K(−) Electron Capture

$$_{4}Be^{7} - 1^{e0} \rightarrow {}_{3}Li^{7} + \gamma + Q$$

The Radioactive Series

As emphasized earlier, some 40 species with different radioactive properties have been identified in nature. With a physical or chemical separation, it is possible to study the radioactive decay and/or growth curves along with specific properties of emitted radiations. It has been found that the natural radioactive elements of high atomic weight fall into three distinct series, (*a*) the thorium, (*b*) the uranium, and (*c*) the actinium series. The first two are named after the longest-life precursors, thorium and uranium, with half-lives of 1.39×10^{10} and 4.51×10^{9} years, respectively.

The parent elements of thorium and uranium undoubtedly had shorter lives and, consequently, no longer exist in any detectable amounts. The parent of the actinium series is not, as was originally supposed, the element actinium, the first member of this series to be discovered, but rather a much longer-life element sometimes referred as actino-uranium (U^{235}) with a half-life of 7.1×10^{8} years. The alpha particle, identified as being identical with the helium nucleus, has a mass of 4 and a Z number of 2. It is evident then that in any disintegration stage where an alpha particle is emitted, the atomic weight of the daughter element must be four units less than the parent, and the Z number two units less.

example: $_{88}Ra^{226} \rightarrow {}_{86}Rn^{222} + {}_{2}He^{4} + Q$

When a beta particle of an electron mass, which is negligible on the atomic weight scale, is produced, the atomic mass will be the same, with a disintegration accompanying a beta particle. The Z number, however, will change, increasing one unit, since a beta decay is accompanied with a transformation of a neutron into a proton.

$$\textit{example:} \quad {}_{82}Pb^{214} \rightarrow {}_{83}Bi^{214} + \text{beta}^- + Q$$

The production of radiation energy as gamma ray may or may not accompany the emission of alpha and beta particles.

In Tables 11.2, 11.3, and 11.4, the published details according to Glasstone (14) and Lapp and Andrews (28) for the thorium, uranium, and actinium series are presented. These tables give the nature of the radiation, the half-lives of the respective members, and some names of the various elements as they have been recorded in the literature.

Radioactive Potassium

The radioactive potassium nuclide K^{40} breaks down in two ways, changing to Ca^{40} by beta-particle emission, and to A^{40} by K-capture. Both processes will be considered, since most of the activity is associated with the production of Ca^{40}, which has been estimated to be 90 per cent; only 10 per cent of the activity is associated with the production of A^{40}. Appropriate corrections must be made when considering the gamma activity or the K^{40} isotope (Figure 11.1).

The element rubidium, which decays only by beta negative, would transfer to Sr^{87} as a result of the decay mechanism. The production of stable elements as the result of decay of the naturally occurring radioactive isotope would then result in a non-series reaction producing stable isotopes, as was the case with K and Rb.

Variations in the Earth's Natural Radioactivity

An estimate of the relative contribution to the earth's radioactivity has been made by Asimov (4), who estimated that, of the 10 most important radioactive naturally occurring elements, only three, K, Th, and U, would occur in the earth's crust to an extent of more than 10 ppm on a weight basis. The other elements, even if each is counted as occurring to the extent of 10 ppm, would be shown to account for not more than 1/1000 of the natural radioactive content of the earth's crust. It is therefore concluded that a kilogram of the earth's crust would have approximately 26 g of potassium, which would produce approximately 800 beta particles per second and about 80 gamma photons per second. The thorium content would be about 0.02 g/kg in the earth's crust. This would produce, per second, approximately 490 alpha particles and 330 beta particles. The amount of uranium in the earth's surface would be about 0.08 g/kg and

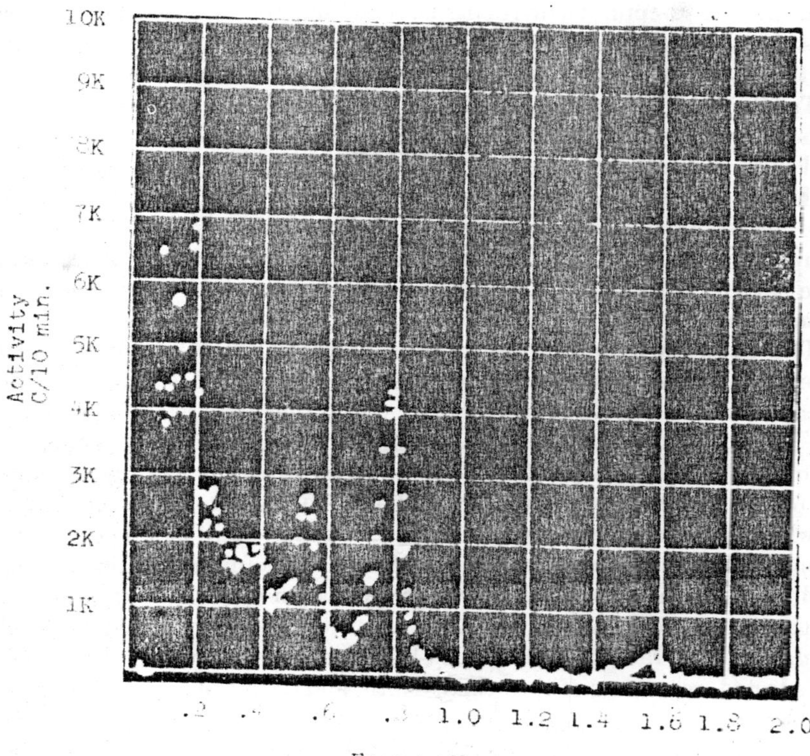

Figure 11.1. Bluegrass-held radioactive fission products as shown by gamma-ray spectrum. The photo peak near 0.8 Mev is caused by $Zr^{95} \rightarrow Nb^{95} \rightarrow Mo^{95}$ (stable); at 0.51 Mev by $Ru^{106} \rightarrow Rh^{106} \rightarrow Pd^{106}$ (stable); at 0.36 Mev by $I^{131} \rightarrow Xe^{131}$ (stable); and at 0.13 Mev by $Ce^{144} \rightarrow Pr^{144} \rightarrow Nd^{144}$ (stable).

would contribute, per second, some 8500 alpha particles and approximately 6200 beta particles. In terms of relative contributions of these three elements to the radioactivity of the earth's crust, uranium would account for 90 per cent of the total radioactivity; thorium, about 5 per cent; and potassium, 4.9 per cent. All other elements would account for less than 0.1 of 1 per cent of the total radioactivity.

Gibbs *et al.* (13) studied the radioactivity of the principal horizons of a number of soil types for relationships of radioactivity to soil properties as a possible method to be used in the identification and classification of soil profiles. After correction for cosmic-ray background, the measurements showed a range in gamma counts from 410 to 2420 cpm and beta counts from 21 to 154 cpm. Their results showed that the radioactivity depends partly on the rock materials from which the profiles are developed and partly on the extent of weathering and leaching of the profiles.

Telfair *et al.* (49) made a study of the natural radioactivity of Miami soils. They concluded that the observed radioactivity varies with soil type and with depth in the soil profile. The gamma activity was found to be high in the surface A_0 layer; this was attributed to fallout from bomb tests. The lowest amount was found in the 5- to 10-inch depth. The activity increased to a high at the 25-inch depth, then fell off with increasing depth. The activity observed was attributed to K^{40} and to the natural series of uranium and thorium.

The naturally occurring series element radon (Rn^{222}) is of particular interest (10) in soils studies, because it follows radium (Ra^{226}) in the natural decay schemes. The decay of a radium atom, therefore, is accompanied by the production of a radon atom, which then decays by alpha emission to polonium^{218}. The half-life of radon is 3.8 days, while the half-life of radium is 1622 years. These two isotopes would, therefore, be in secular equilibrium, in which case for every active atom of radon found there would be an active atom of radium present. And since radium is in equilibrium with uranium and thorium, an analysis for a given soil for radon would give an estimation of the amount of uranium and thorium in the soil. Stout (48), Delwiche *et al.* (9), and Delwiche and Stout (10) have reported finding radon in all soils tested, regardless of the origin, depth, and degree of weathering. These authors reported that a soil developed from limestone contained a higher amount of radon than the carbonate parent material. This observation was explained by the fact that uranium and thorium atoms were associated with the aluminum silicate residues which resisted solution and were not carried away as was the calcium carbonate. This suggests that the amount of radon found in any given soil is associated with the amount of radon found in the parent material of this soil, and would give a ratio representing the change in volume of the parent material during soil formation. This would have an accuracy similar to a determination of the zircon content of parent material as compared to the zircon content of the developed soil, as was done by Marshall *et al.* (31). The determination of zircon is much more time-consuming than would be the determination of radon.

Hansen *et al.* (20) reported some of their experiences related to the recovery of radon. It was presumed by these investigators that the radon would be absorbed by soil as well as the radium or uranium and thorium, but, since radon is a gas, it should be able to escape from the soil and diffuse out into the soil atmosphere. Since these workers anticipated that moisture might be a factor in the release of radon, they studied the effect of adding increments of water to montmorillonite clay and determined the per cent of recovery of the radon. They found that up to 29.8 per cent of the radon could be recovered by adding water to the clay directly. With lower amounts of water, less than this amount of radon was recovered. However, when the clay was suspended in boiling water, all the radon being generated by the clay held, radium was released to the atmosphere. They concluded, then, that clay material which was suspended in boiling water would release all the radon that was in the soil. Their experiences with radium also showed that radium is completely mobile in a soil when determined by leaching with normal ammonium acetate or chloride solutions. This would be expected since radium is a Group 2 element.

Radon of the outer air is derived from soil by diffusion. Kovach (27), who measured the radon content of soil in New York over an entire year, showed that most of the radon that escapes to the air came from a depth of 150 cm or less. Radon escape was usually observed to be lower when soil was frozen.

In a report by Telfair *et al.* (48), the accumulation of radon in sealed cans of soil taken from a Miami silt loam at various depths in the soil profile indicated that the maximum radon accumulation occurred at the sample taken at 25 inches of depth. The sample at the 5-inch depth had the lowest amount of radon, and this radon increased in amounts to the 25-inch sample and then decreased through the profile, indicating that the radium and radon content was closely associated with the sample taken from the lower B horizon.

Radioactive Elements of the Biosphere

Of the natural radioactive elements in the biosphere, those that would be absorbed by plants from the soil and returned to the soil as plant residues (K^{40}, C^{14}, and possibly H^3) would be the most important. The occurrence of C^{14} and H^3 in the atmosphere was studied in detail by Libby (29). These studies resulted in methods for radiocarbon dating, and for dating materials with tritium. Radiocarbon exists in the atmosphere because of a transmutation reaction which is identical with that used in nuclear reactors to make synthetic C^{14}: $N^{14} + {}_0n^1 = C^{14} + \text{proton} + 620$ Kev. The neutrons are provided by cosmic radiation, and the nitrogen is that of the atmosphere of the earth. The probable existence of this as a production mechanism was pointed out by Korff (26). Natural H^3 was discovered in 1951 (17). It is produced by fast neutron reaction (${}_7N^{14} + {}_0n^1 \rightarrow {}_1H^3 + {}_6C^{12}$). The yield of this reaction is very low as compared to that of the C^{14}; however, enough tritium has been produced so that it is found in amounts which allow it to be identified in naturally occurring materials. Hansen *et al.* (20) have estimated that an average green crop will contain about 0.11 mc (millicuries) of radioactivity per acre per year. This would amount to a turnover by good crop land of about 7 mc per square mile. Thus the naturally occurring radioactivity per square mile per crop year would be somewhat smaller than the amounts of synthetic or man-made radioactive nuclides, which have been distributed all over the world. The radioactivity due to fallout and fission-product disposal is greater than 80 mc per square mile.

Radioactive Nuclides Deposited on Soil by Fallout

During the last few years radioactivity in the form of fission products and neutron-activated radionuclides have been added to the soil as a result of the nuclear-weapons-testing program. In some instances, this added radioactivity is small compared to the natural radioactivity which the soil already contains. Gustafson, Marinelli, and Brar (18) in 1958 reported on a study of the natural radioactivity of soil in relation to the fallout. They demonstrated that the fission products from the fallout contributed only 7 milliroentgen per year to the total background dose rate measured at points 3 feet above the ground, as compared with 77 milliroentgen per year contributed by the natural radioactivity of the soil. When the radioactivity is expressed as millicuries per kilogram of soil, the contribution from the fission products would be more noticeable, since many of the fission products are beta emitters and would contribute very little radiant activity measured at an air distance of 3 feet above the surface. Table 11.5 lists the common fission products which are found in soils and the common neutron-activated elements.

The activity of the fission-product isotopes, such as Cs^{137}, Sr^{90}, Zr^{95}, Ce^{144}, etc., is likely to be lower in most soils than the same level of the natural radioisotopes, especially K^{40}. In recent years, much effort has been strongly fixed on the fission-product isotopes, because of the supposition that the activity level alone is not a valid measure of potential biological hazards and because there is a nation-wide concern associated with the hazards to man from such radioactive isotopes in fallout. It may well be that neither of these reasons has much sound scientific basis and that radioactive fallout is much less harmful or interesting than has been supposed. Nevertheless, with our present state of knowledge, revolving around the biological hazard of these isotopes, it is important to know something about the distribution and the large area of contamination which has resulted from the nuclear-weapons-testing program.

Alexander, Hardesty, and Hollister (1) reported on a worldwide fallout testing program for Sr^{90} in soils. They indicated that they wanted to relate the levels in soils to the levels found in plants, animals, human foods, and human beings. Secondly, they wanted to relate the levels in soils to the

TABLE 11.5. MAN-PRODUCED RADIOACTIVE NUCLIDES COMMONLY FOUND IN SOILS

Nuclide*	Half-life	Emission and energy
	Significant fission products	
$_{8}Sr^{89}$	50.4 day	$Beta^-$ 1.47
$_{8}Sr^{90}$	28 yr	$Beta^-$ 0.54
↘		
$_{39}Y^{90}$	64.2 hr	$Beta^-$ 2.26
Y^{91}	59 day	$Beta^-$ 1.55
Zr^{95}	65 day	$Beta^-$ 0.40 0.36, 0.89, gamma 0.72
↘		
$_{41}Nb^{95}$	35 day	$Beta^-$ 0.16, gamma 0.77
$_{44}Ru^{103}$	40 day	$Beta^-$ 0.21, 0.13, 0.71, gamma 0.04, 0.50
$_{44}Ru^{106}$	1 yr	$Beta^-$ 0.04 gamma, 0.51, 0.62, 1.22
↘		
$_{45}Rh^{106}$	2.2 hr	$Beta^-$ 0.79, 1.62, gamma 0.51, 0.22, 1.22
$_{55}Cs^{137}$	30 yr	$Beta^-$ 0.52, 1.18
↘		
$_{56}Ba^{137}$	2.6 min.	Gamma 0.662
$_{58}Ce^{144}$	285 day	$Beta^-$ 0.31, 0.19, gamma 0.034, 0.134
↘		
$_{59}Pr^{144}$	17.3 min.	$Beta^-$ 2.98, gamma 0.692, 2.18
$_{61}Pm^{147}$	2.5 yr	$Beta^-$ 0.23, gamma 0.12
	Neutron activated	
$_{6}C^{14}$	5770 yr	$Beta^-$ 0.156
$_{1}H^{3}$	12.26 yr	$Beta^-$ 0.018
$_{27}Co^{60}$	5.27 yr	$Beta^-$ 0.31, gamma 1.17, 1.33
$_{30}Zn^{65}$	245 day	$Beta^+$ 0.33, gamma 1.11

* See footnote to Table 11.1.

amount produced in the weapons-testing program and to the amounts existing elsewhere in the environment, in order to determine the worldwide material balance. They concluded that rainfall was undoubtedly the principal vehicle in which Sr^{90} is brought to the soil from the air. The spots selected were, therefore, to be sufficiently permeable to absorb all the precipitation that would fall on the land. The sites from which samples were taken, however, were located so that water from higher levels would not enter the soil as a result of runoff. Also, the soil should have a base-exchange capacity so that the Sr^{90} would be held by the surface soil and not be leached into ground water. They looked for areas with a good grass turf, not sheltered by buildings and trees, so that the absorption would be at a maximum and runoff at a minimum. They indicated that vegetation may enhance the removal of Sr^{90} from the air, but they had no opinion as to the amount which might be removed by the vegetative filtering of the air.

Recent investigations, such as the one reported by Russell and Possingham (41) indicated that the contamination of vegetation consisted entirely of particles adhering to the leaf surface. In no instance was the outline of the leaves apparent when they investigated vegetation by preparing radioautographs of the leaves. In this case the material must have been filtered from the air or deposited in rain and not brought from the soil into the grass material.

In investigations by Graham (fall of 1962) in which bluegrass vegetation was collected at weekly intervals at Columbia, Missouri, the amount of fallout material on the bluegrass was the same regardless of whether or not it had rained. This indicated that the effect of vegetation as a filter from air-borne fallout must be of considerable importance, as shown in Figure 11.1. Alexander (2) and his co-workers indicated that the vertical distribution of Sr^{90} is of some concern because it may have a direct effect on the uptake of the isotope by the plant. In the main, they concluded that Sr^{90} remained close to the surface of the vegetative, undisturbed soil in Monroe County, Illinois, where, in 1958, the 0- to 2-inch sample contained 27.2 mc per square mile, while the 2- to 6-inch sample contained 15 mc and the 6- to 16-inch one contained only 5.2 mc per square mile. These were sandy soils and probably the penetration would have been much less had the soil been a silt loam, a clay loam, or one high in organic matter.

These studies in the United States indicated that, as of 1958, the range of Sr^{90} in mc per square mile was from a high value of 58.0 at Springfield, Illinois, to a low value of 14.7 at Burbank, California.

Although Sr^{90} has received much more attention than any of the fallout products, there are many others of considerable interest. Work at the Argonne National Laboratory reported by Gustafson (19) in 1959, indicated that the Zr^{95}, Nb^{95}, Ru^{106}, Ru^{103}, Cs^{137}, Ce^{141}, and Ce^{144} would contribute greatly to the radioactivity of the soil. With the exception of Cs^{137}, these isotopes have half-lives of one year or less; they are, therefore, decaying at a fairly rapid rate, and in the event that nuclear-weapons testing does cease, in a short time these isotopes would be gone from the soils. During the years from 1958 to 1962, however, the short-life isotopes were falling on the soil at a much greater rate, probably ten to a hundred times greater than any of the Sr^{90} values that have been reported.

Of the fission-products isotopes that have been distributed to the soil as fallout, the long-life isotopes Cs^{137} and Sr^{90} have received the most attention. Studies by Anderson (3) revealed that measurements of the Cs^{137} content of milk gave data that were useful in studying fallout as a process. He concluded that the precipitation received in any given area was an important factor in determining the Cs content of milk. This observation is the result of the deposition rate and the accumulation of Cs on the plants and in the soil. Studies by Knapp (24) on the effect of the deposition rate and accumulative soil level of Sr^{90} on the Sr^{90} content of milk indicated that its concentration in milk was associated with its average accumulative level in the soil, as mc per square mile, plus its average deposition during the preceding month, similarly expressed. This author presented a formula for determining the average Sr^{90} content of milk in micro-microcuries per liter as follows:

$$\text{per liter} = \frac{1}{9}\left\{\begin{matrix}\text{average cumulative}\\ Sr^{90}\text{ level}\\ \text{mc/square mile}\end{matrix}\right\} + 2.6\left\{\begin{matrix}\text{average } Sr^{90}\text{ deposited}\\ \text{preceding months}\\ \text{mc/square mile}\end{matrix}\right\}$$

Knapp's data show that the formula worked well for the periods from October 1957 through January 1958, and from October 1959 through May 1960. In some instances the content predicted from the formula was about 30 per cent too low for that measured in the milk. He premises this difference on the fact that, for that particular period, many cows were fed significant amounts of hay and grain which had been stored and accumulated in other places instead of the animals being on pasture.

The amount of Sr^{90} in the milk predicted by Knapp's formula would indicate that it had been deposited on the surface of the vegetation as filtered from the wind, or that it had been soil-derived from the accumulated level of Sr^{90} there. These data would lead one to believe that once the Sr^{90} has become part of the soil system, the percentage turnover in the amount of Sr^{90} per given crop year would be low in relation to the amount of Sr^{90} falling daily from the atmosphere and accumulating on the surface of the vegetation. The contribution from the soil becomes most significant when the soil's cumulative level is ten to a hundred times greater than the amount deposited in a given month.

A study recently reported by Hansen *et al.* (21) on the concentration of I^{131} and Sr^{90} in milk, as moderated experimentally by adjustments in variable farming practices, would lead one to conclude that soil is a much more important influence on the amount of Sr^{90} in milk than one would have been led to believe as the result of Knapp's studies.

Hansen (21) and co-workers in the St. Louis Milkshed pastures near Lebanon, Missouri, reported the following results of their studies. The levels of I^{131} and Sr^{90} in milk were experimentally reduced by a factor of 50 per cent or more throughout the entire pasturing season by grazing cows on abundant, well-fertilized pastures. A relationship between optimal fertilization of the land and decreased radio-nuclide content of the milk was established. Forages from the well-fertilized land contained a lower concentration of these radioisotopes than forages from the unfertilized land. A mechanism of dilution explaining the findings was shown. The principle of optimal fertilization and management of land, they concluded, may prove to be a practical means of exercising control on the limits of public exposure to contamination from I^{131} and Sr^{90} in milk.

Studies by Graham *et al.* (15) have shown that the concentration of Sr^{85} in millimicrocuries per gram of ryegrass was affected greatly by the kind of colloid in the cultures on which the ryegrass was grown. For example, the highest concentration of 61.6 millimicrocuries per gram was obtained in rye grown on a kaolinite-sand mixture, while a low value of 15.9 millimicrocuries was found on rye grown in sand cultures containing montmorillonite. This relationship would be predictable if one studied the distribution coefficient of Sr on montmorillonite as compared with its coefficient on kaolinite. The K_d (distribution coefficient) value for Sr^{85} adsorbed on montmorillonite was 22.3. The value obtained under the same conditions, the same volume mass relationships, and the same soil concentrations was 0.2 for kaolinite.

The results obtained in the same study with organic colloids, such as fibrous peat and sedimentary peat, allowed rather large amounts of Sr^{85} to be observed in the tissue of the rye, and a high percentage of recovery of added Sr to the systems, but did not agree with the distribution coefficient observed for these particular systems. In the same study, information obtained on Cs^{137} and Co^{60} showed a good agreement between the concentration and recovery of a given isotope from a colloid system and the reciprocal of the distribution coefficient of cation when measured for the colloidal exchange system used to grow the plants.

Plant growth as a factor in the dilution of soil-held isotopes has been studied by many workers. Killion (23) reported the effect of nitrogen as a factor in growth on the absorption of radioactive isotopes. Nitrogen tended to increase total uptake and per cent recovery of Cs^{137} and Co^{60} by the plants. This was true for all combinations, except that less total Sr was taken by rye that had received a nitrogen treatment.

The addition of nitrogen to the plant cultures affected the uptake of Sr much less than the uptake of Cs or Co. In fact, the Sr uptake was 27.6 millimicrocuries/g of dry weight observed on rye grown on illite clay with nitrogen in comparison to 39.5 when grown on illite clay without nitrogen. The greatest difference obtained was for Putnam clay, where rye grown with nitrogen revealed 30.5 millimicrocuries/g as compared to 70.1 in rye grown without nitrogen treatments.

Nishita *et al.* (35) found that the soil type was a factor in the variation of Sr^{90} and that relationships which appeared to influence Sr^{90} did not influence Cs and Ru to the same degree.

Factors such as temperature, moisture, and other environmental conditions associated with the growth of plants would, no doubt, have considerable effect on the removal of a radioactive isotope from the soil by a given crop during a given year.

USE OF RADIOACTIVE ISOTOPES IN SOIL RESEARCH AND SOIL TESTING

Study of Fertility Levels of Soil Phosphorus by Tracer Techniques

Despite the fact that a large number of empirical methods have been employed during the past century for estimating the quantity of soil phosphorus available to plants, the establishment of a reliable technique for determining the phosphate status of the soil remains a problem of agricultural importance. The recent development in this field is the use of radiophosphorus P^{32} for the determination of the phosphorus level of soils and for information on the fate of applied phosphate. The two best-known methods for determining available phosphorus in soils are those of Fried and Dean (12) and of Larsen (30). The formula presented by Fried and Dean is as follows:

$$a = \frac{(b)(1 - y)}{y}$$

where a = phosphorus in the soil; b = phosphorus applied to the soil; and y = fraction of phosphorus taken by the plant from the phosphorus applied to the soil.

The formula as presented by Larsen,

$$Y = X\frac{(C_0 - C)}{C}$$

where Y = available phosphorus in the soil; X = phosphorus applied to the soil; $C_0 = k \times$ total phosphorus in the plant; $C = k \times$ phosphorus taken by the plant from X; and k = proportionality constant.

In Larsen's expression, C_0 is defined as the constant of radiophosphorus in X, and C as the constant radiophosphorus isolated from the plant. However, the definitions here come from the relationship: activity per unit P^{32} in the plant equals activity per unit P in the fertilizer, divided by the total P^{31} content of the plant, which is divided by P^{31} in the plant from fertilizer. For any given experiment

$$B = X \quad \text{and} \quad \frac{1 - y}{y} = \frac{C_0 - C}{C}$$

Therefore, $a = y$ and, for all practical purposes, the equations are identities without restrictions on the rate of the application of the phosphorus.

Factors Affecting the a and y Values of Soils

There is some difference of opinion about the isotope dilution concept as to whether equilibration of the P^{32} for the P^{31} might occur inside the plant root cells or the exchange occur outside the plant in the soil system as the result of equilibration of the added phosphorus, both P^{31} and P^{32} exchanging with a definite fraction of the soil-held phosphorus. Experiments on the rate of equilibration by Scheffer and Ulrich (45) indicated that, in spite of the identity of the mathematical expression used, the results can have different meanings due to variations in experimental procedures. The dilution principle holds only in uniformly mixed systems with high activity concentrations of the added labeled ions. With the calculation of isotopically exchangeable phosphorus one must, therefore, establish four main conditions.

(*a*) Thorough mixing of the labeled fertilizer with the whole soil, or a known part of the soil mass, must occur.

(*b*) The fertilizer, or phosphorus salt, used in the experiment must be water-soluble, so that all the labeled ions can take part in the isotopic exchange reactions.

(*c*) Absorption or precipitation of the added water-soluble fertilizer by soil particles should not convert the phosphate ions into a less reactive form. In this case, the rate of the isotopic exchange reaction serves as a measure of reactivity.

(*d*) The ratio between the added fertilizer and the isotopically exchangeable soil phosphorus should be as narrow as possible. The mean error of the latter phosphorus can become serious with high ratios of phosphate in the plant coming from the fertilizer divided by isotopically exchange phosphorus. Scheffer and Ulrich concluded that the determination of isotopically exchangeable soil phosphorus in a pot or laboratory experiment leads to a fairly good agreement with the common chemical or physiological methods for determining available phosphorus, considering the above restrictions. This result is not of great significance, because there is no absolute measure of the nutrients available to plants in the soil.

Experiments on the isotopic exchange pool performed by Rhoades (39) and other authors (42) have clearly shown that the isotopic exchangeable soil phosphorus is not a homogeneous fraction of the soil, but is more or less a heterogeneous fraction. Considered as a whole, therefore, it will not be an exact measurement of phosphorus available to the plant, but will, instead, be loosely correlated with available phosphorus. Perhaps the greatest value of isotope experiments performed with radioactive phosphorus in soil systems would be the elucidation of the nature of the phosphorus-fertility problem in soil systems.

With the development of the isotope exchange techniques, a more precise means of investigation in the field of soil chemistry has been initiated. It has been found possible by means of P^{32} to determine the phosphorus compounds formed in the soil.

Yuan *et al.* (54) extracted three acid sandy soils using the method of Jackson (22) to estimate the water-soluble, aluminum, iron, and calcium phosphate forms. Over 80 per cent of the added phosphorus was retained as aluminum and iron phosphates; less than 10 per cent was in the water-soluble and calcium phosphate forms. The balance not accounted for was considered to be in other forms not mentioned. The P^{32} was added with P^{31}, so that a comparison of results by chemical and tracer methods could be made. In general, both methods gave satisfactory results. However, because of the greater fluctuation of the percentage recovery of added phosphorus by the chemical method as compared with the tracer method, the results of the experiments performed with added P^{32} activity led to a better interpretation of the data. Using the same method, Volk and McLean (50) found, in four acid soils from Ohio, that in most cases 90 per cent or more of the applied phosphorus was accounted for as aluminum and iron phosphates.

Absorbed Cations in Soils Using Radioactive-isotope Techniques

The exchangeable cation content and capacities, and the dissociation of adsorbed cations of soils, have long been useful criteria by which soil scientists have characterized various soil and plant relationships. The ease with which plants may obtain nutrients from the soil has been related rather conclusively to the amount and nature of the soil exchange material. Many methods have been proposed for the determination of exchangeable cation content and capacity of soils. A step common to most methods involves extraction of the soil with a salt solution followed by measurements of quantities of cations adsorbed from the solution and displaced by it. The proposed isotope-exchange methods differ mainly in the selection of the solutions employed, techniques for carrying out the adsorption and displacement reactions, and the manner of a quantitative determination of amounts of cations adsorbed and displaced. The superiority of the information gained with radioisotope techniques is due largely to its avoidance of (*a*) the characteristic of soil that allows for the dissolution of some cations present in the soil which are not in exchangeable form; (*b*) the failure of certain solvent cations to completely displace all the absorbed cations in the soil; and (*c*) the possible fixation of the displaced cations in a nonexchangeable form and the possible nonexchangeable ion being extracted by the soil solution.

Use of extraction methods of measuring absorbed cations in soils has implied or presumed that the quantities measured are capable of obtaining kinetic equilibrium with the cations of the same element in the soil solution. If this equilibrium exists, then a radioactive isotope placed in an aqueous soil suspension should be expected to interchange readily between the soil and the soil solution, and finally be distributed between the two phases in the same manner as the nonradioactive portion of the same element. The utility of this phenomenon was evident to a number of investigators in studies of ion-exchange equilibria and for quantitative measurements of ions adsorbed on the soil.

One of the earlier investigations of the direct measurement of the rate of exchange and the degree of equilibrium between cations of the same element for exchange surfaces was conducted by Paneth (37). In an effort to measure the specific surface of precipitated lead sulfate, thorium B, a radioisotope of lead, was allowed to equilibrate in an aqueous lead sulfate solution with insoluble lead salt. The amount of interchange of the thorium B with the solid lead sulfate was studied under varying conditions of time, shaking, concentration of the lead solution, and amount of lead sulfate salt in the system. Paneth (37) found equilibrium essentially established after 3 minutes' shaking; limited absorption of the isotope continued, however, for long periods of time at a decreasing rate. This observation was explained on the basis of the increasing surface formation by mechanical breakage during shaking and by the slow entrance of the isotope into particle cracks and crevices.

An extension of the work of Paneth was carried out by Kolthoff and Rosenbloom (25). Much of Paneth's earlier work with lead sulfate was repeated with a salt of higher purity. Kolthoff and Rosenbloom suggested that the lead salts used by Paneth contained absorbed lead and that the isotope equilibrated with the absorbed lead, as well as with lead in the crystalline surfaces. This work indicated that kinetic exchange was practically complete after an hour, although after 24 hours a complete equilibrant was not established. It was suggested that lead sulfate molecules under the surface layer were participating in the exchange.

In an additional investigation, Kolthoff and Rosenbloom (25) observed that the equilibration of thorium B with lead sulfate was slowed greatly when Ponceau 4R was added to the system before the isotope was added. The reaction rate of the lead isotope was not affected if the addition of Ponceau 4R followed the addition of the isotope. It was concluded that the absorbed Ponceau 4R inhibited the rate of exchange of thorium B at the lead-sulfate–water interface by forming a monomolecular layer of lead sulfate surface.

In recent years a great many investigations employing radioactive isotopes have been made to study the relationship between solutions and the exchange materials of systems such as soils, clays, and resins. The majority of the studies have involved the radioactive isotopes of phosphorus, calcium, and sodium. Because of the relatively important position of phosphorus in soil chemistry, the fact that the isotope has been produced in large quantities at a reasonable price, and its relatively short half-life accompanied by a strong beta emission which minimizes the disposal and measurement problems, the P^{32} has been found to be a very valuable isotope in soil studies. The techniques that have been used in the equilibration of P^{32} between soil and soil solution are similar to those involved in cation equilibration. The isotope is allowed to equilibrate between the soil and an aqueous solution which may or may not contain added nonradioactive phosphorus. McAuliff (34) and others (39, 42) observed that part of the soil phosphate equilibrated rapidly with the added isotope and that it was followed by a much slower rate of reaction. In most instances, the rapidly reacting forms corresponded rather closely to the other estimates of phosphorus availability to plants. The milligrams of phosphorus exchanged in the first 30 minutes will correlate with the phosphorus available to plants for soils containing medium to high levels of available phosphorus. The rate of equilibration has further been shown (6) to be affected by temperature, with higher rates for that above room temperature, and vice versa.

Olsen (36) defined surface phosphate as the amount of phosphate associated with the solid phase which would readily equilibrate with P^{32} in the same ionic form. McAuliff (34) found marked uniformity of results over a wide range of soil-water ratios and concentrations of P^{31} in the final solutions. Other investigators utilized similar techniques of P^{32} equilibration and have given information regarding exchange properties within the solid phase of soils, and forms of residual phosphorus in calcareous soils (5).

Rhoades (39) found that the amount of soil phosphorus equilibrated with P^{32} in a given time was affected by the concentration and kind of salts in the outside solution. He found a much greater amount of phosphorus in the isotopic pool when he equilibrated the soils with Bray's weak reagent than when the soils were equilibrated with 0.001 *M* calcium chloride.

Equilibration investigations relating to cation exchange in soils with radioactive isotopes were carried out by Wiklander and Gieseking (52). In in effort to determine the exchange ability of potassium and strontium from exchange surfaces as influenced by the degree of saturation and the nature of the complementary ion, varying degrees of saturation of the four ion pairs, potassium-sodium, potassium-barium, strontium-sodium, and strontium-barium, were established on cation-exchange resins. The degrees of exchange ability of the potassium and strontium were determined by the K^{42} and Sr^{90} equilibration in the system. The exchangeabilities of both potassium and strontium decreased with decreasing saturation when sodium was the complementary ion. For both the potassium and the strontium, the per cent replacement approached a limit value at very low degrees of saturation.

In a study of potassium fixation in clay saturated with different cations, Wiklander (51) observed the rate of exchange of K^{42} between exchangeable and fixed potassium on an illite clay to be very slow, even though the system was held for 14 hours at 87°C.

Borland and Reitemeier (8) utilized the radioactive isotope of calcium, Ca^{45}, in kinetic exchange studies on calcium-saturated kaolinite, halloysite, hydrous mica, beidellite, and montmorillonite. One gram of each of the clays was mixed with 25 ml of water containing 0.1 mc of Ca^{45} and equilibrated on an end-over-end shaker for varying lengths of time up to 48 and 120 hours. At the end of the various time intervals, the suspension was centrifuged, decanted, and Ca^{45} activity measurements made. Results of these studies indicated equilibration was essentially complete in one-half to one hour, which was the shortest time interval studied. In addition, a study to determine the effect of increase in calcium concentration in the solution phase upon the final equilibrant was carried out. Exchangeable calcium measurements were observed to be fairly constant through a wide variation of calcium in solution. It was further observed that the exchange calcium removed by neutral normal ammonium acetate leaching was essentially the same as that measured by the Ca^{45} equilibration procedure. This indicated that calcium replaceable by ammonium acetate was in kinetic equilibrium with the calcium in solution. On the basis of these studies with calcium, the investigators claimed direct experimental confirmation of a basic premise of cation exchange, that the cations of an element in solution are in kinetic equilibrium with the exchange cations of the same element. Work by Fisher (11) with exchangeable cations on a number of Missouri soils revealed data which agreed with the data reported by Borland and Reitemeier (8).

Smith, Blume, and Whitaker (47) utilized the findings of Borland and Reitemeier for evaluating the rates of liming materials in soils. This study involved the measurement of the concentrations of Ca^{45} and Ca^{40} in dilute solutions of calcium nitrate before and after equilibration with a sample of soil previously treated with limestone or slag.

In a more detailed study of the determination of an exchangeable cation and the capacity of soil for such by equilibration with Ca^{45}, Blume and Smith (7) compared their findings with those obtained by methods described by Peech (38), in which ammonium acetate leaching was involved. The exchange calcium equilibration was carried out by adding 50 ml of the solution of 100 ppm of calcium to 25 g of soil and shaken for 3 hours. The suspension was then centrifuged, and the Ca^{45} measurements were made on both the supernatant liquid and the equilibration solution. Of the 22 noncalcareous soils studied, those containing less than 10 to 12 meq calcium per 100 g soil generally gave similar results by the two methods. Those soils having greater calcium content tended to show high values by the equilibration method. One soil which indicated a significantly higher exchangeable calcium by the Ca^{45} equilibration method was leached with 350 and 1000 ml of ammonium acetate. A value of exchangeable calcium which more nearly approached that obtained with the Ca^{45} technique was found. These results indicated that the value obtained by equilibration may be approached by ammonium acetate leaching if the ammonium acetate volume is sufficiently large.

The cation-exchange-capacity measurements were also investigated by Blume and Smith by saturating the soils with calcium prior to the Ca^{45} equilibration. Of 12 soils studied, one gave a higher value by the ammonium acetate leaching, six showed no difference, and five gave higher values by the equilibration method. Two of the soils in the latter group were calcareous. Observations generally indicated that those soils giving higher values of exchangeable calcium by the Ca^{45} equilibration method also gave higher values for cation-exchange capacity.

The investigators suggested that some mechanism was operating to allow free interchange of the calcium ions between the soil and soil solution while restricting the entrance of the ammonium ion into the soil complex. These restrictions apply to the rate of entrance rather than to the total amount which could enter, since long-continued leaching with ammonium acetate led to the extractions of nearly as much calcium as that found by the Ca^{45} equilibration method.

Work on exchangeable calcium in soils containing 6.9 to 66.8 per cent calcium as calcium carbonate was reported by Barbier and Tyszkiewicz (5). Four-gram samples of soil were shaken for 24 hours with 100 ml of Ca^{45} calcium chloride solution containing known amounts of stable calcium. The values obtained compared favorably with those obtained by Hissink's method of successive leaching with normal sodium chloride. Tests showed that the calcium of the limestone did not interfere to any important degree with isotopic exchange. It was suggested that the technique of using isotopic exchange is suitable for determining cation-exchange capacity, without the necessity of replacing the calcium by another cation or of previously destroying the calcium carbonate present, if there is sufficient calcium in the equilibrating solution to displace other cations in the exchange complex.

In many of the studies reported on the movement of exchangeable calcium ions between colloidal particles and solution, it was concluded that this exchange is very rapid, that the fixed ions are in a state resembling the ions of solution, and that all exchange cations of the same kind have a determined position and play the same role in systems in equilibrium. It has been suggested that measurements of Ca^{45} in the supernatant liquid could be more accurately carried out if a quantity of calcium equal to that absorbed in the soil was added to the equilibration mixture. This would allow for a greater portion of the Ca^{45} to be in the supernatant liquid, thus facilitating measurements. According to our recent work with Ca^{45}, using liquid scintillation systems, this factor may not be so important since 1-ml portions of solution containing Ca^{45} could be satisfactorily counted in solutions containing dioxane, methyl alcohol, naphthalene, and toluene, with primary and secondary scintillators, so that a count of 100 per minute per ml could be made with accuracy.

Rosenqvist (40) reported a routine procedure for measuring the cation-exchange capacity of Scandinavian soils, using Ca^{45} equilibration procedures which permitted one technician to carry out as many as 25 analyses per day. Two to 5 ml of neutral 0.5 *M* calcium chloride solution containing Ca^{45} was allowed to equilibrate with 0.25 g of soil for 2 to 10 minutes. The suspension was centrifuged and the soil wet three to four times with distilled water to remove Ca^{45} in the solution phase. The soil was placed in a suitable container for counting, allowing correction for self-absorption. Weighing was unnecessary since the samples are considered to be infinitely thick. The measured activity was compared with standards of known concentration.

Measured Activity Compared with Standards of Concentration

Previous saturations of the soil with sodium or calcium did not affect the values obtained in this manner. Saturation with hydrogen, however, gave values approximately 20 per cent lower. Equilibrium with mineral calcium was thought not to occur because of the short equilibration time employed.

Salam and Hashish (44) compared values obtained by various methods for exchangeable calcium in specially prepared calcium-saturated soils containing varying amounts of calcium carbonate and calcium sulfate. They found that the Ca^{45} equilibration procedure gave values very close to the theoretical value on the prepared soils containing 0 to 50 per cent calcium carbonate. For soils containing calcium sulfate, the Ca^{45} equilibration method gave good results when the calcium sulfate was less than 10 per cent, but high values were obtained if the calcium sulfate was greater than 10 per cent. On desert soils containing calcium carbonate and calcium sulfate, the equilibration procedure gave values very near those found for ammonium absorption from ammonium acetate.

With the exception of a few unusual cases, the isotope equilibration procedure employing Ca^{45} generally appeared to give satisfactory estimates of quantities of absorbed calcium in soils. This further implies that absorbed calcium is near kinetic equilibrium with calcium in the associated solution phase. Since corresponding information regarding other cations is not available, it would be of interest to examine these relationships with respect to magnesium, potassium, and sodium.

Ion-exchange Equilibria in Terms of Ion Selectivity

Radioactive isotopes lend themselves to the study of practically all ion-exchange reactions involving the contraction of a single ion-exchange substance and an electrolyte. The availability to plants of divalent and monovalent cations in the soil is associated with ion-exchange reactions. These interreactions of cations in the absorption processes by plants have been studied extensively in recent years. The extent to which ions are absorbed by plants is determined largely by two factors One is the potential of the ion or the energy situation, which has been investigated in terms of mean free-bonding energy by Marshall and co-workers (32), and the energy of exchange studies, as reported by Schofield (45) and Woodruff (53). The second important concept which must be considered is the amount of the ion in the exchange pool. This may be expressed as ppm or pounds per acre. The amount of the exchangeable ion has been previously determined by making an extraction with ammonium acetate or some other strong salt. With radiochemical isotopes, isotopic-exchange studies for the amount of the exchangeable ion have been made. Much information has been obtained by partial exchange, such as would occur in shaking a soil with a weak salt solution. It appears that an analysis of both the energy situation, or the ion potential, and the size of the isotopic exchange pool (or the pounds-per-acre value) may be obtained by one single determination. The analysis may be performed for both major and trace elements on a single colloidal system by equilibration, radiochemical counting, or flame or colorimetric photometer determinations for the concentration of the stable isotope. Mixtures of ions as well as single ions may be analyzed, since gamma-emitting isotope and gamma-ray spectrometer analyses of three ions or more may be made and quantitative results can be expressed. Also, whenever a beta-emitting isotope and a gamma-emitting isotope are available, the double labeling can be carried out by first counting the gamma-emitting isotope on a gamma counter and then counting the two isotopes on a beta counter, making a coefficient of correction for the gamma-emitting isotope as recorded by the beta counter. The results may be calculated in terms of selectivity numbers or distribution coefficients. The various ion-exchange equilibria may be expressed in terms of ion selectivity, which is calculated from the equilibration reaction from the following:

$$\text{Ion selectivity number} = (A/B)_{\text{outside}}\,(B/A)_{\text{inside}}$$

where $(A/B)_{\text{outside}}$ refers to the activity of ions A and B in the solution phase, and $(B/A)_{\text{inside}}$ refers to the activity of ions B and A on the solid exchanger. This value may be calculated as $T - S$ when compared to the activity of a known standard to obtain the total activity, S being the measured activity of the solution phase.

This expression may be rewritten as follows:

$$\text{Ion selectivity number} = \frac{\dfrac{T^B - S}{B_s}}{\dfrac{T^A - S}{A_s}}$$

This relationship would then be the distribution coefficient of ion B divided by the distribution coefficient of ion A. The particular relationship described above is sometimes referred to as the selectivity coefficient. Ion selectivity is a ratio approximation of the ions in solution and the ions on the exchanger which is derived from the Donnan equation. This is a useful expression for studying the distribution of ion species, especially at low concentrations, between external solution and the ion exchanger, when one of the ions is present in excess and the other ion is in tracer amount.

The expression will be referred to as the selectivity number. Marshall and Garcia (31) referred to this relationship as a selectivity number in their clay studies. The same arrangement of ion ratios was used by Russell *et al.* (43), who called it the soil-solution factor.

In their study, Russell *et al.* expressed the concentrations as activity of the radioisotope in counts per minute. This expression, although used for divalent to divalent ions, may be used with investigations as to the proper volume mass relationships for any pair of ions and expressed in suitable units, such as moles, equivalents, or, in the case of radioactive isotopes, as counts per minute or disintegrations per minute. The distribution selectivity number which would result in a value of 1 would therefore indicate that the two ions were similarly partitioned, i.e., the ratio of the amount in solution to the amount on the exchanger would be equal for both ions. A value of 0.5 would show that the distribution value of the top number of ions was half the distribution value for the bottom number of ions. The advantage of giving the expression "distribution selectivity number," in which a distribution coefficient of one ion is divided by the distribution coefficient of other ions, is that it appears to be more objective than the soil solution factor arrangement used by Russell *et al.* This experiment may be carried out rather rapidly and results in data which yield much information about the nature of a given colloidal system.

To carry out experiments on soil suspensions for selectivity numbers, aliquots of previously prepared Ca-saturated soil colloids may be transferred to 30-ml ethylene bottles; enough radioactive Ca^{45}, Cs^{137}, Co^{60}, and Sr^{85}, or other isotopes in solution of known activity, plus an amount of known molar concentration of $CaCl_2$ may be added to each suspension. This would bring the outside concentration of the $CaCl_2$ to some known value, say 0.01–0.001 M. A control blank must be prepared in a similar manner, with distilled water being substituted for the suspension of colloidal material.

The bottles containing the radioactive isotopes and soil colloid would then be stoppered and allowed to come to equilibrium by constant shaking for a period of 48 hours. The suspensions may then be removed by centrifuging, and the clear supernatant liquid can be drawn off for analysis. In the case of mixtures of gamma-ray-emitting isotopes, a single-channel or, preferably, a multichannel analyzer may be employed, using the sodium-iodide crystal scintillation detecting system. When the author has counted double isotopes of gamma-emitting materials and beta-emitting materials, the gamma activity is determined by using a single-channel analyzer. Then the same amount of solution is counted in the liquid scintillation counter, and with the proper settings for discriminators, voltage, etc., the activity of both the gamma-emitting and beta-emitting isotope is determined. Coefficients of correction are made by counting the single gamma-emitting isotope first in the single-channel gamma-ray detector; then, by placing the same amount in the liquid scintillating counting detector, a coefficient of correction may be obtained which allows for a correction of the beta activity when unknowns are counted. With some isotopes, the coefficient of correction is small. However, with isotopes such as Cs^{137} and Ba^{137} (a parent-daughter relationship) the coefficient of correction will be very high.

Comparisons between the activity of the control blank and the activity of the supernatant liquid containing a colloidal exchanger allow for the calculation of the distribution coefficients for each ion. The distribution coefficient is calculated in the following manner. The activity in the control blank is referred to as the total activity used in the experiment, and the activity in the supernatant from the colloid is called the activity in the outside solution. The total of activity on the exchanger would then be the total found in the blank minus that found in the solution. This calls for the calculation of the distribution coefficient as such:

$$K_d = \frac{\% \text{ of ion on exchanger}}{\% \text{ of ion in solution}}$$

This is usually multiplied by a volume mass relationship when distribution coefficients are to be contrasted, but when this information is used to determine a selectivity number, the mass volume relationships cancel out and thus it is not necessary to determine them.

The results of some studies of distribution selectivity by Graham *et al.* (16) are presented in Table 11.6. The values obtained show many interesting differences between samples of soil colloids used in the study. The highest value obtained for Cs selectivity (23.7) over Ca was obtained for Putnam soil colloid. A similar value (21.8) was obtained for illite. The similarity of these two selectivity values is not too surprising, since the Putnam soil is thought by some to behave principally like illite. A different value was obtained for montmorillonite (4.7) and a still smaller value (0.25) for fibrous peat. A value of 0.21 was obtained for kaolin. The two colloids, Putnam and illite, showed a high selectivity of Cs over Ca; montmorillonite was medium; and the two colloids, kaolin and a fibrous peat, were low. This same set of information was not obtained when Co distribution selectivity was considered over Ca. This study revealed a very high value (74.8) for fibrous peat; none of the clays showed a value anywhere near this. Montmorillonite, however, was second with 1.7, and kaolin, the lowest, showed a value of 0.014. Fibrous peat was second to montmorillonite, Putnam third, and

TABLE 11.6. SELECTIVITY NUMBER FOR CESIUM AND CALCIUM, COBALT AND CALCIUM, AND STRONTIUM AND CALCIUM FOR SOME SOIL COLLOIDS

Soil colloid*	Cs/Ca†	Co/Ca†	Sr/Ca†
Illite	21.8	0.1	0.14
Putnam	23.7	0.6	0.25
Montmorillonite	4.7	1.7	1.17
Kaolin	0.21	0.2	0.014
Peat Fibrous	0.25	74.8	0.66

* The colloidal systems were saturated with Ca (pH range 6.5 to 7.3). The Cs, Co, and Sr were added in sub-micro amounts as radioactive isotopes. Equilibrations were carried out in 0.01 *M* $CaCl_2$ outside concentration. Ca^{45} was used as the radioactive tracer for Ca.

† A number of 1.0 shows equal adsorption of ion *B* and ion *A* for the colloid. A number smaller than 1.0 reveals that the Ca ion is held with greater energy than ion *B*. Numbers larger than 1.0 show the colloid adsorbs ion *B* with greater energy than Ca

illite next. Selectivity numbers obtained in the study by Russell *et al.* (43) revealed a set of soils which varied in selectivity of Sr over Ca from 1.25 to 1.85, with a mean value of about 1.37. Although these values are somewhat higher than any obtained in our laboratory, they most closely approximate the values for montmorillonite, which would indicate that the predominant clay in these samples was montmorillonite.

Soil studies associated with trace-element problems have been worked out by employing radioactive trace elements. The great sensitivity that modern counting techniques offer represents advantages not available with color chemistry. This is especially true for short-life isotopes of high specific activity (radioactivity expressed as curies per grain or curies per liter, etc.); those which lend themselves well to trace element studies are Fe, Mn, B, Cu, Zn, Mo, Co, Cs, Sr, and Cl.

Several studies on the use of radioisotopes of trace elements point to the value of tracer isotopes in studies revolving around chelation, saturation of bentonite plant necrosis, and plant nutrition.

To be able to use isotopes in the study of problems of soil and plant nutrition demands a good understanding on the part of the investigator of the phenomenon of isotopic exchange. It is the opinion of the writer and others that the phenomenon of isotope exchange has been oversimplified. It should not always be concluded that information obtained by using calcium will be the same for sodium, or that the stable isotope will always behave the same as the radioactive isotope. The mass difference of radioactive ${}_1H^3$ as compared to stable ${}_1H^1$ is certainly going to make some difference. It is entirely possible that the stable isotope of a given species will move at a different rate than the radioactive one. This would be especially true in biological studies of diffusion rates and may be true as well in physical studies of these rates. The unpredictability of the exchange rate will no doubt reveal properties of importance about biological and physical systems which, eventually, will be assets rather than disadvantages.

REFERENCES

1. Alexander, L. T., *et al.* "Radioisotopes in Soils: Particularly with Reference to Strontium-90," Radioisotopes in the Biosphere, University of Minnesota General Extension Division, 1960.
2. Alexander, L. T., "Strontium-90 Distribution as Determined by the Analysis of Soils," Joint Committee on Atomic Energy, Congressional Hearings, May 5–8, 1959.
3. Anderson, E. C., *Science* 128, 882–886 (1958).
4. Asimov, I., *J. Chem. Educ.* 31, 24–25 (1954).
5. Bärbier, G., and Tyskiewiez, E., *Comptes Rendus des Seances de l'Academie des Sciences, Commission II and IV* 2, 79–82 (1953).
6. Blanchet, R., *Annales Agronomiques, Series A* 10, 5-267 (1959).

7. Blume, J. M., and Smith, D., *Soil Sci.* **77**, 9–17 (1954).
8. Borland, J. W., and Reitemeier, R. F., *Soil Sci.* **69**, 251–260 (1950).
9. Delwich, C. C., *et al.*, *Proc. 2nd International Conf. on the Peaceful Uses of Atomic Energy* (*U. N.*), **18**, 551–556 (1958).
10. Delwich, C. C., and Stout, P. R., *Bull. Am. Meteorol. Soc.* **40**, 285–290 (1959).
11. Fisher, T. R., Doctor's thesis, University of Missouri, 1962.
12. Fried, V. M., and Dean, L. A., *Soil Sci.* **73**, 261–271 (1952).
13. Gibbs, H. S., and McCallum, G. J., *New Zealand J. Sci. Tech.* **37**(3), 354–368, (1955).
14. Glasstone, S., "Sourcebook on Atomic Energy," New York, D. Van Nostrand Co., Inc., 1958.
15. Graham, E. R., and Killion, D. C., *Soil Sci. Soc. Am. Proc.* **26**, 545–547 (1962).
16. Graham, E. R., "Measurement of Cation Selectivity in Soils by Radioactive Isotope Distribution Equilibration Methods," AEC Tech. Progress Report, mimeographed, 1963.
17. Grosse, A. J., *et al.*, *Science* **113**, 1 (1951).
18. Gustafson, L. D., *et al.*, *Science* **127**, 1240 (1958).
19. Gustafson, L. D., *Radiology* **75** (2), 282–288 (1960).
20. Hansen, R. O., *et al.*, "Radioisotopes in Soils," Radioisotopes in the Biosphere, Minneapolis, Minnesota, Univ. of Minnesota, Gen. Extension Division, pp. 23–36, 1960.
21. Hansen, W. G., *et al.*, "Fission Products in Milk," II Memo. Report USPHS, St. Louis, Mo., 1963.
22. Jackson, M. L., "Soil Chemical Analysis," Engelwood Cliffs, N. J., Prentice-Hall, Inc., 1958.
23. Killion, D. D., Master's thesis, University of Missouri, 1962.
24. Knapp, H. A., "The Effect of Deposition Rate and Cumulative Soil Level on the Concentration of Strontium-90 in U. S. Milk and Food Supplies," USAEC TID-13945 (1961).
25. Kolthoff, I. M., and Rosenclum, C., *J. Am. Chem. Soc.* **55**, 2664-2672 (1933).
26. Korff, S. A., *Terrestrial Magnetism and Atmospheric Electricity* **45**, 133 (1940).
27. Kovach, E. M., *Trans. Am. Geophys. Union* **26**, 241–248 (1945).
28. Lapp, R. E., and Andrews, H. L., "Nuclear Radiation Physics," Englewood Cliffs, N. J., Prentice-Hall, Inc., 1954.
29. Libby, W. F., *Phys. Rev.* **69**, 671 (1946).
30. Larsen, S., and Cooke, I. J., *Plant Soil* **4**, 1–10 (1952).
31. Marshall, C. E., and Garcia, G., *J. Phys. Chem.* **63**, 1663–1666 (1959).
32. Marshall, C. E., and Upchurch, W. J., *Soil Sci. Soc. Am. Proc.* **17**, 222–227 (1953).
33. Marshall, C. E., and Haseman, J. F., *Soil Sci. Soc. Am. Proc.* **7**, 448–453 (1943).
34. McAuliff, C. D., *et al.*, *Soil Sci. Soc. Am. Proc.* **12**, 119–123 (1948).
35. Nishita, H., *et al.*, *J. Agr. Food Chem.* **9**, 101–106 (1961).
36. Olsen, S. R., *J. Phys. Chem.* **56**, 630–632 (1952).
37. Paneth, F., and Vorwerk, W., *Z. Physik. Chem.* **101**, 445–479 (1922).
38. Peech, M., *et al.*, U.S.D.A Circ. 757, 1947.
39. Rhoades, C. R., Master's thesis, University of Missouri, 1961.
40. Rosenqvest, J. T., *Proc. 2nd Conf. on Radioisotopes, Oxford, England* **1**, 412–416 (1954).
41. Russell, R. S., and Possingham, J. N., "Section 2, Progress in Nuclear Energy Bio. Sci.," Vol. 3, New York, Pergamon Press, 1961.

42. Russell, R. S., *et al.*, *J. Soil Sci.* **5**, 85–105 (1954).
43. Russell, R. S., *et al.*, *Proc. 2nd U.N.I.C. on the Peaceful Uses of Atomic Energy Isotopes in Agriculture* **27**, 146–148 (1958).
44. Salam, A., and Haskish, S., *Proc. U.N.I.C. on the Peaceful Uses of Atomic Energy Isotopes in Agriculture* **278**, 172–175 (1958).
45. Scheffer, F., and Ulrich, B., *Proc. U.N.I.C. on the Peaceful Uses of Atomic Energy Isotopes in Agriculture* **27**, 149–151 (1958).
46. Schofield, R. K., *Proc. 11th Cong. Pure and Applied Chem. London* **3**, 25 (1947).
47. Smith, D. H., *et al. Agr. Food Chem.* **1**, 67–70 (1953).
48. Stout, R. R., *et al.*, USAEC TID-7512 (1956).
49. Telfair, D., *et al.*, *Science* **131**, 727–728 (1960).
50. Volk, V. V., and McLean, E. D., *Soil Sci. Soc. Am. Proc.* **27**, 53–58 (1963).
51. Wiklander, L., *Soil Sci.* **69**, 261–268 (1950).
52. Wiklander, L., and Gieseking, J. E., *Soil Sci.* **66**, 377–384 (1948).
53. Woodruff, C. M., *Soil Sci. Soc. Am. Proc.* **19**, 167–171 (1955).
54. Yuan, T. L., and Robertson, W. K., *Soil Sci. Soc. Am. Proc.* **24**, 447–450 (1960).

CHAPTER 12

Methods for Chemical Analysis of Soils

William J. Hanna

Department of Soils and Crops
Rutgers—The State University, New Brunswick, New Jersey

This chapter contains a brief outline of some methods for chemical analysis of soils and is designed to complement the subject matter of the chapters preceding it. The methods described have been used satisfactorily, but not exclusively, in the soils laboratories of the New Jersey Agricultural Experiment Station.

It is not the intent of this brief review to cover all the techniques used in soil analysis, for example such widely used methods as differential thermal analysis, and x-ray absorption, diffraction, and fluorescence. Nor is it the intent to cover all the many limitations of the methods discussed and the precautions that should be observed in their use. Before choosing a procedure for a particular determination the reader should consult more detailed instructions for analysis such as those to be found in the works listed as references 1, 2, 3, and 5.

The collection, preparation, and storage of the sample represent the most significant steps in the entire analytical operation. It cannot be overemphasized that no analytical results can be more accurate or representative than the sample examined. For samples on which determinations of trace-element content are to be made, it is particularly important to exert special effort to avoid contamination during and after sampling.

Soils are notoriously heterogeneous materials. If this heterogeneity is to be evaluated in the process of sampling and analysis, then more individual samples, not composited, need to be examined. The analysis of replicate samples is much to be preferred over replicate determinations on the same sample, since laboratory variability is relatively small.

Blank determinations should be made simultaneously with each group of unknowns. The blanks should include all reagents used in the determinations, and the procedures should be carried through for the blank exactly as for the unknown.

In each instance, the choice of a sampling procedure depends upon the particular purpose of the investigation and the analytical methods to be used. A good outline of the principles involved in soil sampling can be found in Jackson (5).

ANALYSES FOR TOTAL CONTENT OF MAJOR CONSTITUENTS

Moisture

The soil sample should be in an air-dry condition after mixing, sieving, and grinding. If stored in a moisture-tight container, the free H_2O content should not vary appreciably during storage. Most of the subsequent determinations will be made on the air-dry sample, but for certain purposes, such as obtaining summation values, a moisture content value is required.

Weigh 2 g soil into a tared, shallow, stoppered weighing bottle and heat with the bottle uncovered for several hours, or overnight, in a drying oven at 110°C. Cover, cool in a desiccator, and weigh. Report percentage loss in weight as moisture.

The dried sample may be held for determination of loss on ignition.

Organic Matter

The most accurate procedure for the determination of organic matter in soils is the dry-combustion method for total C, in which the CO_2 evolved on heating is trapped in some absorbent, and a correction is made for any native CO_3^{-2} present. Since this is a tedious method, it is often felt that one of the faster, wet-oxidation procedures using chromic acid will give results of a desired accuracy. As an approximate method, loss of weight of an oven-dry sample on ignition is a method often employed.

Loss on Ignition

Transfer the dried sample from the moisture determination to a Pt or porcelain crucible and heat slowly over a gas burner or in an electric muffle furnace to about 700°C. Hold at this temperature for $\frac{1}{2}$ hour, and then remove, cool in a desiccator, and weigh. Report percentage loss in weight as "loss on ignition." The result will more clearly approximate the organic matter content of the sample if a correction is made for H_2O lost from clay crystals: subtract 5 per cent of the weight of clay in the sample (particle-size analysis necessary) from the loss in weight, before calculating "loss on ignition."

The residue from this determination may be used later for the sodium carbonate fusion.

Chromic Acid Digestion

Reagents:

Chromic acid, 0.4 N solution. Dissolve 19.616 g oven-dry $K_2Cr_2O_7$ in 1 liter of 1-1 mixture of 96 per cent H_2SO_4 and 85 per cent H_3PO_4 by heating in a water-bath at 95 to 100°C for 2 hours with occasional stirring. This solution is very nearly 0.4 N and is stable for at least 2 months.

Ferrous ammonium sulfate, approximately 0.2 N solution. Dissolve 80 g of $Fe(NH_4 \cdot SO_4)_2 \cdot 6\ H_2O$ in approximately 500 ml distilled H_2O containing 5 ml of 96 per cent H_2SO_4. Dilute to 1 liter and store in a close-stoppered bottle away from light.

Diphenylamine indicator. Dissolve 0.20 g diphenylamine in 100 ml 96 per cent H_2SO_4. Store in amber-glass dropping bottle.

Procedure. Mix each sample of soil thoroughly and grind approximately 5 g to a fine powder with a mortar and pestle. Weigh an 0.300 g sample and transfer to a dry 25 × 200 mm "Kimax" or "Pyrex" test tube. Use a similar amount of pure sea sand as a blank. Add from a burette exactly 10 ml of the chromic acid to each test tube. Heat to just 175°C in approximately 3 minutes. Use sand bath, wax bath, or electric oven. When the contents of the tubes reaches 175°C, remove from heating device and cool tubes at once by dipping in cool H_2O. Rinse contents of tubes into 250-ml beakers with enough distilled H_2O to make approximately 100 ml. Add 5 drops of diphenylamine and titrate with the ferrous ammonium sulfate solution to the sharp green end point. Calculate the percentage C by the following equation

$$\frac{(B - U) \times D \times N \times A \times 100}{B \times W} = \%\,C \qquad (1)$$

where B represents volume of ferrous ammonium sulfate solution in ml required for the blank; U, volume of ferrous ammonium sulfate required for sample; D, ml chromic acid used (10 ml); N, normality of chromic acid (0.4 N); A, meq weight of C (0.003 g); and W weight of sample used (0.3 g).

Since this procedure is not as efficient as the dry-combustion method, results from its use should be multiplied by a factor of about 1.2 to make the results of the two methods comparable. The usual factor for conversion of C to organic matter is 1.724. The percentage of organic matter in the sample is obtained by combining these two factors into a single factor of 2 and multiplying this by the percentage C from equation (*1*).

Nitrogen

Reagents:

Sodium hydroxide, 45 per cent solution. Dissolve about 10 pounds of NaOH flakes in approximately 7 liters of distilled H_2O. Dissolve 120 g K_2S in 500 ml distilled H_2O. Mix the two solutions and dilute to 10 liters with H_2O.

Boric acid, 4 per cent solution. Dissolve 40 g H_3BO_3 in 1 liter of distilled H_2O.

Mixed indicator. Dissolve 0.5 g bromcresol green and 0.1 g methyl red in 100 ml ethanol (95 per cent). Adjust to pH 4.5 with dilute NaOH.

Standard acid, 0.1 N HCl.

Procedure. Digest 5 to 10 g soil in a 500-ml Kjeldahl flask with 25 to 30 ml of 96 per cent H_2SO_4, 0.7 g HgO (or 0.65 g Hg), and 5 g anhydrous Na_2SO_4 for $1\frac{1}{2}$ hours or until clear. Cool and dilute with about 200 ml distilled H_2O. Cool again, add a pinch of zinc dust, and just before connecting to Kjeldahl distillation apparatus, add an excess (about 100 ml) of the NaOH solution. Connect the flask to the Kjeldahl apparatus and mix the contents of the flask thoroughly. Distill about 150 ml into a 250-ml Erlenmeyer flask containing about 50 ml of the boric acid solution. Titrate with the standard acid, using 5 drops of the mixed indicator. To correct for any nitrogen in the reagents, run a blank determination, using about 0.2 g sucrose instead of soil. The percentage nitrogen in the soil is calculated from the equation

$$\frac{T \times N \times .014 \times 100}{W} = \% \text{ N} \qquad (2)$$

where T is the volume in ml of standard acid (less than the blank); N, the normality of the standard acid; 0.014, the meq weight of nitrogen; and W, the weight of sample (g) used.

Sulfur

Reagents:

Barium chloride, 10 per cent solution.

Procedure. Fuse 2 g finely ground soil with 10 g Na_2CO_3 and 0.3 g $NaNO_3$ in a Pt crucible. Do the fusion in an electric furnace to avoid S contamination from gas flame. Remove the melt from the crucible by digesting in a 250-ml beaker with about 75 ml H_2O on steam bath for several hours (preferably overnight). Filter into a 200- or 250-ml volumetric flask, washing the paper with hot H_2O. Cool, make to volume, and remove an aliquot (50 to 100 ml) for S determination, retaining the remainder if desired for determination of Cl.

Dilute the aliquot for S to 150 ml and add sufficient HCl to neutralize the Na_2CO_3 and to bring the solution to about 1 per cent HCl in excess. Heat to boiling and add 10 ml of the $BaCl_2$ solution slowly while stirring. Digest for about 30 minutes, and then cool and filter through Whatman No. 44 paper. Wash with hot H_2O until filtrate is free of chlorides. Transfer paper to Pt crucible and ignite paper and $BaSO_4$ precipitate in electric furnace. Cool and weigh. To insure against inclusion of SiO_2, add 2 drops HF and 1 drop H_2SO_4 (1 + 1), heat carefully until sulfate fumes have ceased, re-ignite, and weigh again. Calculate concentration of S from weight of $BaSO_4$ precipitates and appropriate dilution factor as follows:

$$\frac{\text{Weight of } BaSO_4(g) \times 137370}{\text{g soil equivalent in aliquot used}} = \text{ppm S} \qquad (3)$$

Chlorine

Reagents:

Silver nitrate solution (1 mg Cl equivalent/ml). Dissolve 4.7914 g $AgNO_3$ in H_2O and dilute to 1 liter.

Sodium chloride (1 mg Cl/ml) solution. Dissolve 1.6486 g NaCl in H_2O and dilute to 1 liter.

Potassium chromate, 5 per cent solution.

Sulfuric acid, 0.1 N solution.

Procedure. Adjust the pH of an aliquot from the solution (containing 5 to 50 mg Cl) for sulfate determination, or from a separate $NaCO_3$-$NaNO_3$ fusion, to about pH 8 (just acid to phenolphthalein) with 0.1 N H_2SO_4. Dilute to about 100 ml with H_2O, add 1 ml of the K_2CrO_4 solution as indicator, and titrate to a faint permanent red color with the standard $AgNO_3$ solution.

$$\frac{\text{ml } AgNO_3 \text{ solution} \times 1{,}000{,}000}{\text{g soil equivalent in aliquot used}} = \text{ppm Cl} \qquad (4)$$

Potassium and Sodium

Reagents:

Ammonium carbonate, 5 per cent solution.

Standard solutions of K and Na (50 ppm for K, 20 ppm for Na).

Procedure. Grind together in an agate mortar 0.5 to 1.0 g of soil and an equal weight of NH_4Cl. Add 4 to 5 g $CaCO_3$ powder and mix thoroughly. Put approximately 0.2 g $CaCO_3$ in bottom of a Pt crucible and add mixture from mortar. Clean mortar of all the sample mixture with an additional 0.5 g $CaCO_3$ and add this to the top of the mixture in the crucible. Tap to settle, cover, and heat over very low flame until release of ammonia ceases. Do not let temperature in any part of the crucible get high enough to cause NH_4Cl to be lost. When all ammonia has evolved, increase the flame until the crucible is a dull red, and continue heating at this temperature for 40 to 60 minutes or until the mass is completely fused. Cool and transfer the shrunken fusion to a 150-ml Pt or porcelain evaporating dish. Wash all remains of the fusion from the crucible and its cover into the dish with hot H_2O. Cover the dish with a watchglass and digest on steam bath for several hours until fusion mass is thoroughly disintegrated. If after digesting overnight the mass is not completely slaked, crush any lumps remaining with a stirring rod or pestle. Filter through Whatman No. 2 paper into a 400-ml beaker. Wash the dish and filter with hot H_2O in small portions until the filtrate volume is 300 ml. Add 50 ml $(NH_4)_2CO_3$ solution to precipitate most of the Ca. Concentrate by evaporation to about 50 ml, filter into 250-ml beaker and wash paper and residue with at least 100 ml hot H_2O in small portions. Concentrate volume to 50 ml on steam bath and transfer to 150-ml porcelain evaporating dish. Rinse beaker with hot H_2O and add rinsings to dish. Evaporate contents of dish to dryness and ignite to expel ammonia salts. Dissolve residue in hot H_2O; cool, transfer to 100 ml volumetric flask, and make to volume with H_2O. Determine K and Na on aliquots of this solution on flame photometer after adding Li as internal standard and diluting to an appropriate volume, depending on K and Na content of soil.

Sodium Carbonate Fusion for Al, Fe, Ca, Mg, Mn, P, Si, and Ti

Mix in a Pt crucible 2 g finely ground soil, or the residue from the loss-on-ignition determination, with five times its weight of anhydrous Na_2CO_3. Cover and heat gradually in a muffle furnace, or over a burner, to 900°C. Heat at this temperature for 10 to 15 minutes, occasionally tilting and rotating the crucible with Pt-tipped tongs to insure complete fusion of the sample. Remove from heat and, while the melt is still liquid, rotate with the tongs so that the melt solidifies in a thin layer on the bottom and lower sides of the crucible. Cool and cover the melt in the crucible with a little water and heat over flame to loosen melt, or immerse crucible and melt in H_2O in 400-ml beaker and allow to sit overnight or until melt disintegrates completely. Remove crucible, scrub with a rubber policeman and a little dilute HCl, and rinse into beaker. Cover the beaker with a watchglass and gradually add 100 ml of concentrated HCl followed by 10 ml $HClO_4$ (70%). Stir well, uncover, and heat on steam bath until volume is reduced to less than 50 ml. Transfer to hot plate and continue heating at low level for 15 minutes after first fumes of $HClO_4$ evolve. It is not necessary or desirable to evaporate to dryness. Cool, then take up in a few ml of hot H_2O and filter through Whatman No. 42 paper. Wash beaker and filter with repeated washings of dilute HCl to a volume of 200 ml. Save the filtrate and washings for determination of Al, Fe, Ca, Mg, Mn, P, and Ti. Designate as Solution *A*.

Silicon

Transfer the filter paper containing the dehydrated silica residue to a Pt crucible and dry at 120°C in oven. Ignite slowly, starting with a cold furnace, or using a low flame, until the paper is reduced to a white ash or nearly so. Increase the heat to 900°C and hold at this temperature for 10 to 15 minutes. Cool in a desiccator and weigh. Silicon may be determined from this weight. For more accurate results, the silicon may be volatilized as follows: add 1 ml 1 + 1 H_2SO_4 and 5 to 6 ml HF and evaporate on sand bath to evolution of SO_3 fumes. Cool, then add another ml of HF and evaporate to complete dryness in the sand bath. Ignite as before, cool, and weigh. Calculate the percentage silicon from the loss in weight.

Fuse the residue in the crucible with 1 g Na_2CO_3 as previously described. Dissolve cooled melt in dilute HCl as before and add to Solution *A* for determination of Al, Fe, etc.

Mixed Oxides of Aluminum, Iron, Phosphorus, and Titanium

Reagents:

Ammonium hydroxide, 3 M solution.
Ammonium nitrate, 1 per cent solution.
Hydrochloric acid, 6 M solution.

Procedure. Concentrate the combined filtrates (Solution *A*) from the Na_2CO_3 fusion and Si determination by evaporation to about 150 ml, transfer to 250-ml flask, and make to volume with H_2O. Designate as Solution *B*, remove two 50-ml aliquots, and reserve for determination of P, Fe, and Ti. Wash the remaining 150 ml into a 400-ml beaker and evaporate to 100 ml. Add NH_4OH solution, carefully with stirring, to pH 6.3.

Cover beaker with watchglass, bring to boil on hot plate or low flame, remove from heat, allow precipitate to settle for 2 to 3 minutes, and stir and filter rapidly through Whatman No. 41 paper. Wash beaker and filter several times with hot NH_4NO_3 solution. Return the paper containing the precipitate to the original beaker, add 10 ml HCl solution, macerate the paper with a stirring rod, add 75 ml H_2O, and heat to boiling. Remove from heat, adjust pH to 6.3 with the NH_4OH solution, heat to boiling, and filter as before. Wash precipitate free of chlorides with hot NH_4NO_3 solution; combine the filtrates and save for Ca, Mg, and Mn determinations (Solution *C*). Dry the precipitate and paper and transfer to tared crucible; ignite slowly until the paper is ashed and then raise temperature to 900°C for 10 minutes; cool in a desiccator and weigh. Weight of residue represents combined weights of Al_2O_3, Fe_2O_3, P_2O_5, and TiO_2.

Iron

Reagents:

Jones Reductor. Prepare and maintain as directed in "Methods of Analysis" (1, p. 701) or Jackson (5, p. 311–312).

Potassium permanganate, 0.05 N solution.

Procedure. Add 10 ml H_2SO_4 to one of the 50-ml aliquots reserved from Solution *B* in a 250-ml beaker. Evaporate on hot plate to evolution of white fumes of SO_3. Cool and cautiously dilute with about 150 ml H_2O. Stir and pass through a Jones reductor (1) at about 25 ml per minute. Titrate to first pink color with the $KMnO_4$ solution. Save solution for Ti determination. Calculate percentage Fe as follows:

$$\frac{V \times N \times B}{W \times A} = \%\ \mathrm{Fe} \qquad (5)$$

where, *V* is volume in ml of $KMnO_4$; *N*, normality of $KMnO_4$; *W*, weight of sample taken for Na_2CO_3 fusion; *B*, volume in ml of aliquot taken for Fe and Ti determinations (50 ml); and *A*, volume in ml to which filtrates from fusion were diluted (200 ml).

Titanium

Reagents:

H_2O_2, 30 per cent.

Standard solutions of Ti in 2 N H_2SO_4, 2 to 50 ppm Ti.

Procedure. Evaporate the solution from the Fe determination to about 50 ml. Cool and transfer to a 100-ml volumetric flask. Add 2 ml 30 per cent H_2O_2 and dilute to volume. Read intensity of yellow color in colorimeter at 400 mμ. Prepare a standard curve from standard solutions treated as for unknowns. Plot concentration of Ti against log of light transmission. Calculate percentage Ti in samples from the standard curve.

Phosphorus

Reagents:

Stannous chloride, 0.2 M solution. Dissolve 40 g $SnCl_2 \cdot 2H_2O$ in 100 ml of concentrated HCl. Dilute to 500 ml with freshly boiled H_2O and store under layer of mineral oil in container with provision for dispensing from bottom.

Ammonium-molybdate-HCl solution. Dissolve 25 g MoO_3 in a mixture of 38 ml NH_4OH (sp. gr. 0.90) and 68 ml H_2O. Pour this solution with constant stirring into a mixture of 120 ml HNO_3 and 287 ml H_2O. This is a stock solution and will keep for about 2 months. To prepare solution for use, mix 25 ml of the stock solution with 75 ml H_2O and 25 ml HCl.

Jones reductor. Prepare as directed in "Methods of Analysis" (1, p. 701) or Jackson (5, p. 311–312).

Standard P solution, 0.5 to 10.0 ppm.

Procedure. Pass the 50-ml aliquot reserved for P determination from Solution *B* above through a small Jones reductor into a 100-ml volumetric flask. Rinse the Jones reductor column with several small portions of H_2O to bring the volume in the flask to the mark. Mix and transfer an aliquot to a 50- or 100-ml volumetric flask. Size of aliquot and size of flask are chosen to have the range of P concentration in the final solution between 0.1 and 1.0 ppm. Dilute with H_2O to about half the volume of the flask, add 1 ml ammonium molybdate solution, and mix well. Add 5 drops stannous chloride solution, mix at once, and make to volume with H_2O. Read the intensity of the blue color after 5, but within 20, minutes at 660 mμ. Prepare a standard curve from the standard P solutions to establish the P concentrations in the unknowns

Calcium

Reagents:

Ammonium oxalate, saturated solution.
Potassium permanganate, 0.05 N solution.
Bromphenol blue indicator, 0.04 per cent solution.
Ammonium hydroxide, 8 M solution.
Sulfuric acid, 1 M solution.

Procedure. Concentrate Solution *C*, the combined filtrates from the NH_4OH separation of mixed oxides, to approximately 150 ml; transfer to 200-ml volumetric flask and make to volume. Transfer a 50-ml aliquot to a 250-ml beaker; add 5 drops bromphenol blue indicator and add 8 *M* NH_4OH to the end point at pH 4.0. Bring to boiling on hot plate, add 10 ml of the $(NH_4)_2C_2O_4$ solution, cover with watchglass, and boil slowly for 1 hour or until precipitate is coarsely granular. Filter through Whatman No. 42 paper and wash free of oxalates with H_2O. Pierce the tip of the filter with a stirring rod and wash the precipitate into the original beaker with 1 *M* H_2SO_4. Dilute to about 100 ml with hot H_2O and titrate with $KMnO_4$ to faint pink. Add filter paper and rinsings of funnel into beaker. Continue titration with $KMnO_4$ to pink color, which persists 1 minute. Calculate per cent Ca from volume and normality of $KMnO_4$ and dilution factors.

If flame photometer suitable for Ca determination is available, determine Ca in Solution *C* with this instrument rather than as directed above [see Jackson (5, p. 461) or Snell (13, p. 502)].

Magnesium

Reagents:

Disodium-EDTA, 0.01 M solution.

Eriochrome Black T indicator. Dissolve 0.5 g of the indicator powder and 5 g hydroxylamine-HCl in 100 ml methanol.

Ammonium hydroxide, 8 M solution.

Potassium cyanide, 2 per cent solution.

Procedure. Transfer a 50-ml aliquot of Solution *C* to a 150-ml beaker or 125-ml flask, adjust to pH 10 with NH_4OH, and add 5 drops of the Eriochrome Black T indicator and 1 ml KCN solution. Titrate with Na-EDTA to blue endpoint. This titration includes Ca and Mn as well as Mg. The amounts of Ca and Mn determined separately are subtracted to give a value for only Mg. Save solution for Mn determination. Where adequate equipment is available, Mg may also be determined more conveniently by flame photometer.

Manganese

Reagents:

H_3PO_4, 85 per cent.

Potassium periodate, crystals.

Procedure. Add 15 to 20 ml HNO_3 to the solution saved from the EDTA titration, or to a 50-ml aliquot from Solution *C* if Ca and Mg are determined by flame photometer. Digest on hot plate for 30 minutes while covered; remove cover and evaporate to dryness. Repeat until ammonium salts are decomposed. Add 10 ml H_3PO_4 to residue, warm on steam bath, and add 50 ml H_2O and 0.2 g KIO_4. Digest on steam bath for 30 minutes. Transfer to a 100-ml volumetric flask, cool, and make to volume with H_2O. Read permanganate color at 540 mμ. Calculate Mn content from standard curve prepared from standard solutions treated in a manner similar to that described above.

Aluminum

Calculate the percentage Al or Al_2O_3 in the sample by subtracting from the percentage of mixed oxides the percentage of P_2O_5, Fe_2O_3 and TiO_2 as determined separately, or determine directly as follows [(4), and (5, p. 298–300)].

Reagents:

Sodium acetate buffer solution. Dissolve 25 g NaOH in about 900 ml H_2O containing 125 ml glacial acetic acid. Adjust pH to 4.2 and dilute to 1 liter.

Aluminon-buffer solution. Dissolve 0.50 g aluminon (aurin tricarboxylic acid) in 1 liter of the NaOAc buffer.

HCl, 1 M solution.

Standard solution of Al, 5 ppm. Take 1- to 10-ml aliquots of this solution through procedure as for unknowns to construct standard curve.

Procedure. Transfer to a 100-ml volumetric flask an aliquot of the solution used for P determination which has passed through the Jones reductor. The aliquot should contain no more than 0.1 mg Al. Add sufficient 1 *M* HCl (usually 3 to 6 ml) to the flask to bring pH of solution to 3.8 to 4.0 after the addition of the aluminon-buffer solution.

Heat on steam bath for 20 to 30 minutes; cool, and dilute to about 75 ml with H_2O. Add 15 ml of the aluminon-buffer solution; dilute to volume and allow color to develop for 30 to 40 minutes. Read intensity of color at 520 mμ and determine Al concentration from a standard curve prepared from standard solutions treated as above.

ANALYSES FOR TOTAL CONTENT OF TRACE ELEMENTS

Colorimetry

Cobalt

Reagents:

Carbon tetrachloride. Distill over CaO through all-"Pyrex" glass still and pass through acid-washed filter paper.

Phenolphthalein. One per cent solution in 95 per cent ethanol.

Cupric acetate solution. One per cent solution in redistilled H_2O.

Stock ortho-nitrosocresol solution. Dissolve 6 g hydroxylamine hydrochloride and 15 g $CuCl_2$ in 900 ml redistilled H_2O and add 5 ml meta-cresol. Add 15 ml of 30 per cent H_2O_2 with stirring. Let stand 2 hours at room temperature, then add 25 ml of concentrated HCl and extract with four successive 150-ml portions of petroleum ether in a large separatory funnel. Add an additional 25 ml of concentrated HCl and again extract with four more 150-ml portions of petroleum ether. Wash the yellow petroleum ether extract twice with 0.1 *N* HCl and twice with redistilled H_2O. Extract the *ortho*-nitrosocresol from the petroleum ether with successive 50- to 100-ml portions of 1 per cent cupric acetate solution until the aqueous phase no longer turns deep blood-red. Add 25 ml of concentrated HCl to the aqueous Cu solution and extract the free *ortho*-nitrosocresol with two successive 500-ml portions of petroleum ether, thus transforming the *ortho*-nitrosocresol to the ether phase. Combine the two ether extracts, wash twice with 150- to 200-ml portions of 0.1 *N* HCl and then with several portions of redistilled H_2O. Store in the dark at a temperature near 4°C. *Ortho*-nitrosocresol prepared and stored in this way appears to be stable for at least 4 months without repurification for preparation of the sodium salt.

Sodium nitrosocresol solution. To prepare 100 ml of the sodium salt, extract 100 ml of the petroleum ether solution of the free nitrosocresol with two successive 50-ml portions of Na_3BO_3 buffer, pH 9.1, prepared as subsequently directed. This should yield a solution of such strength that 2 ml or less is sufficient for a determination. As this solution is less stable than the stock solution, prepare only enough for a day's run each time.

Sodium borate buffer. (*a*) *pH 7.8.* Dissolve 20 g H_3BO_3 in 1 liter redistilled H_2O and bring to proper pH with 1 *N* NaOH (approximately 50 ml). (*b*) *pH 9.1.* Add 120 ml 1 *N* NaOH to 1 liter of the foregoing buffer.

Dithizone solution. Dissolve 0.5 g dithizone in 600 to 700 ml CCl_4 (redistilled is unnecessary), filter into a separatory funnel containing 2.5 to 3.0 liters of 0.02 *N* NaOH, shake well, and discard the CCl_4 layer. Shake with 50-ml portions of redistilled CCl_4 until the CCl_4 phase has, as it separates, a pure green color. Add 1 liter CCl_4 and acidify slightly with HCl (aqueous phase changes from orange to colorless). Shake the dithizone into the CCl_4 layer and separate. Store in a cool, dark place.

Hydroxylamine-acetate buffer. Dissolve 10 g hydroxylamine hydrochloride and 9.5 g anhydrous $NaC_2H_3O_2$ in 500 ml redistilled H_2O. The pH should be between 5.0 and 5.2.

Ammonium citrate solution (40 per cent). Dissolve 800 g lead-free citric acid in 600 ml redistilled H_2O. Adjust the pH to 8.5 with redistilled ammonium hydroxide. Dilute to 2 liters and extract with 10-ml portions of dithizone solution in CCl_4 until the aqueous phase stays orange and the CCl_4 phase remains predominantly green. Then extract the solution with CCl_4 until all the orange color is removed.

Ligroin. Obtain ligroin from Eastman Kodak Co. (boiling point, 70° to 90°C). Ask for material purified by permanganate.

Standard cobalt solution. Prepare a stock solution containing 100 μg/ml of Co by heating 5 g of $CoSO_4 \cdot 7H_2O$ in an oven at 250 to 300°C to constant weight (6 to 8 hours sufficient). Weigh out exactly 0.477 g of the $CoSO_4 \cdot 7H_2O$ and dissolve in 50 ml of redistilled H_2O and 1 ml of concentrated H_2SO_4. Make to a volume of 1 liter. Transfer 5 ml of the stock cobalt solution to a one-liter volumetric flask and dilute to volume with redistilled H_2O for the standard solution. This solution contains 0.5 μg Co/ml.

Preliminary Procedure. Weigh 0.1 to 0.3 g soil into a Pt crucible and ignite in a muffle at 400 to 500°C to destroy organic matter. Add a few ml H_2O, 0.5 ml 70 per cent $HClO_4$, and 3 ml HF. Slowly evaporate the mixture to dryness on a sand bath. Take up the residue with a few ml H_2O, 0.5 ml $HClO_4$, and 3 ml HF, and again evaporate to dryness. Take up the residue with 2 ml redistilled HCl and 10 ml H_2O, and heat.

Dithizone Extraction. Transfer the solution or an aliquot to a separatory funnel. Add 5 ml ammonium citrate (40 per cent) to tie up the Fe. Adjust the pH to 8.5 with NH_4OH using phenolphthalein as an indicator. Formation of a precipitate may be prevented by adding more ammonium citrate. Add 10 ml dithizone solution in CCl_4 and shake vigorously for 5 minutes. Draw off the CCl_4 phase into a beaker. It is often necessary to repeat the extraction with an additional 5 ml of dithizone solution. The extraction is complete when the aqueous phase remains orange and the CCl_4 extract remains predominantly green. To the aqueous phase in the separatory funnel add a small quantity (2 to 3 ml) of CCl_4 to remove the remaining dithizone solution. Without shaking, draw off and add to the combined CCl_4 extracts.

Add 5 ml CCl_4 to the aqueous phase. Shake for 2 minutes. Draw off the CCl_4 phase and add to the combined extracts. If the dithizone extraction was complete, the CCl_4 extract at this point will be pure green. Repeat the extraction if necessary.

Add 1 ml $HClO_4$ to the combined CCl_4 extracts and reflux in a covered beaker on a hot plate until all $HClO_4$ has gone off. If free acid remains it will interfere with subsequent steps where pH adjustment is important.

Dissolve the residue in 4 ml 0.01 *N* HCl and transfer to a 120-ml separatory funnel with redistilled water. Add 4 ml of the Na_3BO_3 buffer to bring the pH up to about 7.0.

Determination. Make up the sodium nitrosocresol from the stock solution as given under directions for reagents. Add this solution drop by drop to the contents of the separatory funnel, until all Cu and other elements present have reacted. This point is indicated by the appearance of an orange color distinct from the pink color of the Cu complex which may be present. Add 1 ml in excess.

Add exactly 5 ml of ligroin and shake for 10 minutes. Ligroin stays on the top. Remove the aqueous phase, and to the ligroin add 5 ml of 1 per cent aqueous cupric acetate solution to remove the aqueous phase. Wash the ligroin with redistilled H_2O and finally with 5 ml hydroxylamine sodium acetate buffer to eliminate any Fe (reduces Fe). After shaking, remove aqueous layer in each case. Transfer the ligroin to 25-ml glass-stoppered volumetric flasks. Measure the color intensity at 360 mμ and calculate concentration of Co by comparison with a standard curve prepared from a series of standards treated, as for the unknowns.

Precautions. The hood must be kept spotless to prevent Fe contamination during evaporation.

Cleaning of separatory funnels:

(*a*) Clean those used for ligroin extraction with aqua regia, redistilled H_2O, two or three rinsings of petroleum ether, and finally with redistilled H_2O.

(*b*) Clean those used for regular extractions with H_2SO_4 and finally with redistilled H_2O.

Copper

Reagents:

Dimethylglyoxime, 0.5 g in 100 ml of 95 per cent alcohol previously redistilled in "Pyrex".

Ammonium citrate. 20 g in 100 ml of solution; make just alkaline to cresol red with NH_4OH and add 2 ml in excess. Extract with successive small portions of dithizone (0.01 per cent in CCl_4) until the extracts are green, and then with CCl_4 until the extracts are colorless. Pass through filter paper previously washed with hot dilute HCl.

Thymol blue indicator, 0.04 per cent in water.

Sodium diethyldithiocarbamate, 0.1 per cent solution. Preserve in a brown bottle.

Carbon tetrachloride, reagent grade.

Standard copper solution, 0.01 per cent (100 ppm) stock solution. Dissolve 0.1964 g clear uneffloresced crystals of $CuSO_2 \cdot 5H_2O$ in water, add enough HCl to make the final acidity about 0.1 *N*, and dilute to 500 ml. Dilute 20 ml of this stock solution to 1 liter to give a solution containing 2 μg Cu/ml.

Procedure. Weigh 0.25 g soil that has been ground in an agate mortar to an impalpable powder, or pass through a 100-mesh sieve, and transfer to a Pt crucible. If the soil is exceptionally high in organic matter, ignite in a muffle at 400 to 500°C to destroy organic matter. Add 1 ml redistilled H_2O, 1 ml $HClO_4$, and 3 ml HF. Heat on a sand bath until all fumes are expelled. Cool and again add 1 ml H_2O, 1 ml $HClO_4$, and 3 ml HF. Evaporate to dryness and cool. Repeat this process until virtually all SiO_2 has disappeared (a small trace will do no harm). Add 1 ml $HClO_4$ and evaporate to dryness. Cool and add 2 ml redistilled HCl (constant boiling) and 5 ml H_2O; warm to dissolve residue. Filter through a small paper, and wash with hot H_2O.

Transfer solution to a 125-ml separatory funnel. Add 1 ml dimethylglyoxime solution and 5 ml ammonium citrate solution. Add a few drops of thymol blue indicator and adjust pH to 9.0 to 9.2 with NH_4OH. Add 1 ml carbamate solution and make volume up to 30 ml with H_2O. (A larger volume may be more convenient if the volume of the filtrates is large. In such a case equally satisfactory results are obtained as long as the reference curve is constructed by use of this larger volume of solution.) Add exactly 5 ml CCl_4 and shake for 2 minutes. Allow the CCl_4 layer to settle, dry stem of funnel with filter paper on a glass rod, and run the CCl_4 layer into a suitable cell. Determine the transmittancy of the solution at light wavelength of 400 mμ. Compare to reference curve made by using known amounts of Cu (0 to 50 μg) in solutions of the same volume and acidity as the unknown solution.

Molybdenum

Reagents:

Stannous chloride, 10 per cent solution in 1 + 9 HCl. A freshly prepared solution is required. Dissolve 20 g $SnCl_2$ in 20 ml of concentrated HCl by warming almost to boiling. Then add 180 ml distilled H_2O.

Ammonium thicyanate, 10 per cent solution.

Ferric chloride, 10 mg Fe/ml. Dissolve 49 g $FeCl_3 \cdot 6H_2O$ in distilled H_2O and make up to 1 liter.

Isopropyl ether. Shake, in a separatory funnel, reagent isopropyl ether with one-tenth its volume of a mixture containing one-third each of the $SnCl_2$ and NH_4SCN solutions and water. The ether must be prepared the same day it is used.

Sodium nitrate, 5 N solution.

Standard molybdenum solution, 100 ppm Mo. Dissolve 0.150 g C.P. MoO_3 in 10 ml 0.1 *N* NaOH. Make slightly acid with HCl and dilute with distilled H_2O to 1 liter. An aliquot of this solution is used to prepare a 2-ppm solution each time a series of standards are to be prepared: i.e., 20 ml diluted to 1 liter = 2 ppm = 2000 μg; 1 ml of this solution = 2 μg; 0.5 ml of this solution = 1 μg.

Procedure. Mix 2 g finely ground soil with 100 g anhydrous Na_2CO_3. Place in a Pt crucible and burn off excess organic matter. Then place in a muffle furnace and heat to 950 to 1000°C for 20 minutes. Remove the muffle, tipping crucible lightly and turning it as the melt solidifies. When cold, detach the cake from the crucible, by gently rolling the crucible between the thumb and forefinger, and place in a 250-ml beaker half filled with H_2O. Cover and heat on a steam bath. Break up the cake with a stirring rod as the softening process takes place. Transfer the liquid to another beaker and the solid to an agate mortar. Grind the softened pieces of the cake into a fine slurry. Transfer to the liquid in the beaker and digest on the steam bath for about 1 hour. Filter on a Büchner funnel and wash. Acidify the filtrate with a moderate excess of HCl and evaporate to dryness on the steam bath. Take up with 7 to 8 ml of concentrated HCl and 40 to 50 ml H_2O. Filter off the SiO_2 and wash until the volume is about 100 ml.

Transfer the solution to a separatory funnel and add the following reagents: 1 ml $FeCl_3 \cdot 6H_2O$; 5 ml 10 per cent NH_4SCN; 1 ml $NaNO_3$; and 5 ml 10 per cent $SnCl_2$. Shake the solution. Then add exactly 10 ml isopropyl ether and shake 100 times. Allow the two layers to separate and draw off the aqueous phase. Dry stem of funnel with filter paper. Deliver the ether into a glass-stoppered flask. (If preferred for greater accuracy, extract with two 5-ml increments of isopropyl ether.) Determine the Mo in the ether layer with a colorimeter or spectrophotometer at 475 mμ.

Zinc

Reagents:

Bromcresol green. 0.04 per cent purified with dithizone and stored in paraffin-lined bottle.

Sodium acetate, 0.5 M solution. Free the solution from heavy metals by shaking with successive small portions of dithizone solution, and pass through a small moistened filter paper to remove droplets of CCl_4.

Sodium thiosulfate solution. 50 g ($Na_2S_2O_3 \cdot 5H_2O$) dissolved in 100 ml H_2O and purified with dithizone.

Dithizone, 0.002 per cent in CCl_4 *(reagent grade).*

Sodium sulfide solution. Dilute 40 ml of a 1 per cent Na_2S stock solution with 1 liter Zn-free H_2O.

Sodium thiosulfate-acetate wash solution. Mix 225 ml of 0.5 *M* sodium acetate solution, 10 ml of 50 per cent $Na_2S_2O_3$ solution, and 40 ml of 10 per cent HNO_3, and add H_2O to make 500 ml. Remove Zn by shaking with dithizone.

Acid mixture for digestion. Mix concentrated HNO_3 ("Pyrex" redistilled), 70 per cent $HClO_4$, and concentrated H_2SO_4 in the ratio of 10:0.75:0.25.

Standard zinc solution. Dissolve 0.2 g reagent-grade Zn in concentrated HCl and make up to 1 liter with Zn-free H_2O in a "Pyrex" glass-stoppered volumetric flask. Dilute 10 ml of this stock solution to 1 liter with Zn-free H_2O for a solution containing 2 μg Zn per ml. This should be freshly prepared each time it is used.

Precautions and cleaning of apparatus. 1. Cleaning of separatory funnels before starting the determination is absolutely essential. Shake funnels with 1:1 HCl about 50 times; rotate glass stopcock several times. Rinse out with redistilled H_2O; add 2 drops $NaC_2H_3O_2$ buffer solution and turn stopcock. Rinse with four portions of redistilled H_2O, turning stopcock each time.

2. Avoid Zn contamination by coating reagent bottles with paraffin. (This refers to those bottles used to store Na-containing or alkaline reagents.)

3. The Na_2S solution stored in ordinary glass will lose strength by precipitation of sulfides from the glass. Therefore, make up a new stock solution every three or four weeks.

4. Run a blank through the whole procedure.

Procedure. Weigh 0.1 g of soil that has been ground in an agate mortar to an impalpable powder, and place in a Pt crucible. Proceed as for determination of Cu to the point where virtually all SiO_2 has been removed and the $HClO_4$, except for a drop or two, has been expelled. Add 3 ml of approximately 3 *N* HCl and heat to boiling. Cool and transfer into a separatory funnel.

The volume at this state should be about 15 ml. Add 2 ml of 25 per cent ammonium citrate solution. Adjust the pH to neutrality with approximately 5 *N* NH_4OH, using litmus paper as indicator, and add two drops excess. Extract Zn and other reacting heavy metals with 0.005 per cent dithizone solution. Transfer the Zn to 0.02 *N* HCl solution by shaking with three 5-ml portions of the latter. Add 3 drops of bromcresol green. The color of the solution will now be yellowish green in the acid solution. From a pipette add 0.5 *M* $NaC_2H_3O_2$ until the pH is between 5.0 and 5.5 (a light bluish green). Add 2 to 3 ml $Na_2S_3O_2$ solution, which will form complexes with the interfering trace elements such as Cu, Pb, and Hg. Add from an all-glass burette exactly 10 ml dithizone solution and shake approximately 100 times. If the dithizone turns red or purplish red, addition of dithizone solution must be continued in 10-ml increments until the dithizone remains green on the last extraction. Keep account of the total volume added, since the zinc standard curve is made up on the basis of 10 ml solution. Thus, if, for example, 40 ml of the dithizone solution is required before it remains green, the final results must be multiplied by 4. Separate the dithizone layer, leaving a slight amount in the funnel. Replace exactly 10 ml of the dithizone layer in a clean separatory funnel. Wash once with about 10 ml of the sodium thiosulfate-acetate wash solution, and once with H_2O. Then wash with several successive 5-ml portions of Na_2S wash solution to dissolve the excess dithizone; the last portion of the sulfide wash solution must remain colorless. Zinc dithizonate now separates in the bottom as a pink to red solution. Withdraw this solution and determine color intensity at 540 mμ. Calculate Zn concentration by comparing color intensity readings with a standard curve prepared from a series of standards treated as for unknowns.

Boron

Reagents:

Quinalizarin indicator. Dissolve 5 mg quinalizarin in 1 liter of concentrated H_2SO_4.

Glassware. All beakers, flasks, vials, and other apparatus should be low-boron glass. "Pyrex" glassware cannot be used. Corning No. 7280 is satisfactory low-boron glassware.

Preparation of Color Standards. Dissolve 2.8578 g H_3BO_3 in 1000 ml distilled H_2O. This solution contains 0.5 mg B/ml and serves as the primary (*a*) base stock solution. Prepare a second (*b*) stock solution containing 0.01 mg B/ml by diluting 2 ml of the primary base stock solution to 1000 ml with distilled H_2O, and a third (*c*) stock solution containing 0.001 mg B/ml by diluting 100 ml of second stock solution to 1000 ml.

Transfer varying amounts (see Table 12.1) of the second stock solution for amounts above 0.001 mg B, and of the third stock solution for the others, to boron-free test tubes or glass vials. Glass vials approximately 20 by 100 mm are convenient and satisfactory for this purpose. For the correct concentration of acid when the color is developed, it is necessary to have exactly 1 ml of H_3BO_3 solutions in each vial. This is conveniently obtained by dispensing the boron stock solution from a burette and then adding H_2O from another burette to bring the volume to 1 ml in each case. Now add 10 ml of concentrated H_2SO_4-quinalizarin solution. After 30 minutes the standards are ready for use. These colors are permanent if the vials are tightly sealed to prevent absorption of H_2O.

Procedure. Intimately mix 5 g finely pulverized soil with 15 g anhydrous

TABLE 12.1. DATA FOR PREPARING BORON STANDARDS FROM STOCK SOLUTIONS (*b*) AND (*c*)

Standard no.	Distilled H_2O (ml)	Stock solution (ml)	Boron (mg)
1	1.00	0.00	0.0000
2	0.80	0.20 (c)	0.0002
3	0.60	0.40 (c)	0.0004
4	0.40	0.60 (c)	0.0006
5	0.20	0.80 (c)	0.0008
6	0.00	1.00 (c)	0.0010
7	0.85	0.15 (b)	0.0015
8	0.80	0.20 (b)	0.0020
9	0.75	0.25 (b)	0.0025
10	0.70	0.30 (b)	0.0030
11	0.65	0.35 (b)	0.0035
12	0.60	0.40 (b)	0.0040

Na_2CO_3 ; transfer to a large Pt crucible, place in a cold muffle furnace, and raise temperature to 900 to 950°C. Pour the red-hot melt into a large Pt dish floating on cold water. Immediately rotate the dish to spread the melt into a thin layer. When cold, break the melt up into small pieces by means of a mortar and pestle, and transfer to a 250-ml Erlenmeyer flask. Clean the crucible with hot water and a policeman. Pour all the liquid into the Erlenmeyer flask. The final volume in the flask should be about 50 to 60 ml. Allow to stand overnight, or shake in a shaking machine until all the melt has disintegrated. Slowly and cautiously add 29 ml of concentrated HCl and shake on the machine for 5 to 10 minutes. Add small increments of $CaCO_3$ (reagent grade) until the solution becomes alkaline and all Fe and Al silicates have precipitated out. Excess $CaCO_3$ is essential. Attach the flask to a reflux condenser, the inner tube of which is of boron-free glass. Cautiously bring to a boil over a low flame and reflux for 10 to 15 minutes. After removing flame, loosen the stopper from the flask, and wash down the reflux tube with distilled H_2O. Pour the suspension onto a Büchner funnel containing a large filter paper, fitted up around the sides. Suck the precipitate dry, and wash three or four times with hot distilled H_2O, sucking the precipitate dry each time. Return the liquid into a 250-ml beaker of boron-free glass and evaporate on a slow hot plate to slightly less than 100 ml. Cool, make exactly to 100 ml, and transfer to a boron-free flask.

To develop color, place 1 ml, equivalent to 0.05 g soil, in a comparison vial, and add exactly 10 ml quinalizarin indicator from a burette. The quinalizarin indicator should be added quickly and the vial stoppered immediately. Allow the sample vials to stand 30 minutes to 1 hour, and then determine the B content by comparing with the set of prepared standards, viewing through the sides of the vials placed over a light comparator.

The mg B found in the 1-ml aliquot used × 20,000 gives the ppm B in the sample.

Emission Spectrographic Analysis

Preparation of Sample

Weigh 2.0 g oven-dry and finely ground soil into a Pt crucible. Add 1 ml H_2SO_4 and 3 ml HNO_3 and heat in sand bath to evolution of white fumes. Add, further, 3-ml portions of HNO_3 to destroy all organic matter. If organic matter persists after repeated treatments with HNO_3, add 2 ml $HClO_4$. When all organic matter is destroyed and all HNO_3 and $HClO_4$ have been volatilized, cool and add 2 ml H_2O and 2 ml HF. Allow to stand in hood overnight, heat on sand bath to volatilize the silica. Use additional 2 ml portions of HF as necessary to remove all silica. Add 1 or 2 ml additional H_2SO_4 if required to avoid taking the residue to complete dryness. Transfer the residue from the Pt crucible quantitatively to a 50- or 100-ml volumetric flask, make to volume, and designate as Solution *A*. More details may be found in Mitchell (7) and Shimp (12).

Determination of Elements by $NaNO_3$ Buffer Method

Reagents:

Molybdate solution, 4.0 mg Mo/ml.

Acid-proofed electrodes (graphite electrodes, tops coated with 2 per cent plicene in benzene).

Procedure. An appropriate aliquot (usually 25 ml) of Solution *A* is evaporated to a volume of about 1 ml in a Berzelius beaker on a hot plate. Add 2.0 g $NaNO_3$, 0.5 ml HNO_3, and 0.5 ml of the molybdate solution. Dilute to exactly 10 ml. The concentrations of Al, Fe, Mg, and some other elements are usually too high for satisfactory determination in this solution, in which case dilute to suitable level with a solution that is 20 per cent in $NaNO_3$, 5 per cent in HNO_3 and 0.02 per cent in Mo.

One drop of the $NaNO_3$ solution, properly diluted, is placed on the ends of each of two prepared electrodes. Dry the solution on the electrode thoroughly in an oven at 110°C and hold in the oven or under a heat lamp until removed for immediate arcing. Arc electrodes with 2400 v a-c, 2.2 amperes. Prepare a standard curve from a series of standard solutions in $NaNO_3$ solution as above. Determine Al, Ca, Cu, Fe, Mg, Mn, Ti, etc., as desired.

Determination of Elements by Preconcentration in Al_2O_3-Fe_2O_3

Reagents:

8-Quinolinol solution, 5 per cent in 2.0 N acetic acid.
Ammonium hydroxide: 8 M.
Ammonium acetate: 2 N.
Tannic acid, 10 per cent in 2 N ammonium acetate.
Thionalide, 1 per cent glacial acetic acid.
Powdered graphite mixture, 1 g of graphite containing 0.25 mg GeO and 0.75 mg Pd.

Procedure. An aliquot of Solution *A* containing 20 to 30 mg of Al_2O_3 and Fe_2O_3 is adjusted by the addition of $AlCl_3$ or $FeCl_3$ to have an Al_2O_3: Fe_2O_3 ratio of 4.5:1. Dilute to 160 ml, add 10.0 ml 8-quinolinol solution and enough 8 *M* NH_4OH to bring the pH to 1.8. Adjust to pH 5.1 to 5.2 with 2 *N* ammonium acetate. Add 2 ml tannic acid solution, 2 ml of thionalide solution, and 5 ml 8 *M* NH_4OH. Stir, allow to stand overnight, and filter through Whatman No. 42 paper. Wash 5 or 6 times with H_2O. Dry the filter paper containing the precipitate and ignite in Pt crucible. Weigh the Al_2O_3-Fe_2O_3 ash and mix with twice its weight of the powdered graphite mixture. Transfer 6.0 mg of this mixture to an anode cup electrode and arc to graphite counter electrode with 220 v d-c at 10 amperes. Develop standard curves from a series of standards prepared as for unknowns in Al_2O_3-Fe_2O_3 matrix. Determine concentration of Cd, Co, Cr, Ga, Ni, Pl, Sn, V, Zn, etc., as described where concentrations are in detectable range.

SPECIAL EXTRACTION PROCEDURES

Trace Elements Extracted With 2.5 Per Cent Acetic Acid

Reagents:

Acetic acid, 2.5 per cent solution, pH 2.5.

Procedure. Shake 50 g air-dry soil with 250 ml acetic acid solution for several hours, preferably overnight; filter with suction and leach soil with an additional 250 ml of acetic acid in 50-ml portions. Transfer the filtrate to a 600-ml beaker and evaporate to less than 100 ml. Add 10 ml H_2O_2 (30 per cent) and evaporate to dryness on steam bath.

Retreat with H_2O_2 as necessary to destroy organic matter. Take up the residue in dilute HCl and determine trace element content by the procedures described above.

Hot-Water-Soluble Boron

Special Glassware. "Corning 7280" or other alkali resistant low-boron glassware should be used.

Reagents:

Potassium carbonate, 10 per cent solution.

H_2SO_4, 0.36 N solution.

Quinalizarin-H_2SO_4 mixture. Dissolve 0.005 g quinalizarin in 1000 ml of 98 per cent N_2SO_4. Protect from H_2O in air by use of an automatic dispensing burette with a drying tube.

Standard solutions of boric acid for color standards. Prepare as directed under Boron in Table 12.1.

Procedure. Transfer 100 g of air-dry soil to a 500-ml Erlenmeyer flask. Add 200 ml of boiling H_2O and shake intermittently for 25 minutes. Add approximately 0.5 g of $CaCl_2 \cdot 2H_2O$, shake, and after 5 minutes filter through a Whatman No. 2 paper into a 250-ml beaker. Transfer 40 ml of filtrate to size 2 Coors porcelain crucible, add 3 to 4 drops of 10 per cent K_2CO_3 solution, and evaporate to dryness on a hot plate or a steam bath. Ignite in a muffle furnace at 550 to 600°C for 2 hours. Add 5 ml of 0.36 *N* H_2SO_4 and carefully scrape residue from sides of the crucible with glass rod to insure the solution of all boron. Transfer 1-ml aliquot to a 15 x 80 mm weighing bottle with ground-glass stopper. Add 10 ml of quinalizarin-H_2SO_4 mixture, and stopper, mix, and cool. Compare visually with standards prepared similarly.

Easily Reducible Manganese

Reagents:

Ammonium acetate-quinol solution. Prepare neutral normal ammonium acetate solution as directed for Cation-Exchange Capacity. Dissolve in this solution 2.0 g quinol for each liter, as suggested by Leeper (6).

Procedure. Weigh 25–50 g air-dry soil into a 500-ml Erlenmeyer flask. Add 200 ml NH_4OAc-quinol solution and shake 8 hours or allow to stand overnight. Filter with suction or through large fluted paper. Wash the original flask and the soil in the filter with an additional 200 ml NH_4OAc-quinol solution in four 50-ml portions. Transfer a 50-ml aliquot of the filtrate to a 250-ml beaker, add 15 to 20 ml HNO_3, and digest on hot plate (while it is covered) for 30 minutes. Remove cover and evaporate to about 10 ml. Cool, add 15 ml HNO_3, and 5 ml $HClO_4$. Evaporate cautiously on hot plate until solution is nearly colorless and white fumes are evolved. Repeat digestion with more HNO_3 if organic matter is not destroyed. Cool, add 5 ml H_2SO_4, and heat to remove all $HClO_4$. Cool, add 10 ml H_3PO_4, warm in steam bath, add 50 ml H_2O and about 0.2 g KIO_4. Proceed from this point as directed under Sodium Carbonate Fusion for Total Mn.

Rapid Chemical Tests for Nutrients in Electrodialysis Extracts

Extraction by means of electrodialysis in 0.05 *M* H_3BO_3. Purvis and Hanna (8) have described the electrodialysis apparatus and cell. Water-soluble and exchangeable cations are removed quantitatively by this procedure; and phosphate ions are extracted in amounts that correlate closely with plant availability of this nutrient. Weigh, or measure by volume, 5 g of air-dried, 2-mm sieved soil into the soil compartment of the electrodialysis cell. Add 75 ml of 0.05 *M* H_3BO_3. Dialyze for 20 minutes and drain dialyzate through tube in bottom of cell into 100-ml beaker. This solution contains extracted nutrient elements from the soil at a dilution ratio of 15:1.

Phosphorus

Reagents:

Ammonium molybdate solution. Prepare as directed under Total P.

Stannous oxalate suspension. Add approximately 0.1 g of stannous oxalate (C.P. powder) to 50 ml of H_2O. Shake it into suspension before using. Make fresh suspension daily.

Standard solutions. Dissolve 0.1156 g $Na_2HPO_4 \cdot 12H_2O$ in H_2O and dilute to 1000 ml. This solution contains 10 ppm P. To construct a standard curve on a colorimeter, dilute 5-, 10-, 15-, 20-, 25-, and 30-ml portions of this solution to 100 ml. These solutions contain from 0.5 to 3.0 ppm P. Five ml of these diluted standards are transferred to matched test tubes. The color is then developed in the standards by the procedure described for the phosphate test.

Procedure. Transfer 5 ml of extract from soil to a matched test tube (15 x 120 mm) and add 0.5 ml of ammonium molybdate solution. Shake well and add 0.5 ml of stannous oxalate suspension. Shake again and read on colorimeter within 5 minutes at 660 mμ.

Potassium and Sodium

Reagents:

Standard K solutions. Dissolve 1.9067 g KCl in H_2O and dilute to 1000 ml. This solution contains 1000 ppm K. Dilute 16.7 and 33.3 ml of this solution to 1000 ml for use as standards of 16.7 and 33.3 ppm, respectively.

Standard Na solutions. Dissolve 2.5413 g NaCl in H_2O and dilute to 1000 ml. This solution contains 1000 ppm Na. Dilute 16.7 and 33.3 ml of this solution to 1000 ml for use as standards 16.7 and 33.3 ppm, respectively. Potassium and sodium standard solutions may be combined if desired. Concentrations of 16.7 and 33.3 ppm were chosen to make calculations of pounds an acre easier.

Procedure. Determine K and Na with Perkin-Elmer, or other flame photometer, using 0.05 *M* H_3BO_3 for zero adjustment and either 16.7- or

33.3-ppm standard for full-scale deflection. Direct readings without an internal standard give satisfactory results if an air-propane flame is used.

Magnesium

Reagents:

Thiazol yellow, 0.01 per cent solution. Dissolve 0.2 g thiazol yellow (Eastman Kodak P5977) in 200 ml of H_2O and store in a brown glass bottle. Dilute 20 ml of this solution to 100 ml with H_2O for use.

NaOH, 25 per cent solution.

Starch-compensating solution. Make a thin paste from 2 g soluble starch and a few drops of H_2O. Add 100 ml of boiling water slowly while stirring. Starch solution should be prepared fresh daily. Cool and mix with equal volume of compensating solution made by dissolving 4.4 g calcium acetate in 1000 ml of H_2O containing 10 ml of concentrated HCl.

Standard solutions. Dissolve 1.0135 g $MgSO_4 \cdot 7H_2O$ in H_2O and dilute to 1000 ml. This solution contains 100 ppm Mg. Dilute 1, 2, 3, 4, 6, 8, and 10 ml of this solution to 100 ml to make standards of 1 to 10 ppm.

Procedure. Transfer 5 ml extract from soil to a matched test tube (15 x 120 mm). Add 0.5 ml of starch-compensating solution and 0.5 ml of thiazol yellow solution, and mix. Next add 0.5 ml of sodium hydroxide solution, mix, and match color visually with standards prepared similarly.

NO_3-nitrogen

Reagents:

Sodium hydroxide, 10 per cent solution. Dissolve 10 g NaOH in H_2O and dilute to 100 ml.

Ammonium hydroxide, 1-1 solution. Mix equal volumes of NH_4OH, sp. gr. 0.90, and H_2O.

Phenoldisulfonic acid solution. Dissolve 25 g of white phenol in 150 ml of H_2SO_4 (sp. gr. 1.84). Add 75 ml of fuming H_2SO_4 (13 per cent SO_3) and heat in a boiling water bath for 2 hours.

Standard solutions. Dissolve 0.7221 g KNO_3 in H_2O and dilute to 1000 ml. Dilute 1-, 2-, 4-, 6-, 8-, and 10-ml portions of this solution to 100 ml with H_2O. These solutions contain 1, 2, 4, 6, 8, and 10 ppm NO_3-N respectively.

Procedure. Transfer 1 to 5 ml, depending on NO_3-N concentration, of soil extract to size 000 evaporating dish, add 3 drops of NaOH solution, and evaporate the mixture to dryness on a steam bath. After cooling, add 1 ml of phenoldisulfonic acid and turn the dish by hand to insure that all of the residue is moistened. Add approximately 5 ml of H_2O carefully to avoid splattering, and then add 10 ml of NH_4OH solution. Transfer the contents of dish, with washings, to an appropriate-size volumetric flask (size depending upon intensity of color), and then make to volume with H_2O. After mixing, transfer approximately a 5-ml portion of solution to a test tube and measure color intensity with colorimeter at 420 mμ. Prepare a standard curve from the standard solutions of KNO_3.

NH_4-nitrogen

Reagents:

Nessler's reagent. Dissolve 115 g of HgI_2 and 80 g of KI in H_2O and dilute to 500 ml. Mix with 500 ml of 6 *N* NaOH.

Standard solutions. Dissolve 0.3821 g NH_4Cl in H_2O and dilute to 1000 ml. Dilute 1, 2, 3, 4, and 5 ml of this solution to 100 ml with H_2O. These solutions contain 1, 2, 3, 4, and 5 ppm NH_4-N, respectively.

Procedure. Transfer 5 ml of soil extract to matched test tubes. Add 0.5 ml of Nessler's reagent, mix well, and measure color developed with colorimeter at 420 mμ. Prepare standard curve by treating 5-ml portions of standard solutions in a similar manner.

CATION-EXCHANGE PROPERTIES

Cation-Exchange Capacity

Reagents:

Ammonium acetate, 1 N, pH 7.0. Dilute 290 ml glacial acetic acid with about 2 liters of H_2O. Dilute 350 ml NH_4OH (28 per cent NH_3) with about 2 liters of H_2O. Mix the two solutions, adjust the pH to 7.0 by small additions of NH_4OH or acetic acid as necessary, and dilute to 5 liters.

Ethyl alcohol, 95 per cent, pH 7.0. Adjust to pH 7.0 with dilute NaOH or acid as required.

NaOH, 1 M solution.

Boric acid, 4 per cent solution.

NaCl, 10 per cent solution.

HCl, 0.1 N solution.

Mixed indicator. Prepare as directed under Nitrogen.

Procedure. Weigh 50 g air-dry < 2 mm soil into a 250-ml Erlenmeyer flask. Add 100 ml NH_4OAc solution and shake occasionally for 2 hours. Filter with suction through Whatman No. 2 filter paper on a Coors No. 2 Buchner funnel into 500-ml suction flask. Wash soil from Erlenmeyer flask onto filter with 50-ml portions of NH_4OAc eight times or until filtrate totals 500 ml. Save filtrate, as Solution *A*, for determination of exchangeable cations.

Wash soil on filter with five 50-ml portions of ethyl alcohol to remove excess NH_4^+. Discard the alcohol washings, rinse suction flask, and leach soil with 400 ml NaCl solution, also in 50-ml portions. In all filtering and washing allow each portion to pass through soil before adding next portion, but avoid letting soil dry excessively to form cracks which lead to inefficient leaching of soil.

Transfer NaCl leachings to a 600-ml Kjeldahl flask. Rinse suction flask into Kjeldahl flask; add a few paraffin chips or a few drops of octyl alcohol to prevent excessive foaming, 0.2 g Zn granules, and 25 ml NaOH solution. Connect immediately to Kjeldahl distillation apparatus and distill at least 150 ml into a 250-ml Erlenmeyer flask containing 50 ml boric acid solution. Titrate with 0.1 N HCl, using the mixed indicator. Calculate cation-exchange capacity according to the following equation.

$$\text{CEC (meq/100 g soil)} = \frac{\text{ml HCl} \times N \text{ HCl} \times 100}{\text{wt of soil used}} \qquad (6)$$

Exchangeable Cations

Evaporate Solution *A* from the NH_4^+ saturation of the soil sample to dryness in a 600-ml beaker. Hold the beaker with tongs over a low flame until no more fumes are evolved. Rotate the beaker to insure that all parts are uniformly heated. Increase the flame, and continue heating and rotating the beaker. When there is no further whitening of the residue, reduce the flame gradually in order to slowly cool the beaker. Allow at least one minute from highest heat to none. Treat residue as indicated below.

Total Exchangeable Bases

Reagents:
HCl, 0.1 N solution.
NaOH, 0.1 N solution.

Procedure. Add 50 ml of the standard HCl solution to the beaker containing the ignited residue. Scrub the sides and bottom of the beaker with a rubber policeman; rinse the policeman with a little H_2O into the beaker, cover with watchglass, and digest for about 1 hour at low heat on hot plate. Filter through Whatman No. 30 paper into a 250-ml Erlenmeyer flask. Wash filter 5 or 6 times with hot H_2O. Titrate the excess HCl with the standard NaOH, using methyl red indicator. Reserve the solution for determination of individual bases (Solution *B*). Calculate the total exchangeable bases as follows:

$$\text{Total exchangeable bases (meq/100 g soil)} \qquad (7)$$
$$(N_1 \times V_1) - (N_2 \times V_2) \times 100/\text{wt of sample in g}$$

where N_1 = normality of HCl; V_1, volume of HCl (ml); N_2, normality of NaOH; and V_2, volume of NaOH (ml).

Sodium and Potassium

Procedure. Transfer the Solution *B* from the titration of total exchangeable bases to a 250-ml volumetric flask and dilute to volume. Designate as Solution *C*. Transfer a 50-ml aliquot of this solution to a 100-ml volumetric flask, add Li-internal standard solution, dilute to volume, and determine Na and K with flame photometer. Save the remainder of Solution *C* in the 250 ml-flask for Ca, Mg, and Mn.

Manganese

Reagents:

Standard solutions of Mn, 0.5 to 10 ppm.

Procedure. Transfer a 10- to 50-ml aliquot (depending on Mn content) of Solution *C* to a 100-ml bleaker. Add an amount of H_3PO_4 (85 per cent) equal to $\frac{1}{5}$ the volume of Solution *C* taken, and approximately 0.1 g KIO_4. Heat in water bath for 30 minutes; cool, transfer to volumetric flask (volume depending on depth of color developed), and make to volume and read light transmittance at 540 mμ. Prepare standard curve as above using standard solutions. Calculate exchangeable Mn content of soil from standard curve and appropriate dilution factor.

Calcium

Reagents:

Ammonium oxalate, saturated solution.
Potassium permanganate, 0.05 N solution.
Bromphenol blue indicator, 0.04 per cent solution.
Ammonium hydroxide, 8 M solution.
Sulfuric acid, 1 M solution.

Procedure. Transfer a 50-ml aliquot of Solution *C* to a 250-ml beaker, add 5 drops bromphenol blue indicator and add 8 *M* NH_4OH to the end, point at pH 4.0. Bring to boiling on hot plate, add 10 ml of $(NH_4)_2C_2O$-solution, cover beaker with watchglass, and boil slowly for 1 hour or until precipitate is coarsely granular. Filter through Whatman No. 42 paper and wash free of oxalates with H_2O. Pierce the tip of the filter with a stirring rod and wash the precipitate into the original beaker with 1 *M* H_2SO_4. Dilute to about 100 ml with hot H_2O and titrate with $KMnO_4$ to faint pink. Add filter paper and rinsings of funnel into beaker. Continue titration with $KMnO_4$ to pink color, which persists for 1 minute. Calculate per cent Ca from the normality and volume of $KMnO_4$ used and dilution factors.

If a flame photometer suitable for Ca determination is available, determine Ca in Solution *C* with this instrument rather than as directed above.

Magnesium

Reagents:

Disodium-EDTA, 0.01 M solution.
Eriochrome Black T indicator. Prepare as directed under Total Mg.
Ammonium hydroxide, 8 M solution.
Potassium cyanide, 2 per cent solution.

Procedure. Transfer a 50-ml aliquot of Solution *C* to 150-ml beaker or 125-ml flask and adjust to pH 10 with NH_4OH; add 5 drops of the Eriochrome Black T indicator and 1 ml KCN solution. Titrate with Na-EDTA to blue endpoint. This titration includes Mn and Ca as well as Mg. The amounts of Ca and Mn determined separately are subtracted to give a value for Mg alone. Where adequate equipment is available, Mg may also be determined more conveniently by means of a flame photometer.

OTHER SPECIAL PROPERTIES

pH

Transfer 50 g < 2 mm air-dry soil to a 100-ml beaker, or Dixie No. 44 paper cup; add 50 ml H_2O and stir occasionally for 1 hour. Determine pH of sample with glass electrode and pH meter while stirring. Soils with very high organic-matter content may require more H_2O to become thoroughly wet.

Lime Requirement

Reagents:

Bromcresol green indicator, 0.04 per cent.
Hydrochloric acid, 0.05 N solution.
Calcium hydroxide to p-nitrophenol solution. Dissolve 8.0 g $Ca(OH)_2$ in 3 liters of freshly distilled H_2O. Add 25.0 g *p*-nitrophenol (PNP) and shake or stir occasionally until dissolved.

Procedure. Weigh into a 250-ml Erlenmeyer flask an amount of air-dry soil equal to twice the reciprocal of the approximate cation-exchange capacity of the soil, expressed as meq/g. Add 100 ml of the $Ca(OH)_2$-PNP solution, stopper, and shake for at least 2 hours, preferably overnight. Filter through Whatman No. 2 paper and titrate an aliquot to the bromcresol green end point (a faint green in the yellow of the PNP, approximately pH 5.2) with the 0.05 *N* HCl. Titrate an aliquot of the $Ca(OH)_2$-PNP as a blank. Calculate the lime requirement in pounds $CaCO_3$ an acre as follows:

$$\frac{(B - T) \times V \times N \times 10^4}{\text{wt of sample in g}} = \text{lime requirement} \qquad (8)$$

TABLE 12.2. INTERPRETATION OF CONDUCTIVITY READINGS

Conductivity 1×10^{-5} mhos) (1:2 soil:H_2O ratio)	Relative level	Plant response
0–20	Low	No injury to any plants; may indicate lack of nutrients
20–80	Medium	Optimum level for most plants; usually indicates well-fertilized soil; salt-sensitive plants may be injured
80–150	High	Injury to salt-sensitive plants; germinating seeds or seedlings are likely to be injured
Above 150	Excessive	Definitely injurious to most plants of all ages

where B = standard HCl for blank; T = ml standard HCl for sample; V = ml $Ca(OH)_2$-PNP in aliquot titrate; and N = normality of the HCl (0.05).

This procedure was originally designed to determine lime requirement to pH 7 Schofield (11), but as modified above will estimate lime required to bring most soils to about pH 6.5.

Salt Concentration—Conductivity Test

Weigh a 50-g sample of soil into a 250-ml Erlenmeyer flask. Add 100 ml of distilled H_2O and let stand at least 30 minutes, shaking occasionally. Filter or pour off supernatant liquid and measure conductivity of liquid with a bridge and conductivity cell. The salt concentration in ppm can be approximated by multiplying the conductivity reading expressed as 1×10^{-5} mhos, by a factor of 8. Direct interpretation of conductivity readings may be made as indicated in Table 12.2 [see Richards (10) for more details].

REFERENCES

1. Association of Official Agricultural Chemists, "Methods of Analysis," 7th ed., Washington 4, D. C., 1955.
2. Bear, F. E., *et al.*, *Soil Sci.* 59, 1–109 (1945).
3. Chapman, H. D., and Pratt, P. F., "Methods of Analysis for Soils, Plants, and Waters," Berkeley, California, University of California, Agricultural Publications, 1961.
4. Hsu, P. H., *Soil Sci.* 96, 230–238 (1963).
5. Jackson, M. L., "Soil Chemical Analysis," Englewood Cliffs, New Jersey, Prentice-Hall Inc., 1958.
6. Leeper, G. W., *Nature* 134, 972 (1934).
7. Mitchell, R. L., Commonwealth Bureau of Soil Sci., Tech. Comm. No. 44, Harpenden, England, 1948.

8. Purvis, E. R., and Hanna, W. J., *Soil Sci.* 67, 47–52 (1949).
9. Purvis, E. R., and Higson, G. E., *Ind. Eng. Chem. Anal. Ed.* 11, 19 (1939).
10. Richards, L. A., ed., "Diagnosis and Improvement of Saline and Alkali Soils," U.S.D.A., Washington 25, D. C., 1954.
11. Schofield, R. K., *J. Agr. Sci.* 23, 252 (1933).
12. Shimp, N. F., *et al.*, *Soil Sci.* 83, 51–64 (1957).
13. Snell, F. D., *et al.*, "Colorimetric Methods of Analysis," Vol 2a, Princeton, New Jersey, D. Van Nostrand Co., Inc., 1959.

Index

Related Books

ISBN: 9789387893832

ISBN: 9789388812276

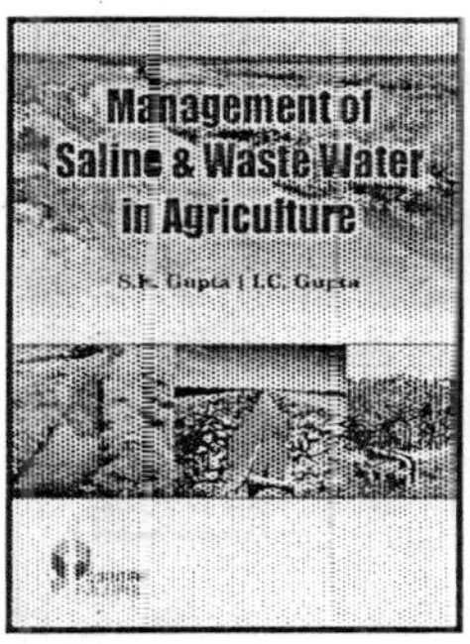

ISBN: 9788172339326

ISBN: 9789388172028

ISBN: 9788172338602

ISBN: 9788172338220

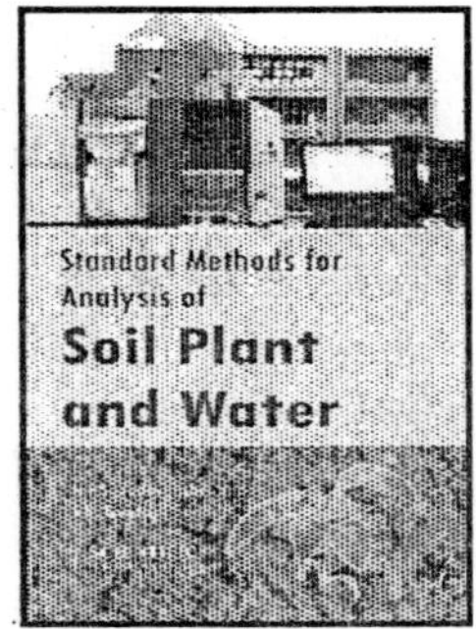

ISBN: 9788172337902

ISBN: 9789383692620

ISBN: 9789383692880

For Full Content & Sample Page Visit Website

To know your local book store call or us on +91-95211-31111

Related Books

ISBN: 9789387307087

ISBN: 9789383692637

ISBN: 9788172335380

ISBN: 9789388043656

ISBN: 9788172336202

ISBN: 9789383692880

ISBN: 9788172337995

ISBN: 9788192686028

ISBN: 9789386347237